Springer Geography

The Springer Geography series seeks to publish a broad portfolio of scientific books, aiming at researchers, students, and everyone interested in geographical research. The series includes peer-reviewed monographs, edited volumes, textbooks, and conference proceedings. It covers the entire research area of geography including, but not limited to, Economic Geography, Physical Geography, Quantitative Geography, and Regional/Urban Planning.

More information about this series at http://www.springer.com/series/10180

Márton Veress · Tamás Telbisz
Gábor Tóth · Dénes Lóczy
Dmitry A. Ruban · Jaroslav M. Gutak

Glaciokarsts

Springer

Márton Veress
SEK, Department of Physical Geography
Eötvös Loránd University
Szombathely, Hungary

Tamás Telbisz
Department of Physical Geography
Eötvös Loránd University
Budapest, Hungary

Gábor Tóth
SEK, Department of Physical Geography
Eötvös Loránd University
Szombathely, Hungary

Dénes Lóczy
Institute of Geography, Department of Physical Geography
University of Pécs
Pécs, Hungary

Dmitry A. Ruban
Southern Federal University
Rostov-on-Don, Russia

Jaroslav M. Gutak
Institute of Mining and Geosystems
Siberian State Industrial University
Novokuznetsk, Kemerovo Oblast
Russia

ISSN 2194-315X ISSN 2194-3168 (electronic)
Springer Geography
ISBN 978-3-030-07330-5 ISBN 978-3-319-97292-3 (eBook)
https://doi.org/10.1007/978-3-319-97292-3

This Springer imprint is published by the registered company Springer Nature Switzerland AG
The registered company address is: Gewerbestrasse 11, 6330 Cham, Switzerland

Preface

This book describes a specific karst type, the glaciokarst. The cause of the peculiarity of glaciokarst that the karsts belonging to this karst type were and are still affected by diverse effects (karstification, glacial erosion, fluvial erosion, frost weathering, mass movements). In a lot of cases, these effects were repeated and alternated. According to this, these areas have one of the most diverse landscapes regarding karst types and because of this it can be a significant area of researchers' interest.

The need for a better knowledge of glaciokarsts is expected to increase in the present and in the future. On our globe, climate change has an increasing dominance. In order to become familiar with former climate changes, the research of glaciokarstic areas may give data which are less possible to be obtained from other non-karstic areas. Since the features that developed on the karst (and thus, glacial erosional features too) are stabile and can be studied well. Thus, ice cover and climate change of the past 1–2 million years can be studied better on karst than on non-karstic terrains. It makes it possible the better understanding of the climatic events of the near past and present.

The book is divided into nine chapters. Chapter 1 deals with the research history of glaciokarst, Chap. 2 involves the general description of this karst type. Glacial erosion taking place on karst is analysed in Chap. 3, which is special in many aspects since the karst features influenced the type of glaciers and the intensity and way of glacial erosion. Chapters 4 and 5 discuss the features of glaciokarstic areas and their genetics. The karst types of glaciokarst are described in Chap. 6 and the geomorphic evolution of glaciokarstic terrains is presented in Chap. 7. The features and events of some sample sites of glaciokarstic areas are dealt with in Chap. 8. A description of the areas of the glaciokarsts of the Earth can be found in Chap. 9 and thus, it gives an overview on the characteristics of physical geography, on geological data, on the history of glaciation and the features occurring there.

Szombathely, Hungary Márton Veress

Acknowledgements

We would like to say thanks for the efforts of all people who contributed to the preparations of this book. Those people are the following: Németh, H., Györe, É., Döbröntei, L., Deák, Gy., Széles, Gy., Richard Maire, Annie-Marie Meyer and Teddy Auly.

Contents

Contributors

Jaroslav M. Gutak Institute of Mining and Geosystems, Siberian State Industrial University, Novokuznetsk, Kemerovo Oblast, Russia

Dénes Lóczy Institute of Geography, Department of Physical Geography, University of Pécs, Pécs, Hungary

Dmitry A. Ruban Southern Federal University, Rostov-on-Don, Russia

Tamás Telbisz Department of Physical Geography, Eötvös Loránd University, Budapest, Hungary

Gábor Tóth Faculty of Natural Sciences, University of West Hungary, Szombathely, Hungary

Márton Veress SEK, Department of Physical Geography, Eötvös Loránd University, Szombathely, Hungary

Chapter 1
History of Glaciokarst Research

Tamás Telbisz and Gábor Tóth

Abstract In this chapter the research history of glaciokarsts is described from 1880 in the following topics: morphological descriptions (landforms on glaciokarst terrains, cave explorations on glaciokarsts), hydrologic and speleological analysis of subglacial and periglacial karst aquifers, new methodologies in glaciokarst research (dating methods, formal stratigraphy, GIS, computer simulations), age of synthesis, anthropogenic effects and climate change on glaciokarsts.

Keywords Glaciokarst research · Morphological descriptions · Hydrologic and speleological investigations · New methodologies in glaciokarst research Age of synthesis · Anthropogenic effects and climate change on glaciokarsts

1.1 Introduction

The history of glaciokarst research can be approached from two directions, either from the study of glaciations or from the study of karstology. The scientific study of Quaternary glaciations began as early as the 1840s by the pioneering work of Agassiz (1840), and over time, more and more scientists joined this research, and after some decades of work it became evident that continental ice sheets and mountain glaciers covered huge areas during Quaternary glacial periods, among others karst terrains as well. Glaciology became a sound and versatile science, but the presentation of its history is out of the scope of this chapter. As the scientific study of glaciokarsts is rather a branch of karstology, in this short history our subject is approached from the direction of karst research.

The history of glaciokarst research can be divided into five periods, nevertheless, these periods have different lengths, and they are partly overlapping (Table 1.1). The first period since the end of the nineteenth century to the 1970s (but occasionally further on) is the time of morphological explorations and descriptions, and processes are interpreted mainly in qualitative terms. In the second period from about 1960 to the end of the 1980s, geochemical and hydrological measurements made glaciokarst studies more quantitative, while periglacial and subglacial environments were more thoroughly explored. The third period is characterized by the

M. Veress et al., *Glaciokarsts*, Springer Geography,
https://doi.org/10.1007/978-3-319-97292-3_1

Table 1.1 Steps of glaciokarst studies

Period	Name	Characteristic approach	Outstanding scientists
1880–1970	Morphological analysis	Study of landforms, typical glaciokarst terrains, qualitative analysis of processes, cave explorations	Corbel, Cvijić, Grund, Horn, Martel
1960–1990	Hydrology and geochemistry of subglacial and periglacial karsts	Analysis of water samples and dissolution processes, speleological exploration of subglacial caves	Atkinson, Dreybrodt, Ek, Ford, Smart
1975 (2000)–	Methodological boom	Speleothem dating by U-series; Cosmogenic dating of superficial and cave sediments; GIS	Atkinson, Ford, Gascoyne, Harmon, Häuselmann, Hughes, Lauritzen, Spötl
2000–	Synthesis	Synthesis of periods and extensions of glaciations; application to glaciokarsts	Audra, Hughes, Maire
2000–	Anthropogenic effects on glaciokarsts	Role of glaciokarst in CO_2 budget; effects of climate change on glaciokarsts; direct anthropogenic effects	Viles, Zeng

application of new methodologies, partly since the end of the 1970s, but the largest boom occurred since the beginning of the 2000s. U-series dating of speleothems was the first among these new methods that really revolutionized views related to the age and development of glaciokarsts. Further improvements are due to cosmogenic nuclide techniques, which are also suitable for dating cave materials, but these methods became available only after the turn of the millennium. Besides cave sediments, cosmogenic nuclide techniques can be used also to infer exposure age of superficial landforms such as moraines or limestone pavements. The integration of spatial data, statistical analysis of morphometrical parameters, recognition of spatial patterns and relationships have become easier due to the widespread distribution of Geographic Information Systems (GIS), which had its impact on glaciokarst studies also since the 2000s. Theoretically, the age of synthesis is presented as a distinct period in this historical overview, however, it basically overlaps the previous period. At the beginning of the twenty-first century, a number of glaciological syntheses have been published. They are based on a wealth of field data and previous research, and they focus the extent and phases of glaciations including both continental ice sheets and mountain glaciers. These syntheses usually serve as inputs or background data to glaciokarst studies, but in many cases, glaciokarst studies themselves contributed to glaciological synthesis with new field data. Finally, studies dealing with anthropogenic effects on glaciokarsts are mentioned, especially those which are related to the present-day global warming. To now, these studies have relatively few importances within the domain of glaciokarst studies, but they mark that climate change, one of the top issues in today's earth sciences, is not negligible from the viewpoint of glaciokarsts, and it is likely that this direction will get more focus in the future.

1.2 Morphological Descriptions

1.2.1 Landforms on Glaciokarst Terrains

For a long time, especially at the beginnings, the exploration and study of glaciokarsts basically went on two parallel paths according to the two main geographic zones in which glaciokarsts are found. Some scientists dealt mainly with *alpine* glaciokarsts, whereas others studied *arctic* glaciokarsts, thus in this short history, we also present these research lines separately at first, but they are tied together afterwards.

The study of *alpine glaciokarsts* is essentially as old as karstology itself. Karst research began in the Classical Karst of Slovenia and Italy at the end of the nineteenth century. The continuation of the Classical Karst, the Alps to the north and the Dinaric Alps to the south (Fig. 1.1) contain a lot of formerly glaciated high karst mountains easily accessible for the early researchers. Recognizing the signs of glaciations, even Jovan Cvijić, the "father of karstology" published several works about glaciokarst terrains in the Balkan Peninsula (Cvijić 1899, 1900, 1903, 1913, 1917, 1920). In his publications, he thoroughly discussed the relationship of karst processes and glaciations, and presented how depressions are jointly formed by these effects. In his work of 1917, he even used a special term, "karstic glaciers". He recognized that in some mountains of the Balkan, notably in the Orjen and Lovćen mountains (in Montenegro), Pleistocene glaciers stretched down to surprisingly low elevations, even below 1000 m asl. Although some of his observations had to be corrected later (see Milivojević et al. 2008), but his principal theses have remained valid till now. Parallel with Cvijić, or following him, other scientists also began to study glaciokarst morphology mainly in the Balkan. Albrecht Penck, who was the doctoral supervisor of Cvijić, analysed the glacial morphology of Balkan Peninsula (Penck 1900), similarly to Grund (1902, 1910) and Sawicki (1911) . In addition, Penck studied the glaciations of the Pyrenees (Penck 1885) and of the Alps (Penck and Brückner 1901) as well, but in those works, glaciokarsts are not emphasized at all.

Publications focussing especially on glaciokarsts of the Alps have been published only later, first about French territories. Allix (1930), for example compared the landforms of the French Préalpes to the Dinaric sample areas, using Cvijić's conceptions, and discussing the preglacial or postglacial formation of cirques. After the forced break of the Second World War, glaciokarst studies became more widespread in space and in thematics. The pivotal works of Bögli about karren morphology were published in the 1960s (Bögli 1960, 1964). He studied the genesis and characteristics of small-scale dissolutional forms, which are partly of glaciokarstic origin. His field work was mostly based on sample areas within the Alps. The first thorough publications about the glaciokarst morphology of the Pyrenees (Miotke 1968; Bertrand and Bertrand 1971; Smart 1986) and of the Rocky Mountains (Ford 1971b) were published some years later. Meanwhile, in the context of the Dinaric Alps, and the glaciation of the Mediterranean region, it was already the time of the first morphological syntheses (Roglić 1961; Messerli 1967).

Fig. 1.1 Typical glaciokarst scenery in the Dinaric Alps, Prokletije Mountains (photo by Telbisz)

The term *arctic glaciokarst* refers to karst terrains, which were affected by the continental ice sheet during the Pleistocene glaciations. Nonetheless, these terrains are not necessarily "sensu stricto" arctic, i.e. their latitudes are in some cases much lower than the Arctic Circle. The classical scene of arctic glaciokarst studies is the British Isles, as it was nearly completely covered by the Pleistocene ice sheet, further on, karst terrains are also numerous here, though not dominant. Moreover, these landscapes were naturally "given" for British earth scientists and karst researchers. The scientific study of British glaciokarsts started roughly at the same time as the description of Dinaric glaciokarsts. Davis already published a paper about the Norber erratics in 1880. Norber is found in Yorkshire, which is one of the best known British glaciokarsts. Hughes (1901) presented a detailed study about Ingleborough, which is also located in Yorkshire. Thus, Yorkshire became a locus classicus of glaciokarst, and it was also the subject of a large number of further studies, namely about the morphology and genesis of limestone pavements, the role of glaciations, and the corrosional erosion rates calculated from the parameters of erratic blocks (Sweeting 1966; Goldie 1973; Vincent 1995). Another extended and thoroughly studied glaciokarst of the British Isles is the Burren (Fig. 1.2), which is situated in western Ireland. It is also famous about the limestone pavements (Williams 1966), but other landforms, such as caves and dry valleys also occur either in Burren, or in Yorkshire. One of the oldest publications (Reid 1887) is related to the development of dry valleys, which are characteristic landforms in some of the British karst terrains, and they are typically formed on Cretaceous chalk limestones in periglacial climate.

Fig. 1.2 Typical glaciokarst scenery in Burren (photo by Mari)

The French Jean Corbel was a broad-minded karstologist, who carried out research in varied karst environments, notably in tropical, alpine (Corbel 1956 1957a) and arctic (Corbel 1952b, c) areas alike. Thus, he had the knowledge and the experience to connect the study of alpine and arctic glaciokarsts (Corbel 1952a, 1957b, 1959). However, in spite of his huge synthesizing work, he is known about a remarkable error in the history of karstology. According to him, karstification is weak below glaciers, but karst corrosion is the most intensive in areas where snow is abundant, since CO_2 is relatively better dissolved in cold than in warm waters. However, several of his statements were later refuted (e.g. by Smith and Atkinson 1976), because the partial pressure of CO_2 is a much more important factor in the karstic dissolution than the temperature, and in turn, CO_2 content is usually several magnitudes higher in tropical soils, thus karstification is the most intensive in warm and wet climates (Jakucs 1977).

The variegated relationship of glaciation and karstification has been synthesized by Ford (1983a) using Canada as an example. This country gives a highly favourable opportunity for synthesis, because there are both alpine and arctic glaciokarsts in Canada. Ford summarized the impact of glaciers on karst terrains and demonstrated examples for destructive, deranging, inhibitive, preservative and stimulative effects.

Naturally, the morphological approach has not been terminated in the 1980s, it is still present, but usually it is completed with several new methods, thus glaciokarst studies became more complex in general since that time.

Somewhat surprisingly, certain glaciokarst terrains were discovered quite recently, that brought even new landform types to light. Of the unknown or recently discovered glaciokarsts, the most famous are the two islands of Diego de Almagro and Madre de Dios in Patagonia (Chile). The islands between 50°S and 52°S latitudes were first explored by a French expedition in 1995. The first trip was followed by nine further expeditions until 2017, with an increasing Chilean participation. The results of the ten expeditions demonstrate that probably most active glaciokarst on Earth is found at the coast of Chile, from both glacial and karstic aspects. Based on the measurements of Maire et al. (1999), the solution rate is 3.5–4 times higher than the highest data measured in the Alps (60 mm/ka). The morphology perfectly records the alternating geomorphological heritage. Intensive glacial effects formed the islands during glacials, and an extremely high dissolution rate was the rule during interglacials. Hydroeolic karren features, unique in the world have been described by the French expeditions and a Hungarian research team (Veress et al. 2006). Due to tectonic movements and glaciokarstification, huge cave systems also evolved, but they have been partly ruined as a result of intense erosion. Since the last deglaciation, 10–12 m of glacio-isostatic uplift have been measured by French researchers.

1.2.2 Cave Explorations on Glaciokarsts

Obviously, cave explorations played a crucial role in the discovery of glaciokarsts. At the beginning (and still now) caving was motivated by nature loving, alpinism, paleontology, sport, etc., however, scientific interest was also an issue almost since the beginnings, that gave birth to speleology. In general, the exploration and investigation of glaciokarst caves is a difficult challenge because of several reasons. First, shafts are common features in alpine caves, and moving in them requires advanced technical skills. Second, in many cases, even the access to glaciokarst caves is difficult, especially in case of arctic caves, which are often situated in remote, rarely populated lands. In case of long and hard caves, it is often necessary to spend several days continuously within the cave. Moreover, glaciokarst caves may be dangerous due to abrupt floods, especially during summer, therefore visiting them is possible only in winter, that, in turn, causes difficulties related to cold temperature, and finally, in case of alpine caves, avalanches may threaten cavers approaching the cave.

The exploration of alpine caves began in the second half of the nineteenth century, in Switzerland, Austria and Italy in the framework of the starting alpinism. It was also the time, when the first caving clubs were founded (Shaw 2004). One of the best known ice caves, the Eisriesenwelt in Austria (Fig. 1.3) was explored in 1879 by a naturalist, Anton Posselt from Salzburg. The Mamut Cave and the Ice Cave of Dachstein Mountains (also in Austria) were discovered some decades later, at the beginning of the twentieth century, with Friedrich Simony playing a key role in the explorations (Pavuza and Stummer 1999).

Fig. 1.3 The Eisriesenwelt Ice Cave in Austria (photo by Egri)

Nevertheless, the most outstanding, pioneering personality of the heroic age of cave explorations was the French Édouard-Alfred Martel, originally an advocate. He participated in many cave explorations, in the French Massif Central at first, then in the French Alps, later in Austria, in the Pyrenees, in the Balkan, and in the United States as well. In many cases, he was the first discoverer of the cave (Audra 2004a, b; Halliday 2004). Beside alpine caves, he could visit arctic glaciokarsts as well, notably in the United Kingdom and in Ireland. He was the first person to descend into the 110 m deep shaft of Gaping Gill, a famous cave in Yorkshire (Judson 2004). He was particularly active not only in explorations, but in scientific descriptions, popularization and in the organization of the speleological public life. It is manifested by the fact that in 1895, he created the French Société de Spéléologie, which was a particularly popular society, which had several foreign members as well (Shaw 2004). In North America, the exploration of glaciokarst caves was somewhat delayed, the first explored alpine cave was the Arctomys Cave in the Rocky Mountains in 1911, and since then, it is the deepest known cave in North America (Halliday 2004).

For a long time, the scientific study of caves generally did not focus on glacial effects, instead, other geologic, hydrographic and hydrologic factors were examined. However, in case of Norwegian glaciokarsts, which are mostly limited in their areal extent, but which were particularly strongly affected by Pleistocene glaciations, the question about the relationship of speleogenesis and glaciations was almost immediately raised. The first review about Norwegian karsts was published by Oxaal (1914), and he already noted the typical situation of caves hanging in high positions at valley sides. Oxaal supposed that cave passages developed during deglaciations, and he emphasized the role of meltwaters in speleogenesis. Horn (1935, 1937, 1947) raised the possibility of *subglacial speleogenesis*, i.e. the formation of cave passages below glacier ice. He also introduced the term *stripe-karst*, which denotes a peculiarity of Norwegian karsts, namely that karsts are usually restricted to relatively narrow marble layers, which are typical in the heavily folded Scandinavian Mountains. These narrow marble bands are commonly bordered by non-karstic metamorphic rocks at both sides. The question, whether the age of Norwegian caves is preglacial, interglacial or postglacial has been the subject of several later studies as well (Lauritzen 1981, 1986). It could be answered only by the new dating methods at the next step of research history.

In parallel with cave explorations, scientific knowledge was increased and different theories about speleogenesis were elaborated. The remarkable differences in the shape, and consequently in the evolution of phreatic and vadose passages were first described during the exploration of Dent de Crolles cave system in 1944. At that time, it was the deepest (−658 m) known cave on Earth (Chevalier 1944a, b).

At the end of the 1960s, Derek Ford led several scientific expeditions into the Castleguard Cave found in the Canadian Rocky Mountains (Ford 1971a, 1983c). This cave is still the longest cave in Canada, and it is especially important, because the passages stretch just below a large present glacier. These expeditions significantly improved our understanding of glaciokarst processes (see mainly in Chap. 5).

Another well documented and glaciokarst focusing expedition was the 1970 British Karst Research Expedition (Waltham 1971). Its aim was to get acquainted with the karstic areas of the High Himalayas so far unknown. At that time, no proper geological map was available, and even the extent of karstifiable rocks was unknown. The main purpose of the research was to explore new cave systems in the region of Annapurna and Kashmir. During the trip, they explored several caves approaching them through sinkholes.

Another significant event was the expedition of French cavers into Greenland in 1983, during which the northernmost glaciokarst of the world was discovered (Loubiere 1987). The cave system, which lies at 80°N latitude in the northeast corner of Greenland, 1100 km from the North Pole, is very difficult to access. The cave is developed in Cambro-Silurian limestones. It was formed in climatic conditions, which are nowhere found in Greenland at present. Based on concretions found in the cave, it was demonstrated that around the Pliocene-Pleistocene boundary, climatic conditions were similar than in the forests of the French Prealps today. Thus, intense karstification was possible at such a latitude during this period.

Since the turn of the millennium, the explorations and scientific studies related to glaciokarst caves have been further increased and deepened, partly due to modern exploration and research techniques, partly due to the growing number of the speleological society, and several new results and theories have been elaborated.

1.3 Hydrologic and Speleological Analysis of Subglacial and Periglacial Karst Aquifers

After the exploration of the surface and subsurface morphology of previously glaciated karst terrains, it was a logical step to go closer to the presently glacier-covered karsts. One of the most important questions was, whether ongoing karst development exists below ice, and if so, how effective it is.

As for permafrost terrains, the general view was that permanently frozen conditions exclude the possibility of karst processes (Brook and Ford 1978; Palmer 1984). However, some permafrost karst terrains are known in Siberia, Svalbard and Canada. Nevertheless, their karstification is usually limited, and probably they were formed in climatic settings different from the present ones (Popov et al. 1972; Salvigsen and Elgersma 1985; Ford 1983a).

It was an important step in the understanding of subglacial karstification to measure the dissolution capacity of water samples collected in karst areas, which are actually glaciated. Ek (1964) was a pioneer of these studies. He collected water samples in the French Alps, and by analysing pH and carbonate content values, he concluded that the carbonate concentration is very low, and that subglacial meltwaters are not aggressive from the viewpoint of limestone dissolution capacity. Ford (1971b), who analysed the geochemistry of water samples collected in the Rocky Mountains, confirmed the statements of Ek. In the following period, further data were collected about subglacial conditions. At the beginning of the 1980s, the Castleguard Cave and the Columbia Icefield above it became a thoroughly studied sample area (Ford 1983b, c). It was revealed that subglacial secondary carbonate precipitations occur both at the rock surface and within the caves. Subglacial calcite precipitates at the rock surface were described by Hallet (1976). Below warm-based glaciers, at the stoss side of glacially polished rocks, water melts due to pressure increase and $CaCO_3$ is dissolved. At the lee side, in turn, water refreezes and calcite precipitates, this is the so-called "regelation-slip process". Maire (1976, 1990) explained similarly his field observations at the Desert de Platé in the French Alps, and at the Tsanfleuron plateau in the Bernese Alps (Switzerland). He also noted that subglacial calcite precipitates often fill karren features in the foreland of glaciers.

Cave precipitations have been studied and explained by Magaritz (1973), Dreybrodt (1982) and Atkinson (1983). In certain conditions, infiltrating subglacial waters are capable of dissolving $CaCO_3$, though only in very low concentrations. When water reaches the cave atmosphere, the temperature increases by some degrees, and water becomes supersaturated, thus speleothem growth may start (Dreybrodt 1982). However, Atkinson (1983) experienced that the above mechanism is not too effective, and the main reason for subglacial speleothem formation is the so-called *common-ion effect*. It means that if water is saturated with respect to calcite, and further dissolution of gypsum or dolomite occurs, then calcite precipitates. An important consequence of this mechanism is that speleothem growth is possible without biogenic CO_2 in glaciated karst or in other bare karsts as well, until the cave temperature is above 0 °C. These findings modified previous views, which stated that speleothem growth is possible only in warm periods, under soil-covered terrains (Ford 1976; Atkinson et al. 1978). However, this latter statement is still valid in general, though not exclusively. Smart (1983) studied the hydrological regime of glaciokarsts, and differentiated supraglacial meltwaters, which have high fluctuations (having both seasonal and daily cycles) and higher dissolution capacity, and subglacial pressure meltwaters, which have negligible dissolution capacity.

Meanwhile, due to the new explorations in Norwegian karsts, several field observations were made. First, it was presented that maze patterns are frequent features in glaciokarst caves of Scandinavia. Second, it was demonstrated that the development of some phreatic passages occurred when flow directions were opposed to the present-day topographic gradients. Based on these facts, Lauritzen (1982, 1983, 1984, 1986) outlined the theory of ice-contact speleogenesis, which takes place when ice and karstifiable rocks form a joint aquifer.

1.4 New Methodologies in Glaciokarst Research

Since the 1970s, several new techniques were introduced into earth sciences in general, and they could be used in glaciokarst studies as well. First, U-series dating methods must be mentioned, that can be applied to carbonate rocks, especially to speleothems. Cosmogenic nuclide methodology became available since the turn of the millennium. These methods provided crucial data that significantly contributed to answer the questions related to the age of speleogenesis, i.e. whether caves are formed during preglacial, interglacial, subglacial or postglacial periods. Naturally, the answer is not of global validity, instead, the age may be different depending on the actually studied cave, but the applied methods were usually similar. Besides the aforementioned techniques, other dating methods, such as luminescence or radiocarbon were also applied on glaciokarsts, but they were less significant.

The widespread use of GIS has become a quasi-standard since around the 2000s. The role of GIS in glaciokarst studies is not restricted to the acquisition of new data, but GIS is also helpful due to its good visualization and analytical capabilities. Besides dating methods and GIS, some other, less significant methodological innovations can be also listed, namely the introduction of formal stratigraphy in glaciological studies, and the application of computer simulations.

Largely due to the new methodologies, it is experienced that the number of glaciokarst publications abruptly increased since the 2000s. However, it may be partly caused also by the transformation of the scientific world, namely that the production of publications is increasingly enhanced in the present-day research system.

1.4.1 *Dating Methods*

1.4.1.1 U-Series Dating of Speleothems

U-series dating of speleothems began in the second half of the 1970s. The first study areas were in the Rocky Mountains, the Nahanni karst (Harmon et al. 1975, 1977; Ford 1976), the British Isles (Atkinson et al. 1978, 1987; Gascoyne et al. 1981; Gascoyne and Ford 1984), thus mostly glaciokarst terrains were investigated first by this methodology. Norwegian speleothems were dated somewhat later by Lauritzen and Gascoyne (1980) and by Lauritzen (1983, 1984), and the glaciokarsts of the Alps followed next with some delay (Audra and Quinif 1997; Spötl et al. 2002a, b; Spötl and Mangini 2007; Holzkämper et al. 2005; Audra et al. 2007; Häuselmann et al. 2008; Luetscher et al. 2011). The measurement limit of the U-series dating method was 350 ka at the beginning, but due to technical improvements, it has been doubled since that time (Dorale et al. 2004).

Given this measurement limit, U-series data can not be directly used to prove Early Pleistocene ages. However, they made it evident that there are several caves in the abovementioned glaciokarst terrains, that survived several glacial cycles, therefore they are *preglacial* in that meaning. Normally, caves are considered truly preglacial if they are older than the oldest Pleistocene glaciation. It is also noted that the above results do not mean that all caves in glaciokarst terrains are older than the last glacial.

Age determinations also made it possible to calculate erosion rates in glaciokarst terrains. In some cases, values were quite low, namely in Canada, where Ford et al. (1981) measured 0.13 mm/ka rate, while in other regions, in Norway, for instance higher values of 0.35 mm/ka were calculated (Lauritzen and Gascoyne 1980).

When speleothem ages are statistically evaluated, i.e. periods of speleothem growth are outlined, then the general approach is that hiatuses mark glacials, whereas high-intensity growth periods mean interglacials (e.g. Atkinson et al. 1978; Gascoyne et al. 1981). Nonetheless, as it was already presented, speleothem formation is possible even in glacial periods if certain conditions are satisfied (Atkinson 1983; Spötl and Mangini 2007), but the growth intensity is obviously much less. Further on, it must be taken into consideration that hiatuses can be caused by other factors as well (Häuselmann et al. 2008).

1.4.1.2 Cosmogenic Dating of Cave Depositions

The principle of cosmogenic nuclide dating is that the amount of certain cosmogenically induced radionuclides exponentially decays after the material is shielded from cosmogenic radiation. The rate of decay is dependent on the isotope. In caves, sediments frequently contain ^{10}Be and ^{26}Al in quartz grains, and they can be used for age determination. The age limit of cosmogenic nuclide method is much larger than that of the U-series technique, it is appropriate in the range of 0.1–5 Ma BP (Häuselmann and Granger 2005; Häuselmann 2007).

Careful measurements in several caves of the Alps manifested that the age of sediments is highly variegated from 0.18 to 5 Ma BP, and it is particularly important that there are several old caves, which started to develop at least in the Pliocene or even earlier (Audra et al. 2007; Häuselmann et al. 2008; Hobléa et al. 2011).

1.4.1.3 Moraine Ages Using Secondary Carbonate Cements

As for the superficial sediments, numerical dating was not possible for a long time. However, on glaciokarst terrains, where the till has some carbonate content, secondary carbonate precipitation is often found in the till, and the age of these secondary carbonates can be determined providing a minimum age for the formation of the till (moraines). U-series dating was applied to glacial deposits in the Apennines by Kotarba et al. (2001), in the Hellenides by Woodward et al. (2004), and in the Dinaric Alps by Hughes et al. (2011). These data made it possible to elaborate a precise chronology of glacial advance and retreat phases, wherever glacial till containing carbonates were preserved.

In case of glaciokarst moraines, where the proportion of carbonate material is high, scientists can use the ^{36}Cl cosmogenic isotopes as well, a technique, which was developed somewhat later, but it turned out to be very useful in carbonate glaciokarsts, and provided high precision data about terrains where previous chronology was insufficient (e.g. Sarikaya et al. 2008, 2014; Zreda et al. 2011; Çiner et al. 2015; Wilson et al. 2013b).

1.4.1.4 Cosmogenic Dating of Limestone Pavements

Limestone pavements are among the most peculiar glaciokarst landforms (Waltham et al. 1997), though some researchers emphasize that they are the products of compound processes, and can not be simply considered as the "automatic consequence" of karstification on glacially eroded bare surfaces (Vincent 1995). The study of limestone pavements was completed by valuable information due to ^{36}Cl cosmogenic isotope exposure age data. Vincent et al. (2010) and Wilson (2012, 2013a) dated erratics in British limestone pavements that gives the age of post-LGM deglaciation, moreover, surface lowering rates could be calculated as well.

1.4.2 Formal Stratigraphy

Literature dealing with glaciokarst terrains has been quickly growing since the turn of the millennium. Occasionally, relatively small spatial units (mountains) are studied with larger detail than earlier. The glaciations of these mountains, especially at lower latitudes of Europe or other continents, were local phenomena, and can not be directly connected to glacial phases of the large glacierized terrains, like the Alps or the continental ice sheets. Hence, Hughes et al. (2005) suggested that in case of locally glaciated mountains, it is recommended to use a formal stratigraphy in the description of glacial (or glaciokarstic) landforms. Glacial sediments and landforms, such as moraines should have a standardized name, which reflects the different glacial phases, because it helps the interpretation and comparison of morphological data. While most of the new methodologies mentioned in the previous points were technical innovations, this latter suggestion means more a change in mind.

1.4.3 GIS, Computer Simulations

Nowadays, GIS practically replaced the former role of maps in earth sciences, moreover, it even completed the traditional map functions by a number of new capabilities. For instance, the acquisition of field data necessary to describe glaciokarst landforms became more effective and more precise with the help of GPS receivers (Hughes et al. 2011; Žebre and Stepišnik 2015a, b). The integration of geologic, topographic, hydrographic and geomorphologic data into a common coordinate system helped the analysis and statistical assessment. Digital elevation models improved the analysis of the altitudinal characteristics of glaciokarst terrains (Telbisz 2010a, b, 2011; Fig. 1.4), the identification of valley networks and larger depressions (Bočić et al. 2015), and more recently, LiDAR data make it possible to investigate relatively small scale landforms, namely the shape of dolines, stream sinks and moraines (Žebre and Stepišnik 2015a; Telbisz et al. 2016). The analysis of relatively small-scale landforms is further supported by the always better quality and availability of aerial and satellite images (Žebre and Stepišnik 2015b). The quality and the content of geomorphological maps about glaciokarst terrains are also continuously getting better due to GIS capabilities (Stepišnik et al. 2009, 2016; Aucelli et al. 2013). Thus, GIS seems to be an essential auxiliary tool in today's glaciokarst studies. Glacial or glaciokarst data are also increasingly available in different GIS formats (e.g. Ehlers et al. 2011).

Computer simulations are not strictly a part of GIS, but here we note that they are also appropriate means in the study of glaciokarst processes. They are not as widespread as GIS tools, because they need special mathematical, physical and computing skills. However, they were applied to simulate the development of maze caves, which are frequent phenomena in Norwegian glaciokarsts, and it was plausibly demonstrated that maze caves could be really formed at both the inflow and outflow sections of stripe-karsts, where the carbonate rocks are in connection with warm-based glaciers (Skoglund et al. 2010).

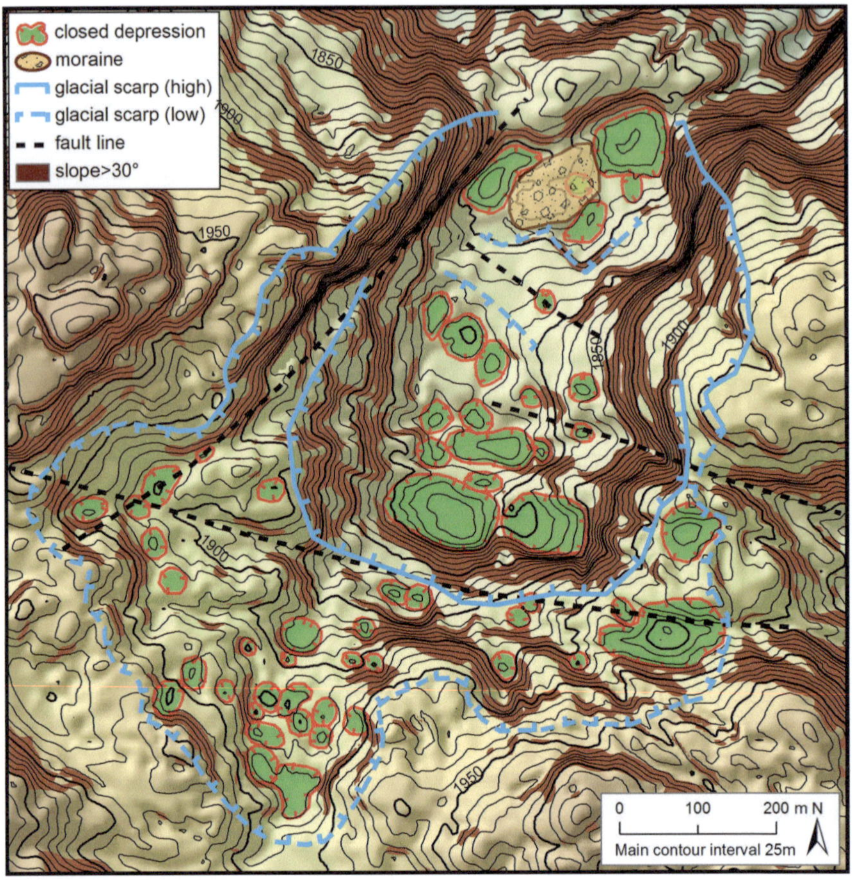

Fig. 1.4 GIS-based geomorphological map of a sample area in Canin Mountains (after Telbisz et al. 2011)

1.5 Age of Synthesis

As a result of more than 150 years of research, the bulk of data about glaciations has grown to extremely large sizes. Hence, it became possible and at the same time necessary to create synthesizing works, which try to briefly and uniformly present the actual knowledge about the spatial distribution of glacial landforms and the chronology of glacial phases. Here, only some of them are highlighted, namely the work of Svendsen et al. (2004) about Europe, of Dyke (2004) about North America, and the global overview of Ehlers and Gibbard (2008), which contain many further citations. Essentially, these glacial syntheses provide input data to glaciokarst research, but occasionally, glaciokarst studies themselves may offer valuable information to glaciology, as speleothems or other glaciokarstic sediments like well-preserved moraines, are suitable for gaining data about climate history. Or simply, if carbonate rocks are predominant in the geological composition of a region, then the glacial synthesis may be largely based on glaciokarsts like in the Mediterranean region (Hughes et al. 2017). Moreover, there are syntheses, which are mostly about glaciokarst features of certain regions, such as the works of Maire (1990) and Audra et al. (2007) about the Alps, or Delmas (2009) and Jiménez-Sánchez et al. (2013) about the Pyrenees.

Beside regional reviews, thematic syntheses are sparse. Notably, the chapters "Alpine Karst" and "Glacierized and Glaciated Karst" of Smart (2004) in the Encyclopedia of Caves and Karst Science, or the book of Veress (2010) about karren features of high mountain, or the review paper of Veress (2017) about glaciokarst depressions can be mentioned. And the present book…

1.6 Anthropogenic Effects and Climate Change on Glaciokarsts

Anthropogenic impacts on glaciokarsts were first studied in the British Isles. Drew (1983) and Moles and Moles (2002) stated that humans contributed to the destruction of vegetation and to the erosion of soils already in the prehistoric times. Nonetheless, later in history, mining exerted a more direct effect on glaciokarsts demolishing occasionally certain landforms (Viles 2003).

Today, the fact of global warming is acknowledged by most researchers (though not everybody), even if the reasons and consequences are not yet fully understood. Viles (2003) examined how much the protected glaciokarst terrains of the British Isles can be affected by the ongoing climate change. He concluded that it is not likely that significant geomorphological changes would occur because of the warming itself. However, he warns that mining and pollution may have significant local effects on glaciokarsts.

Larson and Mylroie (2013) investigated karst terrains from the viewpoint of the global CO_2 budget. In fact, their study focused not primarily on the present climate change, instead, they compared the present climate to the Last Glacial Maximum (LGM) conditions. They calculated that during glacial periods, several million km^2 of continental karst areas were out of the carbon cycle due to the ice sheets and mountain glaciers. On the other hand, similarly large tropical carbonate platforms became subaerial due to the lowering of the sea level, therefore these areas were involved in the carbon cycle during glacial periods.

The hydrological and geochemical properties of water originating from glaciers found on karst terrains were studied by Gremaud et al. (2009) in the Alps, and by Zeng et al. (2012, 2015) in Yunnan (SW China). They tried to quantify how the water budget and the functioning of glaciokarst aquifers are changed due to global warming.

At present, research directions presented in the last section are not among the most significant topics in the study of glaciokarsts, however, it is likely that the examination of anthropogenic effects will grow in the future.

References

Agassiz L (1840) Etudes sur les glaciers. Jent et Gassmann, Neuchatel, p 652
Allix A (1930) Formes glaciaires sur calcaire à plis autochtones. Annales de Géographie 39 (219):299–305
Atkinson TC (1983) Growth mechanisms of speleothems in Castleguard Cave, Columbia Icefields, Alberta, Canada. Arct Alp Res 15:523–536
Atkinson TC, Harmon RS, Smart PL, Waltham AC (1978) Palaeoclimatic and geomorphic implications of $^{230}Th/^{234}U$ dates on speleothems from Britain. Nature 272(5648):24–28
Atkinson TC, Lawson TJ, Smart PL, Harmon RS, Hess JW (1987) New data on speleothem deposition and palaeoclimate in Britain over the last forty thousand years. J Quat Sci 1:67–72
Aucelli PP, Cesarano M, Di Paola G, Filocamo F, Rosskopf CM (2013) Geomorphological map of the central sector of the Matese Mountains (Southern Italy): an example of complex landscape evolution in a Mediterranean mountain environment. J Maps 9(4):604–616
Audra P (2004a) Dent de Crolles Cave system, France. In: Gunn J (ed) Encyclopedia of caves and karst science. Fitzroy Dearborn, New York, pp 583–586
Audra P (2004b) France: history. In: Gunn J (ed) Encyclopedia of caves and karst science. Fitzroy Dearborn, New York, pp 768–770
Audra P, Quinif Y (1997) Une cavité de haute-montagne originale: la grotte Téophile (Alpe d'Huez, France) Rôle des paléoclimats pléistocènes dans la spéléogenèse. Spéléochronos 8:23–32
Audra P, Bini A, Gabrovšek F, Häuselmann P, Hobléa F, Jeannin PY, Kunaver J, Monbaron M, Šušteršič F, Tognini P, Trimmel H, Wildberger A (2007) Cave and karst evolution in the Alps and their relation to paleoclimate and paleotopography. Acta Carsologica 36(1):53–68
Bertrand C, Bertrand G (1971) Le complexe terminal glaciaire du plateau karstique des lacs d'Enol (Picos de Europa, massif Cantabrique central, Espagne du nordouest, province d'Oviedo). Revue Photointerpretation 3:5–18
Bočić N, Pahernik M, Mihevc A (2015) Geomorphological significance of the palaeodrainage network on a karst plateau: The Una-Korana plateau, Dinaric karst, Croatia. Geomorphology 247:55–65
Bögli A (1960) Kalklösung und karrenbildung. Bornträger
Bögli A (1964) Un exemple de complexe glacio-karstique: le schichtreppenkarst. Rev Belge Geogr 88–63
Brook GA, Ford DC (1978) The origin of labyrinth and tower karst and the climatic conditions necessary for their development. Nature 275:493–496
Chevalier P (1944a) Distinctions morphologiques entre deux types d'érosion souterraine. Revue de géographie alpine 32(3):475–486
Chevalier P (1944b) Problèmes et hypothèses d'hydrologie souterraine. Les Études rhodaniennes 19(3–4):228–234
Çiner A, Sarıkaya MA, Yıldırım C (2015) Late pleistocene piedmont glaciations in the eastern mediterranean; insights from cosmogenic ^{36}Cl dating of hummocky moraines in southern Turkey. Quatern Sci Rev 116:44–56
Corbel J (1952a) A comparison between the karst of the mediterranean region and of north western Europe. Trans Cave Res Group Great Br 2:3–25
Corbel J (1952b) Karsts et glaciers en Laponie. Revue de géographie de Lyon 27(3):257–267
Corbel J (1952c) Une région karstique d'Irlande: le Burren. Revue de géographie de Lyon 27 (1):21–33
Corbel J (1956) Le karst du Vercors. Revue de géographie de Lyon 31(3):221–241
Corbel J (1957a) Karsts hauts-alpins. Revue de géographie de Lyon 32(2):135–158
Corbel J (1957b) Les karsts du Nord-Ouest de l'Europe et de quelques régions de comparaison: étude sur le rôle du climat dans l'érosion des calcaires. Revue de Géographie de Lyon 12
Corbel J (1959) Érosion en terrain calcaire (vitesse d'érosion et morphologie). Annales de Géographie 68(366):97–120

Cvijić J (1899) Glacial and morphological studies of the mountains of Bosnia, Herzegovina and Monte Negro (in Serbian). Glas SKAN 57, Belgrade
Cvijić J (1900) L'époque glaciaire dans la péninsule des Balkans. Annales de Géographie 9: 359–372
Cvijić J (1903) New results on the glacial epoch in the Balkan Peninsula (Novi rezultati o glacijalnoj eposi Balkanskog poluostrva). Glas SKAN 65:185–240
Cvijić J (1913) Ledeno doba u Prokletijama i okolnim planinama (Ice Age in Prokletije and the Surrounding Mountains; in Serbian). Glas SKAN 91, Belgrade
Cvijić J (1917) L'époque glaciaire dans la péninsule balkanique. Annales de Géographie 26 (141):189–218
Cvijić J (1920) Geomorfologija, autorizovana skripta po predavanjima. Petrović D (Ur) Ledeno doba i glacijalni oblici Balkanskog poluostrva, pp 434–459
Davis JW (1880) On a group of erratic boulders at Norber, near Clapham, in Yorkshire. Proc Yorks Geol Polytech Soc Geol Soc Lond 7(3):266–273
Delmas M (2009) Chronologie et impact géomorphologique de glaciation quaternaires dans l'est des Pyrénées. Géomorphologie, Université Panthéon-Sorbonne, Paris
Dorale JA, Edwards RL, Alexander EC, Shen CC, Richards DA, Cheng H (2004) Uranium-series dating of speleothems: current techniques, limits, applications. In: Studies of cave sediments. Springer, Boston, pp 177–197
Drew DP (1983) Accelerated soil erosion in a karst area: the Burren, western Ireland. J Hydrol 61 (1–3):113–124
Dreybrodt W (1982) A possible mechanism for growth of calcite speleothems without participation of biogenic carbon dioxide. Earth Planet Sci Lett 58(2):293
Dyke AS (2004) An outline of North American deglaciation with emphasis on central and northern Canada. Dev Quat Sci 2:373–424
Ehlers J, Gibbard P (2008) Extent and chronology of Quaternary glaciation. Episodes 31(2): 211–218
Ehlers J, Gibbard PL, Hughes PD (eds) (2011) Quaternary glaciations—extent and chronology: a closer look. Elsevier
Ek C (1964) Note sur les eaux de fonte des glaciers de la Haute Maurienne (Savoie, France): leur action sur les carbonates. Revue belge de Géographie 88(1–2):127–156
Ford DC (1971a) Alpine Karst in the Mt. Castleguard-Columbia Icefield Area, Canadian Rocky Mountains. Arct Alp Res 239–252
Ford DC (1971b) Characteristics of limestone solution in the southern Rocky Mountains and the Selkrik Mountains, Alberta and British Columbia. Can J Earth Sci 8:585–609
Ford DC (1976) Evidences of multiple glaciation in South Nahanni National Park, Mackenzie Mountains, Northwest Territories. Can J Earth Sci 13(10):1433–1445
Ford DC (1983a) Effects of glaciations upon karst aquifers in Canada. J Hydrol 61(1–3):149–158
Ford DC (1983b) Alpine Karst Systems at Crowsnest Pass, Alberta-British Columbia, Canada. J Hydrol 61(1):187–192
Ford DC (1983c) Concluding discussion. Arct Alp Res 551–554
Ford DC, Schwarcz HP, Drake JJ, Gascoyne M, Harmon RS, Latham AG (1981) Estimates of the age of the existing relief within the Southern Rocky Mountains of Canada. Arct Alp Res 13:1–10
Gascoyne M, Ford DC (1984) Uranium series dating of speleothems. II: results from the Yorkshire Dales and implications for cave development and quaternary climates. Cave Sci 11(2):65–85
Gascoyne M, Currant AP, Lord TC (1981) Ipswichian fauna of Victoria Cave and the marine palaeoclimatic record. Nature 294(5842):652–654
Goldie H (1973) The limestone pavements of Craven. Trans Cave Res Group Great Br 15: 175–190
Gremaud V, Goldscheider N, Savoy L, Favre G, Masson H (2009) Geological structure recharge processes and underground drainage of a glacierized karst aquifer system, Tsanfleuron-Sanetsch, Swiss Alps. Hydrogeol J 17:1833–1848
Grund A (1902) Neue Eiszeitspuren aus Bosnien und der Hercegovina. Globus 78(9):173–174

Grund A (1910) Beiträge zur Morphologie des Dinarischen Gebirges. Geographische Abhandlungen 9(3):230
Hallet B (1976) Deposits formed by subglacial precipitation of $CaCO_3$. Geol Soc Am Bull 87 (7):1003–1015
Halliday WR (2004) America, north: history. In: Gunn J (ed) Encyclopedia of caves and karst science. Fitzroy Dearborn, New York, pp 102–109
Harmon RS, Thompson P, Schwarcz HP, Ford DC (1975) Uranium series dating of speleothems. Nail Speleol Soc Bull 37(2):21–33
Harmon RS, Ford DC, Schwarcz HP (1977) Interglacial chronology of the Rocky and Mackenzie Mountains based upon $^{230}Th–^{234}U$ dating of calcite speleothems. Can J Earth Sci 14(11): 2543–2552
Häuselmann P (2007) How to date nothing with cosmogenic nuclides. Acta Carsologica 36(1): 93–100
Häuselmann P, Granger DE (2005) Dating of caves by cosmogenic nucleides: method, possibilities, and the Siebenhengste example (Switzerland). Acta Carsologica 34(1):43–50
Häuselmann P, Lauritzen SE, Jeannin PY, Monbaron M (2008) Glacier advances during the last 400 ka as evidenced in St. Beatus Caves (BE, Switzerland). Quatern Int 189(1):173–189
Hobléa F, Häuselmann P, Kubik P (2011) Cosmogenic nuclide dating of cave deposits of Mount Granier (Hauts de Chartreuse Nature Reserve, France): morphogenic and palaeogeographical implications. Géomorphol: relief, processus, environnement 17(4):395–406
Holzkämper S, Spötl C, Mangini A (2005) High-precision constraints on timing of Alpine warm periods during the middle to late Pleistocene using speleothem growth periods. Earth Planet Sci Lett 236(3):751–764
Horn G (1935) Űber die Bildung von Karsthöhlen unter einem Gletcher. Nor Geogr Tidsskr 5:494–498
Horn G (1937) Űber einige Karsthöhlen in Norwegen. Mitteilungen für Höhlen und Karstforschung 1–15
Horn G (1947) Karsthuler i Nordland. Norg Geol Unders 165:1–177
Hughes TM (1901) Ingleborough. Part I. physical geography. Proc Yorks Geol Polytech Soc 14:125–150
Hughes PD, Woodward JC (2017) Quaternary glaciation in the Mediterranean mountains: a new synthesis. Geol Soc Lond Spec Publ 433(1):1–23
Hughes PD, Gibbard PL, Woodward JC (2005) Quaternary glacial records in mountain regions: a formal stratigraphical approach. Episodes-News Mag Int Union Geol Sci 28(2):85–92
Hughes PD, Woodward JC, Van Calsteren PC, Thomas LE (2011) The glacial history of the Dinaric Alps, Montenegro. Quatern Sci Rev 30(23):3393–3412
Jakucs L (1977) Morphogenetics of karst regions. Wiley
Jiménez-Sánchez M, Rodríguez-Rodríguez L, García-Ruiz JM, Domínguez-Cuesta MJ, Farias P, Valero-Garcés B, Moreno A, Rico M, Valcárcel M (2013) A review of glacial geomorphology and chronology in northern Spain: timing and regional variability during the last glacial cycle. Geomorphology 196:50–64
Judson D (2004) Britain and Ireland: history. In: Gunn J (ed) Encyclopedia of caves and karst science. Fitzroy Dearborn, New York, pp 341–345
Kotarba A, Hercman H, Dramis F (2001) On the age of Campo Imperatore glaciations, Gran Sasso Massif, Central Italy. Geografia Fisica e Dinamica Quaternaria 24:65–69
Larson EB, Mylroie JE (2013) Quaternary glacial cycles: karst processes and the global CO_2 budget. Acta Carsologica 42(2/3):197–202
Lauritzen SE (1981) Glaciated karst in Norway. In: Proceedings of 8th International Speleological Congress. Bowling Green, KY, USA, pp 410–411
Lauritzen SE (1982) The paleocurrents and morphology of Pikhåggrottene, Svartisen, North Norway. Nor Geogr Tidsskr 36:183–209
Lauritzen SE (1983) Arctic and Alpine Karst symposium. Program and Field Guide, Department of Chemistry, University of Oslo, 89 p

Lauritzen SE (1984) Evidence of subglacial karstification in Glomdal, Svartisen, Norway. Nor Geogr Tidsskr 38:169–170
Lauritzen SE (1986) Kvithola at Fauske; Northern Norway: an example of ice-contact speleogenesis. Nor Geol Tidsskr 66:153–161
Lauritzen SE, Gascoyne M (1980) The first radiometric dating of Norwegian stalagmites–Evidence of pre-Weichselian karst caves. Nor Geogr Tidsskr 34:77–82
Loubiere JF (1987) Observation preliminaires sur les cavites de la region du Lac Centrum (Nord-est Grönland). Karstologia 9(1):7–16
Luetscher M, Hoffmann DL, Frisia S, Spötl C (2011) Holocene glacier history from alpine speleothems, Milchbach cave, Switzerland. Earth Planet Sci Lett 302(1):95–106
Magaritz M (1973) Precipitation of secondary calcite in glacier areas; carbon and oxygen isotopic composition of calcites from Mt. Hermon, Israel, and the European Alps. Earth Planet Sci Lett 17(2):385–390
Maire R (1976) Recherches géomorphologiques sur les karsts haut-alpins du Haut-Giffre et de Suisse occidentale. Thèse 3e cycle, Université de Nice, 456 p
Maire R (1990) La haute montagne calcaire: karsts, cavités, remplissages, paléoclimats, Quaternaire. Karstologia-Mémoires 3:1–731
Maire R, l'Équipe Ultima Esperanza, Pernette JF, Fage LH (1999) Les "glaciers de marbre" de Patagonie, Chili – Un karst subpolaire océanique de la zone australe. Karstologia 33(1):25–40
Messerli B (1967) Die eiszeitliche und die gegenwärtige Vergletscherung im Mittelmeerraum. Geographica Helvetica 22(3):105–228
Milivojević M, Menković L, Ćalić J (2008) Pleistocene glacial relief of the central part of Mt. Prokletije (Albanian Alps). Quatern Int 190(1):112–122
Miotke FD (1968) Karstmorphologische studien in der glazialüberformten Hohenstufe der Picos de Europa, Nordspanien. Selbtverlag der geografischen Gessellschaft, Hannover, 161 p
Moles NR, Moles RT (2002) Influence of geology, glacial processes and land use on soil composition and quaternary landscape evolution in the Burren National Park, Ireland. Catena 47(4):291–321
Oxaal L (1914) Kalkstenshuler i Ranen. Norg Geol Unders 69:1–47
Palmer AN (1984) Objectives and current status of alpine and arctic karst research. Nor Geogr Tidsskr 38(3–4):145–150
Pavuza R, Stummer G (1999) The Dachstein caves: the Dachstein Region—its karst and its caves. In: Mandl GW (ed) FOREGS `99 Vienna. 150 Years Geological Survey of Austria. Field trip guide: Vienna-Dachstein-Hallstatt-Salzkammergut, Wien, vol 113, no 43, pp 101–105
Penck A (1885) La Periode glaciaire dans les Pyrénées. Bulletin de la Societe d'histoire naturelle de Toulouse 19:105–200
Penck A (1900) Die Eiszeit auf der Balkanhalbinsel. Globus 78:133–178
Penck A, Brückner E (1901) Die Alpen im Eiszeitalter: In 3 Bden, Tauchnitz
Popov IV, Gvozdetskiy NA, Chiksishev AG, Kudelin BI (1972) Karst in U.S.S.R. In: Herak M, Stringfield VT (eds) Karst, important karst regions of the northern hemisphere. Elsevier, pp 355–416
Reid C (1887) On the origin of dry chalk valleys and the coombe rock. Q J Geol Soc 43:364–373
Roglić J (1961) Glaciation of the Dinaric Mountains and its Effect on Karst. Papers of VI. International Congress Quat, Warszaw, pp 293-299
Salvigsen O, Elgersma A (1985) Large-scale karst features and open taliks at Vardeborgsletta, outer Isfjorden, Svalbard. Polar Res 3(2):145–153
Sarıkaya MA, Zreda M, Çiner A, Zweck C (2008) Cold and wet last glacial maximum on Mount Sandıras, SW Turkey, inferred from cosmogenic dating and glacier modeling. Quatern Sci Rev 27(7):769–780
Sarıkaya MA, Çiner A, Haybat H, Zreda M (2014) An early advance of glaciers on Mount Akdağ, SW Turkey, before the global last glacial maximum; insights from cosmogenic nuclides and glacier modeling. Quatern Sci Rev 88:96–109
Sawicki L (1911) Die eiszeitliche Vergletscherung des Orjen in Süddalmatien. Verlag von Gebrüder Borntraeger

Shaw T (2004) France: history. In: Gunn J (ed) Encyclopedia of caves and karst science. Fitzroy Dearborn, New York, pp 723–729
Skoglund RØ, Lauritzen SE, Gabrovšek F (2010) The impact of glacier ice-contact and subglacial hydrochemistry on evolution of maze caves: a modelling approach. J Hydrol 388(1):157–172
Smart CC (1983) The hydrology of the castleguard karst, columbia icefields, Alberta, Canada. Arct Alp Res 471–486
Smart PL (1986) Origin and development of glacio-karst closed depressions in the Picos de Europa, Spain. Zeitschrift für Geomorphologie 30:423–443
Smart CC (2004) Glacierized and glaciated karst. In: Gunn J (ed) Encyclopedia of caves and karst science. Fitzroy Dearborn, New York, pp 804–809
Smith DI, Atkinson TC (1976) Process, landforms and climate in limestone regions. In: Derbyshire E (ed) Geomorphology and Ciimale. Wiley, London, pp 367–409
Spötl C, Mangini A (2007) Speleothems and paleoglaciers. Earth Planet Sci Lett 254(3):323–331
Spötl C, Mangini A, Frank N, Eichstädter R, Burns SJ (2002a) Start of the last interglacial period at 135 ka: evidence from a high Alpine speleothem. Geology 30(9):815–818
Spötl C, Unterwurzacher M, Mangini A, Longstaffe FJ (2002b) Carbonate speleothems in the dry, inneralpine Vinschgau valley, northernmost Italy: witnesses of changes in climate and hydrology since the last glacial maximum. J Sediment Res 72(6):793–808
Stepišnik U, Ferk M, Kodelja B, Medenjak G, Mihevc A, Natek K, Žebre M (2009) Glaciokarst of western Orjen, Montenegro. Cave Karst Sci 36(1):21–28
Stepišnik U, Grlj A, Radoš D, Žebre M (2016) Geomorphology of Blidinje, Dinaric Alps (Bosnia and Herzegovina). J Maps 12(sup1):163–171
Svendsen JI, Alexanderson H, Astakhov VI, Demidov I, Dowdeswell JA, Funder S, Gataullin V, Henriksen M, Hjort C, Houmark-Nielsen M, Hubberten HW, Ingólfsson Ó, Jakobsson M, Kjær KH, Larsen E, Lokrantz H, Lunkka JP, Lyså A, Mangerud J, Matiouchkov A, Murray A, Möller P, Niessen F, Nikolskaya O, Polyak L, Saarnisto M, Siegert C, Siegert MJ, Spielhagen RF, Stein R (2004) Late Quaternary ice sheet history of northern Eurasia. Quatern Sci Rev 23(11):1229–1271
Sweeting MM (1966) The weathering of limestones with particular reference to the Carboniferous limestones of northern England. In: Dury GH (ed) Essays in Geomorphology. Heinemann, London, pp 177–210
Telbisz T (2010a) Morphology and GIS-analysis of closed depressions in Sinjajevina Mts (Montenegro) Karst Dev 1(1):41–47
Telbisz T (2010b) Glacio-karst features of the Sinjajevina Mts (Montenegro): an overview and DEM-analysis. Karst Dev 1(1):17–22
Telbisz T, Mari L, Szabó L (2011) Geomorphological characteristics of the Italian side of Canin massif (Julian Alps) using digital terrain analysis and field observations. Acta Carsologica 40 (2):255–266
Telbisz T, Látos T, Deák M, Székely B, Koma Z, Standovár T (2016) The advantage of LiDAR digital terrain models in doline morphometry compared to topographic map based datasets – Aggtelek Karst (Hungary) as an example. Acta Carsologica 45(1):5–18
Veress M (2010) Karst environments: karren formation in high mountains. Springer Science Business Media
Veress M (2017) Solution doline development on glaciokarst in alpine and Dinaric areas. Earth Sci Rev 173:31–48
Veress M, Szunyogh G, Zentai Z, Tóth G, Czöpek I (2006) The effect of wind on karren formation on the island of Diego de Almagro (Chile). Zeitschrift für Geomorphologie 50(4):425–445
Viles HA (2003) Conceptual modeling of the impacts of climate change on karst geomorphology in the UK and Ireland. J Nat Conserv 11(1):59–66
Vincent PJ (1995) Limestone pavements in the British Isles: a review. Geogr J 161:265–274
Vincent PJ, Wilson P, Lord TC, Schnabel C, Wilcken KM (2010) Cosmogenic isotope (^{36}Cl) surface exposure dating of the Norber erratics, Yorkshire Dales: further constraints on the timing of the LGM deglaciation in Britain. Proc Geol Assoc 121:24–31
Waltham T (1971) British Karst Research Expedition in Himalaya – Report, 99 p

Waltham AC, Simms MJ, Farrant AR, Goldie HS (1997) Karst and caves of Great Britain. In: Geological conservation review, vol 12, 358 p
Williams PW (1966) Limestone pavements with special reference to western Ireland. Trans Inst Br Geogr 40:155–172
Wilson P, Barrows TT, Lord TC, Vincent PJ (2012) Surface lowering of limestone pavement as determined by cosmogenic (^{36}Cl) analysis. Earth Surf Proc Land 37:1518–1526
Wilson P, Lord T, Rodés Á (2013a) Deglaciation of the eastern Cumbria glaciokarst, northwest England, as determined by cosmogenic nuclide (^{10}Be) surface exposure dating, and the pattern and significance of subsequent environmental changes. Cave Karst Sci 40(1):22–27
Wilson P, Schnabel C, Wilcken KM, Vincent PJ (2013b) Surface exposure dating (36Cl and 10Be) of post-Last Glacial Maximum valley moraines, Lake District, northwest England: some issues and implications. J Quat Sci 28(4):379–390
Woodward JC, Macklin, MG, Smith GR (2004) Pleistocene glaciation in the mountains of Greece. In: Quaternary glaciations—extent and chronology. Part I: Europe. Elsevier, Amsterdam, pp 155–173
Žebre M, Stepišnik U (2015a) Glaciokarst geomorphology of the Northern Dinaric Alps: Snežnik (Slovenia) and Gorski Kotar (Croatia). J Maps 12(5):873–881
Žebre M, Stepišnik U (2015b) Glaciokarst landforms and processes of the southern Dinaric Alps. Earth Surf Proc Land 40(11):1493–1505
Zeng C, Gremaud V, Zeng H, Liu ZH, Goldscheider N (2012) Temperature-driven meltwater production and hydrochemical variations at a glaciated alpine karst aquifer: implication for the atmospheric CO2 sink under global warming. Environ Earth Sci 65:2285–2297
Zeng C, Liu Z, Yang J, Yang R (2015) A groundwater conceptual model and karst-related carbon sink for a glacierized alpine karst aquifer, Southwestern China. J Hydrol 529:120–133
Zreda M, Çiner A, Sarıkaya MA, Zweck C, Bayarı S (2011) Remarkably extensive glaciation and fast deglaciation and climate change in Turkey near the Pleistocene-Holocene boundary. Geology 39(11):1051–1054

Chapter 2
General Description of Glaciokarsts

Márton Veress and Dénes Lóczy

Abstract This chapter presents the glacier types associated with karsts and the types of glaciokarst. Among the latter, the glacier/karst interactions and planform types are analysed in detail. The conditions of glacier formation are investigated and the geomorphological zones and subzones of glaciokarst are overviewed. When presenting the properties of glaciokarst, the balance of subglacial waters and epikarst characteristics are described, glaciokarst types are distinguished by the age and mode of origin and the modes and rates of transformation of landforms are identified.

Keywords Glacier types · Glaciokarst types · Geomorphological zones Geomorphological subzones · Epikarst

2.1 Introduction

Karst transformed by ice is called glaciokarst (Sweeting 1973; Maire 1990)—even if that concerns low and vegetated terrains (Sauro 2009). The term glaciokarst is also used for karst-like features of glaciers (Andreichuk 1992). However, Marlyudov (2006) proposes the term glacial karst for such features. Karst denuded by glacial erosion, having a bare surface and mainly rising above its environs is regarded a typical glaciokarst where processes only characteristic of this karst type are active at present. For this reason and because of glacial erosion, glaciokarsts show a wide range of particular landforms and landform assemblages (Kunaver 1983, 2009; Maire 1990; Smart 2004). Smart (2004) basically identifies two varieties of recent glaciokarsts: “glacierized karst”, which is covered by ice presently, and “glaciated karst”, which had been covered by ice before. Glaciation may have developed in the area of non-karstic rocks around the karst. When glaciers did not reach the karst, glacier erosion did not take place. The ice, however, can affect karst evolution even in this case when the meltwater input is high and this promotes ponor and swallow cave development. In addition, the karst surface can receive considerable amounts of till of non-karstic origin. In this case, the solution capacity

M. Veress et al., *Glaciokarsts*, Springer Geography,
https://doi.org/10.1007/978-3-319-97292-3_2

of the water reaching the karst surface is not reduced. It is advisable to call this kind of karst semi-glaciokarst, as described from Tasmania (Kiernan et al. 2001).

It has been not only the better knowledge of the complex multifactorial interactions concerning this karst type that glaciokarsts became popular among researchers (Bögli 1964; Barrére 1964; Miotke 1968; Ford 1979, 1983, 1984, 1987, 1996; Smart 1986; Ford and Williams 1989, 2007; Veress 2012a; Telbisz 2010; Telbisz et al. 2005; Veress and Lóczy 2015), but the understanding of the conditions of glaciation of their formation also proved to be challenging (Van Husen 2000; Hughes 2004; Hughes and Woodward 2009; Menkovič et al. 2004; Woodward et al. 2004; Hughes et al. 2006, 2010, 2011; Žebre et al. 2013; Adamson et al. 2014, Žebre and Stepišnik 2015a).

The glaciokarst areas of the Earth primarily reflect the influence of Pleistocene glaciations, but older glaciokarsts are also known, from Australia (Grimes 2012; Playford 2002, 2009), the British Isles (Clayton 1966) and from Turkey (Bayari et al. 2003).

The general properties of glaciokarsts are the following:

- Compared to other karst types, on glaciokarsts more numerous geomorphic agents are or were active.
- In addition to karst landforms, on glaciokarsts glacial features are often present or even dominate. The two kinds of landforms often occur superimposed. The age of landforms of the glaciokarst fall into four classes: of pre-glacial (1), glacial (2), interglacial (3) and recent origin (4). Those of classes 1 and 3 are transformed, destroyed or infilled to large extent. The frequency and distribution of features in the four age classes can be variable—even in the different parts of the same glaciokarst.

2.2 Glacier Types

There are several glacier typologies in international literature (Allen 1970; Embleton 1980; Selby 1985; Matthews 1987; Knight 1999; Bennett 2003; Rau et al. 2005; Hambrey and Fitzsimons 2010; Bindschadler et al. 2011). The properties of karst (location, extension, original coveredness or the lack of it) may influence the glacier type developing in its area. The latter, on the other hand, reacts the glaciation of the karst (the extent, distribution and pattern of glacial erosion). Therefore, here follows an overview of the most common glacier types in karst regions.

- Cirque glacier, which could have developed on steep slopes, but on karst it most commonly occurs in karst depressions (Maire 1977; Nicod 1978; Ford 1979; Smart 1986). Essentially the cirque glacier does not have an ablation area, neither its movement nor its erosion is considerable, at its termination, however, moraines may occur. Such glaciers are found in highly variable environments.

- Valley glaciers are the most common glacier type in karst regions. The glacial valleys here are constituted of two sections: the cirque in the accumulation area and the glacial trough in the ablation area. Valley glaciers in non-karstic areas are formed in river valleys. The valley glaciers of karst areas, however, are seldom developed in river valleys since the latter are less typical on karst. They only appear in fluvial valleys if the karst used to be covered and the river valley was inherited over the bedrock. (Because of intensive denudation, the traces of the cover sediments cannot be detected.) In lack of valleys of fluvial erosion the glacier formed in a karst depression. This could explain why there is often no (remarkably distinct) cirque in the glacial valleys. (Originally, each of the depressions could have been glacial cirques.) At the same time, numerous valley glaciers developed on steep slopes are coupled with (a group of) cirques (Menkovic 1994; Zebre et al. 2013). It is also common that on the margins of glacial troughs, cirques are aligned (Fig. 2.1) or the glacial troughs radiate from the cirque (which evolved from a former karst depression). Examples are found in the Totes Gebirge (Alps, Austria), where glacial troughs directed in three directions originate from the cirque (formerly a depression) below Windgößl peak.
- Glacier network, where, with the breaching of arètes and connecting neighbouring snow accumulation basins, the glaciers merge with one another creating an ice cap nature. This glacier type developed in some parts of the Eastern Alps (Husen 2000).

Fig. 2.1 Cirque developed from a doline at the valley margin (Durmitor Mts, photo by Veress). 1. Margin of a former doline, 2. cirque, 3. margin of glacial trough connected to the cirque, 4. alluvial fan, 5. step

- Pediment glacier, created by the merging of valley glaciers on the lower terrains surrounding the karst. Thus, sections of the neighbouring lower karst could also evolve into glaciokarst. The valley of pediment glacier is hardly observable or does not exist at all. Therefore, ice covering such a terrain did not destroy the depressions. Examples can be cited from the Durmitor (Dinaric Mountains, Montenegro), where the piedmont glaciers (Djurovič 2009) only partially destroyed the preglacial dolines.
- Outlet glacier, a branch of the ice sheet, which often reaches beyond the karst area.
- Pleistocene ice sheets, which covered parts of continents and thus, reached beyond some karst regions.
- Ice caps, which totally covered the higher sections of karst areas or plateaus. This is particularly striking for glaciers in marine environment, typically in the Dinaric Mountains and on the karsts of Greece (Hughes 2004; Hughes et al. 2006, 2011; Adamson et al. 2014).
- In the rock niches with circular margins on plateau edges small-size glaciers form from snow accumulated during a blizzard. These are niche or cliff glaciers (Groom 1959).

As the intensity of the glaciation changes, at the same site on the glaciokarst glaciers of different types could have alternated. Thus, ice sheets or glacier networks could be replaced by a valley or occasionally cirque glaciers.

The nature of glacial erosion depends on glacier type. For valley glaciers, linear glacial erosion is typical, while ice sheets and ice caps exert free glacial erosion. On karst-free glacial erosion operates in the case of pediment glaciers (no glacial trough develops). Erosion is of local character in the case of cirque glaciers and, consequently, this process is mainly restricted to karst depressions.

Wall icing and hanging glaciers are also differentiated (Star and Langenscheidt 2015). In case of the former type, the ice is located on the steep slope of a mountain, while in the latter case the glacier tongue terminates high over its environment.

2.3 Types of Glaciokarst

Glaciokarsts can be classified from different aspects: the presence of meltwater, karstic rocks, distance from the Equator, the relative positions of the glacier and karst (or with regard to glacier/karst interactions), the vicinity of the karst, relative positions of glaciokarsts, the rate of glaciation or glaciokarst orography. The classification according to the presence of meltwater distinguishes between warm, cold and polythermal glaciers. There is meltwater at the floor of warm glaciers, there is not at cold glaciers and polythermal glaciers have warm as well as cold sections (Glasser and Hambrey 2001; Bennett and Glasser 2009; Davies 2014). Most of the glaciokarsts are built up of limestone, but they also occur on marble (Patagonian archipelago, NE United States), dolomite (Dolomites), gypsum

(Canadian Rockies) (Ford 1979; Maire et al. 1999; Veress et al. 2006; Cooper and Mylroie 2015). According to structure, they are of Alpine or continental types (Ford and Williams 2007), geosyncline or tabular type (Veress 2016).

According to their environment, glaciokarsts are found in the marine or terrestrial environment. In different environments, there are different amounts of precipitation and ice cover of the karst.

2.3.1 Classification by Glacier/Karst Interaction

Water tracing experiments confirm that the meltwater from glaciers flows into the karst (Smart 2004). The precipitations in the cavities of the karst under the glacier point to water movement and solution (Smart 2004). With regard to the glacier/karst interaction, Pyrenean and Canadian glacier/karst types of Alpine karst are identified for glaciers where meltwater is present (Ford and Williams 2007). In the Pyrenean-type glacier is located on higher parts of the karst and, thus, the meltwater inflow into the karst can re-emerge in karst springs. This type of karst is presented by Maire (1990) and Audra (2000). The Canadian-type glaciers can develop at lower elevations and reach the karst springs, prevent or inhibit water outflow from the karst, which is only possible at its termination (Fig. 2.2). For the Canadian types, the opportunities for karst development are more limited since the glacial influence directly affects the karst system (Ford and Williams 2007). According to Ford (1979) and Smart (2004), glacial till can block springs and for this reason water outflow can stop here too and karst water can be impounded. This impoundment is of lesser extent and only reaches the level of till cover. If, however, impoundment happens, karst water can appear on the till-mantled valley floors in the interglacial periods too. The impoundment is strengthened by the increasing melting of the ice or by the partial or complete infilling of the caves of the karst.

As a consequence of impoundment, karst water appears in the ice (Ford 1979). According to Ford (1979), if the ice cover reaches beyond the karst, karst water uniformly develops in the ice, if not, this is only limited. In the Canadian-type glacier karst water development is due to the impoundment of karst water, while in the Pyrenean type this is rather due to the rising karst water levels driven by the inflow of meltwater. There is a gradual transition between glaciers of both types. (During major glaciation, the same glacier advances and then, at the beginning or end of the glacial or during minor glaciation it retreats.) If the glacier is located on a plateau, the karst spring is not covered. In this case, the karst water table, rising from the spring level, finally crosses the rock surface. The vadose zone develops between the spring and the intersection point and a larger distance from the intersection, the phreatic zone follows (Ford 1979).

If the steepness of karst water table is identical, the smaller is the height difference between the highest point of karst water table and the base level, the karst water level compared to the spring, the closer is the intersection site of the limestone surface and, thus, the narrower is the vadose zone (Fig. 2.3a). The greater is

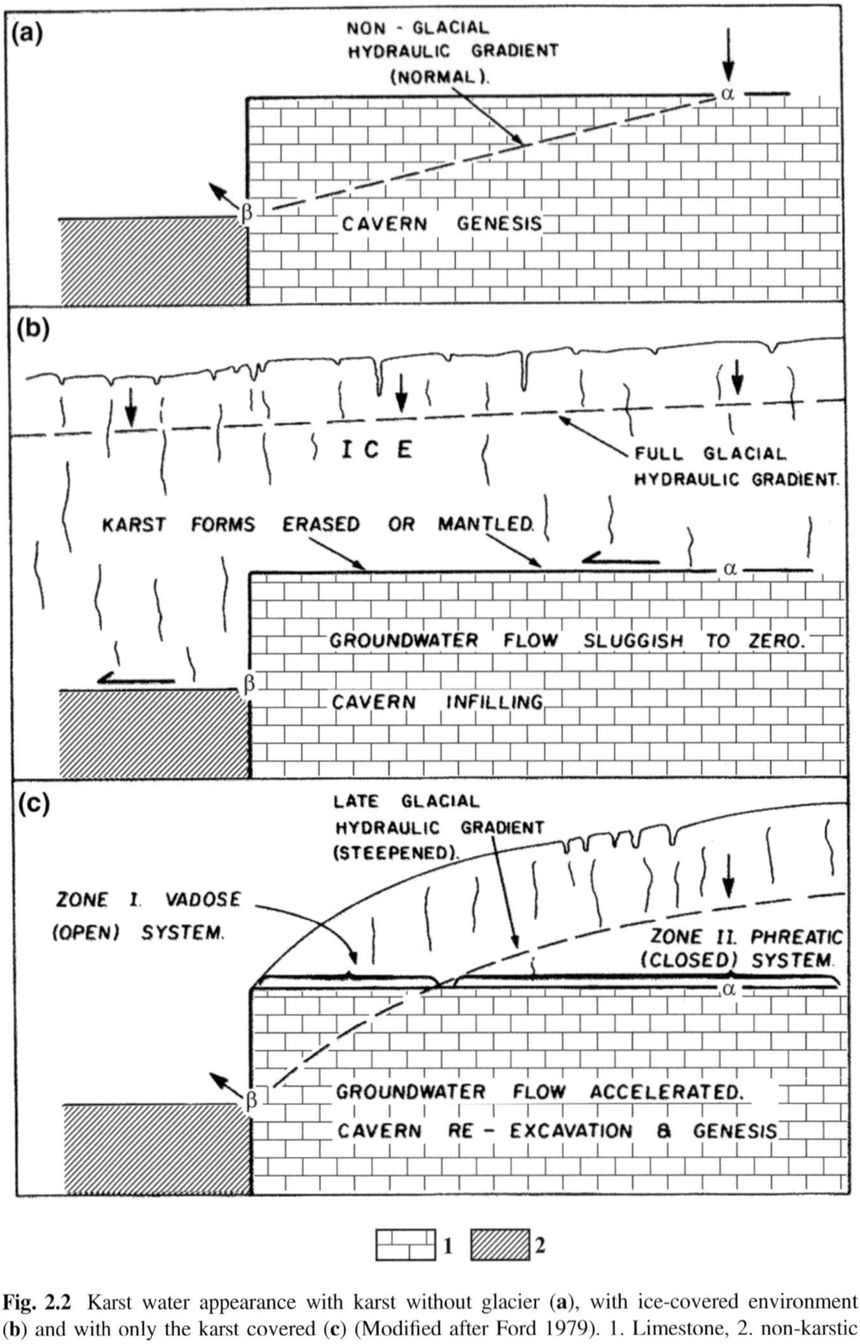

Fig. 2.2 Karst water appearance with karst without glacier (**a**), with ice-covered environment (**b**) and with only the karst covered (**c**) (Modified after Ford 1979). 1. Limestone, 2. non-karstic rock

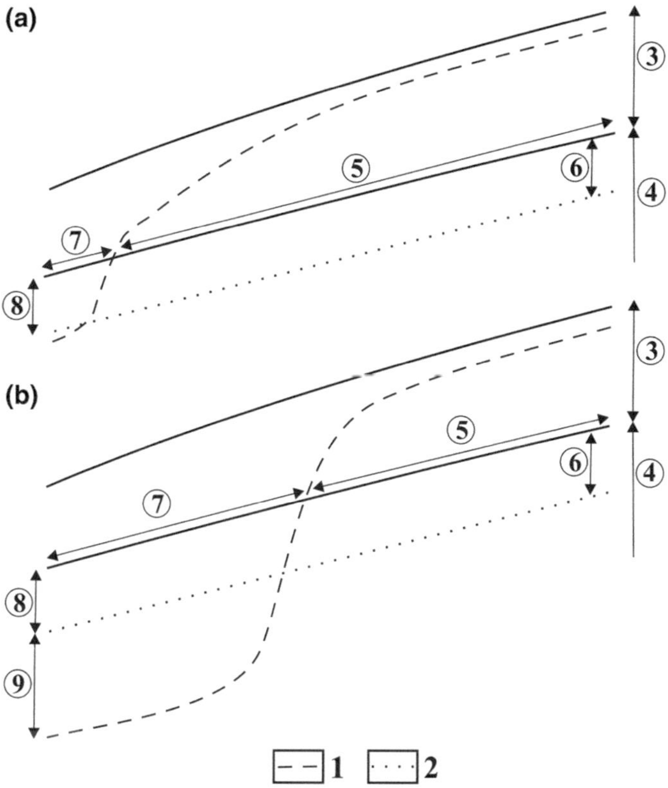

Fig. 2.3 Theoretical relationship on temperate alpine karst between the steepness of karst water table and the extension of vadose and phreatic zones with low (**a**) and high (**b**) steepness. 1. Karst water table, 2. preglacial karst water table, 3. glacier, 4. karst, 5. phreatic zone, 6. phreatic zone transformed from vadose zone, 7. vadose zone, 8. permanent vadose zone, 9. vadose zone transformed from phreatic zone

the height difference, with the same steepness of the karst water table, the farther is the intersection with the limestone surface and the vadose zone becomes wider (Fig. 2.3b). If the height difference is the same, the karst water table intersects with the limestone on the karst with less steep karst water table and in this case, the vadose zone becomes wider. The more cavities are found in the rock, the steepness of the karst water table is more moderate since with numerous cavities friction is reduced and karst water flow is more rapid. Consequently, the widest vadose zone can develop on karst with large cavities.

The extension of the vadose and phreatic zones determines the nature of cavity formation. In the former case shafts and shaft systems develop, while in the phreatic zone horizontal passages form. The extension of both zones may change at the expanse of each other, in the vadose zone (or in part of it) phreatic cavity formation, in the phreatic zone (or in part of it) vadose cavity formation took place.

On temperate alpine karsts, the zones can change into each other during the alternation of glacials and interglacials or during the change of glacier/karst type. The preglacial vadose zone with a lower karst water table (Fig. 2.4a) could transform into phreatic in the glacial because of the rising karst water table. Because of changing steepness of karst water table, in the vadose zone close to the spring, where the karst water table sinks below the former karst water table, a new vadose zone develops from the phreatic zone or the vadose zone becomes wider downwards at the expanse of the phreatic zone (Fig. 2.4b). In the interglacials or in the

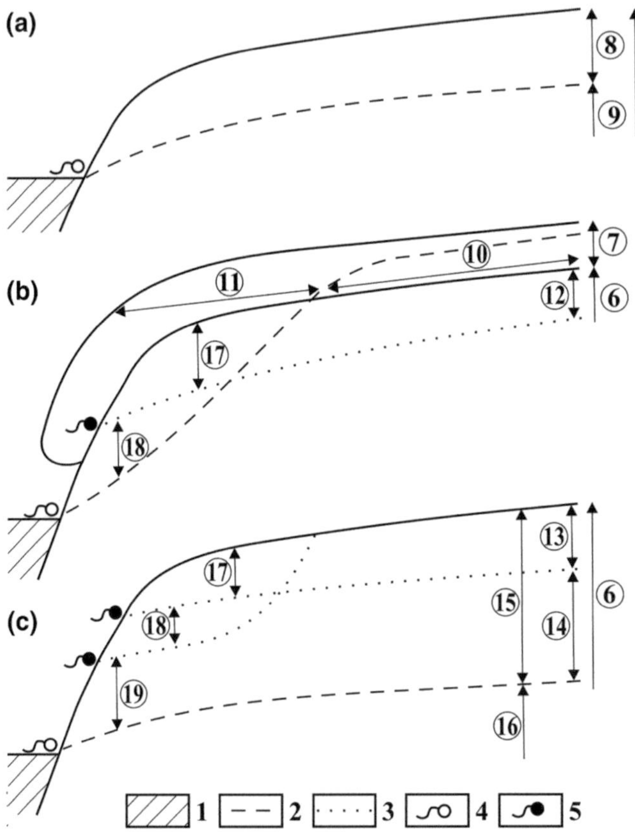

Fig. 2.4 Theoretical opportunities of the mutual transformation of vadose and phreatic zones on temperate alpine karst during glacials and interglacials. **a** Preglacial phase, **b** glacial phase, **c** interglacial or recent phase. 1. Non-karstic impermeable rock, 2. karst water table, 3. former karst water table, 4. active karst spring, 5. inactive karst spring, 6. karst, 7. glacier, 8. vadose zone before the glacial, 9. phreatic zone before the glacial, 10. phreatic zone in the glacial, 11. vadose zone in the glacial, 12. vadose zone transformed into phreatic in the glacial, 13. vadose zone in the interglacial transformed from the glacial phreatic zone (preglacial vadose), 14. vadose zone transformed from the glacial phreatic zone (preglacial phreatic), 15. vadose zone in the interglacial or at present, 16. phreatic zone in the interglacial or at present, 17. permanent vadose zone, 18. zone transformed from phreatic to vadose in the glacial, 19. zone transformed from phreatic to vadose in the interglacial or at present

Holocene, if the base level sank, in the former phreatic zone of the glacial a vadose zone develops, the upper part of which used to be the preglacial vadose zone and the lower part used to be the preglacial phreatic zone (right in Fig. 2.4c). Under the new vadose zone formed in the glacial (Fig. 2.4b) a further vadose zone evolved from part of the former phreatic zone (left part of Fig. 2.4c).

On glaciokarst, hanging karst springs are common in the side or in the valley heads of the main valleys (Ford 1996). During the water supply by meltwaters, the karst water level gets above the base level of erosion. As a result, horizontal cavity formation develops above the base level too. Thus, cavities above the base level may be sites of springs at present.

In the case of the Canadian glacier karst type no vadose zone emerges (Fig. 2.5a) on low-elevation karst, but if the height difference is great (on plateaus), it does (Fig. 2.5b). Therefore, the phreatic and vadose zones can develop in the Canadian type too (Fig. 2.6a).

When the Canadian type is transformed into Pyrenean type, a complete vadose zone appears above the karst water table of the Pyrenean type in the former phreatic zone of the Canadian type, which was the vadose zone in the preglacial. Below the

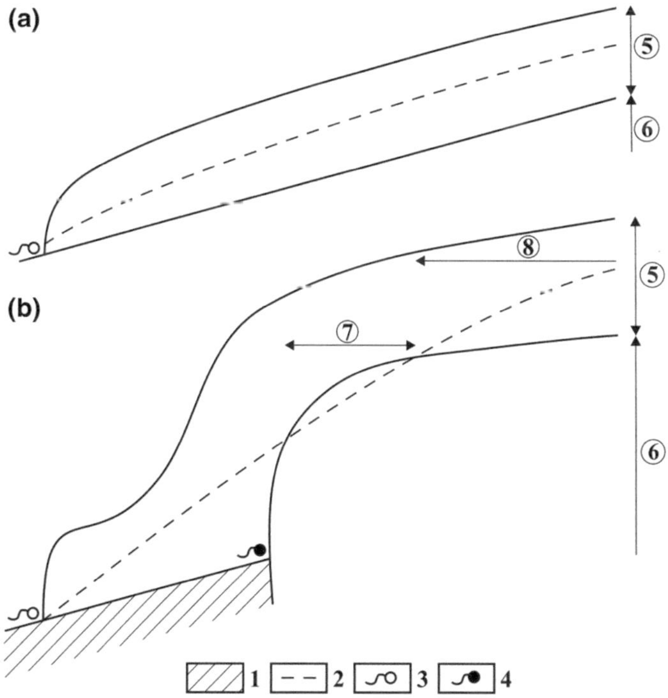

Fig. 2.5 Theoretical relationship between the appearance of the vadose zone and the elevation of the karst for the Canadian glacier-karst type. **a** Low relative relief, **b** high relative relief. 1. Non-karstic rock, 2. karst water table, 3. active spring, 4. inactive spring, 5. glacier, 6. karst, 7. vadose zone, 8. phreatic zone

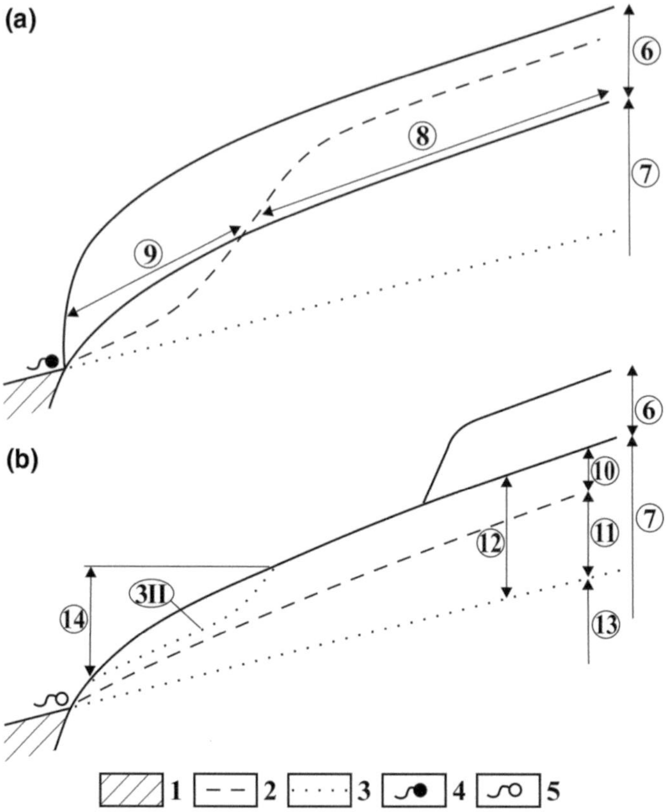

Fig. 2.6 Theoretical opportunities for the mutual transformation of vadose and phreatic zones during glacier retreat. **a**. Canadian glacier-karst type, **b**. pyrenean glacier-karst type. 1. Non-karstic rock, 2. karst water table at the time when the glacier existed, 3. karst water table preceding the existence of glacier, 4. inactive karst spring, 5. active karst spring, 6. glacier, 7. karst, 8. phreatic zone, 9. vadose zone, 10. vadose zone transformed from phreatic (one-time periglacial vadose), 11. unchanged phreatic zone (preglacial vadose), 12. preglacial vadose zone, 13. permanent phreatic zone, 14. permanent vadose zone

karst water table of Pyrenean type, the former preglacial vadose zone remains to be a phreatic zone, irrespective of the change of type (Fig. 2.6b).

Pleistocene glaciations varied in intensity. In the Eastern Alps the intensity of the Riss glacial, compared to the Würm glacial, was lower (Husen 2000); in the Durmitor the glaciations were ever lower intensity in the ever younger glacials (Djurovič 2009). Similarly, in the Orjen Mountains (Dinaric system, Montenegro) intensity was reduced from the older glacials to the younger (Adamson et al. 2014). This induced the formation of shorter glaciers in later glacials. For this reason, the extension of the vadose and phreatic zones on the karst could have changed not only between glacials and interglacials, but also in the subsequent glacials.

2.3.2 Classification by Distance from the Equator

With increasing distance from the Equator and reducing elevation above sea level, in mountains (including karst regions) the number of vertical geographical zones is becoming less and less. At the same time, moving towards the poles, with the same elevation, the extension of higher zones is increasing. Thus, in cold belt glaciokarsts, geographical and also geomorphological zones occur at lower elevations and occupy larger areas. The glaciokarst in this belt are called geomorphologically limited from below. On the glaciokarsts of this type, moving towards the pole, cold-based glaciers will be more and more typical. The statement was even more valid during the glaciations. On glaciokarsts, vertical zonation is best developed in the temperate belt—although during the Pleistocene glaciations also here the glacial zone was dominant. At present, below the occasionally present glacial zone, typical periglacial and temperate fluvial geomorphological zones are found. On the glaciokarsts of this belt, during the glaciations (and also today in the glacial zone), warm-based glaciers are predominant.

The distribution of recent karst landforms on glaciokarsts shows a vertical zonation: below the timberline there are dolines, above that karren (Bauer 1962; Jennings and Bik 1962; Miotke 1968; Jakucs 1977; Ford and Williams 2007); on tropical karsts (in New Guinea) doline and pinnacle terrains, while there are karst hills at lower elevations (Jennings and Bik 1962). This zonation, however, is not always clear, but modified by slope angle, the presence and thickness of cover deposits, water drainage capacity (Ford and Williams 2007). Such factors influence the distribution of karren or the various doline types (Ford and Williams 2007). Through the intensity and duration of solution, zonation is regulated by soil depth and quality (dependent on temperature and ground slope) and the amount and quality of precipitation. The amount and quality of precipitation do not only depend on elevation, but also on morphology, exposure relative to rainfall bringing winds and the distance from seas. Former glacial erosion and accumulation can modify or even fundamentally change zonation (Ford 1979). Thus, if glacial erosion generated flat surfaces, karren develops, while if, at the same elevation, there are dolines and cover deposits, subsidence dolines are formed.

In the tropical belt, a complete series of geomorphological zones can develop. Zonation here is limited from above since karst regions are located at relatively lower elevations than mountains of other (e.g. volcanic) origin. In principle, the glaciers on the glaciokarst of the tropical belt are warm-based glaciers. Glaciokarsts are rather rare in the tropical belt, but examples can be cited from New Guinea (Jennings and Bik 1962; Verstappen 1964; Loeffler 1971; Brook et al. 1977) or Australia, where the Judbarra karst derives from an older, Permian, glaciation, and can be regarded a paleo-glaciokarst (Playford 2002, 2009). The low-lying Judbarra karst presents no zonation. In the tropical belt, most of the karsts have not experienced glaciation in the Pleistocene. Tropical karsts can be placed into a glacial environment in two ways: through large-scale uplift of the karst (see New Guinea) or plate tectonic movements bring it into an environment where, in spite of lower elevation, the tropical form assemblage is glaciated. An example is the Pyrenees, where the Pinnacle (mogote) karst was glaciated (Auly 2000, 2005, 2008).

2.3.3 Classification by Position as Compared to Each Other

According to their position relative to each other, glaciokarsts can be near each other, in this case, they are only separated by valleys, like the glaciokarst of the Northern Limestone Alps (Dachstein, Totes Gebirge and others). Glaciokarst units far from each other are separated by lower, non-glaciated terrains (Dinaric Mountains). Finally, some glaciokarst are not only far from each other, but also separated by non-karstic terrains. In the first, and particularly in the second case, the glaciokarst units do not necessarily have an independent karst water flow system. Especially in the second case, different glaciations for the different units have to be taken into account. In the third case, the glaciokarst units show an independent flow system and the geomorphic evolution of the units and their environments are fundamentally different.

2.3.4 Classification by the Extent of Glaciation

Preglacial karst features (particularly the surface features) may be absent or present on glaciokarsts. They are missing because they have not been formed or because they are removed by glacial erosion if no glacial erosion took place inside the karst features (see Chap. 3). This situation is the more probable, the smaller the karst features were the more intensive the glacial erosion.

The extent of glaciation and the destruction of the karstic landforms can be variable among glaciokarsts. In the case of regional (areal) glaciation, ice covered the entire karst region. According to the degree of glacial erosion, complete, partial and selective denudation can be identified in this case. Complete glacial erosion destroys the whole karstic landform assemblage. With partial glacial erosion, only part of the karstic landforms is destructed. Selective glacial erosion means that part of the landforms is totally destroyed, but others survived. The reason for this can be that the intensity of erosion was high (while at other places it was lower). Semi-glaciation denotes that situation when some section of the karst was not affected by glaciation as the ice did not cover the entire karst. Finally, glacial action can be manifest in accumulation, too. In this case, karstic landforms did not suffer erosion, but filled up or were buried.

2.3.5 *Classification by Morphology*

According to morphology, glaciokarst types depend on what types of glacier(s) developed in the area, in what dimensions and density, with what duration and how many times the glaciation was repeated.

- Plateau glaciokarst (Fig. 2.7a-b)

The karst area is only slightly dissected, side slopes are either steep or gentle. It is the most common karst type. Its surface can be dissected by depressions, marginal horns and glacial erosion features. On the lower plateaus of temperate glaciokarsts, landforms of glacial erosion are of small dimension, today the bare, unvegetated surfaces are absent or of subordinate extension. There appeared glaciers of various types on these plateaus:

- Cirque glaciers, where the ice only filled single depressions (dolines or uvalas) and did not (considerably) reach over their margins. Examples are glaciers in parts of the Biokovo (Dinaric Mountains) (Telbisz et al. 2005) and in the Rax (Alps). At present, karstification affects the once ice-filled depressions (often through covered karst formation) and terrains outside them.
- Valley glaciers developed on the plateau. Glaciers on smaller and lower plateaus were unable to modify the plateau character. The glaciers only covered part of the plateau, did not reach over its margins, did not carve deep troughs, but the cirque and the trough are remarkably distinct. The actual karstification of the plateau is negligible and takes place in the area of the one-time glacier as well as beyond that. Since the glaciers were restricted to the plateaus, their slopes, which are adjacent to valleys bordering the karst, are reshaped by mass movements. Such a glaciokarst is the Schneeberg (Austria).
- Ice cap was formed on a higher, extended or marine-environment karst. Outlet glaciers radiated from the ice cap (e.g. Orjen Mountains—Adamson et al. 2014). Probably the Dachstein (Alps, Austria) is of the same type, but there a lower number of outlet glaciers left the plateau and, consequently, the margins rise more markedly above the environs. The steepness of bordering slopes is also emphasized by the large dimensions of the valleys which embrace the plateau. This type is usually of higher elevation, and in its area, the extension of bare karst can be great. Karstic processes (first of all, karren formation) occurs anywhere on the plateau, but it is particularly typical in the depressions. In this variety, impoundment of karst water could also happen.

- Semi-plateau glaciokarst (Fig. 2.7c).

Either because of the large extension of the karst or because of long-lasting glaciation, large-scale valley glaciers are developed on the karst. They dissected the plateau to an extent that the plateau character was partly or entirely lost. There are hanging valleys above the floors of the bordering glacial troughs. However, even the floors of these higher located valleys are high enough to accommodate bare or only partly vegetated terrains of considerable extension. The bordering deep valleys

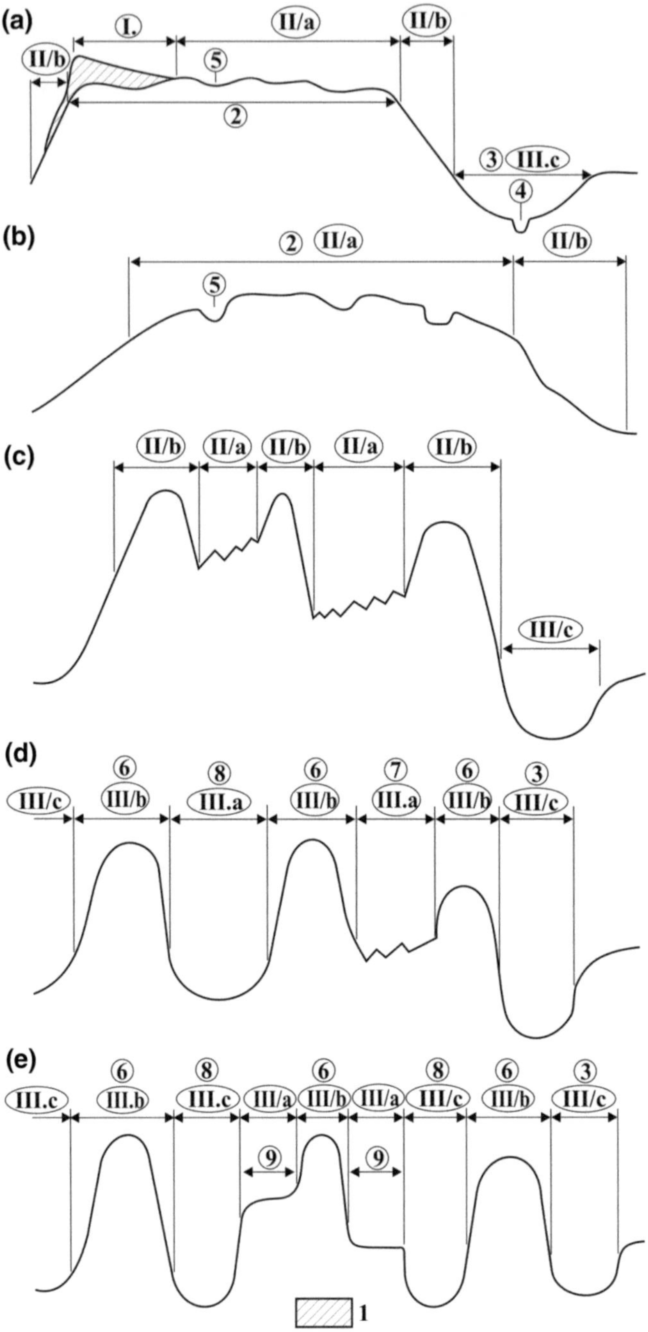
(a)
I.
II/a
II/b
II/b
5
2
3 III.c
4
(b)
2 II/a
II/b
5
(c)
II/b
II/a
II/b
II/a
II/b
III/c
(d)
III/c
6 III/b
8 III.a
6 III/b
7 III.a
6 III/b
3 III/c
(e)
III.c
6 III.b
8 III.c
III/a
6 III/b
III/a
8 III/c
6 III/b
3 III/c
9
9
1

◂**Fig. 2.7** Morphological types of glaciokarst (according to dissection). 1. Glacier, 2. plateau, 3. deeply incised valley transformed by fluvial erosion, 4. channel, 5. solution doline, 6. Arête, 7. glacial trough dissected by steps, 8. glacial trough without steps, 9. hanging valley in higher position. **I** Glacial zone, **II** periglacial zone, **II**(**a**) subzone with solution, **II**(**b**) subzone with frost shattering and mass movements, **III** temperate fluvial erosion zone, **III**(**a**) subzone with solution, **III**(**b**) subzone with frost shattering and mass movements, **III**(**c**) subzone with fluvial erosion. **a**–**b** Plateau glaciokarst, **c** semiplateau glaciokarst, **d** mountain glaciokarst, **e** complex glaciokarst

with steep slopes are wall-like boundaries of the dissected karst area. Actual karstification primarily affects the floors of the valleys dissecting the karst, but sometimes karren formation also happens on the arêtes between valleys (e.g., in the Totes Gebirge).

- Mountain glaciokarst (Fig. 2.7d)

Valley glaciers deepened their valleys to the local base level. Valley floors are found at the level of karst water or of valleys of the karst margin. The floors of the large inner valleys of the karst join the bordering valleys at the same level. In the valleys there are permanent watercourses. Thus, the valleys develop by fluvial erosion. Karst processes are active on top of the arêtes separating the valleys or on the floors of less deep glacier valleys without watercourses (Julian Alps, Slovenia). For this type, in glacials karst water could be impounded.

- Rock fortress glaciokarst

The karst was heavily dissected by valley glaciers. Partly for this reason, plateau-like surfaces did not survive in the area. Also, the valleys deepened to the bordering terrain, which is also karstic. The one-time valley glaciers could reach beyond the karst and created piedmont glaciers. Karstification is present both in the higher regions (mostly on valley floors) and on the bordering lower terrain. In the former, the proportion of bare surfaces is high. The bordering lower terrain lies below the timberline (e.g. Durmitor). If the karstic rock is bordered by non-karstic rocks, the karst water is impounded in the piedmont glacier. In case of the Durmitor, where the 800-m-deep Tara canyon represents the local base level (without glacier, i.e. no ice cover at the water outlet), the impounded karst water level probably did not reach the surface of the limestone.

- Island glaciokarst

A glaciokarst type formed close to the poles, in marine environment, on an island. The valley glaciers reached the sea and, at present, the karst valleys join fjords. The main karst process is karren formation since bare surfaces are widespread and appear anywhere on the karst. The length of the glaciers is influenced by the expansion and altitude of the island. At this type, ice caps and parts of the ice cover could predate valley glaciers. Karst water impoundment could take place in ice too. Such a glaciokarst type is found on the islands of the Patagonian archipelago, such as Diego de Almagro (Veress et al. 2006; Maire et al. 2009).

- Crested glaciokarst

The karst forms a narrow crest surrounded by valley glaciers. Glaciers must have contributed to the transformation of the former plateau into a crest. With the exception of the narrow crest sections, karstification takes place everywhere, but most typically in glacier valleys. A good example of this type is the Hochschwab (Alps, Austria), where primarily the glaciers formed on the northern edge of the former plateau (Plan and Decker 2006) resulted in narrowing the plateau into a crest.

- Composite glaciokarst (Fig. 2.7e)

The karst is heavily dissected by deep valleys. Between valleys, plateau details of higher position accommodate younger valley glaciers. Karstification extends over the latter areas (e.g. the Dolomites). In the larger and deeper valleys fluvial erosion prevails.

- Ice-cover glaciokarst

The extensive ice sheet reached over the karst area. The karst was eroded uniformly (although the extent of erosion was minimal in the case of cold-based glaciers). Major denudation happened where the ice cap functioned as a warm-based glacier (on the ice cap margins) or where outlet glaciers reached out from the ice sheet. Subsequently, after ice retreat, valley glaciers executed large-scale erosion and dissection of the surface (e.g. the British Isles). In ice sheets, locally ice streams with velocities higher than in their environs developed (Davies 2014) with glacier valleys under them (Smart 2004), even if the ice sheet worked as a cold-based glacier. Such valleys show no karstic preformation.

2.4 Glacier Formation on Glaciokarst

Snowline elevation (Equilibrium Line Altitude) not only influence ice accumulation, but also the position and character of the geomorphological zones. The arrangement of glacial features, however, also depends on the type and size of the glacier.

Snowline elevation is controlled by the distance from the Equator, the amount of precipitation, the elevation and morphology of the mountains.

Moving away from the Equator, because of lower temperatures, the climatic snowline shifts to ever lower elevations, but not uniformly in both hemispheres. In the northern hemisphere, it reaches the sea level at latitude 83°N, while in the southern hemisphere at latitude 65–66°S. During glacials, the snowline was replaced to lower elevation at the same site: in the Alps during the Pleistocene glaciations to 1200–1500 m. Slope exposure also affects temperature and snowline elevation. On the poleward side of the mountains, temperatures are lower than at the same elevation on the side towards the Equator. The snowline on the northern side of the Alps is at 2400 m, while on the southern side at 2700 m.

With increasing precipitation, the snowline is lowering. Therefore, the position of the snowline is modified by the vicinity of the tropics and seas and the prevailing winds. (The latter two often exert a joint influence.) With regard to distance from seas, glaciers of marine and terrestrial environments are distinguished (Bennett and Glasser 2009). For island glaciokarsts or coastal glaciokarsts, because of the proximity to seas, the snowline lies lower. In the Dinaric Mountains, the elevation of the snowline reduces in the direction of the Adriatic Sea parallel with increasing precipitation (Menkovič et al. 2004). The differences in precipitation amounts are also reflected in the different glaciations of the coastal ranges of the Dinaric Mountains. In the Biokovo, where precipitation is less than in the Orjen or Velebit, the reconstructed ELA (equilibrium line altitude) is 200–300 m higher (Belij 1985; Bognár et al. 1991; Hughes et al. 2010; Stepišnik and Žebre 2011) than in other mountains. For this reason, glaciation in the Biokovo was more moderate than in the other mountains mentioned (Žebre et al. 2013). The effects of temperature and precipitation on snowline are exerted separately: in humid areas the amount of snow, while in continental regions temperature is the more important control. With the growing size of continents and the resulting decline of precipitation, the elevation of snowline increases. In the Caucasus the ELA is 100–650 m higher than in the Alps.

The influence of prevailing winds on the snowline is great if the strike of the mountains is perpendicular to wind direction. In this case, the windward side is more humid than the leeward. This is a typical situation for the temperate belt, but accompanies other wind directions, too. In the Durmitor wind direction in the older, more intense, glacials were southern, the snowline lay at 1400 m in the southern part of the mountains and at 1600 m in the northern part (Djurovič et al. 2010). In the second, younger and less intense glacial stage, there was no uniform snow line in the mountains, but appeared at different elevations within smaller areas (individual mountain blocks) (Djurovič et al. 2010). It was often lower on northern slopes and showed no regularity. However, in the Pleistocene glaciation, the zone of westerly winds shifted to the north, in the Mediterranean karst mountains the prevailing westerly winds could also have modified the position of the snowline.

Mountain morphology affects both precipitation and temperature conditions. In the basins, the foehn effect may reduce precipitation and raise the snowline. Narrow valleys and karst depressions may favour temperature reduction and, consequently, snow accumulation. The cooling effect of snow accumulated in depressions further enhances the drop in temperature. Depressions are places of wind shelter and promote snow accumulation. Cold air is trapped in them. (Even in dolines below 1000 m altitude in the temperate belt, summer frosts are common.) Snowmelt is slowed down by the shorter duration of sunshine in the depressions than on flat terrain. In addition, snow accumulation prefers wind-sheltered crests, concave slopes with northern exposure (Smart 1986).

The growing height of the karst not only involves declining temperatures, but also increasing precipitation amounts. The former causes the appearance of snowline at a certain elevation, while the latter the lowering of the snowline. The conditions of ice formation could also change in time as a result of the rise of the bearing area. Because of their rapid elevation, mountains may have reached an elevation of several hundred metres during the Pleistocene. Thus, glaciation was not only affected by the lowering of the snowline, but also by the elevation of karst areas too.

Ice can develop in the valley head of a river valley and in a karstic depression. River valleys can only deepen on the karst if it is covered by impermeable, non-karstic rocks since in this case a watercourse can be formed. Consequently, the karst was a buried karst when valley formation began.

On impermeable covered karsts, there are two ways of inheritance:

- The inheritance of the valley was not followed by karstification on the valley floor. There is no karstification if the valley floor is located in the level of the karst water table or if, because of the proximity to the pole, climatic conditions do not allow appropriate solution. The properties of glacier valleys formed in this manner are the following:
  - The density of glacier valleys is high, the valley of the main glacier has numerous tributary valleys. (Valley density reflects the formerly high stream density.) The density, alignment and pattern of glacier valleys were controlled by the one-time watercourses, in the same way as on terrains built of non-karstic rocks.
  - Since the ice accumulated in the valley heads of erosional valleys, the glacier valley can be clearly subdivided into a cirque and a trough section.
  - Since they were formed in non-karstic depressions, the cirques are not closed landforms—although, with subsequent karstification, the cirque can develop into a closed depression, a cirque doline (Barrére 1964). Irrespective of rock quality, a closed cirque also develops if the dip of the strata is directed towards the centre. In this case, the *k* factor, which describes shape, has a high value (Haynes 1968).
  - Transversal scarps may dissect the glacial trough. The scarp emerges where the former river valley floor had a steeper gradient. Since, the steeper sections of the valley floor are further steepened by glacial erosion.

Similar glacier formation was characteristic of part of the Totes Gebirge, where the density of glacier valleys (or cirques) is high (Fig. 2.8).

- Valley inheritance was followed by karstification on the valley floor. This is possible if the karst water level is under the floor of the inheriting valley. Solution dolines and ponors may develop on the floors of inherited valleys (Jakucs 1977; Hevesi 1980; Veress 2016). Ponors emerge if only the lower section of the incising valley cuts through the cover deposit, while the upper section does not. In this case, the valley floor is deepened partly into a non-karstic cover deposit, partly into a karstic rock. At the termination of the non-karstic cover deposit, ponor develops on the valley floor. Solution dolines are formed if the inheritance happens without ponor formation. A dry valley with solution dolines on its floor is produced. During glaciation, the inter-valley

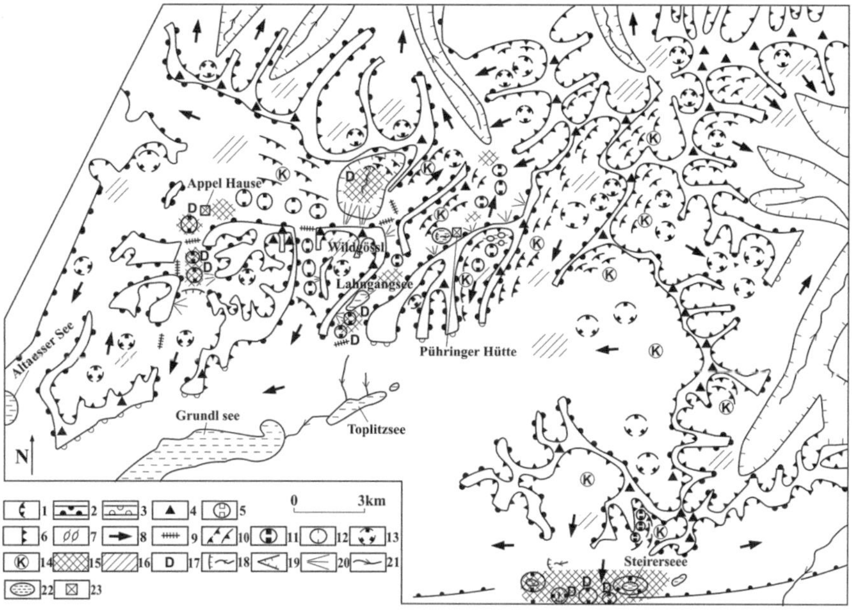

Fig. 2.8 Geomorphological map of part of the Totes Gebirge (based on tourist map) (Veress 2016, modified). 1. Cirque valley, 2. trough on plateau, 3. old trough between plateau sections, 4. horn, 5. rock basin, 6. step, 7. terrain with roches moutonnées, 8. direction of ice movement, 9. glacier bifurcation, 10. gentle slope of trough on bedding plane, 11. giant solution doline of interglacial age, 12. preglacial giant solution doline, 13. assumed interglacial or preglacial giant solution doline, 14. karren, 15. covered karst (mostly with subsidence dolines), 16. assumed covered karst, 17. depression of superficial deposit, 18. ponor, 19. erosion valley, 20. debris fan, mountain collapse, 21. watercourse, 22. Lake, 23. tourist hostel

ridge narrows into an arête. If inheritance over the karstic rock was of small scale and karstification was of high intensity, valley shape is controlled by the dolines (uvalas) and cannot be obliterated by glaciation. At the beginning, all the dolines are sites of ice accumulation (cirque glaciers). Since the dolines are arranged in rows (along the axis of the erosional valley), this favours the merging of ice masses flowing out from the dolines. Since there have been several similar depressions of accumulation, which became part of the ablation area, the resulting glacier has no marked cirque. Valley density is also low as there are few or no tributary valleys to inherited valleys.

This type of glacier formation is characteristic of the Durmitor, where, e.g. Lokvice or Velika kalica are valleys with doline rows (Fig. 2.9).

On non-covered, bare karsts ice accumulation takes place in karst depressions, best suited for the process. Smart (1986) claims that in the Picos de Europa region cirques are formed in karst depressions which lie at high elevations, on N or NE slopes, sheltered from the wind by crests. Such cirques are widespread in the area. Although Smart (1986) calls the landforms transformed from karst depressions

Fig. 2.9 Preglacial giant solution dolines (Durmitor Mountains, lower part of the Lokvice Valley, photo by Veress). 1. Preglacial giant doline, 2. ridge, 3. alluvial fan, apron, 4. glacial trough, 5. lower surface with moraines which borders the mountain

cirque depressions, it is more appropriate to give them the name cirque of karst origin since the name cirque doline (Barrére 1964) is reserved for a cirque transformed into a closed landform through karstification, as it has already been mentioned. Features of similar origin have been described from the karst areas of the Alps first by Fels (1929), then Nicod (1978), Maire (1977) and Ford (1979). The depressions with most common ice accumulation are dolines and uvalas. Ice can accumulate in a single depression or simultaneously in several depressions. The first option probably more frequently happens. Ice may accumulate in a single depression, but flow out of it in several directions, creating glacial troughs extending in different directions as it is observed in the Totes Gebirge, for instance at the depression below the mentioned Wildgöszl peak.

On (details of) plateaus, in the karstic depressions of which cirque glaciers developed, thus the ice did not leave the depressions, no troughs were formed. A good example is the glacier on the northern slope of the Triglav (Julian Alps), which existed until very recently. However, this former cirque glacier was also capable of erosion (Fig. 2.10). Where it was allowed by local conditions, cirque glaciers accumulating till could also be generated, while in sites, where it was not possible only less developed cirque glaciers or niche glaciers occupied the karst depressions. The latter did not form moraines (or marked accumulational landforms) or carve erosional features, neither did the bearing depression transform by erosion. Therefore, their one-time existence is difficult or even impossible to prove. Examples can be cited from the Biokovo, where the ELA was found at 1515 m elevation (Žebre et al. 2013), but on the 1300–1400-m-high sections of the plateau no glacial influence can be identified unequivocally. The most probable explanation is that the minor ice and niche glaciers in the depressions, if existed at all, were incapable of glacial erosion. Traces of glacial erosion can be the corridor-like features, described by Telbisz et al. (2005), which occur on dividing walls between dolines. (Their glacial origin is explained in Chap. 3.) Glaciation is also indicated

Fig. 2.10 Remnant of the cirque glacier below Triglav (**a**) and the surface eroded by the former glacier (**b**). (photo by Veress) **b**: 1. Uncovered karst detail of cirque, 2. covered karst detail of cirque, 3. debris fan, 4. subsidence dolines, 5. Shafts, 6. unfilled shafts

by the nivation niches on the eastern side of the Troglav peak, mentioned by Telbisz et al. (2005) , and the cirque remnants of the Prisika valley at about 1100 m elevation as well as the remnants of erratic blocks at 900 m.

Karst depressions without linked troughs also refer to cirques and ice patches, but formerly they demonstrably had an ice fill. Such a doline-size closed depression is the Podcajtoige, in the Snežniki area (Žebre and Stepišnik 2015b), which used to be filled with ice, but it is not reshaped considerably by glacial erosion.

With regard to the above statements, the following cases of ice formation associated with karst depressions are identified:

- Ice accumulated in a single depression (cirque glaciers or niche glaciers). The depression is transformed to a small extent.
- Ice accumulated in a single depression, but flowed out of it. If the ablation area was small, only moraine accumulation took place, if it was larger, a major trough developed.
- The ice flowing out from the depression reached other glacier tongues and established a small-scale glacier network. This can happen if the depressions are close to each other, but the intensity of glaciation is low. In the Durmitor such a glacier formation could happen in the Valoviti do area.
- Leaving the depression, ice produced a continuous ice sheet (ice cap). This can happen if glaciation is intensive and the density of depressions is high. This kind of glacier development was characteristic of the Orjen (Adamson et al. 2014), the Šneznik and the Gorski Kotar plateaus (Žebre et al. 2016).

Ice can probably also accumulate on steep slopes, on plateau margins, occasionally in rock niches with circular edge, hollows of rock and snow avalanches and depressions of snow erosion (cliff glacier). In this case, the ablation area is also located on the steep slope of the karst. Such glaciers have many cirques in clusters aligned in a row along individual troughs or at the margin of a plateau. Such an evolution is characteristic of some parts of the Totes Gebirge (Fig. 2.8). It is probable that snow accumulation and ice development also take place in lack of depressions on leeward slopes. In several cases, such leeward slopes can be observed at present too where snow accumulation occurs and the snow survives permanently in the present climate too. An example of this is the eastern slope of the ridge that closes the Schneeberg plateau from the west. In such a case a relatively small amount of ice is also enough for the formation of the glacier snout since the snow is accumulated in non-deep and non-closed depressions.

2.5 Geomorphological Zones of Glaciokarsts

The number of geomorphological zones is controlled by the elevation and position of the glaciokarst (distance from the Equator). In the development of geomorphological zones, the distribution of ice and the vegetation (the timberline) are most influential. Their properties are not only dependent on elevation (and the related factors already described), but also rock quality, exposure and morphology. On glaciokarst, like in non-karstic mountains, the closer they lie to the Equator and the higher is the elevation, the more geomorphological zones can be identified. Comparing two glaciokarst in the same distance from the Equator and with the same number of geomorphological zones, the one with the higher snowline (for instance, because of less precipitation) has broader geomorphological zones.

On the glaciokarsts, a primary and a secondary zonation can be distinguished. If not modified by morphology, the primary zonation would extend over the whole karst. The secondary zonation is modified by morphology and causes that the alignment of the zones does not coincide with the plateau margins, but follows the contours instead. On the floors of the glacier valleys the lower geographical (and also geomorphological) zone is widening towards the upper end of the valley. This can be explained by the relative closedness of glacier valleys which have more favourable climatic conditions in their interior, particularly if they are open towards the Equator, as it derives from their altitudinal position. On the slopes of the glacier valleys, however, the upper zone is predominant (Fig. 2.11). The reason for this is that steep slopes and resistant karstic rocks result in the lack of vegetation on the slopes and the widespread distribution of mass movements.

In the high-mountain karsts (alpine karsts) a zonation of karst landforms can be established: at the bottom the zone of dolines is followed by the karren zone and then the periglacial zone with increasing elevation (Ford 1979; Jennings and Bik 1962). Ford (1979) identified the following zones in the Canadian Rocky Mountains: "boreal forest", "glaciated tundra" and "extraglaciated tundra". The latter zone includes areas which were not glaciated in the glacial(s) either, in spite of their possible location above the glaciers. Zonation may develop locally, too. According to Nicod (1972), on the European glaciokarsts (Alps and Pyrenees), in large dolines below the timberline tundra emerges after frost pockets had appeared in the area. In karst areas, vertical zonation can partly be overprinted by solution governing karstification. The intensity of karstification does not necessarily decrease with elevation, but rather the landform assemblage changes, motivated by the following reasons:

- Adjustment to the glacial form assemblage and its change which partly shows zonation. For instance, the surface is either covered by till or bare (in the first case subsidence dolines are typical, in the second schachtdolines).
- Adjustment to changes or lack in vegetation (e.g. karren of special types on bare surfaces).
- Adjustment to changes in the character of solution. Thus, for instance, with increasing elevation, the time solution becomes longer.

In the changes of solution intensity, the factors responsible for zonation (temperature and precipitation) are also influential. To the effect of these factors the amount of CO_2 originating from the soil declines at higher elevations. Other factors can also modify or even increase the rate of solution. In the periglacial zone the intensity of solution can increase to the effect of the solution induced by snowmelt

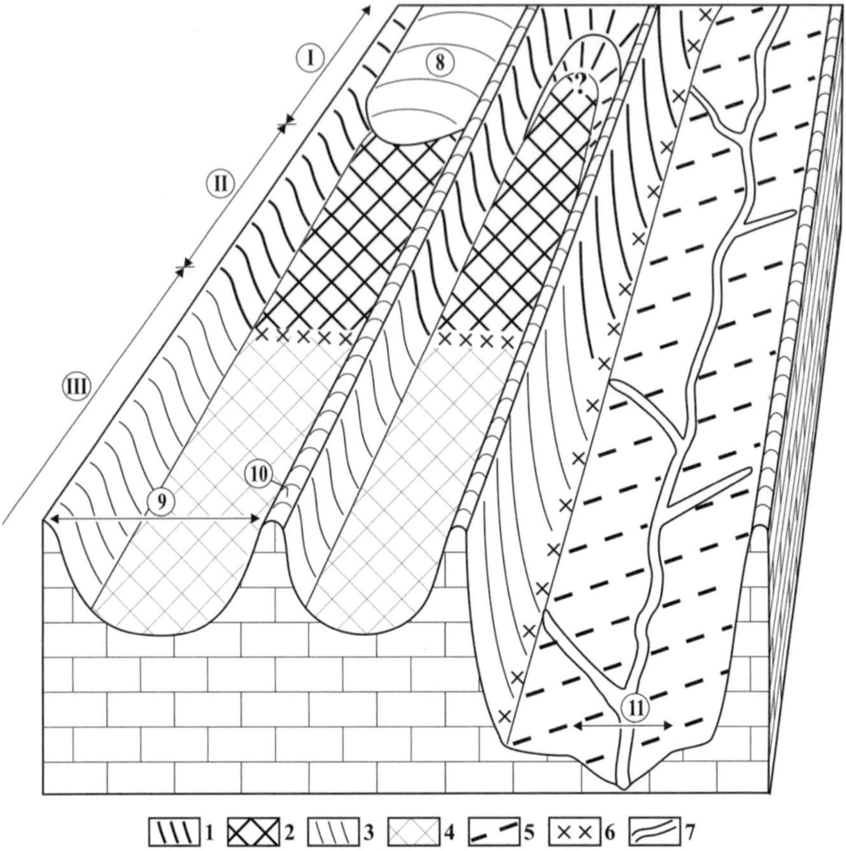

Fig. 2.11 Zones on glaciokarst. **I** Glacial zone, **II** periglacial zone, **III** temperate fluvial erosion zone. 1. Subzone with frost shattering and mass movements (in the periglacial zone), 2. subzone with solution (in the periglacial zone), 3. subzone with frost shattering and mass movements (in the fluvial erosion zone), 4. subzone with solution (in the fluvial erosion zone), 5. fluvial erosion (in the fluvial erosion zone), 6. timberline, 7. watercourse, 8. glacier, 9. glacier valley, 10. arête, 11. river valley

(the intensity of which depends on numerous factors, including local snow accumulation or the daily duration of snowmelt). This is perhaps the reason why solution on bare surfaces is not weaker than on soil-mantled surfaces (Bauer 1964; Delannoy 1983). In the periglacial zone, however, because of the pattern of soils and vegetation, solution can be rather variable and mosaical. In these sites solution generated by snowmelt and CO_2 of soil origin mutually strengthen each other and other factors (e.g. slope conditions) may have a modifying effect. On steep slopes, the soil is often missing (CO_2 production is lower), at the same time, with increasing slope angle runoff velocity grows along with the intensity of solution (Trudgill 1985; Dubljanszkij 1987).

In spite of the above-described facts, the geomorphological zone determines the type of processes on the karst. Because of the properties of the karstic rock (water conduction), the influence of fluvial erosion is limited at the elevation where it is predominant on other rocks. Slope inclination may also modify the intensity of the individual effects or their mutual proportions within the zones. Thus, within the periglacial zone, on the steep slopes mass movements (avalanches, collapses) will prevail and solution becomes subordinate. (Solution only produces karren of certain type.)

The typical geomorphological zones, primarily with respect to temperate belt glaciokarsts, are the following (Fig. 2.11).

- In the glacial zone active glacial landforms occur. Solution due to meltwaters takes place even under ice (Maire 1990; Bennett and Glasser 2009) since the subglacial water which reaches the karst are unsaturated (Atkinson 1983; Smart 2004). According to Smart (2004), solution is efficient on the karst if the meltwater contains little sediment. Meltwater from ice exerts a solutional effect, not only under the ice, but, flowing out form below the ice, on ice-free surfaces. Solution by meltwaters is allowed and enhanced by low (ever lower) water temperature and its high pressure (Ford 1979; Bennett and Glasser 2009), and the incorporation of atmospheric air into the glacier (Ek 1964; Metcale1984). At lower temperatures CO_2 is more soluble. Karstic solution produces heat, which is absorbed by the ice (inducing icemelt). Therefore, water temperature remains low and CO_2 uptake is continuous. The heat derived from intensive solution, however, contributes to meltwater generation. At higher pressure less equilibrium CO_2 is needed (Jakucs 1977; Ford and Williams 2007), and its surplus can be present in the system as agressive CO_2. Solution is increased by high flushing rates, when the water rapidly flows in the glacial system and there no time is available for saturation (Bennett and Glasser 2009) before it reaches the bedrock. Also, CO_2 uptake from atmospheric air can be uninterrupted (Ek 1964; Metcale 1984). Solution is further enhanced by turbulence, when large amounts of fine material are transported and, thus, a large surface is created (Bennett and Glasser 2009). Because of the turbulence, however, the bedrock is also intensively dissolved since the boundary layer on the rock is regenerated through turbulence and, thus, further Ca ions are released from the rock (Curl 1966; Trudgill 1985). It is to be noted that there is a view that the meltwater is not under pressure (Lliboutry 1979), or in the case when the passage system of the karst is yet underdeveloped (Röthlisberger 1972; Shreve 1972).

Naturally, in the glacial zone, old karst features can be destroyed or transformed by glacial erosion. (Only landforms under cold-based glaciers are exceptions.) Below ice, meltwater erosion and, in parallel, features of solution origin may come about. The latter may survive, but may also be destroyed by glacial erosion. Solution and glacial erosion may affect the surface jointly too (see Chap. 3).

- In the glaciokarsts, the periglacial zone is most widespread. On karsts of higher latitude and lower altitude (e.g. islands) it can be exclusive. In this zone vegetation, which affects solution as well as periglacial processes, is not continuous but patchy (dwarf pine or herbs) or missing. With increasing altitude, even the gentler slopes become devoid of vegetation. According to age, the landforms in this zone can be periglacial (mostly karst remnants and transformed features), glacial, erosional and accumulational as well as interglacial (also mostly transformed karst forms) and recent postglacial landforms. The latter can be periglacial, non-karstic or karstic landforms. The periglacial zone can be subdivided into two subzones.

In the subzone of solution (1) widespread solution prevails. In variable extension and intensity, however, pluvial erosion, periglacial geomorphic evolution (frost weathering) and fluvial erosion (on moraine surfaces) may also be present. The subzone occupies the relatively lower portions with gentler slopes, glacier valleys, floors of depressions and lower plateaus within the zone. In the lower region of the subzone soil covered, as well as soil and till-covered (or frost-shattered debris covered) surface parts alternate with bare surfaces. In the upper region the soil is missing and surfaces mantled with till and frost-shattered debris alternate with bare surfaces. In the subzone, the karst features predating the Last Glacial are being destroyed through the bordering slopes becoming gentler and the infilling of depressions. At the same time, new karstic or non-karstic forms may emerge, too.

Solution is caused by snow meltwater and rainwater. Since the amount of snow precipitation may exceed several thousands of millimetre, solution by snowmelt is predominant. As it was mentioned, the intensity of solution is increased by the slow melting of snow (lasting for several months) and contact with rock for several hours in the summer half-year. Snow traps develop (at steep slopes and karst depressions) which promotes the local accumulation of meltwater. At the same time, with the large extension of snow patches, solution affects remarkable areas. Solution intensity is enhanced by the low temperature of meltwater, which not only allows the uptake of more CO_2, but increases the viscosity of water, reduces the velocity of flow and makes the duration of solution longer. According to Kestin et al. (1978), the kinematic viscosity of water at 10 °C is 1.3063, at 2 °C 1.6736 and at 0 °C-os 1.7915 $mm^2\ s^{-1}$. Added to the rate of solution, higher flow velocity increases the variation in concentration between flowing water and the boundary layer and ion transport is also increased (Dubljanszkij 1987). Therefore, compared to gentle slopes created by glacial erosion, on steeper slopes the intensity of solution may grow. The solution effect of snow water also depends on what snow it comes from. Fresh superficial snow has the highest CO_2 content (4.5 mg l^{-1}),while the older, crystalline snow has a lower CO_2 content (1.05 mg l^{-1}) (Küfmann 2014).

On karsts, particularly on glaciokarsts, snow karst is identified if much snow and meltwater feeds the karstic reservoir (Ford and Williams 2007). Snow karst can develop in glaciated areas, but also on karsts without glaciation (Ford and Williams 2007). A special case of the latter is “karst en vagues” (Peritaz 1996), where a series of asymmetrical depressions come about. The snow accumulated on the sides of depressions results in their deepening and increasing asymmetry.

As it was mentioned, in the periglacial zone of the karst, soil patches may be present. Here, in the soil under snow patches, the intensity of soil microbial life and CO_2 production rise even at around 0 °C temperature (Mariko et al. 1994) Also, higher CO_2 production can result from the fact that the evergreen vegetation of soil patches does not photosynthesize if covered with snow, but only dissimilate (Bauer 2005; Veress 2012b). The meltwater from snow patches enhances biological activites in vegetation patches (and thus, in soil microbial life). This promotes CO_2 production.

In the other subzone (2) with mass movements and frost weathering, solution is missing or subordinate. Gelisolifluction, frost heaving, frost shattering, thermokarst processes, nivation and mass movements independent of ground ice occur. Because of the properties of the karstic rock, some of them (e.g. the effects of ground ice) are of subordinate importance. On non-morainic terrain frost shatttering and mass movements independent of ground ice (rock and snow avalanches) are significant. It is to be noted that nivation is not only important here, but also in the other two zones. Snow erosion is composed of several partial processes and can be slow or rapid (Thorn and Hall 2002). With slow nivation, there is remarkable frost shattering and sheet wash under snow patches (French 2003). Rapid snow erosion includes the impact of snow avalanches. Meltwater from snow patches exerts solution on the bordering slopes or reworks the fine-grained material.

- The temperate fluvial erosion zone is subdivided into three subzones. The solution subzone (1) covers the glacier valley sections below the timberline and above the base level as well as the dry valleys. In the second subzone (2) erosion is predominant and till-lined troughs and dry valleys, valley sections, occur. On the floors of the latter fluvial erosion may temporarily be active in rainy spells. In the valley heads of dry valleys springs issue and water only seeps away at some distance. In this subzone, the features of fluvial accumulation are also characteristic, often in the major karst depressions, poljes (Adamson et al. 2014; Žebre and Stepišnik 2015b). The third subzone (3) with steeper slopes presents frost shattering and mass movements. Even if in a limited extension and of lower intensity, essentially, the processes of the periglacial zone work on the steeper slopes.

In the solution subzone, where frost shattering and mass movements are of lower intensity, the periglacial and glacial landforms are less destroyed. (Recent solution features develop on them.) There are two varieties of the karst of the solution subzone: one is with former glaciation and the other without. In the latter, karst landforms, which occasionally show periglacial impact, retained their karst processes to our days. Such areas appear on warm temperate (Mediterranean) karsts (in the Dinaric Mountains: Šneznik and Gorski Kotar (Žebre and Stepišnik 2015b). In the erosion subzone, periglacial karst features, if ever present, have mostly been removed by now and, at present, the glacial landforms are being reshaped, mainly by fluvial erosion. In this zone, the solution is due to rainwater, meltwater and the water rich in CO_2 which percolates through the soil.

The modes of solution are strictly different in the individual zones. In the glacial zone, snow meltwater can be effective, in the periglacial zone meltwater and rainwater, but, in the zone of fluvial erosion, in addition to rainwater, the solution effect of snow meltwater can be present. If glaciation on the karst had several centres, the periglacial zone forms several, ring-like stripes around the glaciated areas. The zone of temperate fluvial erosion, however, embraces the preglacial rings and can even intrude among them.

Solution is sometimes modified or intensified by the wind on glaciokarsts, primarily in the periglacial zone. A good example is the Patagonian archipelago, including the Diego de Almagro Island, where the wind blows incessantly, occasionally with 150–200 km h^{-1} velocity (Zamora and Santana 1979). Here, landforms elongated in wind direction (roche moutonnée-like features) are observed. It is also discernible that because the higher rate of solution, windward slopes are steeper and the leeward slopes are gentler (Maire et al. 2009; Veress et al. 2006). The role of wind in the governing solution is further attested by non-karstic blocks with remnants of karstic rocks on the leeward side (Fig. 2.12—Maire et al. 2009).

The higher rates of the solution on the windward side are explained as follows: The wind increases pressure on water, the wind keeps water on the slope surface and, thus, increases the duration of solution (Veress et al. 2006). Veress et al.

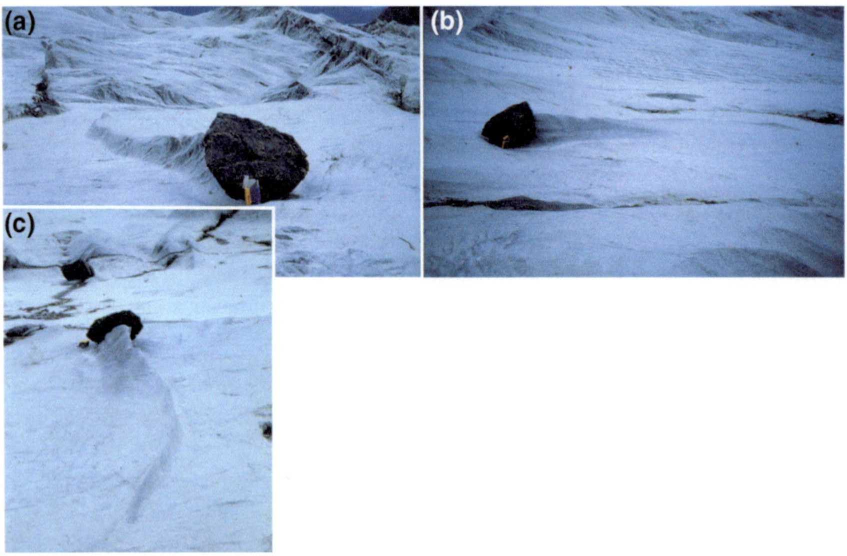

Fig. 2.12 Streamlined elevation of marble in the wind shelter of metamorphic block on glacially polished surface, due to lower rate of solution, Diego de Almagro Island, Chile (photo by Veress). **a** Viewed from wind direction, **b** viewed from right angles to wind direction, **c** viewed from direction opposite to wind

(2006) claim that the wind also promotes solution by adding rainwater yield on the windward slopes. To the effect of the wind, the raindrops acquire a component in the direction of the wind and, consequently, the amount of rainwater fallen in unit time will be larger. The wind induces higher flow velocity, causes turbulence, vapourizes raindrops, increases their total surface. As a result, more air and CO_2 are mixed with water.

As the snowline shifted to lower elevation in the glacials, the geomorphological zones were also displaced downwards. This involved that in the zone of fluvial erosion of the temperate belt (below the timberline) a glacial zone developed. In the warm temperate (Mediterranean) belt, however, a periglacial zone emerged.

In the different glaciokarsts, the geomorphological zones acquire a different character. The characteristics of zonation are presented below on temperate glaciokarsts.

On the plateau glaciokarsts, well-developed broad periglacial zones are found with a solution (karstic) subzone (Fig. 2.7a, b).

On the semi-plateau glaciokarst, an internal and an external zonation appears within the karst. The internal zonation is present on the slopes of the glacier valleys, overwhelmingly with mass movements and frost shattering on the arêtes, while on the valley floors karstification is prevalent. In the Alps this latter occurs on (sections of) the floors of glacier valleys above the timberline, which lies at 1600–1800 m elevation (although timberline altitude can be modified by a range of factors). External zonation develops on the marginal slopes of the karst, where the subzone with mass movements and frost shattering of the periglacial zone is replaced by fluvial erosion zone (Fig. 2.7c).

On mountain glaciokarsts, the arêtes and valley slopes carved by former glaciers belong to the subzone with mass movements and frost shattering of the periglacial zone. Fluvial erosion is active on the lower sections of large-scale valleys. The lower parts and floors of glacier valleys with floors in higher position are located within the solution subzone of the periglacial zone (Fig. 2.7d).

There are several levels of erosion on composite glaciokarsts. The area of arêtes, horns and other elevations carved by Pleistocene glaciers and their upper slopes as well as the valley slopes belong to the subzone with mass movements and frost shattering of the periglacial zone. The floors of the large valleys are in the fluvial erosion morphological zone. The higher plateau surfaces are part of the solution subzone of the periglacial zone, where the solution is predominant (Fig. 2.7e).

2.6 The Characteristics of Glaciokarsts

- The appearance and extension of glaciokarsts depend on the tectonic situation, distance from the Equator and their environment (i.e. the amount of precipitation arriving from the sea). Part of the glaciokarsts on Earth are located close to the pole, while others are found on young, folded and overthrust structures (so-called geosyncline karst), which have been uplifted rapidly into higher elevations. The above factors jointly control the number of geomorphological zones, their extention and, thus, the intensity of solution. The extension and nature of the glaciokarst equally depend on the preglacial karstic (occasionally erosional) geomorphic evolution and the types of glaciers.

- Independent from their origin, glaciokarst features, explained by the stability of the carbonate rocks, are stable. They survive for longer periods and only show minor alteration (in lack of fluvial erosion). As a consequence, newer and newer features develop on existing landforms. Because of ice advance and retreat, superimposed accumulational and erosional features can develop (Smart 2004).
- According to age, glaciokarst landforms are preglacial, interglacial, glacial (subglacial) or recent (postglacial) landforms (Ford 1979). Older karst systems are inherited (Ford 1996). Preglacial surface and subsurface landforms (caves) are distinguished. Preglacial caves were found in the Picos de Europa area by Smart (1986). For their origin, Ford (1979) identified karstiglacial, glaciokarstic and mixed origin landforms. In the case of karstiglacial formations, karstic processes have modified the closed glacial landforms, while in the case of glaciokarstic formations karstic landforms were transformed by glacial processes. The mixed forms were affected by several glacial and karstic influences. Recent forms are either karstic or non-karstic. Recent karstic and recent non-karstic features may occur on both karstiglacial and glaciokarstic landforms.
- At present, bare karstic and covered-karst patches or stripes, developed at accumulations of till (often of calcareous content) or recent debris, alternate on the glaciokarst. The covered karst is either cryptokarst or concealed karst (Veress 2016). In the former case, the cover is impermeable, while in the latter case it is permeable. On glaciokarst, because of the calcareous content of the cover, covered (concealed) karst is different from the characteristic features of other covered karsts. On the concealed karst of glaciokarst, the karstic processes are more complex (see Chap. 6) than in case of non-glaciokarsts. On glaciokarst allogenic karst (contact karst) also occurs, where the waters flowing down from the impermeable cover reach the karstic rock, but sometimes with allogenic karst details where erosion exhumed impermeable non-karstic rocks. Buried-like karst may also occur. At these sites, there is a lack of superficial karst phenomena. However, caves may occur since caves that developed in the surroundings of them may expand below buried karst too.
- The connection between glacial and fluvial systems on the karst is limited (Colhoun et al. 2010).
- The change of base level or that of the altitude of the karst affects the position of karst water. This is also valid for glaciokarst. As a result of the subsidence of the base level, the karst water circulation may acquire a lower position and a greater width (Smart 2004). The elevation of the phreatic zone and the width of the vadose zone changes too. In case of coastal karsts, the position of karst water is mainly affected by eustatic-level changes, while in case of karsts with a terrestrial environment, the position of karst water is influenced by isostatic

movements caused by ice loading (karst subsides if ice cover develops, while it is uplifted if the ice melts), the denudations or infillings of the surrounding non-karstic terrain. If the karst is uplifted and if the elevation of base level does not change, the elevation of karst water and the karst water level does not change. However, the phreatic zone develops at a new site (where a newer cavity formation may begin), the vadose zone widens. As a consequence, the former phreatic zone and its cavities (inactive cavities) constitute a part of the vadose zone (Fig. 2.13i(a)). If the base level subsides and the elevation of the karst does not change, the width of the vadose zone increases too, but the altitude of the newer phreatic zone changes (decreases). The cavities of the former phreatic zone also become parts of the vadose zone (Fig. 2.13ii(b)). If the karst subsides and if the elevation of base level does not change, the elevation of karst water does not change, the phreatic zone develops at a newer site at the expense of the vadose zone, the former phreatic zone gets into the deep karst (stagnant zone), its cavities become filled with calcite. In case of those having a marine contact, precipitation ceases (Fig. 2.13iii(a)). If the base level acquires a higher position and the elevation of the karst does not change, the altitude of karst water and that of karst water level also increases. The phreatic zone will also be of a higher position, the former phreatic zone also gets into the deep karst (Fig. 2.13iii(b)).

- During glacier retreat large amounts of water reach the karst and generate fractures in the area isostatically uplifting after pressure release (Ford 1983). The water input raises the karst water table, but its slope also grows. To the former effect, the level of cavity formation rises, while, to the latter karst water flow becomes faster. The growth of water amounts generates the expansion of corrosion. The higher water input results in the activation of paleokarst horizons and paleokarst surfaces.
- Either in glacials or interglacials, there is water input into the glaciokarst and only a low proportions of it constitutes output by surface runoff. This is not only explained by the high water conductivity of karstic rocks, but also by the morphology of glacier valleys. Cirques are often closed depressions. In the glacial troughs karst depressions, rock basins are common and sometimes the trough itself is a closed form.
- According to Ford (1983), if the ice is thicker than 1000 m, the hydraulic gradient is high both in the ice and in the karst. If the ice is thin, however, the value of the gradient is low.
- Under glaciers, sites with meltwater, frozen ground and percolation may alternate (Ford 1983). Meltwater under the ice can flow in any direction (Ford 1983). In lack of allogenic karst, water conduction in the interglacials is diffuse, while in the glacials can be either diffuse or point like. According to Smart (1986), concentrated conduction is evidenced by the pattern of the main shaft systems in the Picos de Europa area. Subglacial waters are drained through preglacial ponors, dolines or subglacial ponors (Ford 1979; Smart 2004). Drainage is point-like at shafts and at large-scale cracks. Epikarst develops in the area of diffuse drainage.

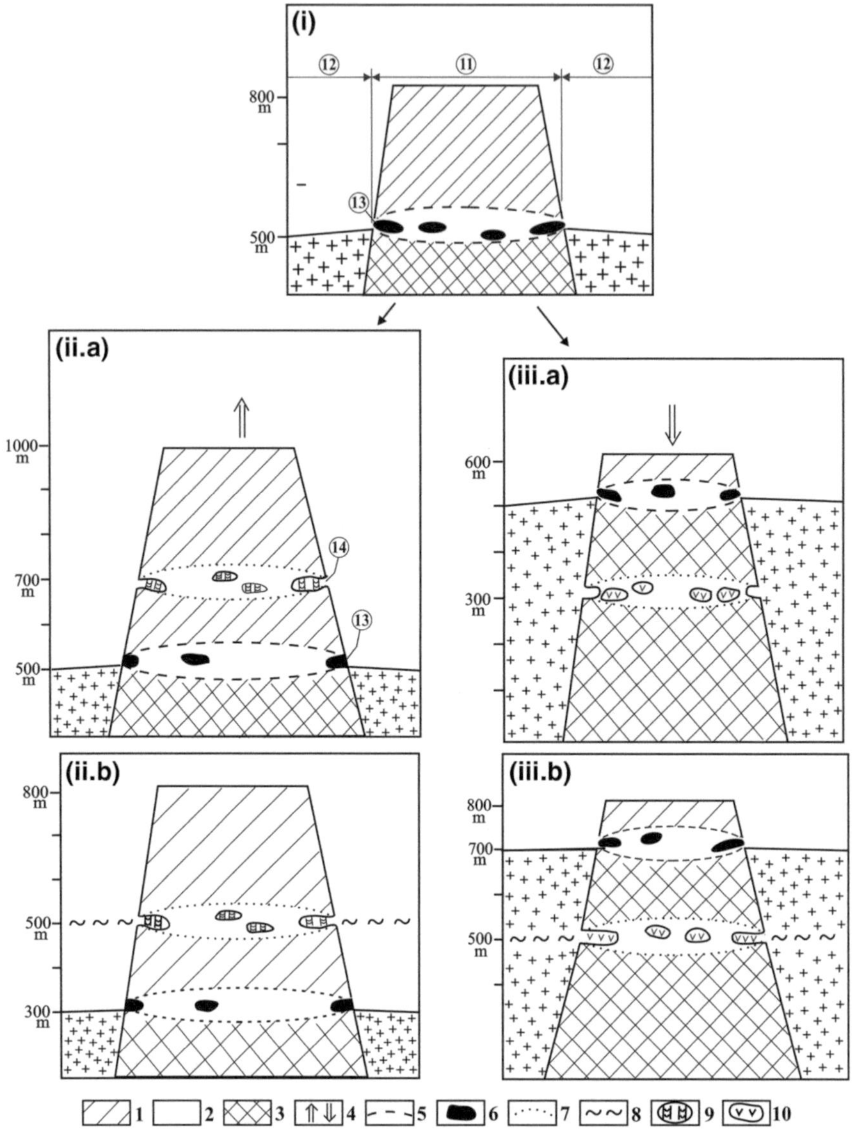

Fig. 2.13 Change in the relative and absolute altitude of karst water zone (modified after Veress 2004). **i** Initial state, **ii** growing height difference between karst surface and base level due to the uplift of karst (**ii**(**a**)) or the erosion or subsidence of the base level (**ii**(**b**)), **iii**. decreasing height difference between the karst surface and the base level due to subsidence of the karst (**iii**(**a**)) or the rise of base level or infilling at the base level (**iii**(**b**)). 1 Vadose zone, 2. phreatic zone, 3. stagnant zone, 4. subsidence and uplift of karst, 5. karst water table, 6. active cavity filled with karst water, 7. former karst water table, 8. former surface, 9. inactive cavity with dripstones, 10. inactive cavity filled with calcite, 11. karst plateau, 12. non-karstic rock (base level), 13. active corrosional spring cave, 14. inactive corrosional spring cave

- Meltwaters leaving the ice carry out a significant gemomorphic evolution in the foreground of the ice (valleys and gorges develop and the already existing ones develop more intensively and fluvioglacial sediments are formed). According to Woodward et al. (2008), the amount of meltwater increases in the glacial to the percolation retarding effect of the ice and the debris. Thus, in the glacial fluvial erosion and reworking are predominant in the foreground of the ice, while in the interglacial karstification becomes significant. However, the meltwater also gets into the karst in areas covered with ice. Therefore, the amount of meltwater leaving the ice depends on the maturity of the epikarst and on the inclination of the bearing terrain.
- Since the ice exerts hydraulic pressure on the karst, water can flow upwards through vadose shafts (Smart 2004).
- According to Smart (2004) the hydraulic gradient is not controlled by surface topography, but, in the case of ice sheets, it can be controlled by the slope of the ice surface. Therefore, karst springs can issue under the ice. Smart (2004) cites examples from Svalbard. In such cases, the base level of the karst is controlled by the ice.
- There is epikarst under warm-based glaciers. (Epikarst also includes karren.) This kind of epikarst is different in several respects from the epikarst of karst which are not glaciokarsts: In the epikarst of glaciokarst water can be present in both liquid and solid states (changing seasonally or even diurnally). The cavities and passages of epikarst under ice can be filled with water, partly because of meltwater recharge and partly because, with a reduced number of cavities, drainage is also reduced (Williams 2008). If the karst water table rises (e.g. springs are covered with ice), the epikarst is filled up with water from below. Lauritzen (1984) claims that the minimum height of the karst water table in the glacier is controlled by the elevation of the upper ends of those passages from where water input to the ice mass is still possible (at the lower ends of these passages input of meltwater under pressure is possible). The water level may have an external control when water arrives from ice-marginal lakes into the passages of the bedrock and then into the ice mass (Lauritzen 1986). If the epikarst is filled up from below, the solution may be modified and its intensity can be even higher, partly because of karst water flow, partly because of mixing corrosion. Water flow is made possible by water input into the ice or output from there on the slopes bordering the plateaus even if there are no impermeable strata in the rock sequence. Former outlet sites are caves in the bordering slopes without impermeable rocks below them. If the karst water table rises further, water flow is replaced into the ice and this reduces or even stops solution. If the evolution of the epikarst through solution comes to an end, it is partly (Ford and Williams 2007) or completely destroyed by ice. From the surface eroded by ice, the traces of solution (karren) may be absent. In lack of impoundment, the vadose zone may be thick if the height difference between the surface and the local base level is high.
- In the literature there is few data on the relation between the glacier and the ground ice. According to Ford's observation (1996), the glacier thaws the

permafrost situated below it because of the insulating effect of the ice. The melting of ground ice may become more intensive when the glacier retreats. In this case, a cold-based glacier may turn into a warm-based glacier. The ice sheet may also contribute indirectly to the melting of permafrost. As a result of the loading of the ice sheet, the crust becomes fractured. This may contribute to the formation of magma. (Magma can heat up the permafrost from below.) During the process, a cold-based glacier can even become a warm-based glacier. Warm-based glaciers may also occur on permafrost patches (the ice covers a terrain with non-continuous permafrost), the ice of which patches may melt partly or completely later by the meltwaters of the glacier. At such places of the karst system, the character of water flow or circulation may greatly depend on whether the shrinking of the permafrost took place vertically or horizontally.

- Because of their cooling effect, warm-based glaciers can create ground ice or permafrost in their environment. Its expansion and durability greatly depend on the cooling effect of the ice and on the distance from the Equator. The thickness of the permafrost can be significant. Thus, in the caves of the Low Tatras such as in the Mesacny Tien cave, it reached a depth of −285 m (Žák et al. 2012). If permafrost develops in the surroundings of ice, the karst system becomes of a duplex nature. The area covered with ice will be autogenic-like karst where there is diffuse water transmission into the karst. The environment of the ice where permafrost exists functions as allogenic karst. Here, allogenic karst patches are separated from each other by autogenic karst sections (where there is no permafrost). Sinkholes develop at the edge of allogenic karst patches.
- On glaciokarsts there was no planation of the surface. A countereffect is the high elevation of the area, the erosion properties related to climate and the joint operation of both ice and karstification towards the dissection of the surface.
- According to Ford (1983), the effect of ice on the karst is generally mainly destructive, inhibitive, preservative or stimulative.
- On glaciokarsts geomorphic evolution happens during glaciation or in a period without glaciation (in interglacials or today). During glaciation the processes are glacial, glacial-karstic or only karstic. In periods without glaciation, karstic influences (solution) are active and supplemented with periglacial and fluvial action.
- On glaciokarsts, fluvial erosion on the surface is temporally (times of snowmelt) and spatially limited (at lakes or till-covered surfaces). In the glacials, it can be active on ice-covered surfaces, although with limited intensity since the hardness of the debris traansported by water does not exceed the hardness of the bedrock. In the interglacial or at present, in till-covered areas (if the till is impermeable) it is effective or appears where the floor of the glacier valley reached down to the local base level. Lakes are common on glaciokarsts, developed either in plugged karst depressions (Ford 1979; Smart 2004) or in rock basins. Their water induces at least local fluvial erosion. It can be drained through the ponor of the lake basin, for instance, at Lake Elm (Bauer and Zötl 1972) or in the case of the rock basin on Diego de Almagro island (Maire et al. 2009). Lakewater contributes to the erosional evolution of caves on the

glaciokarst. Overflow from the lake occasionally reaches the karst through ponors, shafts and karren. Before conducted into depth, overflowing water may produce erosional channels even on the uncovered karst. In the periglacial zone, the intermittent watercourses (rivulets) are concentrated and capable of transporting debris even on bare karsts (Smart 1986).

- On glaciokarsts, karstification and frost shattering operate jointly and alternatingly. The rate of frost shattering is decisive for the survival and condition of karstic landforms.
- On glaciokarsts, snow meltwater exerts erosion by means of sediment transport concentrated in certain locations (Smart 1986).
- Both in the interglacials and glacials, the debris (which is mostly limestone) may considerably reduce the solution of the bedrock. According to Williams (1966), there is no solution on the bedrock if the moraine is thick and composed of limestone debris. The explanation of this phenomenon is the following: the water percolating through the calcareous cover deposits dissolves the limestone and, if the cover is thick enough, become already saturated when reaches the bedrock (Williams 1966; Trudgill 1972, 1985). With respect to saturation, the grain size and the thickness of the debris cover are of great importance. The water reaching the bedrock is capable of solution only if the cover is thin or composed of rocks other than limestone.
- During the glacials, glacial erosion interrupted (where no meltwater accumulated) or modified karstification (on terrains with meltwater).
- In the erosion of the surface under ice, glacial and non-glacial (karstic) processes alternate (Smart 1986) or operate jointly. Smart (1986) claims that the glacial erosion of the karst surface is more intensive, but surface karstification is a process of longer duration and equals the former process in efficiency. Smart (1986) contradicts that in cave formation the rates of glacial and interglacial processes are similar and, because of the longer duration of the latter, cave formation is mainly concentrated in the interglacial(s). Geomorphic evolution is rather different in glacials and interglacials. In glacials, as it has already been referred to, glacial erosion, fluvial erosion (in a limited way below the ice, more expanded in periglacial areas) and solution have a joint or alternating effect. In the interglacials, solution, frost weathering, mass movements and limited fluvial erosion shape the surface. In glacials, the intensity of solution does not show zonality, but it does in the interglacials (at a lower altitude, the amount of CO_2 increases because of the appearance of soil). In the glacials, the epikarst is truncated to the effect of glacial erosion, but it redevelops in a lower position with the lowering of the saturation level. In the interglacials, the solution is the main geomorphic process. Under the glaciers (in glacial periods) the rate of solution may reach or even surpass the intensity of solution observed in the interglacials on bare surfaces. Part of the explanation is that subglacial waters are under pressure, partly because the duration of the solution by meltwater between the ice and bedrock is probably longer (continuous in the summer half-year).

- On glaciokarsts, the rate of surface erosion is lower than that of subsurface erosion. Smart (1986) estimates subsurface erosion 5–10 times faster than surface erosion. According to him, all this can be explained by large-scale subsurface erosion. In the nature or extent of subsurface denudation (cave formation), there is no difference between glacials and interglacials (Smart 1986), although in glacials point-like water conduction is more intensive or at the end of glacials when solution intensifies, more water enters the karstic reservoirs and accelerates erosion.
- The relationship between glacial erosion and karst evolution is complex. The following varieties are observed.
  - The landforms were not destroyed and no new landform developed because the ice was stagnant. This situation refers to cold-based glaciers. Preglacial features are preserved.
  - Preglacial forms or forms of glacial age are partly destroyed. Such partly truncated landforms were described from the British Isles (Rose and Vincent 1983; Vincent 2009). Kunaver (2009) also mentions truncated karren and fossil schachtdolines.
  - Landforms are generated by solution under ice, e.g. shafts (Ford 1984), or regenerated by solution and glacial erosion (see Chap. 3).
  - Preglacial landforms are totally destroyed.
  - Glacial erosion and solution jointly eroded the surface, but locally with different efficiency.
  - Preglacial landforms were disrupted by glacial erosion into segments (e.g. caves), transformed because of the simultaneous erosion of all their parts. However, the shape of the features changed too: they widened and/or their slopes became gentler. During this, symmetrical features turned into asymmetrical. The landforms may have deepened and minor features emerged upon them.
  - The landform is destroyed and its remnant is partly infilled.
  - The landform is totally infilled or occasionally buried, its depression character ceased, but its hydrological functions were preserved (Ford 1983). Since there is water input and sediment transport into the karst, the depression is regenerated. It happens that the depression is not infilled, only buried (e.g. the karren).
  - According to Smart (2004), the ice sheet destroyed the surface karst (by erosion and accumulation by burial), where warm-based glacier conditions were predominant over a longer period.

- The size of superimposed landforms on the glaciokarst and the transformation of landforms on which younger features develop largely depend on the degree of influences active during glacials and interglacials. The glaciers of intensive glacials destroyed the older landforms to a larger extent or transformed them and to a lesser extent if glaciation was less intensive. The intensity of interglacial effects (karstification) depended on the location of the karst and the elevation of its sections. During the interglacials, on higher terrain the rate of karstification could be lesser than at lower position, because of the shorter ice-free period or because on higher terrain mainly periglacial processes (frost shattering) were active. On higher terrain, karren are the predominant karst landforms, which were eroded to larger extent in the next glacial. For this reason, by origin, the landforms (and surfaces) on glaciokarst are multiphase and poligenetic. The possible cases are the following:

- Preglacial karstification followed by multiphase glaciation and recent karstification.
- Preglacial karstification followed by multiphase glaciation with intercalating interglacial karstification and then recent karstification. The landforms of interglacial karstification can be preserved or destroyed by later glaciation(s).
- Multiphase glaciation, interglacial karstification and then recent karstification.
- Multiphase glaciation followed by recent karstification.

- On glaciokarsts, cirques are often less distinct from glacial troughs. On the lower part, there is no counterslope section in all cirques.
- On glaciokarsts, landforms can be closed to various degrees. Thresholds of various heights close the depressions.
- In the glacial troughs, valley watershed is common. The valley floor slopes towards the lower end of the cirques or the trough. The origin of the watershed is probably due to karstification before glaciation. The original slope of valley floor sections could not be changed by glacial erosion either.
- Glacial troughs often fork out, which points to the bifurcation of former glaciers. This was probably allowed by the very gentle slope of the preglacial karstic surface.
- Where valley glaciers formed after ice sheet retreat, the glacial form assemblage is of double nature: an almost planated terrain is dissected by glacial troughs.
- Relative relief on karst plateaus is low. Therefore, glacial erosion is of low rate and on the gentler surfaces the depth of glacial erosional forms is smaller. It is common that, even on plateaus relatively elevated above their environment, the slopes of troughs are low, lower than in the mountains of non-karstic rock. In the latter case, the glaciers were formed in river valleys, which had a relatively steep longitudinal profile from the valley heads to the mountain margin. The steep profile generated high rates of erosion for the glaciers and resulted in several transversal scarps on the floor as compared to the karst. Glacier valleys of small depth came about even if the height difference between the mountains and their environs is low. An example is the Zijovo Mountains (Dinaric Mountains), where relative relief was low, glacier valleys are less deep than, for instance, in the Durmitor, where relative relief was higher (Djurovič et al. 2010).
- Holocene karstification mostly affects glacially eroded surfaces and landforms. The process has been active for a short time and the created landforms are of small dimension. The transformation of the surface, such as the dissection of glacio-karstic landforms or the planation of the ice-shaped surfaces, is of moderate scale. The minor abrasional features which increase the roughness of the glacially modified surface (striations), however, are destroyed by solution

in the Holocene (Kunaver 2009; Žebre and Stepišnik 2015b). The surfaces smoothed by ice become even smoother (Fig. 2.14). Thus, the abraded rough surface becomes smooth through surface solution by the uniform sheet of water. According to the presence or absence of moraine or its thickness, the following cases are possible (Fig. 2.15).

- The glacially formed surface was not buried and the solution by water flow produced smooth surfaces (Fig. 2.14b). At the same time, solutional features (karren) emerged on the flat surface, which this way became rougher by karstification (Fig. 2.15c).
- The glacially formed surface is covered by thick till under which, according to Kunaver (2009), minor abrasional features (glacier striations) or features of subglacial solution (grooves) can be preserved. Since the till can be so thick that the infiltrating water reaches the bedrock saturated (Fig. 2.15).
- With erosion the cover is thinning. Under the cover a double set of landforms emerges. On the one hand, the water reaching the bedrock is capable of solution and generates karren (at first, mainly microkarren), on the other, abrasional features are retained on the surfaces between karren (Fig. 2.15c). Because of the cover, the water arriving at the bedrock can hardly flow, but mostly infiltrates (Veress 2010, 2016). Where there is no water flow, abrasional features are preserved.

Fig. 2.14 Flat surface detail formed by glacial erosion and further smoothed by solution (valley below Mt. Tragl, Totes Gebirge, photo by Veress)

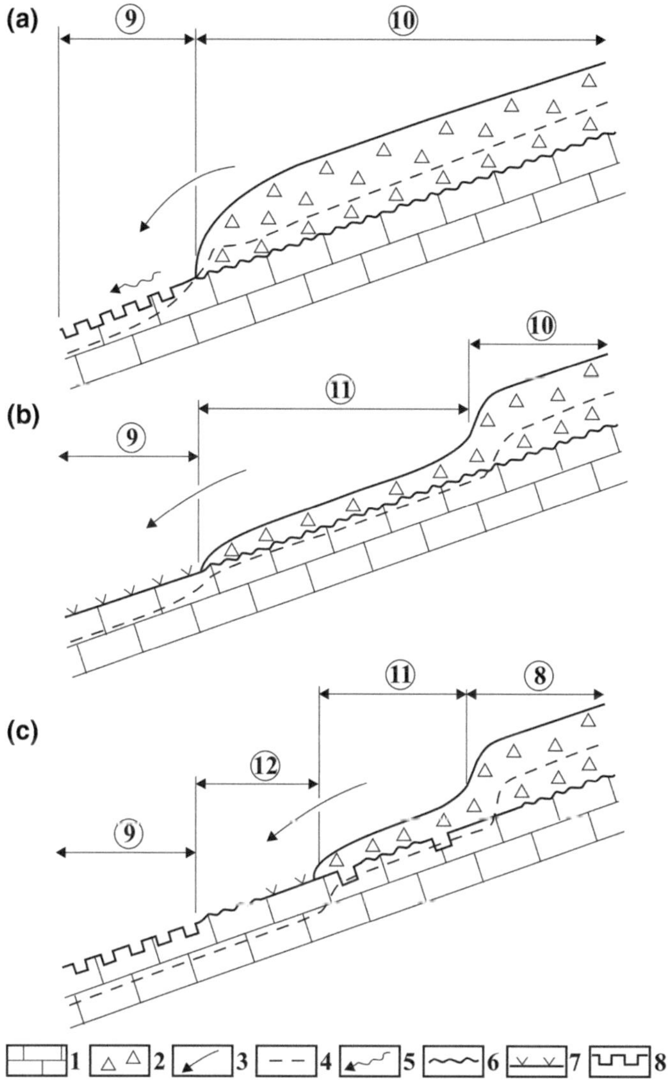

Fig. 2.15 Karren formation on till-covered terrain. 1. Limestone, 2. till, 3. till transport, 4. saturation level, 5. water flow on limestone surface, 6. surface shaped by glacial erosion (microforms, such as striations), 7. surface smoothed by solution, 8. Karren, 9. surface of old karren formation, uncovered even during glaciation, 10. terrain with thick till cover and landforms of glacial erosion, 11. glacially eroded surface with karren, 12. recently exhumed surface with karren formed under cover and without cover and traces of glacial features. **a** slope covered with thick till, **b** till cover partly thinning, **c** growing extension of bare surface, because of the removal of till

  - With the total removal of the cover deposits, young bare surfaces are exposed. On the bare surfaces abrasional landforms may be preserved or destroyed. In the first case, water flow on the surface of the bare bedrock is retarded (for instance, because the density of already existing karren is high or because the bearing slope is short). In the latter, well-developed water flow takes place on the bare surface and the smoothing of the abrasional surface happens on exhumed sections (Fig. 2.15b, c). Experiments show that the process may take 10 years (Sweeting 1966).

- Glacial erosion has produced a double set of landforms. On the one hand, it created minutely dissected surfaces and further enhanced relative relief. On the other, planated the floors of glacier valley to flat surfaces (Fig. 2.14). Both influences of glacial erosion can be manifested jointly: as a consequence, a series of smooth surfaces may appear on the floor of glacial troughs and their surrounding slopes (see Chap. 3).
- Part of the sediments of glaciokarst surfaces were transported into karst passages (Ford 1979; Smart 1986, 2004; Adamson et al. 2014). Therefore, glaciofluvial action (the formation of outwash plains) is subordinate (Žebre and Stepišnik 2015a). Repeated water inflow removes the sediments of karst passages partly or entirely and this contributes to the rejuvenation of caves (Smart 2004).
- The limestone debris of the cover is recharged by frost shattering. Material removal from the cover on the surface is limited (lack of watercourses and presence of closed landforms). Because of recharge and limited material removal, solution is partly or totally restricted to the cover deposit. The material removal on the surface and the length of transport depend on the emergence and number of conduction sites and the development of cavities.
- Part of the landforms on the glaciokarst are arranged in zones. Certain landforms, however, can develop in any geomorphological zone, independent of elevation (see Chap. 4). Preglacial and interglacial karst features, bearing an identical genetic origin, may show zonality because the forms occuring at different elevations are transformed to various degrees. Djurovic et al. (2010) distinguish between different types of karst depressions according to the extent of transformation and the processes affecting them: glacial depressions (in higher position, with cryogenic-nivation processes and surface solution); glacial-karstic depressions (at lower position and with warmer climate, karstification on the floor); glacio-fluvial depressions (at lowermost position, with watercourses, blind valleys on their floor, intermittent lakes in recently formed bottom depressions).
- The activity of karst landforms transformed by glacial erosion is ceased. After ice retreat such features are not activated, but, in their area, karstification can restart and new, usually smaller, features are generated. The type of karstification may also change (e.g. karstification on bare surface replaced by covered karst formation).
- In many glacier valleys on glaciokarsts, relatively few moraines are found (e.g. in the Totes Gebirge). This is partly due to the dissolution of debris and partly to

transport into karst passages (Žebre and Stepišnik 2015a). Partly till is not visible since buried under younger debris. The reason can also be that relatively small amounts of rock were removed in karst areas by glacial erosion as compared to mountains built up of other rocks.

- The paraglacial environment (Benn and Evans 1998) is also present on glaciokarst, which involves material transport of high intensity (Ravanel and Deline 2008). In the environment (or area) of former glaciers these processes are: mud flow, fluvial aggradation, debris flow and fluvial reworking (Church and Ryder 1972). However, fluvial aggradation and fluvial reworking have a limited effect on glaciokarst. The paraglacial state exists until following the retreat of ice, the rate of sediment yield drops to rates typical of unglaciated areas (Ravanel and Deline 2008) and thus, it can spread over the interglacial (Church and Slaymaker 1989).

2.7 Conclusions

Some morphological properties (karst depressions) which are only typical of karst regions contribute to glacier formation on the karst. Ice accumulation takes place in the valley heads of fluvial valleys of the covered karst, karst depressions of inherited fluvial valleys or in depressions of bare karst. The mode of glacier development, the density of glacier valleys and the morphology of the trough depends on the presence or absence of karstification after inheritance. The type of glaciers related to bare karst depressions is governed by the density of karst depressions and the intensity of glaciation. In karst depressions small cirque glaciers develop. If the extent of glaciation increases, the development of valley glaciers has a higher probability. If both the intensity of glaciation and density of depressions are high, ice caps may come about.

The steepness of karst water table depends on glacier type, cavity formation in the karst and the elevation of the karst above base level. Steepness governs the extension of the vadose and phreatic zones on the ice-covered terrain. In the vadose zone, a phreatic zone may develop and, in the phreatic zone, a vadose zone may emerge. For this reason, on the karst, cavity formation of the vadose zone (vertical cavities) and of the phreatic zone (horizontal cavities) can be present in the same place.

Glaciokarsts can be classified from various aspects: by elevation, distance from the Equator, their position as compared to each other and their environments. A morphological classification is possible according to the rate of erosion and extension distinguishing between plateau, semiplateau, mountain, cliff castle, crested, complex glaciokarsts, glaciokarsts with ice sheet and island glaciokarsts. Even with the same elevation, because of different degrees of dissection, the pattern and extent of the geographical and geomorphological zones on these glaciokarst types are different as well as their position as compared to each other and proportion of their subzones (and, thus, their karstification).

On glaciokarsts geomorfological zones are subdivided to subzones. There is a subzone both in the periglacial zone and in the temperate fluvial zone where karstification is predominant. In the periglacial zone such subzone is represented by the glacier floors, while in the erosional subzone those valleys where the floor did not reach the base level. In both subzones, the steep slopes are eroded by periglacial processes (mass movements, frost shattering, nivation). Out of the karstic processes, at most only karren formation is represented on these slopes. Fluvial erosion can be active in the till-covered surfaces of both zones, particularly if the till is (almost) impermeable. This latter situation also occurs in dry valleys partly infilled with sediments and in valleys which incised to the base level.

On glaciokarsts, under warm-based glaciers, epikarst develops which is different from the epikarsts of other karst regions in several properties: solution (primarily on valley floors of the periglacial zone) is caused by meltwater (in the glacials from ice, in the interglacials from snow). In the glacials, the epikarst is continuously filled up with water, which is temporarily frozen. It may be filled up with karst water from below. In the interglacials, the epikarst is intensively destroyed by glacial erosion.

On glaciokarsts with warm-based glaciers, the manner of drainage and geomorphic evolution varies between glacials and interglacials. In times of glacials, glacial erosion and solution, while in interglacials, solution (which is of various intensity depending on altitude), frost weathering and mass movements are the main geomorphic processes. The character of subsurface karstic evolution is primarily erosional at point-like inflows, but in glacials it is more intensive since point-like inflow into the karst is more considerable.

The age of surface karst landforms is preglacial, interglacial, glacial or postglacial. The forms of glacial age are of glacial, glacial-karstic or erosional origin. The preglacial landforms have been destroyed by glacial or periglacial processes. Present-day karstification takes place on periglacial landforms transformed glacially and on glacial and glacial-karstic landforms and surfaces.

On glaciokarsts surface fluvial erosion is restricted both in space and time. Here, the landforms created by both glacial erosion and karstic processes show specific properties. The development of glacial erosional forms and the type and pattern of glaciers (and also the pattern of the landforms created by them) were influenced by the former karst, its set of landforms and the karstic surface. In the case of the karstic features, the specific character is due to the fact that karstification was influenced by meltwaters from ice and the karst landforms were transformed by glacial erosion, which also affected preglacial forms and those developed under ice. Both preglacial and glacial landforms have been transformed by recent periglacial processes and recent karstification.

References

Adamson KR, Woodward JC, Hughes PO (2014) Glaciers and rivers: Pleistocene uncoupling in a Mediteranean mountain karst. Quat Sci Rev 94:28–43

Allen JRL (1970) Physical processes of sedimentation. Allen & Unwin, London, p 248

Andrejchuk VN (1992) Some aspects of glaciokarst study. Problems of Pseudokarst. Abstracts 58–62. Perm, Russian

Atkinson TC (1983) Growth mechanisms of speleothems in Castleguard Cave, Columbia Icefields, Alberta, Canada. Arct Alp Res 15:523–536

Audra P (2000) Le karst haut alpin du Kanin (Alpes juliennes, Slovenie-Italie). Karstologia 35 (1):27–38

Auly T (2000) La région d'Arudy: les relations glacio-karstiques, élément fondamental de l'organisation morphologique. T.L.G.P.A. 18:33–61

Auly T (2005) Comparison entre les bassins de Lourdes et d'Arudy: Aspects géomorphologiques. Actes de colloque: «Lucrarile seminarului geografic Dimitri cantemir: 100 de ani de invatamant geografic iesean», vol 25. Iasi, Roumanie, pp 7–21

Auly T (2008) Quelques morphologies de rapport karst/glaciaire dans les Pyrénées (France). In: Tyc A, Stefaniak K (ed) Karst and cryokarst. University of Silesia Faculty of Earth Sciences, University of Wrocaw Zoological Institute, Sosnowiec-Wroclaw, Pologne, pp 129–154

Barrére P (1964) Le relief karstique dans l'ouest des Purénées centrales. Revue Belge de Geographie. Ed. Soc. Roy. Geogr. Special Publ., Karst et Climats Froids 88(1–2):9–62

Bauer F (1962) Nacheiszeitliche Karstformen in der osterreichischen Kalkalpen. In: Prodeedings of the 2nd International Congress of Speleology, Bari, pp 299–328

Bauer F (1964) Kalkabtragungsmessugen in der Östereichischen. Erkunde 18:95–112

Bauer N (2005) A növényzet egy újabb lehetséges hatása a magashegységi karros térszínek fejlődésére (A new possible effect of plant life on the development of high mountain karrs reliefs). Karszt és Barlang 2000–2001 évf.:17–20 (in Hungarian)

Bauer F, Zötl J (1972) Karst of Austria. In: Herak M, Stringfield VT (eds) Karst, important karst of the Northern Hemisphera. Elsevier, Amsterdam, London, New York, pp 225–265

Bayari S, Zreda M, Çiner A, Nazik L, Törk K, Özyurt N, Klimchouk A, Sarikaya AM (2003) The extent of Pleistocene ice cap, glacial deposits and glaciokarst in the Aladaglar massif: central Taurids range, southern Turkey. In: XVI INQUA Congress, Reno, Nevada, 23–30 July 2003. Abstracts with Programs. Abstract 40–9, p 144

Belij S (1985) Glacijalni i periglacijalni reljef južnog Velebita. Beograd, Srpsko geografsko društvo, 68 str

Benn DI, Evans DJA (1998) Glaciers and glaciation. Arnold, London, p 734

Bennett MR (2003) Ice streams as the arteries of an ice sheet: their mechanics, stability and significance. Earth-Sci Rev 61(3–4):309–339

Bennett MR, Glasser NF (2009) Glacial geology: ice sheets and landforms. Wiley-Blackwell, Chichester UK, p 385

Bindschadler R, Choi H, Wichlaez A, Bingham RG, Bohlander J, Brunt KM, Corr H, Drews R, Fricker HA, Hall M, Hindmarsh RCA, Kohler J, Padman L, Rack W, Rotschky G, Urbini S, Vornberger P, Young N (2011) Getting around Antarctica: new high-resolution mappings of the grounded and freely-floating boundaries of the Antarctic ice sheet created for the international polar year. Cryosphere 5:569–588

Bögli A (1964) Le Schichttreppenkarst. Un example de complexe glaciokarstique. Revue Belge de. Geographie 88(1–2):64–82

Bognar A, Faivre S, Pavelić J (1991) Tragovi oledbe na Sjevernom Velebitu. Geografski glasnik 53:27–39

Brook D, Chapman P, Wilde KA (1977) Caves and karst of the Hindenburg ranges. Geogr J 143 (1):27–41

Church M, Ryder JM (1972) Paraglacial sedimentation: consideration of fluvial processes conditioned by glaciation. Geol Soc Am Bull 83:3059–3072

Church M, Slaymaker O (1989) Disequilibrium of holocene sediment yield in glaciated British Columbia. Nature 337:452–454
Clayton KM (1966) The origin of the landforms of the Malham area. Field Stud 2:359–384
Colhoun EA, Kiernan K, Barrows TT, Goede A (2010) Advances in quaternary studies in Tasmania. In: Bishop P, Pillans B (eds) Australian Landscapes, vol 346. Geological Society of London, Special Publication, pp 165–183
Cooper MP, Mylroie JE (2015) Glaciation and speleogenesis. Springer, Berlin, Heidelberg, New York, 142 p. https://doi.org/10.1007/978-3-319-16534-9
Curl RL (1966) Scallops and flutes. Trans Cave Res Group Great Br 7(2):121–160
Davies D (2014) Antarctic glaciers. www.antarcticglaciers.org/modern-glaciers/glacial-processes/
Delannoy J (1983) Le complexe souterrain du Plateau du Sornin (Vercours, France), Le Gouffre Berger et le Scialet de la Fromagere. Karstologia 2:3–12
Djurovič P (2009) Reconstruction of the pleistocene glaciers of mount Durmitor in Montenegro. Acta geogr Slov 49(2):263–289
Djurović P, Petrović AS, Simić S (2010) The overall impact of pleistocene glaciation on morphological diversity of uvalas at Durmitor and Žijovo. Serb Geogr Soc 90:17–34. https://doi.org/10.2298/GSGD1001017D
Dubljanszkij VN (1987) Teoreticseszkoje modelirovanije dinamiki formirovanija gidrotermo-karsztovüh polosztyej. Metodi i izucssenyija geologicseszkihjavlenyij. Novoszibirszk, pp 97–111
Ek C (1964) Note sur les eaux de fonte des glaciers de la Haute Maurienne: leur action sur les carbonates. Rev Belge Geogr 88:127–156
Embleton C (1980) Glacial processes. In: Embleton C, Thornes J (eds) Processes Geomorphol. E. Arnold, London, pp 272–326
Fels E (1929) Das Problem der Karbildung in den Ostalpen. Petermanns Mitt Ergänzungsh 202:1–84
Ford DC (1979) A review of alpine karst in the Southern Rocky Mountains of Canada. Bull Nat Speleol Soc 41:53–65
Ford DC (1983) Effects of glaciations upon karst aquifers in Canada. J Hydrol 61:149–158. https://dx.doi.org/10.1016/022-1694(83)90240-8
Ford DC (1984) Karst groundwater activity and landform genesis in modern permafrost regions of Canada. In: LaFleur RG (ed) Groundwater as a Geomorphic Agent. Allen & Unwin, London, pp 340–350
Ford DC (1987) Effects of glaciations and permafrost upon the development of karst in Canada. Earth Surf Process Land 12:507–521. https://doi.org/10.1002/esp.3290120508
Ford DC, Williams PW (1989) Karst geomorphology and hydrology. Unwin Hyman, London, p 601
Ford DC (1996) Karst in a cold climate. In: McCan SB, Ford DC (eds) Geomorphology sans Frontières. Wiley, Chichester, pp 153–179
Ford DC, Williams PW (2007) Karst hydrogeology and geomorphology. Wiley, Chichester, p 562
French HM (2003) The development of periglacial geomorphology 1: up to 1965. Permafrost Periglac Process 14(1):29–60
Glasser NF, Hambrey MJ (2001) Styles of sedimentation beneath Svalbard valley glaciers under changing dynamic and thermal regimes. J Geol Soc Lond 158:697–707
Grimes KG (2012) Surface karst features of the Judbarra/Gregory National Park, Northern Territory, Australia. Helectite 41:15–36
Groom GE (1959) Niche glaciers in Bünsowland, Westspitsbergen. J Glaciol 3(25):368–375
Hambrey MJ, Fitzsimons SJ (2010) Development of sediment-landform associations at cold glacier margins. Dry Val Antarct Sedimentol 57:857–882
Haynes VM (1968) The influence of glacial erosion and rock structure on corries in Scotland. Geograf Ann 50A:221–234
Hevesi A (1980) Adatok a Bükk-hegység negyedidőszaki ősföldrajzi képéhez (Data to the Quarternary Paleogeographical Features of the Bükk Mountains). Földtani Közlemények 110 (3–4):25–36 (in Hungarian)

Hughes PD (2004) Quaternary glaciation in the Pindus Moutains Northwest Greece. Ph.D. thesis, University of Cambridge 341 p
Hughes PD, Woodward JC (2009) Glacial and periglacial environments. In Woodward JC (ed) The physical geography of the mediterranean. Oxford University Press, Oxford, 700 p
Hughes PD,Woodward JC, Gibbard PL, Macklin MG, Gilmour MA, Smith GR (2006) The glacial history of the Pindus Mountains, Greece. J Geol 114:413–434
Hughes PD, Woodward JC, van Calsteren PC, Thomas LE, Adamson KR (2010) Pleistocene ice caps on the coastal mountains of the Adriatic Sea. Quat Sci Rev 29(27–28):3690–3708
Hughes PD, Woodward JC, Calsteren PC, Thomas LE (2011) The glacial history of the Dinaric Alps, Montenegro. Quat Sci Rev 30:3393–3412
Husen VD (2000) Geological processes during the quaternary. Mitt Österr Geol Ges 92:135–156
Jakucs L (1977) Morphogenetics of karst regions. Adam Hilgar, Bristol, p 284
Jennings JN, Bik MJ (1962) Karst morphology in Australian New Guinea. Nature 194:1036–1038
Kestin J, Sokolov M, Wakeham A (1978) Viscosity of liquid water in the range −8 °C to 150 °C. J Phys Chem Ref Data 7(3):941 948
Kiernan K, Lauritzen S-E, Duhig N (2001) Glaciation and cave sediment aggradation around the margins of the Mt. Field Plateau Tasmania. Aust J Earth Sci 48:251–263
Knight PG (1999) Glaciers. Stanley Thornes, London, p 272
Küfmann C (2014) Solution dynamics at the rock/snow interface during the ablation period in the subnival karst of the Wetterstein Mountains (Northern Calcareous Alps, Germany). Z Geomorfol 58(1):37–57
Kunaver J (1983) Geomorphology of the Kanin Mountains with special regard to the glaciokarst. Geografski Zbornik 22:197–346
Kunaver J (2009) The nature of limestone pavements in the central part of the southern Kanin plateau (Kaninski podi) Western Julian Alps. In: Ginés A, Knez M, Slabe T, Dreybrodt W (eds) Karst rock features. Karren sculpturing, carsologica, 9. Zalozba ZRC, Institut za raziskovanje krasa ZRC SAZU, Postojna, Ljubljana, Slovènia, pp 299–312
Lauritzen S-E (1984) Evidence of subglacial karstification in Glomdal, Svartisen. Nor Geogr Tidesskr 38(3–4):169–170
Lauritzen S-E (1986) Kvithola at Fauske, northern Norway: an example of ice-contact speleologenesis. Nor Geogr Tidesskr 66:153–161
Lliboutry LA (1979) Local friction laws for glaciers: a critical review and new openings. J Glaciol 23:67–95
Loeffler E (1971) The pleistocene glaciation of the Saruwaged Range, territory of New Guinea. Aust Geogr 11(5):463–472. https://doi.org/10.1080/00049187108702584
Maire R (1977) Les karsts haut alpins de Plate, du Haut-Giffre et de Suisse Occidentale. Rev Geog Alpine 65:403–423
Maire R (1990) La haute montagne calcaire. Karstologia-Memoires 3 La Ravoire, 731 p
Maire R, Pernette JF, Fage LH (1999) Les "glaciers de marble" de Patagonia, Chili. Un karst subpolaire oceanique de la zone austale. Karstologia 33:25–44
Maire R, Jaillet S, Hobléa F (2009) Karren Patagonia a natural laboratory for hydroaeilien dissolution. In: Ginés A, Knez M, Slabe T, Dreybrodt W (eds) Karst rock features. Karren Sculpturing Carsologica, 9. Zalozba ZRC, Institut za raziskovanje krasa ZRC SAZU, Postojna, Ljubljana, Eslovènia, pp 329–348
Mariko S, Bekku Y, Koizumi H (1994) Effux of carbon dioxide from snow covered forest floors. Ecol Res 9:343–350
Matthews SW (1987) Ice on the world. Nat Geogr 171(1):78–103
Mavlyudov BR (2006) Glacial karst, why it's important to research. Acta Carsologica 35(1):55–67
Menkovič LJ (1994) Glacial traces in the Djeravica area. Prokletije Mountains (in Serbian). Geogr God 30:139–146
Menkovic LJ, Markovic M, Cupkovic T, Pavlovic R, Trivic B, Banjac N (2004) Glacial morphology of Serbia, with comments ont he Pleistocene Glaciation of Montenegro, Macedonia and Albania. In: Ehlers J, Gibbard PL (eds) Quaternary glaciations: extent and chronology. PART 1: Europe. Elsevier pp 379–384

Metcale RC (1984) Field pH determinations in glacial melt waters. J Glaciol 30:106–111
Miotke FD (1968) Karstmorphologische Studien in der glazialuberformten Hohenstufe der Picos de Europe' Nordspanien. Jahrbuch der geographischen Gesellschaft zu Hannover Sonderhef, 4:161 p
Nicod J (1972) Pays et Paysages du Calcaire. Paris. Presses Universitaire de France, 244 p
Nicod J (1978) Phenomenes glacio-karstiques du Devoluy meridional. Rev Geog Alpine 66: 149–167
Perritaz L (1996) Le „karst en vagues" des Ait Abdi (Haut-Atlas central, Maroc). Karstologia 28 (1):1–12
Plan L, Decker K (2006) Quantitative karst morphology of the Hochschwab plateau, Eastern Alps, Austria. Z Geomorphol 147:29–56
Playford PE (2002) Palaeokarst, pseudokarst, and sequence stratigraphy in Devonian reef complexes of the Canning Basin, Western Australia. In: Keep M, Moss SJ (eds) The sedimentary basins of Western Australia, vol 3. Petroleum Exploration Society of Australia, Symposium, Perth, W.A., pp 763–793
Playford PE (2009) Guidebook to the geomorphology and geology of Devonian reef complexes of the Canning Basin, Western Australia. Geological Survey of Western Australia, Record 5, 72 p. http://www.dmp.wa.gov.au/GSWApublications/
Rau F, Mauz F, Vogt S, Khalsa SJS, Raup B (2005) Illustrated GLIMS glacier classification manual, version 1.0. GLIMS Regional Centre, 'Antarctic Peninsula': GLIMS (Global Land Ice Measurement from Space), NSIDC, 36
Ravanel L, Deline P (2008) La face ouest des Drus (massif du Mont-Blanc) évolution de l'instabilité d'une paroi rocheuse dans la haute montagne alpine depuis la fin du petit âge glaciaire. Géomorphol Relief Process env 4(this issue)
Rose L, Vincent P (1983) Some aspects of the morphometry of grikes a micture model approach. In: Paterson K, Sweeting M (eds) New directions in karst, Proceedings of the Agnlo French karst symposium, Geo Books, Norwich pp. 497–515
Rothslisberger H (1972) Water pressure in intra and sub-glacial channels. J Glaciol 11:177–203
Sauro U (2009) Glaciokarst landforms of the lower Adige and Sarca valleys. In: Gines A, Knez M, Slabe T, Dreybrodt W (eds) Karst rock features. Karren sculpturing, carsologica, 9. Zalozba ZRC, Institut za raziskovanje krasa ZRC SAZU, Postojna, Ljubljana, Eslovènia, pp 323–328
Selby MJ (1985) Earth's changing surface: an introduction to geomorphology. Clarendon Press, Wortley, p 607
Shreve RL (1972) Movement of water in glaciers. J Glaciol 11:205–214
Smart C (1986) Origin and development of glacio-karst closed depressions in de Europa, Spain. Z Geomorphol N.F. 30(4):423–443
Smart C (2004) Glacierized and glaciated karst. In: Gunn J (ed) Encyclopedia of caves and karst science. Fitzroy Dearborn, New York, London, pp 388–391
Star A, Langenscheidt F (2015) Landforms of high mountains. Springer, Verlag, Berlin Heidelberg. https://doi.org/10.1007/978-3-642-53715-8
Stepišnik U, Žebre M (2011) Glaciokras Lovčena (E-GeografFF 2). Ljubljana, Znanstvena založba Filozofske fakultete, Oddelek za geografijo, 82 str. http://geo.ff.uni-lj.si/sites/default/files/glaciokras_lovcena_0.pdf. Citirano 5 Feb 2013
Sweeting MM (1966) The weathering of limestones with particular reference to the carbon: ferous Limestone of northern England. In: Dury GH (ed) Essays in geomorphology. Heinemann, London, pp 177–210
Sweeting MM (1973) Karst landforms. Columbia University Press, New York, p 362
Telbisz T (2010) A montenegrói Sinjajevina karsztfennsík felszínalaktani vizsgálata terepi és térinformatikai módszerekkel. Karsztfejlődés XV:85–101
Telbisz T, Dragašice H, Nagy B (2005) A horvátországi Biokovo-hegység karsztmorfológiai jellemzése terepi megfigyelések és digitális domborzatelemzés alapján. Karsztfejlődés X:229–243
Thorn CE, Hall K (2002) Nivation and cryoplanation: the case for scrutiny and integration. Progress Phys Geogr 26:553–560

Trudgill ST (1972) The influence of drifts and soils on limestone weathering in N.W. Claire, Ireland. Proc Univ Bristol Speleol Soc 13:113–118
Trudgill ST (1985) Limestone geomorphology. Longman, New York, p 196
Veress M (2004) A karszt (The karst) BDF Természetföldrajzi Tanszék, Szombathely, 215 p (in Hungarian)
Veress M (2010) Karst environments. Karren formation in high mountains. Springer, Dordrecht, Heidelberg, London, New York, p 230
Veress M (2012a) Glacial erosion and karst evolution (karren formation on the surfaces formed by glaciers). In: Veress B, Szigethy J (eds) Horizons in earth science Research, vol 8. Nova, New York, pp 1–94
Veress M (2012b) Morphology and solution relationships of three karren slopes in different environments (Totes Gebirge, Eastern Alps). Z Geomorphol 56(Suppl 2):47–62
Veress M (2016) Covered karst. Springer, Berlin, Heidelberg, New York, 536 p. https://doi.org/10.1007/978-94-017-7518-2
Veress M, Lóczy D (2015) Depression of superficial deposit. In: Veress B, Szigethy J (eds) Horizons in earth science research, vol 13. Nova, New York, pp. 37–64
Veress M, Szunyogh G, Zentai Z, Toth G, Czopek I (2006) The effect of the wind on karren formation on the Island of Diego de Almagro (Chile). Z Geomorphol 50(4):425–445
Verstappen HTh (1964) Karst morphology of the star mountains (Central New Guinea) and its relation to lithology and climate. Z Geomorph NF 8:40–49
Vincent P (2009) Limestone pavements in the British Isles. In: Ginés A, Knez M, Slabe T, Dreybrodt W (eds) Karst rock features. Karren sculpturing, carsologica, 9. Zalozba ZRC, Institut za raziskovanje krasa ZRC SAZU, Postojna, Ljubljana, Eslovènia, pp 267–274
Williams PW (1966) Limestone pavements with special reference to western Ireland. Trans Inst Br Geogr 40:155–172
Williams PW (2008) The role of the epikarst in karst and cave hydrogeology: a review. Int J Speleol 37(1):1–10
Woodward JC, Macklin MG, Smith GR (2004) Pleistocene glaciation in the mountains of Greece. In: Ehlers J, Gibbard PL (eds) Quaternary glaciations—extent and chronology Part 1: Europe. Elsevier, Amsterdam, pp. 155–173
Woodward JC, Hamlin RHB, Macklin MG, Hughes PD, Lewin J (2008) Glacial activity and catchment dynamics in norteast Greece: long-term river behaviour and the slackwater sediment record for the last glacial to interglacial transition. Geomorphology 101:44–67
Žák K, Richter DK, Filippi M, Živor R, Deininger M, Mangin A, Scholz D (2012) Cryogenic cave carbonate—a new tool for estimation of the Last Glacial permafrost depth of the Central Europe. Clim Past Discuss 8:2145–2185
Zamora F, Santana A (1979) Charactersitics climaticas de la costa occidental de la Patagonia entre las latitudes 46°40′ y 56°30′ s. An Inst Patagon 10:109–154
Zebre M, Stepišnik U (2015) Glaciokarst landforms and processes of the southern Dinaric Alps. Earth Surf Process Landf. https://doi.org/10.1002/esp.3731
Žebre M, Stepišnik U (2015b): Glaciokarst geomorphology of the Northern Dinaric Alps: Snežnik (Slovenia) and Gorski Kotar (Croatia). J Maps. https://doi.org/10.1080/17445647.2015.1095133
Žebre M, Stepišnik U, Fabekovič G, Grij A, Koblar S, Kodelja B, Pajk V, Stefanic K (2013) Pleistocenska poledenitev Biokova. Delai 39:141–155
Žebre M, Stepišnik U, Colucci RR, Forte E, Monegato G (2016) Evolution of a karst polje influenced by glaciation: the Gomance piedmont polje (northern Dinaric Alps). Geomorphology 257:143–154

Chapter 3
Glacial Erosion on Karst

Márton Veress

Abstract The chapter presents the movement of warm-based and cold-based glaciers. The factors influencing the slide and flow of glaciers are described. The geomorphic action of glaciers is discussed with a focus on karst. On karst, glacial erosion is characteristic in karst depressions, in stratified limestone, as well as in troughs which developed on similar rocks. As regards depressions, the denudation of slopes the inclination of which is identical with (downstream slopes) or opposite to (upstream slopes) the direction of glacier flow, the destruction of the thresholds of uvalas (all of them) and the glacial erosional transformation of depressions partly filled with non-moving ice are analysed. The formation of stepped surfaces of stratified limestone terrains as well as the relationship between troughs with stepped surfaces, valley directions and the dip direction of beds are presented.

Keywords Warm-based glacier · Cold-based glacier · Polythermal glacier
Asymmetric depression · Depression with a composite slope
Dividing wall · Step · Separation of beds · Dissolution along fracture
Dissolution along bedding plane · Asymmetric glacier valley

3.1 Introduction

Based on the response to the shear stress, the movement of ice may happen in three ways: with internal deformation or flow, with basal sliding and with subglacial bed deformation. The shear stress of glaciers, which comes from the weight of the ice eventually, can be summarized in the equation (Bennett and Glasser 2009):

$$\tau = pg(s - z) \cdot \sin \alpha,$$

where τ is the shear stress,
p is the density of ice,
g is the acceleration due to gravity,

M. Veress et al., *Glaciokarsts*, Springer Geography,
https://doi.org/10.1007/978-3-319-97292-3_3

α is the surface slope of the glacier,
s is the surface elevation,
z is the elevation of a point within the glacier.

The level of shear stress (τ) depends on the ice thickness ($s - z$) and on the surface slope (α) of the ice (or of the floor). However, it is also affected by the quality of the floor. The value of shear stress is lower if the floor is composed of deformable material (Bennett and Glasser 2009). In this case, ice moves forward together with the material of the floor (see below). Deformation results in sliding and plastic flow of ice. On the upper section of the glacier, where the ice is more rigid, it folds, at even more rigid sites, it is dissected along fractures into blocks, which are thrust above each other along slip planes.

Warm-based, cold-based and polythermal glaciers are differentiated (Bennet and Glasser 2009; Davies 2014). Warm-based glaciers have meltwater at their base, while cold-based glaciers do not. Cold-based glaciers have a frozen base, rock fragments are frozen to the base and there is permafrost in the rocks. The upper section of a polythermal glacier acts as a warm-based glacier, while it behaves as a cold-based glacier at its margins (Hambrey et al. 1999; Glasser and Hambrey 2001, 2003).

The water of the floor of warm-based glaciers mainly originates from the melting of ice (Bennett and Glasser 2009; Davies 2014). Melting is primarily caused by the pressure increase of ice (Benn and Evans 2010; Cuffey and Paterson 2010), but it can also be caused by the Earth's geothermal heat (Davies 2014), and the water can also directly originate from the surface waters of the glacier (Hooke and Le 1989). The increase in pressure is controlled by ice thickness (Davies 2014). The thicker the ice, the more meltwater may be produced, where the glacier is thinner, no or only slight melting occurs. Thus, based on ice thickness, warm-based or cold-based glacier section can alternate within one glacier caused by pressure change. There are typical meltwater glaciers in the Alps (Goodsell et al. 2005). Taking the whole Earth into account, temperate glaciers and tropical glaciers belong to this type, while some glaciers under cold climate belong to cold-based glaciers.

Warm-based glaciers show basal sliding because of the presence of water between the ice and the base, while cold-based glaciers flow because this water is absent and their base is frozen. The sliding velocity depends on the level of shear (in case of a double level shear, the sliding velocity is eight times larger), on the quality of the base, whether it is rigid or plastic (Bennett and Glasser 2009). The degree of sliding and flow depends on the distance from the base (Fig. 3.1), but sliding and flow also modify each other (Summerfield 1991). The sliding and flow of warm-based glaciers are presented in Fig. 3.2a, the flow of cold-based glaciers can be seen in Fig.3.2b, while on a plastic base, the movement of warm-based glaciers originating from the deformation of the plastic base, as well as their sliding and flow are presented in Fig. 3.2c (Boulton 1993). Sliding is slowed down by the irregularities of the base. In case of high roughness, ice is in contact with the base at several places, which increases the braking effect, but the debris between the ice and the base has a similar effect too (Bennett and Glasser 2009). The presence and pressure of the water between the ice and the base reduce friction. It decreases even in the case of some mm thick water (Davies 2014). However, it is also influenced by the degree of water supply. In case of maritime glaciers, there will be more meltwater because of a larger water supply, thus, at these glaciers, the movement of ice is faster than at continental glaciers (Kerr 1993). In karst areas, point-like water drainage is common under glaciers. Meltwater is concentrated at these sites and accelerates sliding. At sites from where waters flow towards drainage places, the amount of water will be smaller, which reduces sliding.

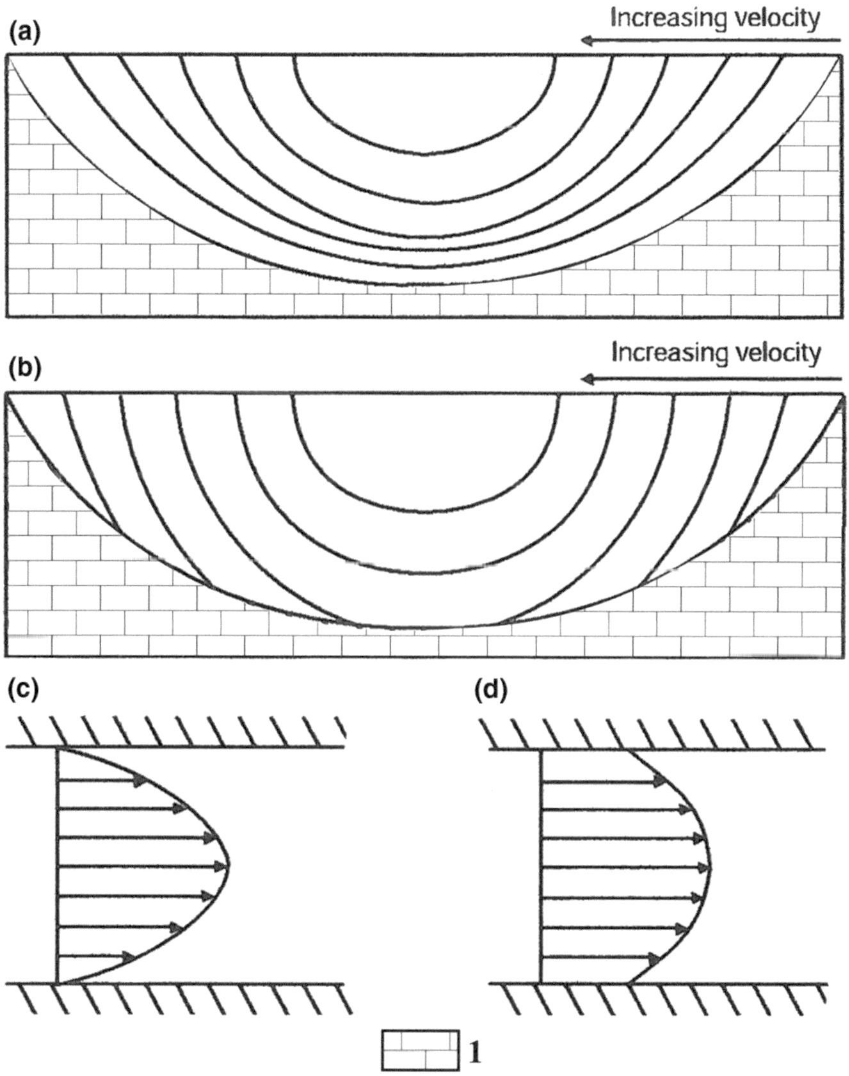

Fig. 3.1 The pattern of ice movement in a glacier valley **a** flow pattern in case of a cold-based glacier, where there is no sliding, **b** flow pattern where there is sliding (in case of a warm-based glacier), **c** flow in plan view in case of a cold-based glacier, **d** flow pattern where there is sliding (Summerfield 1991, modified) 1. limestone

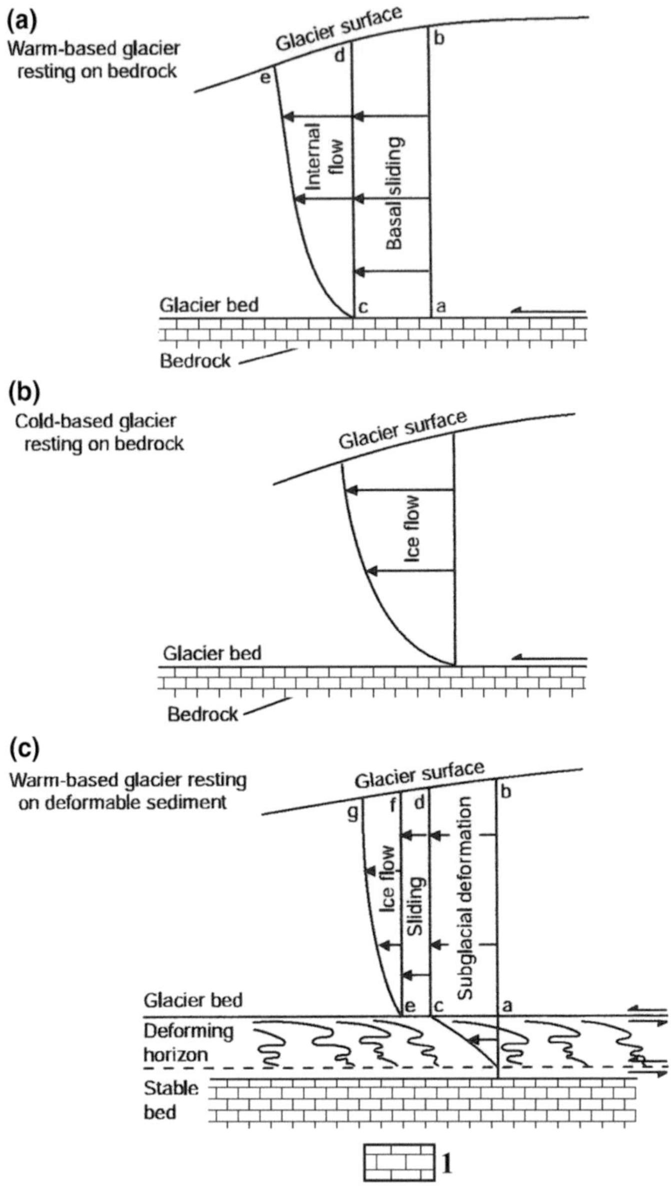

Fig. 3.2 The distribution of speed in case of a warm-based glacier (**a**), in case of a cold-based glacier (**b**), and in case of a warm-based glacier where the floor is deforming sediment (**c**) (Boulton 1993, modified) 1. limestone

Glacial erosion is bound to warm-based glaciers (or glacier parts) since they move (slide) as compared to their base, while cold-based glaciers do not perform glacial erosion as this movement is absent. Nor can dissolution be expected under cold-based glaciers as meltwater is absent. Sliding velocity is affected by several factors. The relationship between sliding and ice temperature is indicated by the fact that if ice temperature is −1 °C, at the same level of shear stress, the extent of sliding can be 1000 times larger than at −20 °C (Bennett and Glasser 2009). Basal sliding is modified by other factors too. Thus, as it has already been mentioned, sliding increases if the number of ice–base contact points is small, but it is also dependent on the amount and pressure of water between the ice and the base as well as on the amount of debris. If there is a lot of debris, the rate of sliding decreases (Bennett and Glasser 2009).

Recently, geomorphic action is supposed to be possible in case of cold-based glaciers too, which is expressed mainly by debris entrainment and transfer (Atkins et al. 2002, 2009; Glasser and Hambrey 2003; Lloyd Davies et al. 2009), and by the scour of bedrock bumps and the plucking of smaller ones (Davies 2014). It is probable that this process does not result in significant erosion. However, cold-based glaciers (ice sheets) may behave partly or completely the same way as warm-based glaciers. Thus, in the Antarctic ice sheet there are zones (ice streams) where the movement of ice accelerates compared to its environment (Davies 2014). The calving of outlet glaciers indicate that they behave partly or totally as warm-based glaciers. But the ice sheets of glacials (or their margins and then more and more increasing parts) were also probably warm-based glaciers at the beginning of glacials when the ice sheet was moving forward. The marginal part of the ice sheet becomes warm-based glacier and the warm-based glacier section shifts during the failure of the ice sheet (Bennett and Glasser 2009). The ice sheet may behave as a warm-based glacier at the end of the glacials when a lot of meltwater is produced because of melting of the ice sheet.

A cold-based glacier can be transformed into a warm-based glacier if its pressure increases because of large ice thickness (Murphy et al. 2015). However, it can also be triggered by the fact that the crust becomes fractured by which magma masses can get close to the surface from depth. The effective normal pressure (N) has an important role in the destructive work of the glacier, which is the force per unit area imposed vertically by a glacier on its bed (Bennett and Glasser 2009). This is summarized by

$$N = p \cdot g \cdot h,$$

where

p is the density of ice,
g is the acceleration due to gravity,
h is ice thickness

Its value is decreased by the pressure of meltwater. As regards the erosional role of the effective normal pressure, there are two approaches (Bennett and Glasser 2009). According to Boulton's model (Boulton 1974), the effectiveness of erosion increases with growing pressure. This cannot grow infinitely because the growing amount of debris increases friction, which slows down sliding. We have seen that meltwater decreases the effective normal pressure, but in those rocks where water can be drained into the karstifying rock, this pressure will not drop at all or it will decrease to a small extent only. According to Hallet's model, erosion is independent of effective normal pressure.

The erosion of glaciers can take the form of plucking, abrasion and erosion or solution by meltwater. The destructive activity of warm-based glaciers takes places in the following ways:

- Plucking:
  - The glacier plucks the floor. According to Bennett and Glasser (2009), plucking or quarrying is the removal of blocks of the bedrock. This consists of two stages: the fracturing of bedrock (1), and the entrainment of these fractured pieces and their detachment from the floor (2). Water freezes in the cracks of the rock and the cracks widen. The rock block between the crack and the cavity of the bedrock is removed. (Blocks can rotate at that place where there is a cavity in the bedrock.) Meltwater freezes at the cavity when pressure drops. The block freezes to the cold patch that developed by freezing. The moving ice entrains the block (Fig. 3.3).
  - At the same place under the glacier, pressure force changes on the base. This promotes the physical weathering of the rocks on the floor. The larger amount of meltwater "props up" the ice, ice pressure may decrease, and then when the amount of meltwater decreases because of karstic drainage, in this case, the ice pressure increases.
- Abrasion operates in the following way:
  - With the parts frozen into ice, the sliding ice scratches the base (abrasion). Polishing is proportional to sliding velocity. The effectiveness of polishing

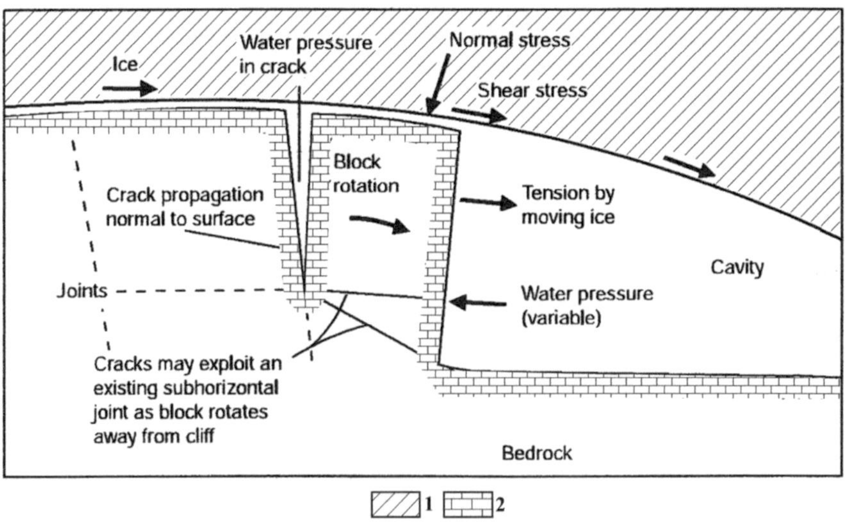

Fig. 3.3 Processes causing plucking (Bennett and Glasser 2009, modified) 1. ice, 2. limestone

increases if the hardness of debris is larger than that of the basement. The effectiveness of polishing depends on the amount of debris. It will be more intensive in case of a larger amount of debris. We have already mentioned that with increasing amounts of debris, friction also grows and reduces sliding. But its effectiveness also depends on the shape of debris, in addition to this, to what extent it is supplied by since the existing pieces are used up fast (Bennett and Glasser 2009).

  - The polishing effect of the fine-grained material (silt), the degree of which is said to be larger than the destruction by frozen rock fragments (Bennett and Glasser 2009) is especially strong with more load.
  - During sliding, the rock fragments situated between the ice and the base rotate. During this rotation not only the base but also the rock fragments are denuded, this latter one contributes to the formation of fine-grained material.
  - The movement of ice is discontinuous. When sliding begins, the rock fragments pushed by the glacier crash into the bedrock bumps and destroy them.

- The destruction of ice intensifies at bedrock bumps. This can be the following:
  - Sliding can be of two kinds at bedrock bumps (enhanced basal creep and regelation slip). Enhanced basal creep occurs in case of larger bumps, while regelation slip is characteristic of smaller bumps. The larger the bump, the greater the pressure and plasticity on the upstream side increasing the flowing tendency. Pressure increases on the upstream side of smaller bumps. To the effect of pressure increase, meltwater is generated (which flows around the bump) and reaches the downstream slope of the bump. Here, the meltwater freezes because of pressure decrease. The meltwater infiltrates into the cracks of the rock and thus, frost weathering occurs on this slope of the bump (Rastas and Seppälä 1981). A cold patch develops on this slope. The ice of the patch freezes to the base. As previously mentioned, the frozen ice removes the loosened rock pieces during the movement of the glacier. This phenomenon is called plucking (Fig. 3.4).
  - The load of the rock block passing through the bumps creates cracks on the base. At the termination of cracks, a shear surface develops, along which rock pieces are removed from the base (Fig. 3.5, Walden 2002).

Various effects can be mentioned which are strongly connected to the activity of ice, complementing and modifying its activity. Thus, these effects are snow erosion, erosion and dissolution by meltwater, mass movements, the erosion and redeposition exerted by (rock and snow) avalanches. The erosion of meltwater: its effectiveness is increased by the growth of flow velocity (the flow will be turbulent), cavitation (Bennett and Gasser 2009) and the amount of debris. In carbonate areas, the dissolution of meltwater: dissolution effect may be significant mainly under wet (maritime) glaciers since a lot of precipitation is added to the meltwater (Bennett and Glasser 2009).

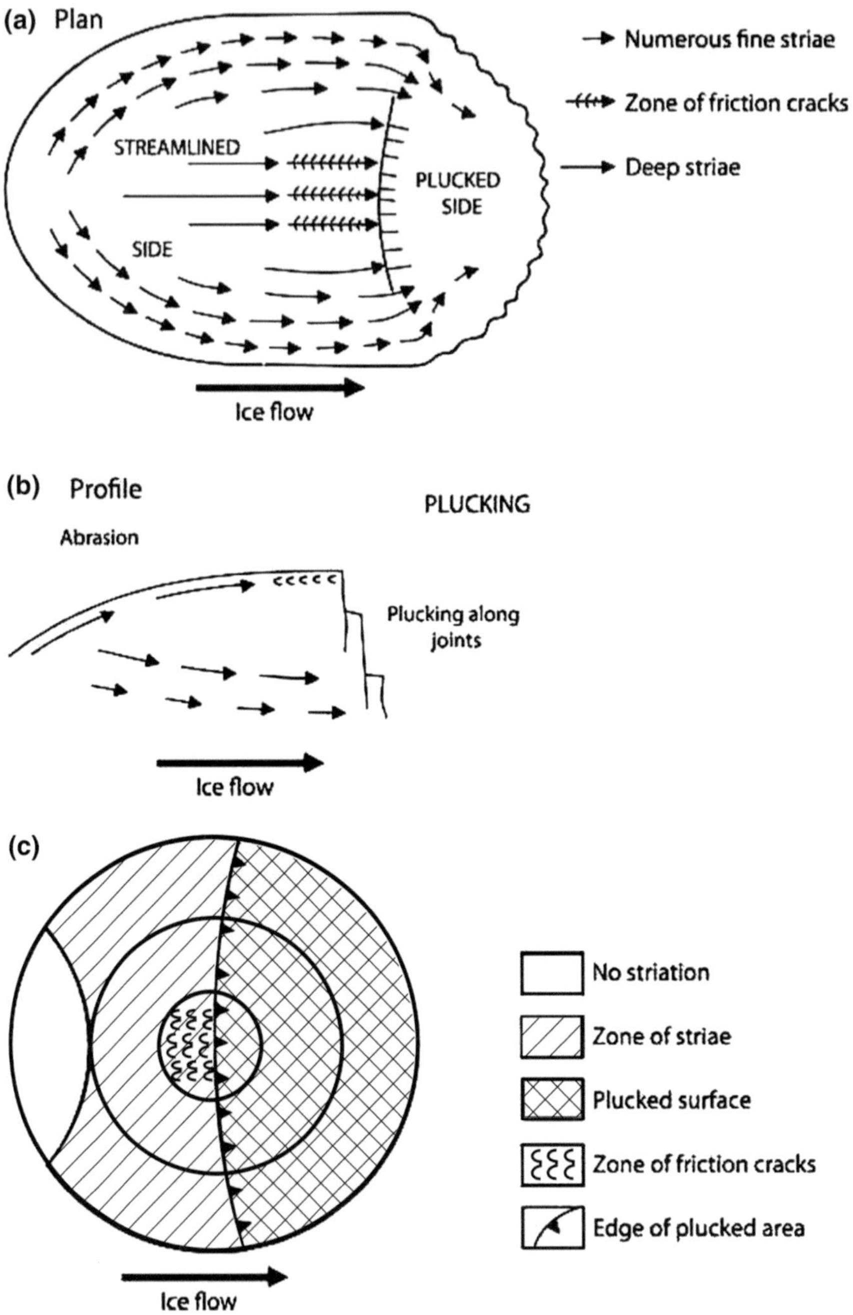

Fig. 3.4 Denudation of the surface of a roche moutonnée by abrasion and plucking (Chorley et al. 1984, modified) **a** and **b** roche moutonnées in plan and side view, **c** model on the distribution of glacial erosional micro features on a roche moutonnée

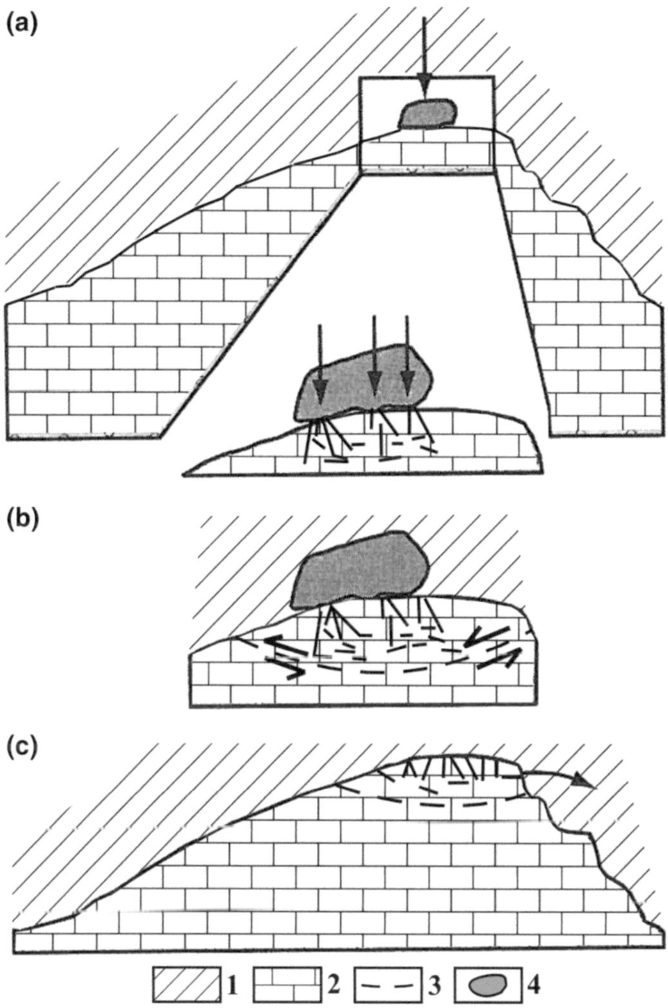

Fig. 3.5 Denudation of the roof level of roche moutonnées; **a** rock under rock blocks becomes fractured, **b** a shear surface develops at the fractures, **c** a rock block becomes separated at the shear surface (Walden 2002, modified) 1. ice, 2. limestone, 3. shear surface, 4. rock block

Glacial erosion creates glacial features and glacial landscapes. The features can be of small, medium and large size (Bennett and Glasser 2009). During the destructive activity of mainly valley glaciers, cirques, troughs, rock basins, transversal steps and longitudinal steps, roche moutonnées, whalebacks and striae develop independently of rock quality. Transverse and longitudinal steps are not identical with the above-described steps of stepped surfaces. The former features occur individually and are large sized (thus, transverse steps go transversely the valley floor, their vertical expansion can be several 10 m), while the latter consist of a series of steps, where the vertical expansion of scarp fronts (which are separated by surfaces with bedding planes of different heights from each other) can range from some tens of centimetres to some metres. In the area of the ice sheet (ice caps), glacial plains and roche moutonnées develop.

The development of cirques happens by the retreat of eroded slopes, during which the length of the arch of the wall increases (Gordon 1977). The retreat is caused by several factors (glacial erosion, frost action, mass movements). During the deepening of the cirque floor, a closed feature can also develop (Haynes 1968). V-shaped valleys are transformed into U-shaped troughs because glacial erosion increasingly shifts towards the valley floor (Harbor et al. 1988). The side slope of roche moutonnées being aligned parallel to ice movement, as it has already been mentioned, becomes steeper and uneven because of frost weathering and the capacity of ice to entrain rock blocks (Rastas and Seppälä 1981). Transverse steps develop at valley floors being steep preceding glacial erosion or at the boundary of rocks with different resistance. Rock basins also develop by selective denudation, due to overdeepening. Fluvioglacial features, gorges under ice (sichelwannen), potholes, channels, like N-(Nye) channels (Hooke and Le 1989; Bennett and Glasser 2009), as well as outwash plains and alluvial fans (Lepirica 2008; Adamson et al. 2014; Zebre and Stepišnik 2015a, b) accumulate from subglacial meltwater.

According to their position, accumulation features can be lateral moraines (morainal bank), surface moraines, basal moraines, terminal moraines and end moraines (Mollard and Jones 1984). According to their deposition, they can be primary moraines (they are directly connected to glacier activity) and secondary moraines (Dreimanis 1989).

3.2 Glacial Erosion on Karst

Cold glaciers (or the non-moving ice lenses of depressions) most effectively contribute to the infilling of karst depressions. However, warm glaciers do not only fill the depressions, but they also transform them. Glacial erosion has many specific characteristics on karst and its effectiveness does probably not lag behind the erosion occurring on other rocks. Its characteristic features are the following:

- The transported debris is almost exclusively limestone. Its hardness is great, although it is identical with the hardness of the bedrock.
- The amount of debris is probably not too large and thus, it does not significantly slow down sliding. Its supply is continuously ensured by the pieces removed from the floor (see below).
- Meltwater is not extremely abundant since the water partly percolates away in the karstic rock where cavities are formed. This ensures large effective normal pressure.

On karst, glaciers can develop isolated, only in some depressions (such a glacier was formed, e.g. on the Austrian Rax plateau). However, these can coalesce later. Glaciers formed by coalescence can cover the terrain with depressions and uvalas continuously, in a larger expansion too, adjusted or not adjusted to the topography (Figs. 3.6, 3.7).

Glacial erosion can be modified by the characteristics of the karst. Therefore, geomorphic evolution may also be specific. The particular feature development of glacial erosion on karst is a result of karst depressions and rock structure as well as the dissolution of the rock.

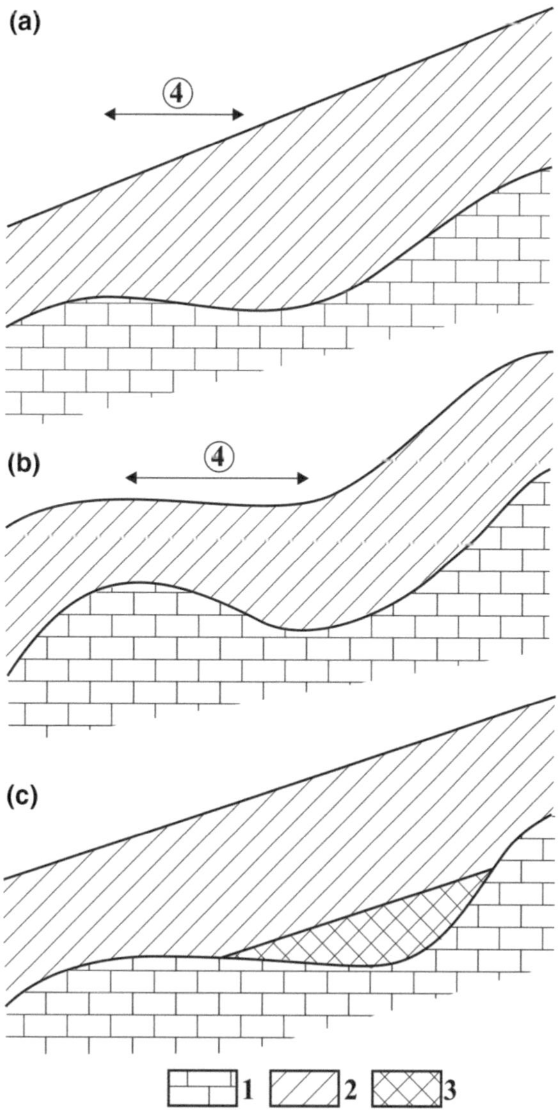

Fig. 3.6 Various glaciers over karst depressions **a** ice moves to the floor of depression, **b** the glacier does not completely fill the depression, **c** there is non-moving ice on the depression floor, 1. bedrock, 2. moving ice, 3. non-moving ice filling the depression, 4. the zone of glacier extrusion

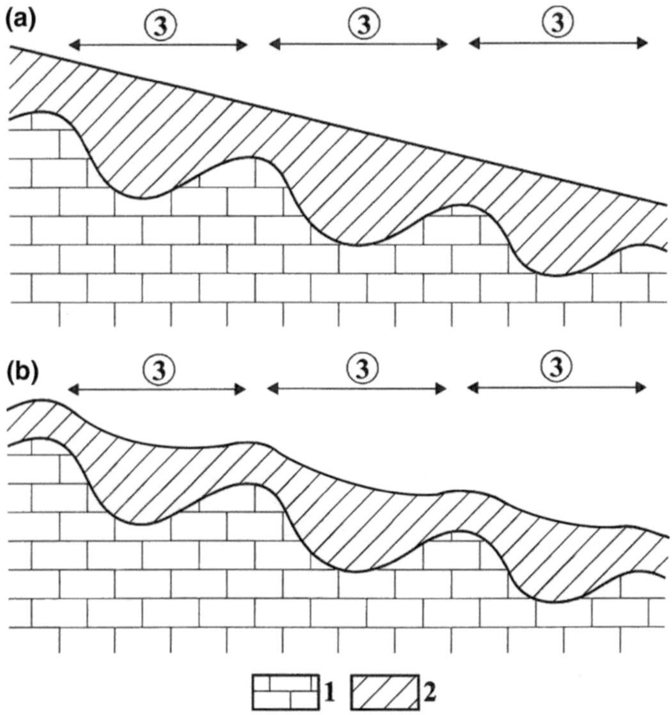

Fig. 3.7 Glacier on a terrain with uvalas **a** the surface of the thick glacier evenly slopes towards the termination of the glacier, **b** a thin glacier bends inwards at depressions, 1. limestone, 2. ice, 3. constituting doline

According to Ford (1983, 1996) Ford and Willliams (2007), glaciers exert their effect on glaciokarst in the following ways:

- Erasure, when the ice planates the surface.
- Dissection, when the ice dissects the bearing rock into blocks.
- Infilling, when the till fills up the karstic features.
- Injection, when the till penetrates into the passages of the karst.
- The bedrock is shielded from postglacial dissolution by limestone debris.
- The bedrock is sealed from dissolution by the impermeable cover.
- Meltwater is superimposed on the karst.
- The deep injection of meltwater into the karst.

Collapses commencing in glacials can be mentioned among other effects, when the existing features regenerate. As previously mentioned, glacial erosion can be modified by rock structure, karst features such as karst depressions (most often dolines, but ponors and poljes can also be under the influence of glacial erosion) and the thresholds separating the depressions.

3.2.1 *Surface Formation in the Depressions of Karst*

Karst features (mainly depressions) can undergo the following changes:

- truncation,
- partial truncation,
- deepening,
- change of shape: the slope becomes gentle or steeper,
- widening (lengthening),
- the dividing walls between depressions become rounded, they are denuded partially, but to a different extent,
- the development of slopes with various gradients.

Opinions on the degree of glacier erosion regarding the denudation of karst features vary. According to Kunaver (2009), even small-sized karst features (karren too) undergo the only truncation, while according to Djurovič et al. (2010), Smart (2004), Djurovic (2009) glacial erosion destroyed most preglacial features. However, the complete truncation of larger features such as karstic depressions is not probable since the dolines of troughs the diameter of which is often several hundred metres could not develop in a postglacial way. For this, the duration of postglacial dissolution was too short.

3.2.1.1 Erosion in Depressions

On glaciokarst, depressions also developed preceding Pleistocene ice cover (Fells 1929; Bauer 1952; Zötl 1963; Smart 1986). Depressions promoted the development of glaciers, and glaciers shaped their valleys by passing through these depressions (Cvijič 1913; Menkovič 1994). Depressions were transformed to a different degree and in various ways by moving ice. The ice passing through karstic depressions has a connection with three slopes of various positions and it exerts its effects on them. These are the following:

- the slope of the depression is aligned parallel with the direction of ice movement,
- the slope of the depression is aligned opposite to the direction of ice movement,
- the direction of the slope is perpendicular to ice movement.

The first two of cases are dealt with in this chapter. The degree of ice movement (v), which, for simplicity's sake, we consider to be proportional to sliding (flow ignored), one of its components presses ice to the base, the other is perpendicular to this and gives the extent and direction of shifting. We call the former one loading component or pressure component triggered by the weight of ice (v_1), we consider it to be the same as the "effective normal pressure (N)" described by Bennett and Glasser (2009), while the latter is the actual shifting or the horizontal component of ice movement (v_2, Fig. 3.8a). The loading component is only vertical if the ice

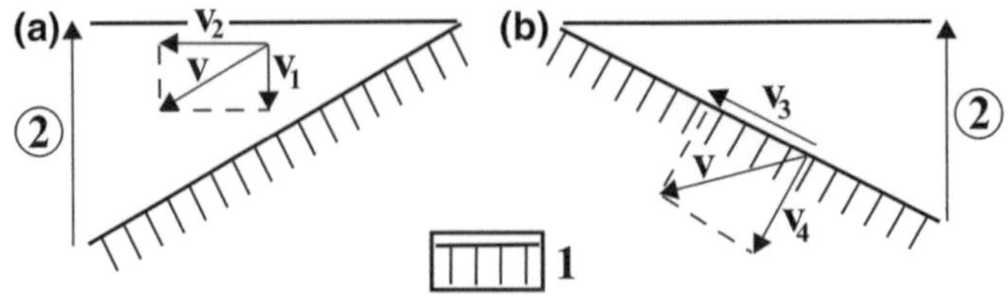

Fig. 3.8 Components of ice speed if the direction of glacier movement and the dip angle of the floor is identical (**a**) and if it is different (**b**) 1. floor, 2. ice, v the degree of the speed of the glacier and the direction of its movement; v_1 direction and degree of pressure component triggered by the weight of ice, v_2 horizontal component of ice movement, v_3 component of ice movement forced by the floor, v_4 pressure component forced by the floor

movement (sliding) corresponds to the direction of the slope. Such places will be the downstream slopes of the depressions. The loading component is smaller on these slopes than on the upstream slope and at a lower angle with the slope than with the direction of ice movement, the steeper is the slope.

However, on the upstream slope, it will not be vertical, but perpendicular to it (v_4, Fig. 3.8b) which is therefore called pressure component forced by its floor. In this case, the actual shifting forced by the floor (v_3) is parallel to the slope, which results in an ice movement opposite to dip direction. If the loading component is at a higher angle with the slope than to the direction of ice movement, during its movement the ice would go away from the base, while if it has a lower angle, it would penetrate into the base. There are no data or observations for such ice movements (apart from the case when one part of the base is vertical, which will be discussed below), and therefore the actual shifting has to be regarded parallel to the slope.

If the shifting (sliding) and the slope angle are identical, the loading component encloses a higher angle with the surface on the upstream slope of the doline than the loading component of the downstream slope (Figs. 3.8a, b). The rate of plucking is probably higher in the case of a loading component with a higher angle. Although the actual shifting of the ice (the horizontal component of ice motion) is larger on the downstream slope, its destructive effect is small since it removes the ice from the base. Thus, both the polishing and the scratching (striation) effects will be weaker than on the upstream slope of the doline where shifting (component forced by the floor) is parallel to the surface. In addition, here the polishing effect will also be larger because of the larger loading component. Therefore, on the whole, the denudation of the upstream slope of the doline will be larger than on its downstream slope.

More rapid denudation of upstream slopes can be experienced if we compare the gradient of the downstream slopes of the paleodolines of glaciokarst with the gradient of their upstream slopes. For this, the slopes of some paleodolines in the Durmitor (the Dinarides) were compared. In order to determine the movement direction of glaciers, the glacier reconstruction developed by Djurovič (2009) was used. It can be seen that the upstream slopes are gentler than the downstream slopes, which is attributable to the fact that the former was destroyed to a greater extent during glacial erosion (Fig. 3.9).

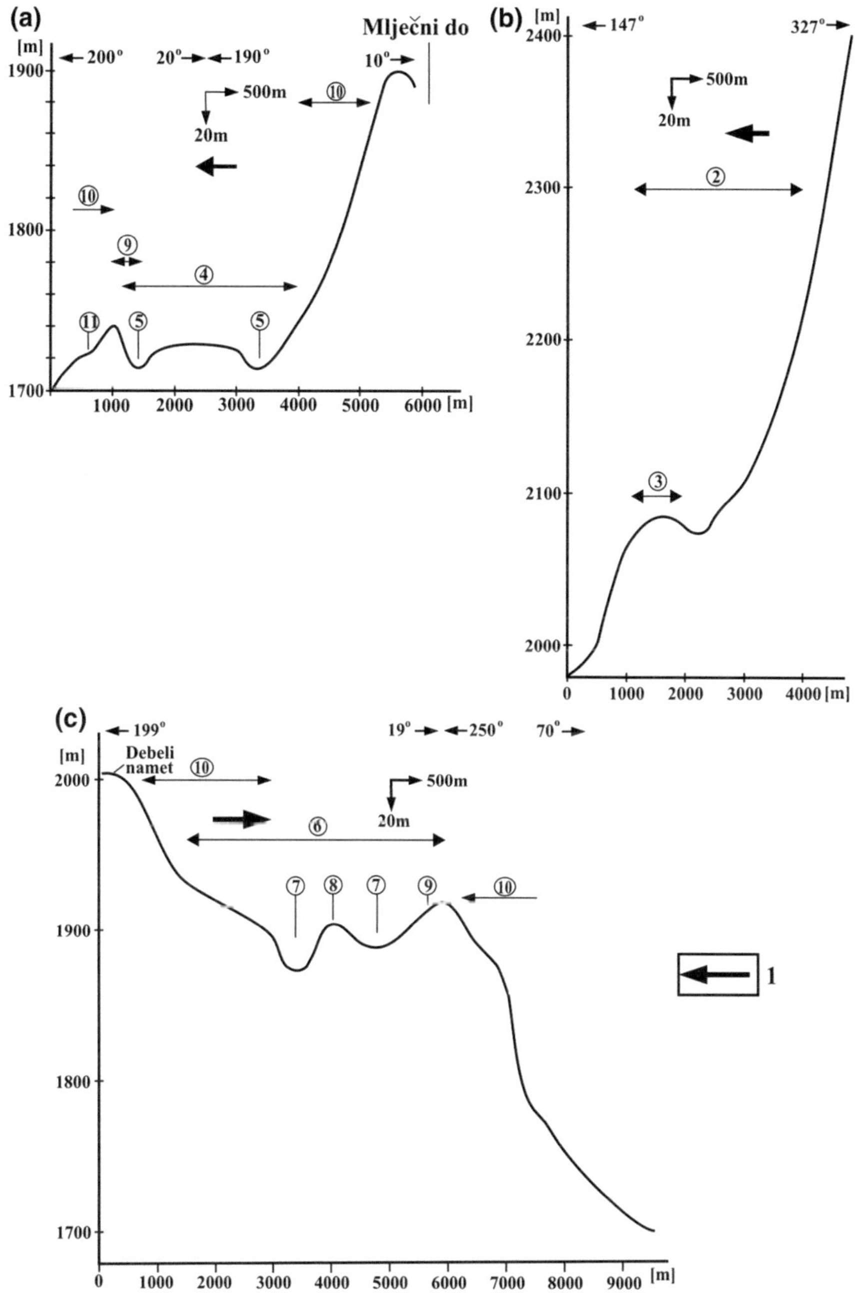

Fig. 3.9 Longitudinal section of dolines and uvalas (Durmitor) **a** Urdeni do, **b** V. korita, **c** Velika Kalica, 1. direction of ice movement, 2. cirque, 3. cirque threshold, 4. doline, 5. inner doline, 6. uvala, 7. constituting doline, 8. denuded dividing wall between constituting dolines, 9. upstream slope, 10. downstream slope, 11. road

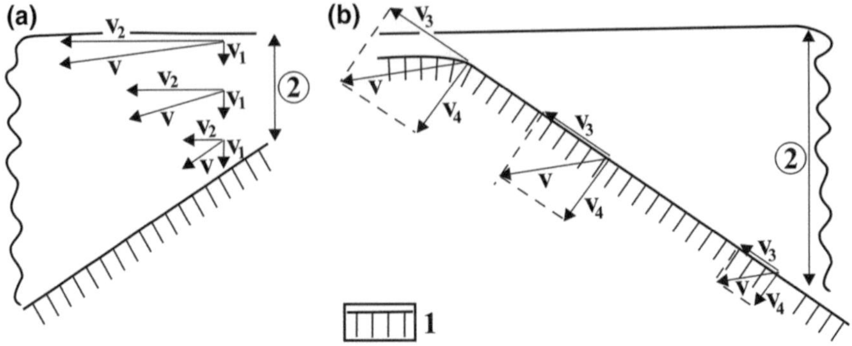

Fig. 3.10 The change of ice speed components in case of different ice speeds if the dip direction of the floor is similar to ice movement (**a**) and if it is different from it (**b**) 1. floor, 2. ice, v. the degree of the speed of ice-flow or glacier flow and the direction of its movement; v_1 pressure component triggered by the weight of ice, v_2 horizontal component of ice movement, v_3 component of ice movement forced by the floor, v_4 pressure component forced by the floor

If the sliding velocity of the ice increases (in case of points being farther from the base), the value of shifting increases (v_2 Fig. 3.10a). This does not alter the degree of destruction on the downstream slope (only debris transport will be larger), while on the upper parts of the upstream slope both components will be larger because of shifting to an increasingly larger extent, v_3 and v_4 in Fig. 3.10b. Therefore, the rate of denudation increases on the upstream slope towards the upper margin of the doline slope (Fig. 3.10b).

As compared to the doline, the ice motion in the doline depends on the angle between the doline slope and the inclination of the glacier surface. It can be seen in Fig. 3.11 that ice is only able to leave the doline if the angle enclosed by the slope and the ice surface is larger than 90°. On the doline slope affected by erosion, since its inclination changes, the value of the pressure component forced by the floor changes [from $v_2(\alpha_1)$ to $v_2(\alpha_2)$], and the value of the actual shifting (component forced by the floor) also changes [from $v_1(\alpha_1)$ to $v_1(\alpha_2)$] (Fig. 3.12). With the decrease of the slope angle, the pressure component forced by the floor decreases while the actual shifting component increases. Therefore, it is probable that the doline slope with an increasingly smaller inclination is destroyed at a lower rate. In case of an unchanged slope angle, depending on the inclination of the glacier, the values of both components also change (Fig. 3.13). With the growth of the surface inclination of a glacier, the value of the pressure component forced by the floor increases [from $v_2(\alpha_1)$ to $v_2(\alpha_2)$].

3.2.1.2 Erosional Transformation of Depressions

Glacial erosion can be observed in the depressions of several glaciokarsts (mainly dolines). The geomorphic activity of ice in dolines can be of two types: degradation

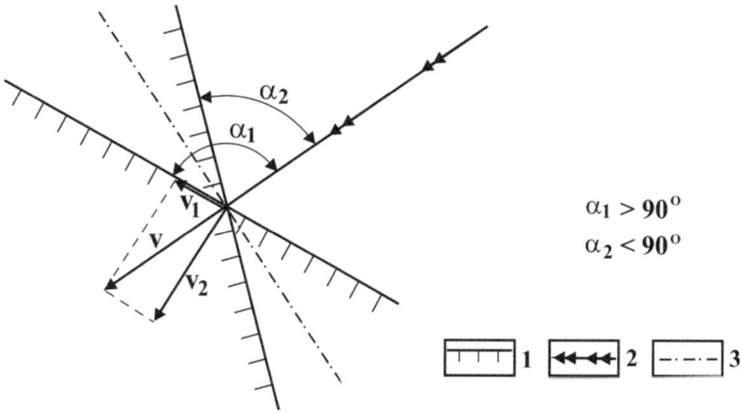

Fig. 3.11 Ice movement determined by the inclination of the bearing slope 1. floor, 2. inclination of ice surface, 3. imaginary surface being perpendicular to ice movement, α dip of the bearing terrain; if $\alpha_1 > 90°$, the glacier is still able to move on the upstream surface, if $\alpha_2 < 90°$, the glacier is not able to move on the upstream surface since v_1 cannot be determined, v the degree of glacier flow speed and the direction of its movement; v_1 component of ice movement forced by the floor, v_2 pressure component forced by the floor

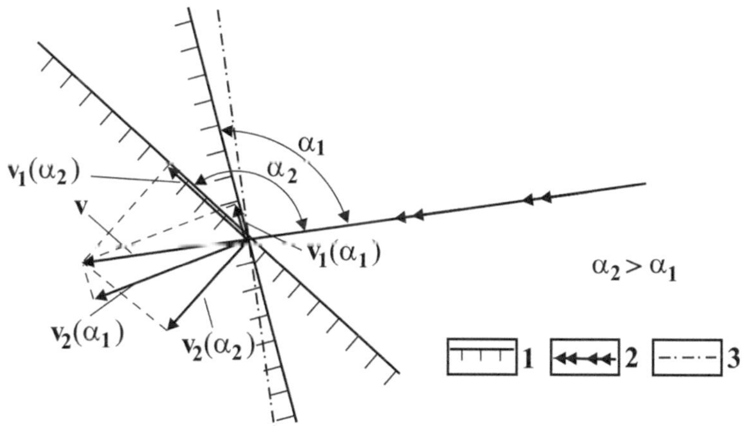

Fig. 3.12 Relation between the shifting of ice and the change of the pressure component if the dip of the bearing terrain decreases 1. floor, 2. inclination of ice surface, 3. imaginary surface being perpendicular to ice movement, α dip of the bearing slope, v direction of glacier movement and the degree of its speed; $v_1(\alpha_1)$ component of ice movement forced by the floor in case of a α_1 slope angle, $v_2(\alpha_1)$ pressure component forced by the floor in case of a α_1 slope angle $v_1(\alpha_2)$ component of ice movement forced by the floor in case of a α_2 slope angle, $v_2(\alpha_2)$ pressure component forced by the floor in case of a α_2 slope angle

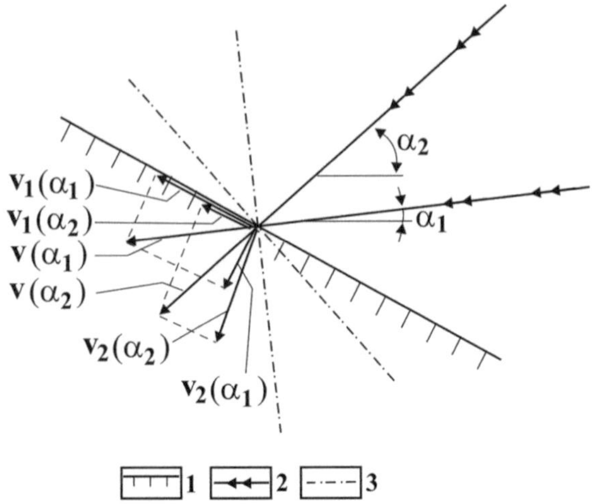

Fig. 3.13 Changes of ice movement and pressure effect in case of different ice movement directions 1. floor, 2. direction of ice movement, 3. surface being perpendicular to ice movement, α the angle of the direction of ice movement as compared to horizontal, $v(\alpha_1)$ degree of glacier-flow speed and its direction in case a movement with an α_1 dip direction, $v(\alpha_2)$ degree of glacier-flow speed and its direction in case of a movement with an α_2 dip direction, $v_1(\alpha_1)$ component of ice movement forced by the floor in case of an ice movement with an α_1 dip direction, $v_2(\alpha_1)$ pressure component forced by the floor in case of an ice movement with an α_1 dip direction, $v_1(\alpha_2)$ component forced by the floor in case of an ice movement with an α_2 dip, $v_2(\alpha_2)$ pressure component forced by the floor in case of an ice movement with an α_2 dip direction

and accumulation. The exact degree of erosion is unknown. It depends on the duration of erosion, the thickness of ice, the inclination of the terrain bearing the doline, the factors influencing sliding (thus, e.g. on the amount of meltwater and on where and to what extent the karst drainage system drains the water).

The deepening of depressions is probably not significant. Deepening is limited by the fact that during the process, by the lengthening of slopes, the plain floor can be completely consumed. Deepening is also controlled by shear stress. The ice is only able to move upstream until the degree of sliding becomes balanced with the shear stress reduced by friction. (During deepening, the ice has to carry out increasingly larger work in order to move and erode.) Deepening is also restrained by the fact that on a gentle doline floor, the movement of ice is probably slow which favours accumulation over destruction.

Depressions mainly widen (lengthen) along the axis of ice movement by their slopes becoming gentle. As it has already been referred to, the downstream slope is degraded, but only to a lesser degree than the upstream slope. The degradation of the doline slopes perpendicular to ice movement is probably not of a great degree. This can be attributed to the fact that the loading component is small on these doline slopes. This refers to the fact that the preglacial uvalas of glacier valleys are wider at their constituting dolines, while they are narrower at merging (Fig. 2.4). Thus, the original karst morphology is preserved better in the case of doline slopes perpendicular to ice movement. Degradation can be different on terrains with dolines and on those with uvalas. In case of an isolated doline, the destruction of the downstream slope is smaller, while it is large on the upstream slope and it is increasingly larger upwards (Figs. 3.8, 3.10b, 3.14b, 3.15i). Therefore, dolines lengthening downstream develop, or if they have been elongated downstream, they will lengthen further. During the process, mainly the upstream slope of the depression becomes gentle and by this, the doline will be increasingly asymmetric (Fig. 3.15i). If the doline is surrounded by a valley, the terrain between the valley and the doline becomes lower too because of slope lowering. An example of such a doline is Urdeni do in the Durmitor Mountains (Fig. 3.9). In the case of an uvala, the asymmetric nature is more significant, since both slopes of the dividing walls between constituting dolines, thus, the downstream slope too, are denuded intensively and thus, the dividing walls can also become lower (Fig. 3.15ii). Since the dividing wall acts as a bump and the meltwater from the upstream slope destroys the downstream slope by frost weathering and glacial erosion by plucking increases. However, at the thresholds, the ice leaving the doline accumulates and is

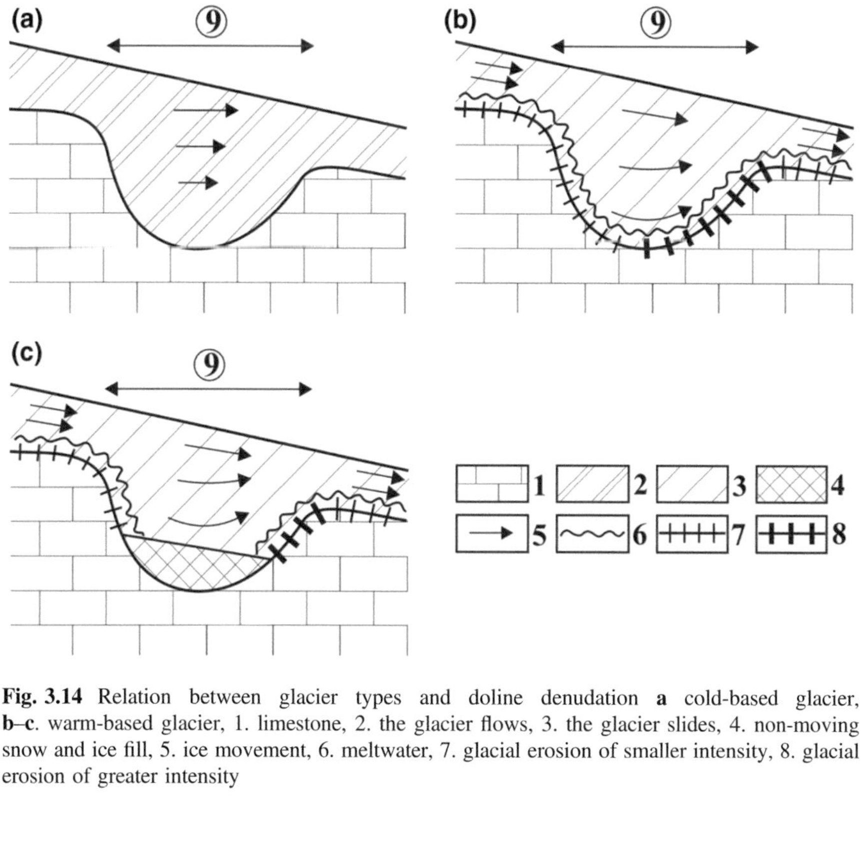

Fig. 3.14 Relation between glacier types and doline denudation **a** cold-based glacier, **b–c**. warm-based glacier, 1. limestone, 2. the glacier flows, 3. the glacier slides, 4. non-moving snow and ice fill, 5. ice movement, 6. meltwater, 7. glacial erosion of smaller intensity, 8. glacial erosion of greater intensity

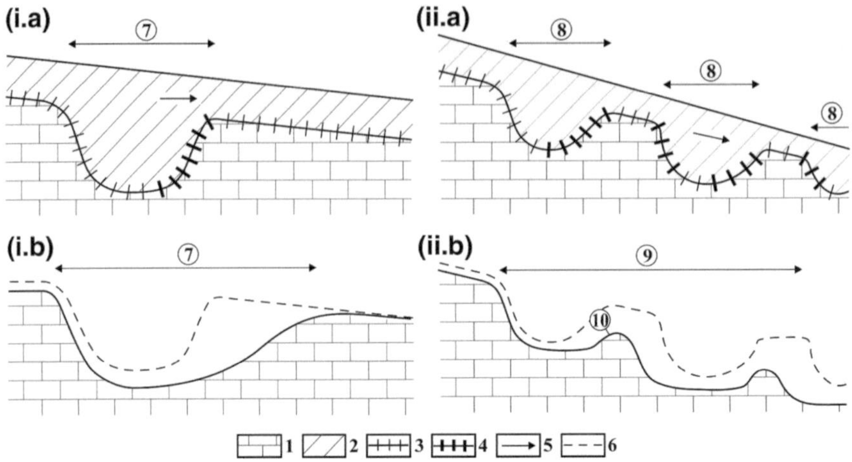

Fig. 3.15 Transformation of an individual depression (**i**), and of an uvala (**ii**) during glacial erosion **a** original state, **b** mature state, 1. limestone, 2. ice, 3. glacial erosion of smaller intensity, 4. glacial erosion of greater intensity, 5. direction of ice movement, 6. former surface, 7. individual depression, 8. constituting depression, 9. large-sized depression which coalesced during glacial erosion (former uvala), 10. roche moutonnée or the stump of a dividing wall1. saturation level, 2. former limestone surface, 3. ice, 4. meltwater, 5. limestone, 6. former denudation surface, 7. later denudation surface, **a** the great amount of meltwater forms a well-matured epikarst zone, **b** meltwater is drained into the cavity system of the epikarst zone, **c** because of glacial erosion, the epikarst thins out, meltwater is drained into the thin epikarst into a lesser degree, **d** therefore, the amount of water increases, the epikarst thickens again to the effect of the dissolution effect of a greater amount of meltwater

pressed together. This also induces meltwater accumulation. The destruction of the dividing wall will be of such degree that only its truncated parts remain (Figs. 3.9, 3.15ii, 3.16, 3.17c) and thus, large-sized (being several 100 m) depressions can develop too. The result of this process can also be observed in the Durmitor Mountains. Thus, the dividing walls between the constituting dolines of the Lokvice valley or that of Velika kalica (former preglacial uvalas) have been almost completely destroyed (Fig. 3.16). The narrowing dividing walls can also be destroyed locally (Figs. 3.18b 3.19). Because of local destruction, corridors may develop on the dividing walls. Such features can be mentioned from the Durmitor Mountains and the Biokovo Mountains too (Telbisz et al. 2005). It is probable that in this case, the dividing wall collapses in some sections because of ice pressure. However, dissolution by meltwater can also play a role in the development of corridors (giant grikes). The dividing walls lower, are destroyed locally or become rounded.

All in all, the degradation of dividing walls can happen in the following ways:

- The roof level of the dividing wall lowers by uniform destruction.
- The dividing wall narrows by the destruction of its slopes.
- Both processes take place simultaneously.

- The pressure of the ice knocks down the dividing wall.
- The ice pressure breaks through the dividing wall locally.
- Meltwater dissolves the dividing wall, the giant grike is widened further by ice pressure.

If the ice does not move on the doline floor, the doline section under the level of ice in standstill is not destroyed (karst features of the floor, the karren also survive), while the slope over this level is being destroyed, by this a slope section with a low inclination develops on the upper part of the doline where karren were also destroyed. Dolines with a double cross section develop (Figs. 3.6c, 3.14c, 3.20).

The type of glacier on karst affects the spreading of erosion. When cirque glaciers develop, the erosion is directed by depressions (their size and type), erosion will be local (Fig. 3.21a–c). If a valley glacier develops in the depressions (uvalas), erosion is directed similarly, uvalas are transformed into great-sized depressions, where constituting dolines are hard to be recognized (Figs. 3.16, 3.21d). The ice caps can cover doline rows. Glacial erosion is not directed by depressions. In the area of the ice cap, the depressions become truncated to a great extent by surface denudation (Fig. 3.21e).

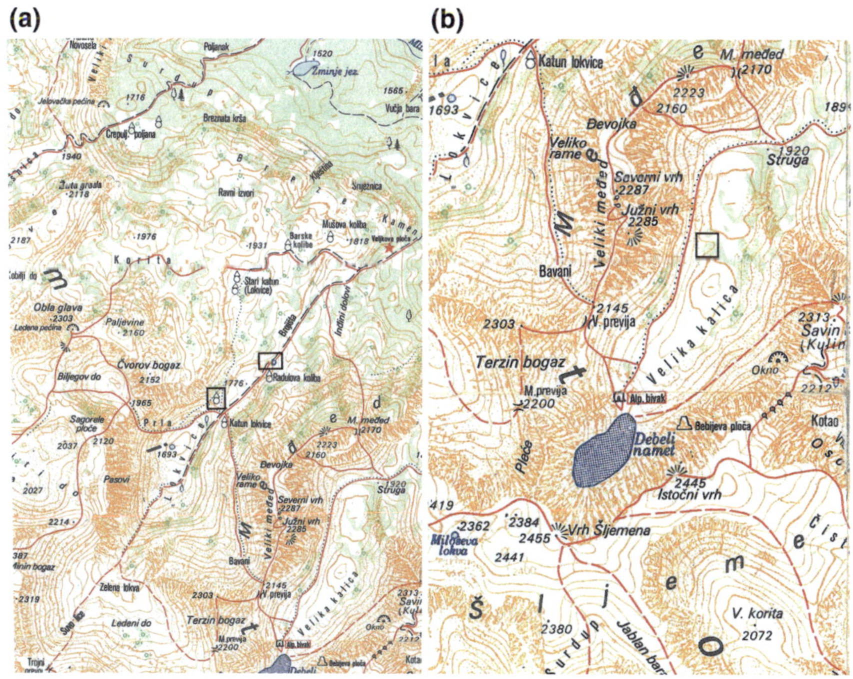

Fig. 3.16 Maps of Lokvice (**a**) and Velika kalica (**b**) (map extracts of the tourist maps of Durmitor: parts marked with a square show the remnants of dividing walls based on contour lines)

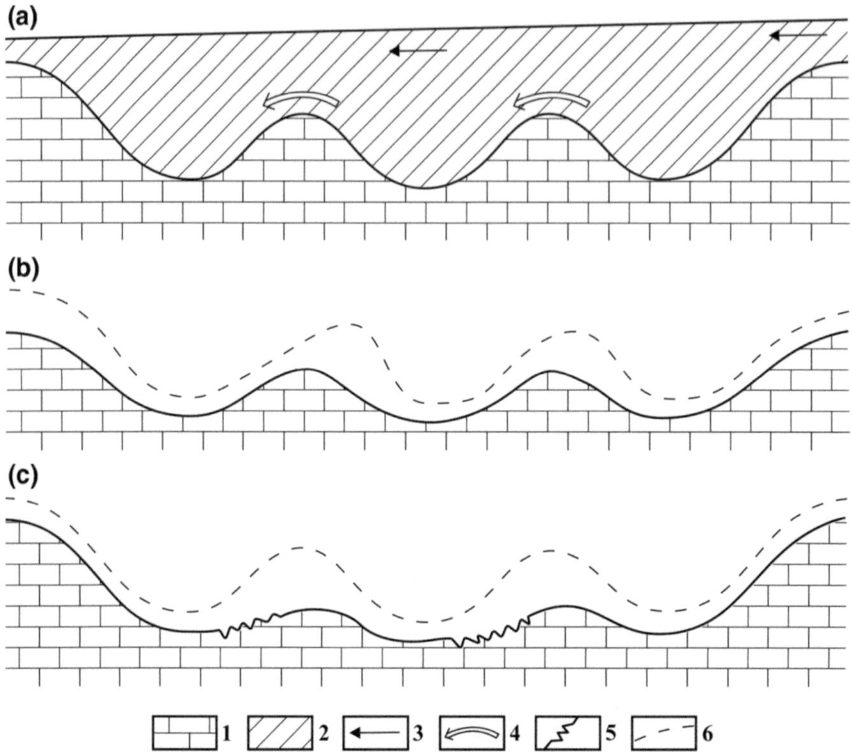

Fig. 3.17 Denudation of the slopes of constituting dolines to a different degree 1. limestone, 2. ice, 3. direction of ice movement, 4. moving ice flows around the bump and the formation of meltwater, 5. denudation of the dividing wall between constituting dolines with plucking, 6. original surface, **a** uvala subject to glacial erosion, **b** upstream slope of constituting dolines is of greater dip and therefore, the denudation of dividing walls can be attributed to the denudation of this slope, **c** truncation of dividing walls is caused by the denudation of both slopes

A moraine can reach depressions at least in three ways. The primary moraine is basal moraine or englacial moraine transported by ice (Fig. 3.22c). The secondary moraine can originate from the englacial, terminal or end moraine (fluvioglacial deposit) of the ice if the front of the ice is at the margin of the ice (Fig. 3.22a). It can come from the basal moraine of the glacier if the glacier front is farther from the depression (Fig. 3.22b). In the latter cases, the moraine material, which can be secondary automatically if it is reworked by the water moving within the ice, (Jones and Arnold 1999) is redeposited again. The sorting of the reworked material increases, its grain size decreases since the finer material is transported away, but also because the size of grains can decrease by dissolution (limestone) or by weathering (non-karstifying rock). Thus, the moraine with various grain size, in which material with sandy and clayey grain sizes may occur too, (Hambrey and Fitzsimons 2010) can become increasingly finer.

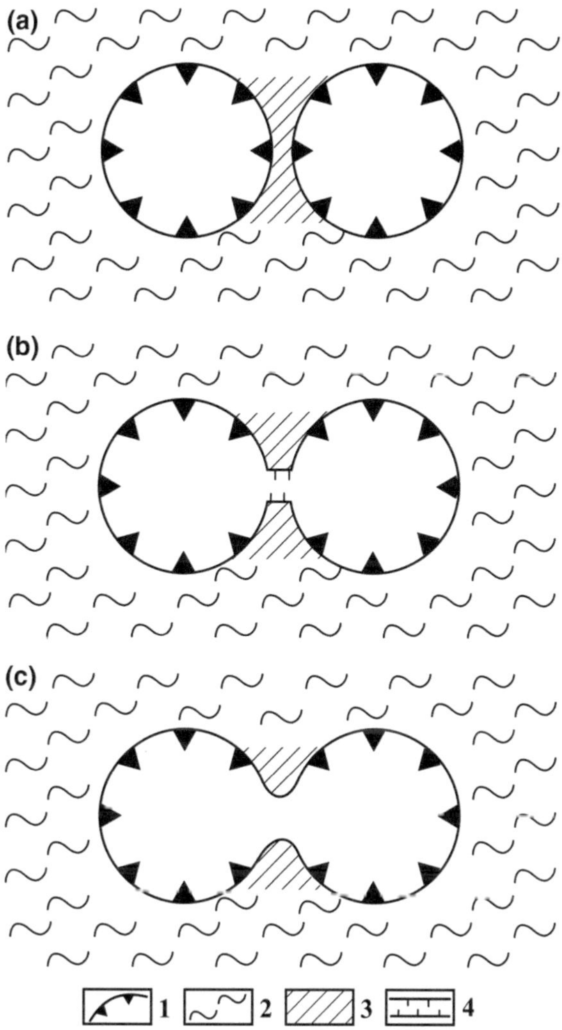

Fig. 3.18 Phases of the coalescence of constituting dolines (in plan view) 1. margin of doline, 2. higher terrain surrounding dolines, 3. dividing wall between constituting dolines, the surface of which is lower than the surrounding terrain, 4. corridor of the dividing wall, **a** constituting dolines are separated from each other by a uniform dividing wall, **b** dividing wall being broken through by ice pressure, **c** middle part of the dividing wall denuded to the floor of constituting dolines

3.2.2 Geomorphic Processes in the Glacier Valleys of Karst

3.2.2.1 Glacial Erosion Forming Stepped Surfaces

A series of steps developed in the troughs of glaciokarst. The steps are dissected into scarp fronts (surface with heads of bed) and exposed bedding planes (surface with bedding planes, Fig. 3.23). The steps and their karren formation have been studied by several researchers (Bögli 1964; Barrére 1964; Williams 1966' Sweeting 1973; Kunaver 2009; Vincent 2009; Veress 2010). The karren, bare surfaces with different inclinations are called limestone pavements and such a karst is called

Fig. 3.19 Local destructions (corridors) on karstic dividing walls: **a–b** local destruction on the dividing wall between Mlječni-do and Urdeni-do (Durmitor), **c–d** local destruction between Todorov-valley and Valoviti-do (Durmitor, photo by Veress) 1. corridor on the dividing wall, in figure **d** the covered karst part of Todorov valley can be seen with steps, hogbacks, roche moutonnées and there is also a giant grike

pavement karst (White 1988). According to Goudie (1990), limestone pavements are confined by exposed, plain and steep slopes, which are dissected by dissolution features. Limestone pavements include bedding plane surfaces (Clayton 1981; Kunaver 2009), karren surfaces intersecting beds (Kunaver 2009) and karren surfaces where the series of bedding planes is interrupted by scarp fronts (Kunaver 2009). The development of limestone pavement is associated with glacial erosion (Clayton, 1966).

Bögli (1964), Williams (1966) classified the series of bedding planes. Bögli (1964, 1978) distinguished Schichttreppenkarst (the beds are horizontal) and Schichttrippenkarst (the beds are inclined). Schichttreppenkarst was called stepped karst by Bögli (1964). Williams (1966) differentiated flat even pavement (idealized pavement), inclined pavement, stepped pavement for karren surfaces with various inclinations. (According to him, the terrain where the inclination of beds is greater than 44° cannot be called pavement.) Kunaver (2009) proposed a new classification and distinguished four categories (Fig. 3.24). Category i includes even flat limestone pavements where the inclination of bedding planes (or beds) is smaller than 10°. Cuesta-like limestone pavements are members of category ii, where the

Fig. 3.20 Dolines with complex slopes, **a** Valoviti-do (Durmitor), **b–c** Surutka (Durmitor), **d** Urdeni-do (photo by Veress) 1. doline margin, 2. border between gentler and steeper slopes of a doline, 3. direction of ice movement, 4. roche moutonnée

inclination of beds is 10°–45° and they are confined by surfaces with different inclination. The flat and sloping stepped limestone pavements belong to category iii where the inclination of the bearing surface is larger than that of the beds. Finally, category iv involves flat and sloping inverse limestone pavements, where the inclination of beds is opposite to the slope of the bearing terrain.

Their classifications show that these surfaces are confined by bedding planes and scarp fronts (heads of bed). In case of the ones belonging to category ii, the slope of scarp fronts is opposite to the inclination of the bearing surface, while that of bedding plane surfaces is identical with surface inclination. In the case of the members of category iii, the scarp fronts are overhanging, and these front slopes turn towards the lower part of the slope.

Limestone pavements are extremely widespread on the British Islands (Williams 1966; Vincent 2009). They occur in an extension of 29 km^2, in Great Britain, while in an extension of 290 km^2 in Ireland. However, they are widespread in Canada too, in Winnipeg they extend to 3500 km^2 (Ford 1996). Thus, not surprisingly, the limestone pavements of the British islands have been investigated by many researchers such as Goldie (1996), Vincent (1983), Williams (1966), Vincent (2009) also explained their evolution by glacial erosion. He distinguished several varieties based on the fact in what way and to which surface the glacial erosion

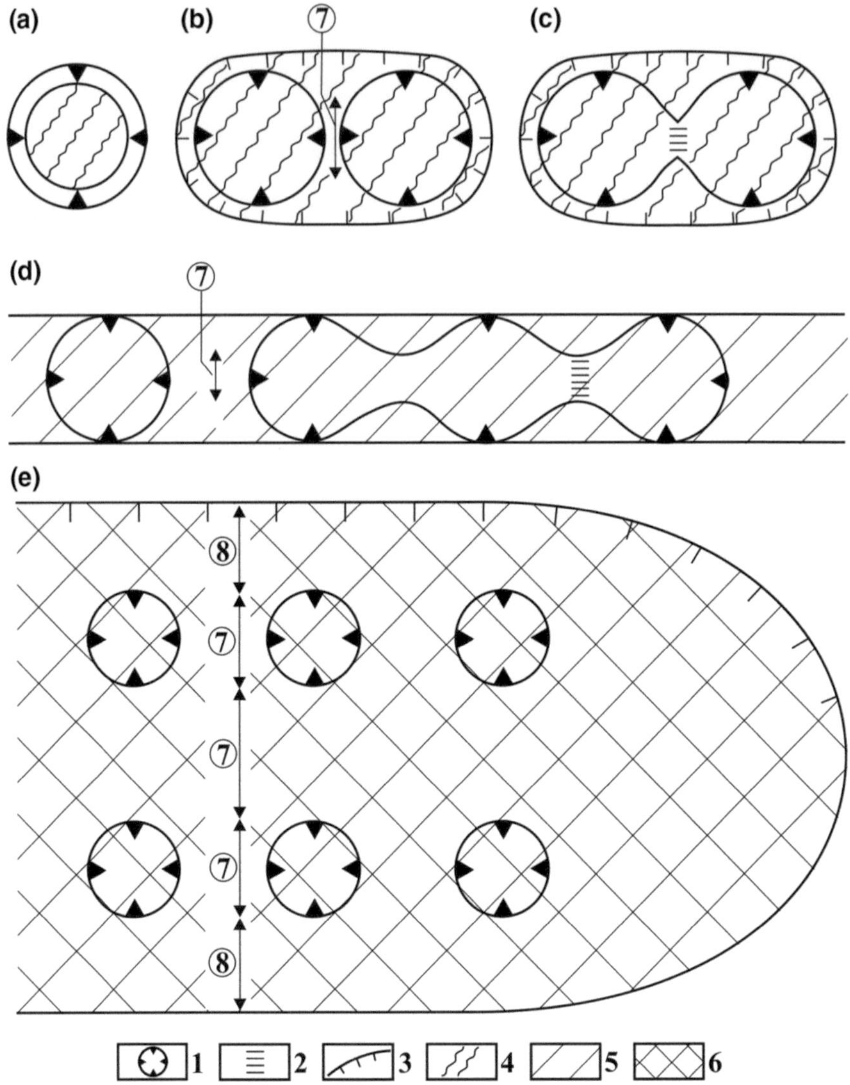

Fig. 3.21 Types of glacier erosion on karst (in plan view) 1. karstic depression, 2. truncated karstic dividing wall, 3. ice margin, 4. local glacial erosion, 5. glacial erosion dependent on karst features, 6. glacial erosion independent of karst features, 7. erosion between depressions, 8. erosion near depressions, **a** cirque glacier develops in a depression, **b** cirque glacier develops in several depressions, **c** cirque glacier develops in an uvala, **d** valley glacier develops in the uvala, **e** ice sheet or ice cap develops on karst

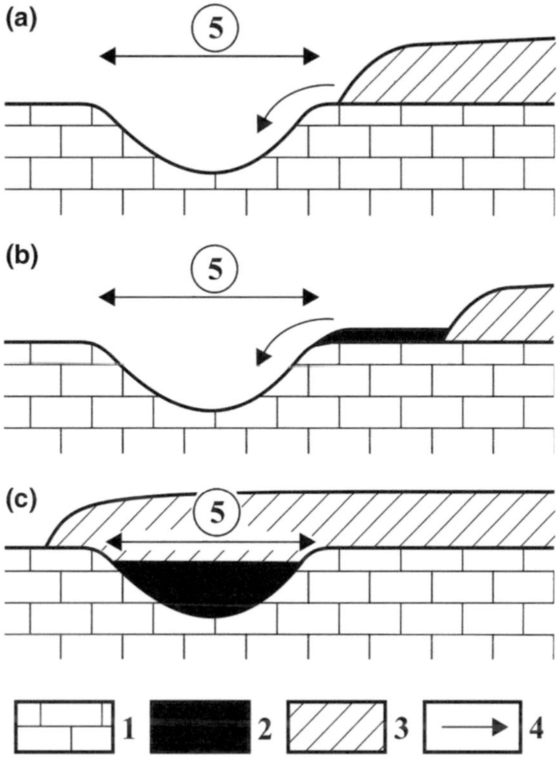

Fig. 3.22 Ways of formation of till in karstic depressions 1. limestone, 2. till, 3. ice, 4. till redeposition by rainwater and stream, 5. karstic depression, **a** the deglaciating englacial moraine of ice redeposits into karstic depression by rainwater, **b** the basal moraine of retreating glacier accumulates into karstic depression by a stream, **c** the englacial moraine of the glacier is transported into the karstic depression by ice

affected the rock. He also took into consideration the exhumation of limestone pavements. The varieties are the following:

- "Glacially plucked joint-dominant pavement" where glacial erosion destroyed the rock by plucking.
- "Glacially abraded calcrete-dominant pavements without paleokarst" where glacial abrasion exerted an effect to the hard, resistant "clay Wayboard".
- "Glacially exhumed calcrete-dominant pavements with paleokarst where glacial erosion exposed karstified (paleokarstic) surface and features
- "Glacially truncated paleokarst", where glacial abrasion partly truncated preglacial karst features.
- "Marine exhumated paleokarst pavements", where exposure happened by maritime abrasion.

Stepped surfaces which are the equivalents of inclined pavements in Williams's classification (1966), also in themselves are the basic morphological elements of glacier valleys. Their significance is intensified by the fact that the karren formation of glaciokarst mainly happens on the bedding planes of the steps here (see Chap. 4). The bedding planes can be inclined towards the axis of the valley (in this case the

Fig. 3.23 Stepped surfaces of glaciokarst (photo by Veress) **a–b** Step systems (Totes Gebirge, Austria), **c** step formed on a bed with greater dip (Totes Gebirge), **d** step developed on a bed with smaller dip (Asiago plateau, Italy)

head of the bed is inclined towards the margin of the valley), but also towards the margin of the valley (in this case the head of the bed is inclined towards the axis of the valley). In this latter case, the steps are like cuestas. Stepped surfaces evolve during the partial denudation of beds to the effect of glacial erosion. This is proved not only by the fact that steps are predominant in the majority of troughs of the glaciokarst, but also by the fact that the stepped morphology can be recognized on the trough floor which was covered with ice some decades ago. The Hallstatt glacier can be mentioned as an example where steps occur on areas being covered with ice some decades ago (Fig. 3.25). Because of the short duration of the unglaciated period, no other action could have created these features.

On various glaciokarst, their distribution is very different. Thus, they are widespread in the Totes Gebirge, however, it may also happen in these mountains that they are absent in some valleys. For example, this is true for the valley bearing Lake Lahngang. In another case, some sections of the same valley have stepped surfaces, while their other parts lack them (e.g. the Seven Lakes Valley, Julian Alps). On other glaciokarst, they rarely occur or are totally absent (e.g. the Durmitor). It seems that they did not develop in those valleys in which rock basins or older dolines are found on valley floors.

According to Clayton (1966), limestone pavements in Great Britain developed in the last glacial. He states that due to the effect of glacial erosion, the rock was destroyed to the bedding plane of beds in a way that ice removed the disintegrated

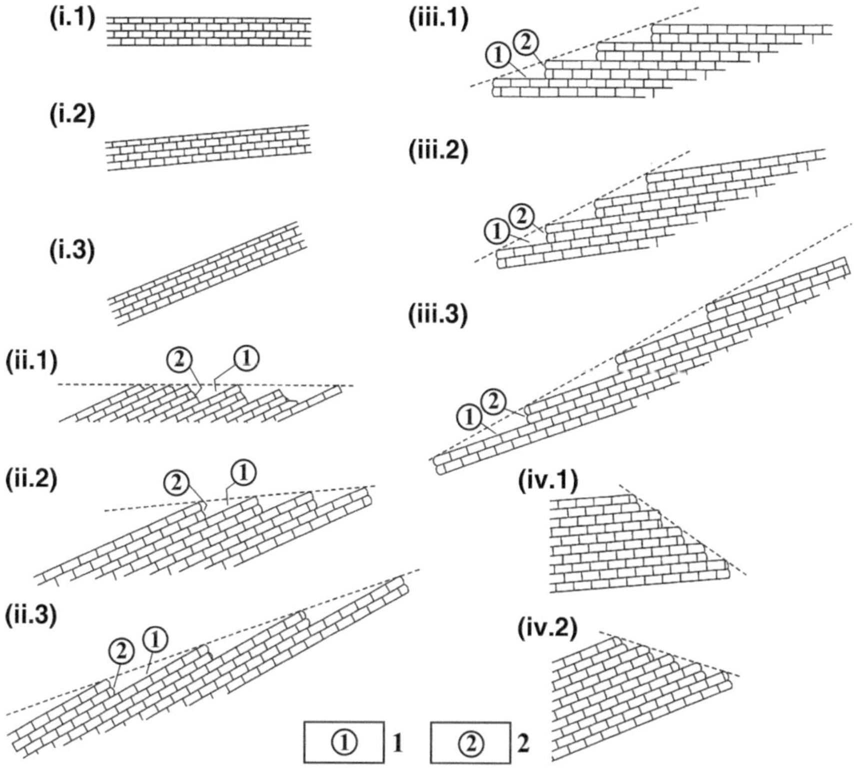

Fig. 3.24 Groups of limestone pavements (stepped surfaces) according to Kunaver from Kanin Mountains (Slovenia, Kunaver 2009, modified) **i** the even flat and sloping limestone pavements, **ii** the flat and sloping cuesta-like limestone pavements, **iii** the flat and sloping stepped limestone pavement, **iv** the flat and sloping inverse limestone pavements, 1. bedding plane, 2. scarp front (head of bed)

material. According to Clayton (1966), limestone pavements did not develop in places without ice cover.

Stepped surfaces develop by the peeling of bed parts to the combined effect of the dissolution by meltwater and glacier erosion. This process takes place in two phases:

- In the first phase, larger or smaller proportion of the bed is partly or completely separated from other parts of the bed and from the bedrock by dissolution, since epikarst develops under the ice and karren features are formed on the rock surface. Some karren features (such as grikes, pitkarren, Schichtfugenkarren) which expand below the surface and drain water into the karst function as part of the epikarst.
- In the second phase, the ice moves away from this part of the bed.

Since there are larger or smaller rock blocks at the termination of the steps, it is probable that ice removes only smaller units from the bedrock at a time and this process is repeated several times. Steps that developed on beds remained in situ are rounded off to some extent by ice. It is also likely that separation can be displacement (the bed unit became completely detached from the bedrock before glacial effect), and tearing (the bed unit did not become completely separated from its environment during glacial effect).

Peeling is promoted if

- joints developed between the beds during rock development (along these joints the rock can become separated more easily and at these places, the initial water movement is faster, which favours a more intensive dissolution),
- dissolution takes place along bedding planes between the beds as well as along fractures in the beds.

Dissolution can take place preceding glacier development in the epikarstic zone. It is likely to be more significant when it occurs with glacial erosion simultaneously. Bennett and Glasser (2009), Ford and Williams (1989, 2007) described dissolution under ice from glacier valleys. In this case, meltwater penetrates into the beds along fractures, then from the fractures between the beds. Water penetration along bedding plane can be observed on glaciokarst at present too. Schichtfugenkarren and rinnenkarren can develop along bedding planes (Veress 2010; Fig. 3.26).

Fig. 3.25 Steps on the valley floor of Hallstatt-glacier (Dachstein, Austria) becoming deglaciated no more than 100 years ago (photo by Veress) bedding plane, 2. head of bed, 3. concretion, 4. moraine

Fig. 3.26 Karren formation of the bedding plane between beds (Seven Lakes Valley, Julian Alps, Slovenia, photo by Veress) 1. exposed bedding plane, 2. head of bed (scarp front), 3. rinnenkarren, 4. grike

The intensity of the process is ensured by the large amount of meltwater being present permanently at the ice-rock interface. Karst water and karst water level can also develop in the glacier (Ford and Williams 2007). This occurs if the karst water table rises. This can be caused by decrease in the discharge of karst springs (Ford and Williams 2007), by the backflow of water under pressure from the passages of karst (Lauritzen 1984), by the inflow of surface waters (Lauritzen 1986) and by feeding by meltwater. If the karst water level acquires at an elevated position above the bedrock surface, dissolution will be more continuous on the bedrock. The epikarstic zone of the bedrock surface will be replaced by the phreatic zone. The degree and size of back-swelling depend on the discharge of water supply, on the maturity of the passage system and on the extent to which the outlets (karst springs) of karst water are covered with ice. Dissolution can take place if meltwater is able to dissolve. Five factors contribute to solution capacity:

- First, the large amount of water can be mentioned, and second low water temperature (in this case, the solubility of CO_2 increases).
- The third factor is the pressure of meltwater. In case of a greater pressure, less equilibrium CO_2 is needed and thus, the surplus can appear as aggressive CO_2-in the system.
- Flow velocity is the fourth factor. As it has already been mentioned, in case of a quicker flow, dissolution intensity increases. In order to maintain solution capacity, it is an important factor that surface waters arrive at the ice floor (such

water contains air), on the other hand, there must be a well-developed air circulation in the glacier which ensures that the air reaches the meltwater.
- Finally, the insulating effect of ice can be mentioned as a fifth factor, which keeps the temperature of meltwater low.

The maturity of the epikarst (the size and density of its karren features, the thickness of the dissolved zone) depends on the amount of percolating water with solution capacity. If more water percolates, karren formation and cavity development increases. The epikarst zone develops and thickens, its conductivity increases because of dissolution. Therefore, a greater amount of water percolates. As a consequence, the amount of meltwater decreases at the ice-base and thus, the sliding of ice too. Although the effective normal pressure increases, the glacier is able to entrain and transport a smaller amount of rock fragments. The epikarst does not grow thinner (it becomes dissected into debris at most), and consequently, meltwater arrives at the lower surface of the epikarst in a saturated state. The downward extension of the epikarst slows down or ceases. As a consequence, a smaller amount of water is lost by seepage (the effective normal pressure decreases) thus, the amount of meltwater increases, which results in more intensive sliding. Because of stronger glacial erosion, the epikarst grows thinner. As a result, and because of the larger amount of meltwater, the epikarst extends downwards since in this case, the saturation level is situated below the lower surface of the epikarst. In spite of its downward denudation, the epikarst can thicken until the saturation level reaches its position at the lower surface of the epikarst.

The effective normal pressure can even grow if because of the increase of water level, water appears not only at the ice–rock interface, but also in the ice. In this case, the weight of ice can increase because of a larger amount of water. Frost weathering can intensify the denudation of the epikarst. In winter, under glaciers or glacier sections where the meltwater or the meltwater that reached the epikarst freezes, dissolution is interrupted, but frost effect occurs. This can intensify the loosening of bed parts partly separated by dissolution.

After all, with increasing intensity of glacial erosion, the thickness of epikarst decreases, while with the dissolution effect of meltwater it thickens to the saturation level. However, the saturation level can acquire a deeper position (a newer saturation level develops) if the upper part of the epikarst is denuded. Its temporary thickness depends on the comparative rates of glacial erosion and dissolution. If glacial erosion is more rapid, the epikarst thins out, if dissolution is more rapid, it thickens. In the case of a balance between glacial erosion and dissolution, the thickness of epikarst does not change, but the epikarst extends downwards. During the process, newer and newer bed units are denuded and thus, more and more bedding planes are exposed (Fig. 3.27).

The epikarst, which can only develop under warm-based glaciers, can be partly or completely removed by glacial erosion, but it can regenerate. Williams (2008) described an epikarst truncated by glacial erosion. In areas without ice cover at present, the maturity of epikarst (paleoepikarst) depends on its thickness during the retreat of ice. Thus, it depends on the nature of the relationship between glacial

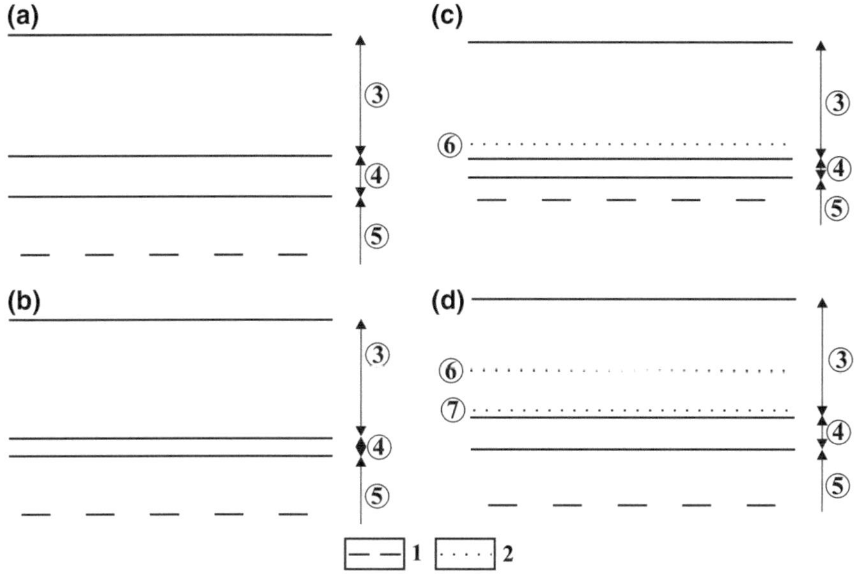

Fig. 3.27 Relation between epikarst and glacial erosion1. saturation level, 2. former limestone surface, 3. ice, 4. meltwater, 5. limestone, 6. former denudation surface, 7. later denudation surface, **a** the great amount of meltwater forms a well-matured epikarst zone, **b** meltwater is drained into the cavity system of the epikarst zone, **c** because of glacial erosion, the epikarst thins out, meltwater is drained into the thin epikarst into a lesser degree, **d** therefore the amount of water increases, the epikarst thickens again to the effect of the dissolution effect of a greater amount of meltwater

erosion and solution capacity during deglaciation. During this period, the epikarst could be of maximum thickness, truncated or completely denuded. The truncated epikarst could become thicker following the disappearance of ice cover. Rose and Vincent (1983) described truncated grike karren from the British Islands. These features probably represent truncated epikarst. While the thickening epikarst is represented by deeper, younger grike karren in the areas studied. It is probable that only certain karren features of relict epikarst develop such as Schichtfugenkarren with grikes and such rinnenkarren which develop on bedding planes between beds (Fig. 3.26). Karren on bedding planes exposed on the surface (kamenitzas, trittkarren rinnenkarren) are not parts of relict epikarst, because if they developed at all, they were denuded by glacial erosion during the period of ice cover.

The development of stepped surfaces takes place in two ways.

- On the one hand, the water reaching the bedding plane and percolating along fractures creates grikes in strike direction (giant grikes, grikes) (Fig. 3.28). The water percolating into the grike and reaching the bedding plane that is the closest to the surface creates Schichtfugenkarren by percolating downslope. (Thus, epikarst under glaciers develops.) Even this partial separation favours the displacement (tearing) of the bed unit by ice. Displacement occurs more easily if further grikes also develop. Such a grike with strike direction can be the one

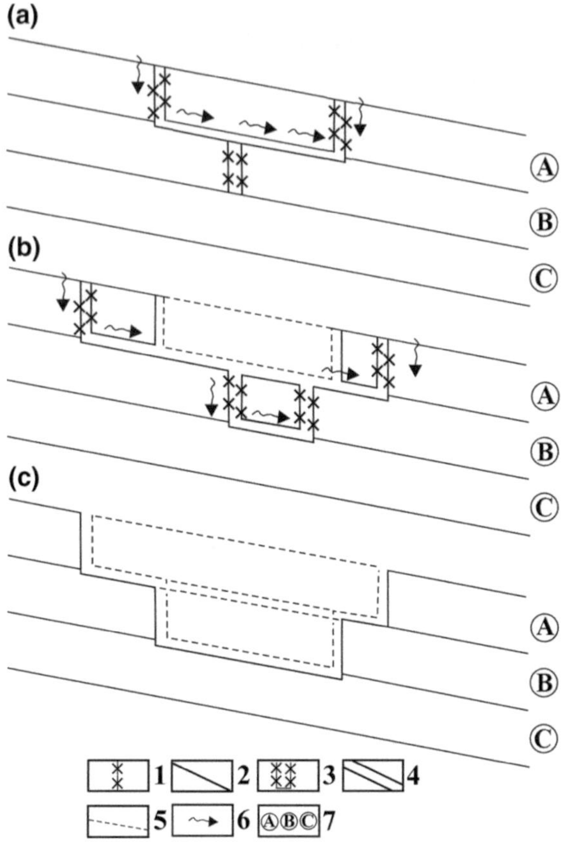

Fig. 3.28 Separation of bed units by dissolution and glacial erosion 1. fracture, 2. bedding plane, 3. grike, 4. Schichtfugenkarren, 5. denuded unit of the bed, 6. water percolation and dissolution, 7. identification mark of bed, **a** dissolution, **b** bed unit of the bed marked A is transported away by ice, **c** a unit of the bed marked B is also transported away by the glacier

which borders or intersects the Schichtfugenkarren on its lower part in a downslope direction, in addition, those grikes in a downslope direction (transversal grikes), which surrounds Schichtfugenkarren. With the development of the latter, the bed unit becomes completely separated from its environment, the displacement of the block happens without tearing too. With the displacement of the block of the uppermost bed, a section of the bedding plane of the bedrock is exposed, by which a scarp front develops.

- On the other hand, at the head of the exposed bed (at the scarp front) the meltwater reaches directly the bedding plane of the bedrock where a Schichtfugenkarren develops too (Fig. 3.28b). The newer block is separated, the scarp front retreats and the exposed bedding plane widens.

Continuously, with the partial denudation of newer and newer bed sections, newer and newer bedding planes can be exposed. An essential precondition for the exposure of newer bedding planes is the retreat of scarp fronts that developed earlier thus, the denudation of bedding planes with higher elevation. As a result of the process, rows of steps are formed (Fig. 3.28c).

Fig. 3.29 Schichtfugenkarren (Totes Gebirge): the Schichtfugenkarren developed preceding the denudation of the bed unit in front of the head of bed: on the one hand, this is proved by the fact that the Schichtfugenkarren has a higher position than the bedding plane part in the foreground of the photo; on the other hand, on this latter one there are rinnenkarren which develop on bedding plane thus, following the denudation of the bed unit, (photo by Veress) 1. scarp front (head of bed), 2. Schichtfugenkarren, 3. bedding plane, 4. rinnenkarren, 5. giant grike

The displacement of bed units and the partial denudation of the epikarst are referred by paleokarren (Fig. 3.29) and truncated paleodolines. The Schichtfugenkarren presented in Fig. 3.29 could only develop when there was water drainage towards its direction. This could occur when on the bedding plane part in front of the step bearing the Schichtfugenkarren no rinnenkarren existed and thus, when the covering bed unit existed in this place. No present water drainage to the Schichtfugenkarren is possible also because the surface of the exposed bedding plane section is of lower elevation than that of the Schichtfugenkarren. The dissolution of the bedding plane section could also occur following the denudation of the bed section (Fig. 3.30). It can be seen that dissolution occurred on the step too (it is dissected by karren features), but the effect of glacial erosion is also observable, the step is slightly rounded off. The above-mentioned facts are also proved by truncated solution dolines. Such a doline can be seen in Fig. 3.31. Although this doline is located at an altitude of 1800–1900 m, it has a small depth and gentle slopes. Its gentle slopes indicate that it is not a recent doline since recent dolines have steep (perhaps vertical) slopes at this altitude. This small depth could develop by the denudation of its upper part thus, by the partial denudation of the beds bearing the epikarst.

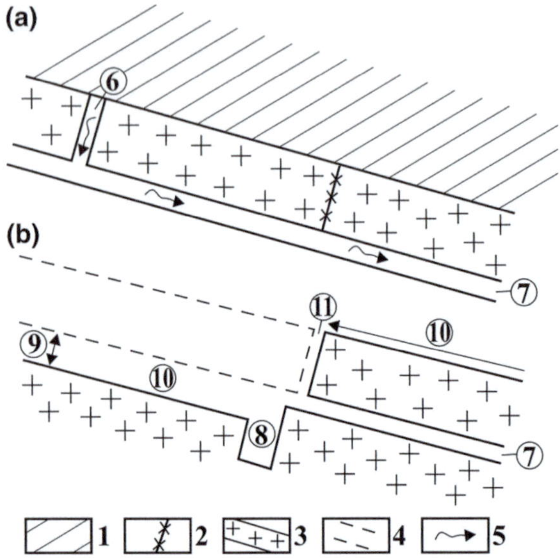

Fig. 3.30 Explanation for the development of the Schichtfugenkarren presented in Fig. 3.29 1. ice, 2. fracture, 3. bed, bed unit, 4. denuded bed unit, 5. water percolation and dissolution, 6. grike, 7. Schichtfugenkarren, 8. giant grike, 9. bedding plane unit denuded by dissolution, 10. bedding plane, 11. head of bed, **a** original state, **b** present state

Fig. 3.31 Truncated doline (valley under Tragl peak, Totes Gebirge, photo by Veress)

Stepped surfaces do not develop in areas being not covered by ice, under cold-based glaciers and if the slopes are extremely steep or no karren formation occurs since for instance meltwater is not able to dissolve. Neither do they develop if the beds are extremely thick or the rock is non-stratified. No Schichtfugenkarren develop in thickly stratified rock also because the ice is likely, not able to displace the thick (of a 5–10 m thickness) bed units. Probably, they do not develop either if the floors of grikes do not reach the uppermost bedding plane during their deepening. If the saturation level of the percolating waters is above the bedding plane, no Schichtfugenkarren develop. Therefore, plucking does not take place along bedding planes. In this case, surfaces without bedding planes can develop on the rock and abrasion comes into prominence over plucking. In case of a given solution capacity, there is a greater chance for this, the thicker the bed is and the larger its inclination is.

With the peeling of more and more extended bed units, more and more extended exposing bedding planes develop. The peeling of beds is affected by bed thickness and the dip angle of beds. The thicker the bed and the larger its inclination, in order to expose a bedding plane section with the same width, the larger work the ice has to carry out in order to displace bed units. Thus, in the case of thick beds with great inclination, exposing bedding planes with small expansion (width) and tall scarp fronts develop (Fig. 3.32ia), while in case of thin beds with a small inclination,

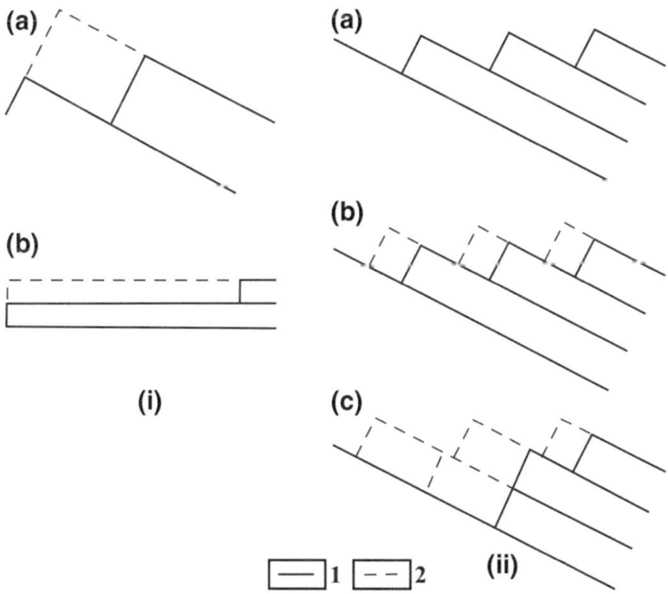

Fig. 3.32 Factors determining the expansion of stepped surfaces (**i**) and the denudation of stepped surfaces (**ii**), **i**(**a**) the beds are thick and of large dip, by this bedding planes with small width are exposed, **i**(**b**) the beds are thin and of small dip, by this bedding planes with large width are exposed, **ii**(**a**) original state, **ii**(**b**) the scarp fronts retreat to a similar extent during their denudation and therefore, the width of bedding planes and the height of scarp fronts do not change, **ii**(**c**) the denudation of scarp fronts of beds with a lower position is faster and therefore, a wide bedding plane and a high scarp front develops, 1. bedding plane, 2. denuded bed unit

wide bedding planes and short scarp fronts develop (Fig. 3.32ib). The different denudation of beds situated above each other can also affect the extension of exposed bedding planes and scarp fronts. If the denudation retreat of the beds situated above each other is the same, the size of bedding planes and scarp fronts does not change (Fig. 3.32iib). If the retreat of the lower ones is faster, the exposed bedding plane of the lowest bed will be wide, and a tall scarp front is formed over it (Fig. 3.32iic). The denudation of the tall scarp front that developed in this way and that is composed of the heads of several beds cannot take place any more thus, the surface with bedding planes does not widen further.

The gradient of bedding planes and heads of bed depends on the inclination of beds. If the bed(s) is (are) horizontal, the slope that is constituted by the head(s) of bed is vertical, while the slope with bedding planes has an inclination of 0°. With increasing inclination, the slope with heads of the bed decreases and the surface slope with bedding planes increases. In case of a 90° dip of bed, the slope with heads of bed is horizontal and the slope with bedding planes is vertical, hogbacks develop. According to Smart (2004), in case of a greater dip of beds, roche moutonnées develop. The nappe structures of the Northern Calcareous Alps are constituted by beds with a dip of 0°–30°. It can be concluded that the slopes with heads of the bed may have an inclination of 60°–90°, while the slopes with bedding planes are mainly 0°–30°. The length of the slopes with heads of bed is between some tens of centimetres and some metres (if the beds are thicker or there are several beds, the slopes are longer), while the length of slopes with bedding planes can be some metres or even 100–200 m.

The pattern of scarp fronts is mainly straight, but it can be meandering, saw-toothed or bay-like. The width of the slope with bedding planes located at their base can be the same, but it can change too. At the base of the scarp fronts, grikes and faults are characteristic, but debris originating from the bed are also common. Their origin is various, some of them are moraines, others are of frost weathering, collapse origin or some of them resulted from karren formation.

In areas exposed by glacial erosion, not only dissolution, but also precipitation can take place. Such concretions were studied for example by Hubbard, Hubbard (1998) at the Transfleuronglacier. According to Hallet (1976), veneers develop during the freezing of meltwater. These features can be spicule, labe and veneers (Smart 2004). In case of the development of veneers, smooth surfaces do not develop by scouring, but they develop by veneer development. Patchy features can be observed on the surface of bedding planes on the valley floor of the Hallstatt glacier that became deglaciated not more than 100 years ago and in karren features (Fig. 3.25). Their colour is white and differs from the colour of the bedrock. Their development is probably very young and it can be traced back to the fact that meltwater froze at the termination of the retreating glaciers because of the retreat of the glacier and, thus, because of the thinning out of ice. On surfaces becoming deglaciated earlier, the colour of the precipitation veneer (its thickness can be even 1–2 cm) does not differ from the colour of the bedrock at all. Two general conditions can be mentioned for the development of precipitations. One of them is that the glacier should be a warm-based glacier, the other condition is that the rate of meltwater drainage into the karst should remain below the rate of supply.

3.2.2.2 The Morphology of Stepped Valleys

The side slope of troughs can be a steep cliff, while its lower part or floor is stepped. When the inclination of bed is identical with the angle of valley slope, the bedding planes dip towards the axis of the valley, while the heads of bed dip towards the margin of the valley. When the inclination is opposite, the bedding planes dip towards the margin of the valley and the heads of bed dip towards the axis of the valley (Fig. 3.33b_3).

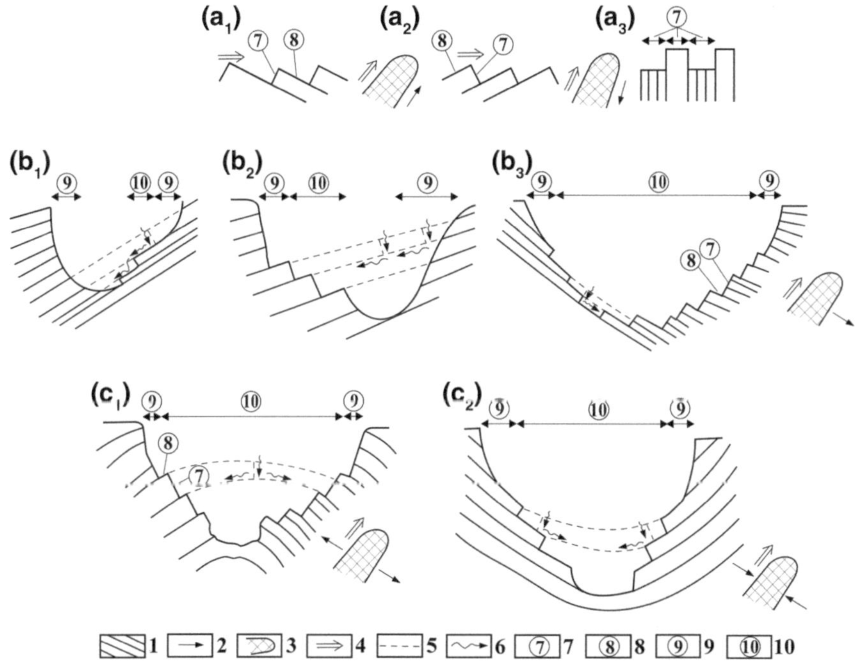

Fig. 3.33 Steps and glacier valley cross sections (Veress 2016, modified) 1. beds, 2. direction of bed dipping, 3. glacier in plan view, 4. direction of glacier movement, 5. former bed unit, former fracture, 6. water percolation, 7. scarp front, 8. exposed bedding plane, 9. steep valley slope with cliffs, 10. gentler valley slope and valley floor with steps **a** valley developed in dip direction, $\mathbf{a}_1$ direction of ice movement was identical with the dip direction of beds, $\mathbf{a}_2$ direction of ice movement was opposite to the dip direction of beds, $\mathbf{a}_3$ the beds are vertical, **b** the valley developed in the strike direction of beds, $\mathbf{b}_1$ asymmetric valley cross section, on one side of the valley the exposed bedding planes dip towards the centre of the valley, $\mathbf{b}_2$ asymmetric valley cross section, on one side of the valley, the exposed bedding planes dip towards the margin of the valley, $\mathbf{b}_3$ asymmetric valley cross section, on one of the valley slopes, the exposed bedding planes dip towards the centre of the valley, while on the other valley slope, they dip towards the margin of the valley, **c** direction of the valley is identical with the strike direction of the fold, $\mathbf{c}_1$ valley developed on an anticline, $\mathbf{c}_2$ valley developed on a syncline

The strike of the steps of stepped surfaces can be perpendicular to the trough (Fig. 3.33b) or identical with its direction (Fig. 3.33a). Perpendicular steps with strike direction develop if the direction of glacier movement was identical with the dip direction of the beds. If the direction of glacier movement was of dip direction, the inclination of bedding planes is identical with the slope of the valley floor (Fig. 3.33a_1), while if the glacier was moving opposite to dip direction, the dip of the bedding planes is opposite to the slope of the valley floor (Fig. 3.33a_2). In this case, steps do not necessarily develop on the floor of troughs, but only if the dip of the glacial erosional surface is identical with the dip of beds (Fig. 3.34a). However, steps do not develop at all if the dip of the glacial erosional surface cutting the beds is smaller than the dip of the beds (Fig. 3.34b), while if the slope of the glacial erosional surface is larger than the dip of the beds, only a surface with heads of bed develops (Fig. 3.34c). If the beds are of vertical position, hogbacks develop which are perpendicular to the direction of the valley (Fig. 3.33a_3).

On glaciokarst, glacial troughs can be symmetric or asymmetric. Troughs with asymmetric cross section occur more frequently. The asymmetry is due to stepped surfaces developing in valleys if the strike direction of scarp fronts and bedding planes situated above each other is identical with the direction of the valley. This occurs if the valley direction and the dip direction of beds meet at an angle of 90°.

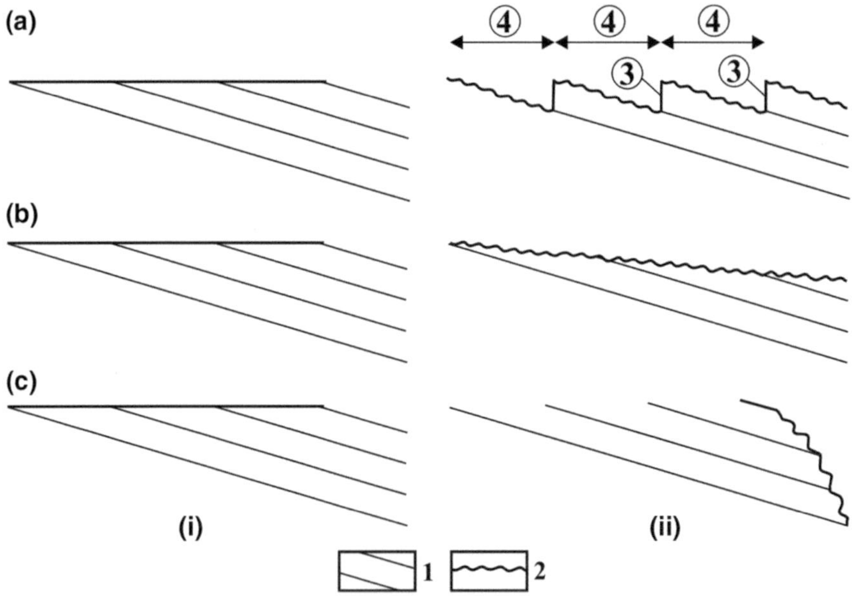

Fig. 3.34 The position of the scoured surface developed by glacial erosion as compared to the beds if the direction of ice movement was identical with the dip direction of beds **i** surface before glacial erosion, **ii** surface that developed by glacial erosion if the dip of the erosional surface is the same as the dip of beds (**a**), if it is of smaller dip (**b**) and if it is of a larger dip (**c**), 1. bed, 2. ice erosional surface, 3. head of bed, 4. bedding plane

However, stepped surfaces can also be formed in symmetric valleys if, as it has already been mentioned, the valley direction is identical with or opposite to the dip direction of beds or the valley developed in an anticlinal or synclinal way.

If the denudation of beds was more rapid at the valley sloping opposite to the dip direction of beds, bedding planes dipping towards the axis of the valley developed or survived during denudation (Fig. 3.33b_1). If the denudation of beds was stronger at valley slopes having an identical dip direction with the dip direction of beds, then bedding planes dipping towards the margin of the valley developed (Fig. 3.33b_2). If the denudation of beds was the most intensive on the valley floor on the valley slope where the direction of the valley slope and that of the beds is similar, then the bedding planes dip towards the axis of the valley, while where it was the opposite, the bedding planes dip towards the margin of the valley (Fig. 3.33b_3).

A symmetric trough is formed if the direction of the glacier was identical with the dip direction of the beds, but also in the case if it was opposite to it or the glacier developed in the area of an anticline or syncline and its direction was identical with the strike direction of the fold. If the glacier developed in an anticlinal, steps developed, the bedding planes of which dip towards the margins of the valley (Fig. 3.33c_1). If the valley developed in a syncline, the bedding planes of the steps dip towards the axis of the valley (Fig. 3.33c_2).

3.3 Conclusions

Glacial erosion was analysed on glaciokarst. Partly, this is similar to erosion and feature development occurring on other non-karstic rock. The difference is manifested in two ways: at the depressions of karst and at glacier valleys where the rock is well-stratified and fractured and dissolution can take place.

On various slopes of depressions, the loading component and the actual shifting are different. Therefore, mainly the upstream slopes of depressions and the ridges separating constituting dolines (dividing walls) are denuded, but no denudation occurs on depression floors where no sliding takes place.

Because of the above-mentioned facts, the depressions of glaciokarst become elongated downslope of ice motion (or they get more elongated) during glacial erosion and their cross section will be asymmetric. The constituting dolines of uvalas coalesce to a much more significant degree, by this there is a greater chance for the formation of large-sized depressions from uvalas, their diameter can even reach or exceed 1 km. Those dolines on the floor of which no moving ice was present, developed into dolines with complex slopes.

On well-stratified, fractured rock, stepped surfaces (limestone pavements) developed. These surfaces are made up of sequences of steps where the steps are bordered by scarp fronts and bedding planes. In the glacier valleys (troughs) of the karst, stepped morphology is characteristic and widespread. The steps can be transverse (if the dip of beds was identical with the movement of ice) and longitudinal (if the direction of ice movement enclosed an about 90° angle with the dip of bed). Stepped surfaces have a great significance in the present karstification (karren formation) of glaciokarst. Their frequency, size and position determine valley morphology. Because of their presence, troughs mainly have an asymmetric cross section (exceptions are those valleys which have the same direction as the strike of folds being developed in an anticlinal or synclinal way and also those ones where the valley direction was identical with or opposite to the dip direction of beds).

The development of stepped surfaces is caused by the mutual effect of meltwater and glacial erosion. Thus, their development can take place under warm-based glaciers if the meltwater has a dissolution capacity. During dissolution, grikes and Schichtfugenkarren develop which contribute to the fact that on the valley floor and on the lower part of valley slopes the units of beds become partly or completely separated from their environment (some bed units separate from other units of the bed or from the bedrock). Autochthon bed unit remnants form a stepped surface.

The process is controlled by glacial erosion and dissolution capacity together. The amount of meltwater getting into the karst does not only depend on the glacier, but on the epikarst too. Thus, the maturity of epikarst also affects the intensity of glacial erosion. However, the maturity (thickness) of epikarst does not only depend on dissolution, but on glacial erosion too. Glacial erosion destroys the epikarst, meltwater regenerates it by dissolution. If the former one is more significant, the epikarst becomes thinner, if the latter, it becomes thicker. In case of a balance between them, its thickness does not change, but the epikarst expands downward. During the process, newer and newer bed units are denuded and thus, newer and newer bedding planes are exposed. Since the epikarst continually regenerates, it could not only develop, but also survive, which, if it was not transformed by later dissolution, preserves its state being present during deglaciation.

References

Adamson KR, Woodward JC, Hughes PO (2014) Glaciers and rivers: Pleistocene uncoupling in a Mediteranean mountain karst. Quatern Sci Rev 94:28–43

Atkins CB, Barrett PJ, Hicock SR (2002) Cold glaciers erode and deposit: evidence from Allan Hills, Antarctica. Geology 30(7):659–662

Atkins CB, van der Meer JJM, Barrett PJ, Hicock SR (2009) Evidence for glacial activity in the Allan Hills, Antarctica. Quat Sci Rev 28(27–28):3124–3137

Barrère P (1964) Le relief karstique dans l'ouest des Pyrénées centrales. Revue Belge de Geographie 88(1–2):9–62

Bauer F (1952) Zur Verkarstung des Sengseng-ebirges in Oberösterreich. Mitt. Höhlenkomm 7–14

Benn DI, Evans DJA (2010) Glaciers and glaciation. Hodder Education, London, p 802

Bennett MR, Glasser NF (2009) Glacial geology: ice sheets and landforms. Wiley-Blackwell, Chichester UK, p 385

Bögli A (1964) Un exemple de complexe galciokarstique: Le Schichttreppenkarst. Revue Belge de Geographie 88(1–2):64–82

Bögli A (1978) Karsthydrographia und physische Speläologie (Karst hydrology and physical speleology). Springer, Heidelberg, New York, Berlin 292 p

Boulton GS (1974) Processes and patterns of glacial erosion. In: Coates DR (ed) Glacial geomorphology, Proceedings of the Fifth Annual Geomorphology Symposia, Binghampton, Allen & Unwin, London, pp 41–87

Boulton GS (1993) Ice ages and climatic change. In: Duff P, Mcl D (eds) Holmes' principles physical geology, 4th edn. Chapman and Hall, London, pp 439–469
Chorley RJ, Schumm SA, Sugden DE (1984) Geomorphology. Methuen, London 605 pp
Clayton M (1966) The origin of the landforms of the Malham area. Field Stud 2:359–384
Clayton K (1981) Explanatory of the landforms of the Malham area. Field Stud 5:389–423
Cuffey KM, Paterson WSB (2010) The physics of glaciers, 4th edn. Academic Press, London, p 704
Cvijič J (1913) The ice age in the Prokletije and surrounding mountains (in Serbian). Glas Srpske Kraljevske Akademije Nauka, Belgrade. XCI, 149 p
Davies D (2014) Antarctic glaciers. www.antarcticglaciers.org/modern-glaciers/glacial-processes/
Djurovič P (2009) Reconstruction of the pleistocene glaciers of Mount Durmitor in Montenegro. Acta geographica Slovenica 49(2):263–289
Djurović P, Petrović AS, Simić S (2010) The overall impact of Pleistocene glaciation on morphological diversity of at Durmitor and Žijovo. Serbian Geographical Society 90:17–34. https://doi.org/10.2298/gsgd1001017d
Dreimanis AA (1989) Tills: their genetic terminology and classification. In: Goldthwait RP, Matsch CL (eds) Genetic classification of glacigenic deposits. Balkema, Rotterdam, pp 17–83
Fels E (1929) Das Problem der Karbildung in den Ostalpen. Petermanns Mitt., Ergänzungsh 202:1–84
Ford DC (1983) Effects of glaciations upon karst aquifers in Canada. J Hidrol 61:149–158
Ford DC (1996) Karst in a Cold Climate: Effects of Glaciation and Permafrost Conditions Upon the Karst Landform Systems of Canada. In: Mc Cann SB, Ford DC (eds) Geomorphology sans frontières. Wiley, Chichester, pp 153–179
Ford DC, Williams PW (1989) Karst geomorphology and hydrology. Unwin Hyman, London, p 601
Ford DC, Williams PW (2007) Karst hydrogeology and geomorphology. Wiley, Chichester, p 562
Glasser NF, Hambrey MJ (2001) Styles of sedimentation beneath Svalbard valley glaciers under changing dynamic and thermal regimes. J Geol Soc, London 158:697–707
Glasser NF, Hambrey MJ (2003) Ice-marginal terrestrial landsystems: Svalbard polythermal glaciers. In: Landsystems Glacier (ed) Evans DJA. Arnold, London, pp 65–87
Goldie HS (1996) The of Great Asby Scar, Cumbria. U. K. Environ Geol 28(3):128–136
Goodsell B, Hambrey MJ, Glasser NF (2005) Debris transport in a temperate: Haut Glacier d'Arolla, Valais, Switzerland. J Glaciol 51(172):139–146
Gordon JE (1977) Morphometry of in the Kintail–Affric–Cannich area of northwest Scotland. Geogr Ann 59A:177–194
Goudie A (1990) The landforms of England and Wales. Blackwell, Oxford, p 394
Hallet B (1976) Deposits formed by subglacial of $CaCO_3$. Bull Geol Soc Am 87:1003–1015
Hambrey MJ, Fitzsimons SJ (2010) Development of sediment-landform associations at cold glacier margins, dry valleys, Antarctica. Sedimentology 57:857–882
Hambrey MJ, Bennett MR, Dowdeswell JA, Glasser NF, Huddart D (1999) Debris entrainment and transfer in polythermal valley glaciers. J Glaciol 45:69–86
Harbor JM, Hallet B, Raymond CF (1988) A numerical model of landform development by glacial erosion. Nature 333:347–349
Haynes VM (1968) The influence of and rock structure on corries in Scotland. Geogr Ann 50A:221–234
Hooke R, Le B (1989) Englacial and subglacial hydrology: a qualitative review. Artic Alpine Res 21:221–233
Hubbard B (1998) Bedrock surface roughness and the distribution of subglacially precipitated carbonate deposits implications for formation at Glacier de Transfleuron, Switzerland. Surf Process Land 23:261–270
Jones H, Arnold N (1999) Modelling the entrainment and transport of suspended sediment in subglacial hydrological systems. Glacial Geol Geomorphol 9:20. http://boris.qub.ac.uk/ggg/papers/full/1999/rp091999/rp09.html. 2003.05.10
Kerr A (1993) Topography climate and ice masses: review. Terra Nova 5:332–342
Kunaver JJ (2009) The nature of limestone pavementsLimestone pavements in the central part of the southern Kanin plateau (Kaninski podi) Western Julian. In: Ginés A, Knez M, Slabe T, Dreybrodt W (eds) Karst Rock Features—Karren Sculpturing Zalozba ZRC, pp 299–312. Institut za raziskovanje krasa ZRC SAZU, Postojna. Carsologica, 9. Ljubljana, Eslovènia

Lauritzen S-E (1984) Evidence of subglacial Karstification in Glomdal Svartisen. Norsk Geografisk Tidesskrift 38(3–4):169–170
Lauritzen S-E (1986) Kvithola at Fauske, northern Norway: an example of ice-contact speleologenesis. Norks Geografisk Tidesskrift 66:153–161
Lepirica A (2008) Geomorphological characteristics of the massif Prenj. Acta Carsologica 37 (2-3):307–329
Lloyd Davies MT, Atkins CB, van der Meer JJM, Barrett PJ, Hicock SR (2009) Evidence for cold-based glacial activity in the Allan Hills, Antarctica. Quaternary Science Reviews 28(27–28):3124–3137
Menkovič LJ (1994) Glacial traces in the Djeravica area, Prokletije mountains (in Serbian). Geografski gadisnjak 30:139–146
Mollard JD, Jones JR (1984) Airphoto interpretation and the Canadian landscape. Canadian Government Publishing Centre, Hull, Québec, p 415
Murphy PJ, Faulkner TL, Lord TC, Thorp JA (2015) The Caves of Giggleswick Scar—examples of deglacial speleogenesis. Cave and Karst Science. 42(1):42–53
Rastas J, Seppälä M (1981) Rock jointing and forms on roches moutonnées, SW Finland. Ann Glaciol 2:159–163
Rose L, Vincent P (1983) Some aspects of the morphometry of grikes: a micture model approach. In: Paterson MM (ed) New direction in Karst, proceedings of the Anglo-French Karst Symposium. Geo Books, Norwich, pp 497–514
Smart C (1986) Origin and development of glacio-karst closed depressions in de Europa, Spain. Zeits f Geom N I 30(4):423–443
Smart C (2004) Glacierized and glaciated karst. In: Gunn J (ed) Encyclopedia of Caves and Karst Science, Fitzroy Dearborn, New York, London, pp 388–391
Summerfield (1991) Global Geomorphology: an introduction to the study of landforms. Longman Scientific & Technical, Harlow XIV: 537 p
Sweeting MM (1973) Karst landforms. Columbia University Press, New York, p 362
Telbisz T, Dragašice H, Nagy B (2005) A horvátországi Biokovo-hegység karsztmorfológiai jellemzése terepi megfigyelések és digitális domborzatelemzés alapján. Karsztfejlődés 10:229–243
Veress M (2010) Karst environments, Karren formation in high mountains. Springer, Dordrecht Heidelberg London New York 250 p
Veress M (2016) Covered Karst, Springer, 536 p. https://doi.org/10.1007/978-94-017-7518-2
Vincent P (1983) The dissolving landscape: karren. Geograph Mag 55:508–510
Vincent P (2009) Limestone pavements in the British Isles. In: Ginés A, Knez M, Slabe T, Dreybrodt W (eds) Karst Rock Features—Karren Sculpturing Zalozba ZRC. Institut za raziskovanje krasa ZRC SAZU, Postojna. Carsologica, 9. Ljubljana, Eslovènia, pp 267–274
Walden J (2002) Glaciers and glacial processes I & II. Lecture Notes—School of Geography and Geosciences, University of St. Andrews, St. Andrews, UK. http://www.st-andrews.ac.uk/www_sgg/personal/jw9link/jwteach/glacial/ppt2003.06.10
White WB (1988) Geomorphology and hydrology of karst terrains. Oxford University Press, Oxford, p 463
Williams PW (1966) Limestone pavements with special reference to western. Trans Inst Br Geogr 40:155–172
Williams PW (2008) The role of the in karst and cave hydrogeology: a review. Int J Speleol 37 (1):1–10
Žebre M, Stepišnik U (2015a) Glaciokarst landforms and processes of the southern Dinaric. Surf Process Land. https://doi.org/10.1002/esp.3731
Žebre M, Stepišnik U (2015b) Glaciokarst geomorphology of the Northern Dinaric Alps: Snežnik (Slovenia) and Gorski Kotar (Croatia). J Maps. https://doi.org/10.1080/17445647.2015.1095133
Zötl J (1963) Zur Morphogenese des Ennstales. Mitt Naturwiss Ver Steiermark 93:155–160

Chapter 4
Karst Landforms of Glaciokarst and Their Development

Márton Veress

Abstract In this chapter, the karst landforms of glaciokarst are presented which are the following: karren, giant grikes, shafts, karst depressions such as giant depressions (dolines, uvalas), small-sized solution dolines, schachtdolines, subsidence dolines, ponors and poljes. We describe their distribution and frequency, their relation to glacial erosional features as well as the relation between each other, their size, morphology, varieties, evolution, development and development age.

Keywords Karren · Giant grike · Shaft · Giant solution doline and uvala
Small-sized solution doline · Schachtdoline · Subsidence doline
Ponor · Polje

4.1 Introduction

This chapter presents the karst landforms of glaciokarst. These can be features transformed by ice (of preglacial, interglacial and glacial age) and features that were not transformed by glacial erosion (of postglacial age). The majority of features and thus, those being transformed are negative features (karren or karstic depression), but karstic relict landforms also occur among them which are mainly positive features. Pure glacial features and other features of non-karstic origin have already been dealt with in Chaps. 2 and 3 thus, their detailed description is only mentioned where it is necessary for the description of karstic features. On glaciokarst, mainly glacier valleys are being karstified at present. The possible karst features of a glacier valley, their relation to each other and their approximate distribution are presented in Fig. 4.1.

The karst features of glaciokarst can be classified and described in many ways. Thus, according to the type of the feature, the development and the development age of the feature, its development environment and its size. In the following, an outline of the features can be found based on the types of the features. We describe their shape and present their morphology, varieties, development age, genetics, size, frequency, distribution and development environment as well as the role of glacial erosion in their development and transformation.

M. Veress et al., *Glaciokarsts*, Springer Geography,
https://doi.org/10.1007/978-3-319-97292-3_4

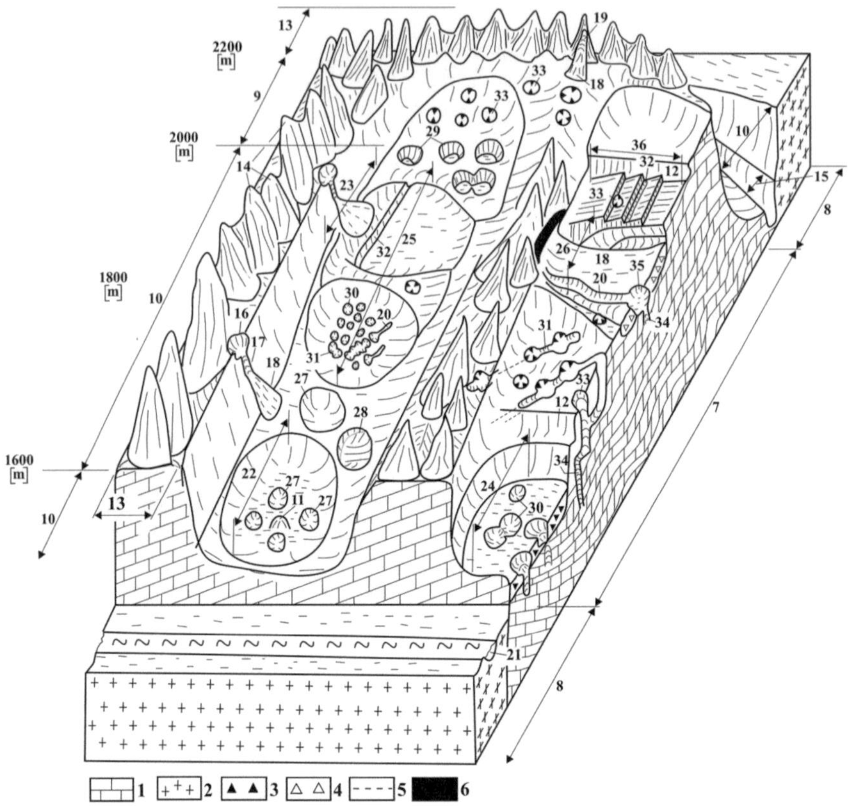

Fig. 4.1 Main features in the glacially transformed karstic high mountains (Veress 2016a, modified). 1. Limestone, 2. older metamorphic basement, 3. till, 4. colluvial debris, 5. fault, 6. siliceous interbedding in limestone, 7. Klippen (plateau), 8. eroded portion of nappe, 9. cirque, 10. trough, 11. roche moutonnée, 12. transverse step, 13. arête, 14. horn, 15. river valley, 16. niche with circular rim, 17. niche of frost shattering, 18. talus cone, 19. path of rock avalanche, 20. furrow of covered karst depression, 21. river, 22. paleodoline, 23. asymmetric paleodoline, 24. paleodoline with moraine, 25. paleouvala, 26. partially filled paleodoline, 27. small-sized solution doline, 28. asymmetric small-sized solution doline, 29. schachtdoline, 30. subsidence doline, 31. subsidence uvala, 32. giant grike, 33. shaft system, 34. passage, chimney, shaft system in cross section, 35. ponor, 36. schichtrippenkarst

On glaciokarst, based on development age, features being developed in the preglacial, interglacial, glacial and postglacial can be differentiated, while according to their evolution and/or development there are karstiglacial, glaciokarstic, mixed, glacial, covered karstic features, as well as features of mass movement, frost weathering, fluvioglacial, fluvial and snow erosional features. The distribution, the density and the size of various features depend on the degree of the former karstification of glaciokarst and on the nature and degree of glaciation. Karstiglacial features preserved their original state better, while glaciokarstic features and mixed features can be denuded or transformed to various degrees. Adjacent features could even coalesce. However, depressions can be lined, filled or buried too.

Both in the periglacial and temperate fluvial zone, karst surfaces can be covered by superficial deposit. On glaciokarst, based on coveredness, uncovered terrains (bare karst), terrains covered by soil (soil-covered karst) and terrains covered by soil and/or debris (till or frost weathering material) may occur (Veress 2016a). The latter is the covered karst which is cryptokarst if the cover is impermeable and it is concealed karst if the cover is permeable (Veress 2016a). Allogenic karst also occurs on glaciokarst (Veress 2016b). This type develops where there is an impermeable cover on the karst or non-karstic rock becomes wedged in karstic rock and the streams of non-karstic terrain flow towards the karst.

Glacial erosional features and fluvioglacial features can still develop today if the glacial zone is present, while former glacial features and glacial sediments are denuded in the periglacial zone. While in the latter, but also in the temperate fluvial zone, features of snow erosion, features of mass movement and frost weathering features are still developing. On glaciokarst, fluvial geomorphic activity is present to a limited extent (in case of certain karst features and on morainic terrains). Gorges which are widespread on bare karst are exceptions. Their number and density are larger than that of those occurring on karst areas which belong to other karst types. This can be explained by the abundant amount of former (also subglacially, Lepirica 2008) and present meltwater, but the overflowing water of the lakes of glaciokarst also provides favourable conditions for their evolution and development. Glaciokarstic and mixed features are polygenetic (and polymorph) and since their former karstic development ceased, they are paleokarstic. They reached their present state and form after a long time (the initiation of their development can even precede the Quaternary). Their development was not continuous, it was interrupted from time to time. Their development rate could alternate.

In older karstic features, the nature of younger karstification changes (thus, a ponor develops instead of a doline), the process only affects a part of the original feature, but in this case it may have an effect at several places (for example solution dolines, or subsidence dolines develop in older depressions). Paleokarstic features may also be absent. Either because they did not develop at all or they were destroyed. Their destruction could mainly happen by filling (see below).

The most common paleokarstic features or old features are former solution dolines, uvalas, shafts and poljes. Less common are ponors, karren and remnants features of karst (thus, karst hills and the different types of karstic inselbergs).

Recent, postglacial features mainly develop in glacial erosional features (in glacier valleys and in the transformed karstic depressions of the floor of glacier valleys), but they also occur on the free glacial erosional surfaces of glaciokarst or on plateau parts exempt from glacial erosion. Some features (for example subsidence dolines) are more common in paleokarst depressions. However, the characteristics of superficial deposit (thickness, composition) and hydrological conditions (degree and duration of water supply) have a significant effect on the distribution and development of covered karst features. However, subglacial features are often in a perched position (Ford 1996). The reason for this may be that the meltwater gets into the karst at the higher parts of the floor of the glacier valley or because the subglacial water under pressure leaves the karst at a more elevated part as compared to its entering site (Lauritzen 1986).

The direct and indirect effects of ice have a different role in the development of the various karst features of glaciokarst. A direct effect is when the meltwater of ice creates the karst feature. Such features can be, for example some karren features or shafts. An indirect effect is when the surfaces (features) or sediments created by ice create a favourable opportunity for karstification. Here again, karren or subsidence dolines can be mentioned as examples. Finally, such a feature development can be mentioned where the glacier or the environment created by the glacier has no role in feature development. Such features can be, for example postglacial solution dolines.

4.2 Karren

Karren are the most typical and most widespread features of glaciokarst. They are mainly dominant in that part of glacier valleys which belongs to the periglacial zone, however, subsoil karren are also present in the fluvial erosional zone of the temperate belt. Regarding their development, karren features are primarily recent (postglacial) features that developed in the Holocene, but they can also be glaciokarstic features (these karren were partly destroyed by glacial erosion), or they can be mixed, when karren were transformed and partly destroyed to the effect of ice, but they regenerated and continued their development to the effect of meltwater. There is a great opportunity to survive for karren developing along bedding planes (schichtfugenkarren) and for karren (grike) going transversely the beds being destroyed by glacial erosion and penetrating into the surviving beds (Ford 1996; Rose and Vincent 1983). Although according to Sauro (2009), large-sized kamenitzas could also survive glacial erosion. However, karren with small depth (rinnenkarren) are destroyed during glacial erosion.

But karren features could mainly survive because they could continuously regenerate under ice. Thus, because of the continuity of karren formation, although the majority of karren features were partly or completely destroyed because of continuous glacial erosion, karren which developed in the moment of the disappearance of ice survived. Subglacial karren can also become conserved. This occurs when the moraine of the surface is injected into karren features (Ford 1996).

On karst, karren and epikarst are in a close relationship with each other. Some karst features (grikes) can form a part of epikarst. This is much more significant on glaciokarst, where meltwater connects the karren and the epikarst. However, this relationship is also expressed by the fact that there is a close relation between the denudation of the surface and the two latter factors too (see Chap. 3). On glaciokarst, karren of glacial age (and their remnants) and also karren of postglacial age such as grikes, pits and schichtfugenkarren can constitute the epikarst.

Karren form karren fields, which are areas of large extension (Bögli 1980) where karren dominate (Monroe 1970). According to Ginés (2009), limestone pavement is also a type of karren fields. Bögli (1964) differentiates three karst types on glaciokarst: Rundhocker karst (the area of roche moutonnées), Schichttreppenkarst (the bedding planes of horizontal beds) and Schichtrippenkarst (the bedding planes of oblique beds). Since karren dominate roche moutonnées and bedding planes, these karsts are virtually karren fields too.

Karren features of glaciokarst are widespread on erosional features being developed by glaciers (Bögli 1964). They are mainly characterized in troughs and on features that developed in them: mostly on the stepped surfaces of valleys, on the side slopes of troughs, on transverse steps, on the side walls of rock basins, on the side slopes of karstic depressions (Veress 2010), on roche moutonnées (Veress 2012a; Stepišnik et al. 2010), but they also occur on erratic blocks, on large blocks of moraines (Veress 2010), on the blocks of rock fall (Djurovič et al. 2010) and on the cemented material of moraines (Žebre and Stepišnik 2015a). However, they can also develop outside glacier valleys, on limestone pavements that were formed by free glacial erosion (Sweeting 1973; Vincent 2009).

As regards the above-mentioned glacial erosional features, limestone pavements and mainly their parts with bedding planes are the features that experienced karren formation for the most part (Fig. 4.2). The spatial position of surfaces with bedding planes and the surfaces with heads of bed (scarp front) (thus, their dip direction and thus, their position in the valley) and thus, the direction of the expansion of karren terrains were determined by the relation between the direction of ice motion and the spatial position of beds. The expansion and inclination of the terrain with bedding planes and the terrain with heads of bed depend on the thickness and inclination of beds (Fig. 3.32). However, this also depends on the characteristics of the limestone constituting the bed. On the British Islands, those beds have a greater chance to constitute a bedding plane (or such surfaces survive from the former surface), which contain less fractures and in which the rock is more massive (Doughty 1968). To the joint effect of glacial erosion, karren formation and the stratification of the limestone, the floor of the glacier valley, partly its side (mainly the trough) is a mosaic-like system of surfaces with bedding planes of different expansions, inclination and relative height. The expansion and inclination of some mosaics and thus, that of the bedding planes (the heads of bed too) do not only depend on the thickness and the inclination of beds, but on glacial erosion and on karren formation under the ice too. Therefore, on glacial erosional surfaces (mainly in troughs), karren have a mosaic-like pattern where the expansion of certain karren patches can be extremely different.

The extent and quality of karren formation of bedding planes can be various. Partly because the expansion of the bedding plane determines the amount of water flowing on a bedding plane and partly because soil and Pinus mugo develop on the upper, marginal part of a significant amount of the bedding planes (mainly on the lower parts of the periglacial zone). The differences are also enhanced by the fact

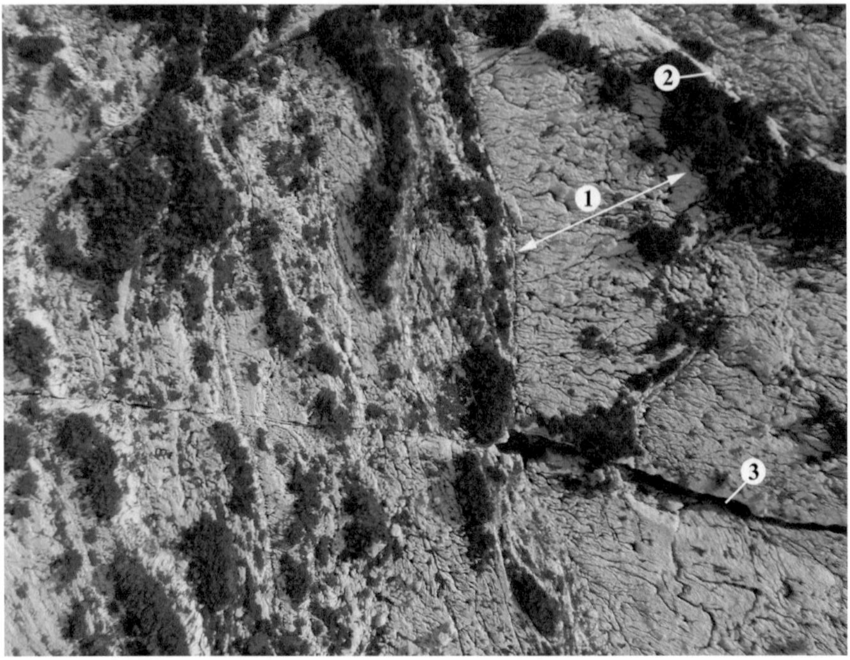

Fig. 4.2 A series of steps with karren (valley under Tragl peak, Totes Gebirge, aerial photograph, photo by Kalmár). 1. Bedding plane with rinnenkarren, 2. scarp front, 3. giant grike

that surficial water flow was partly or completely absent on a number of bedding planes during present karren formation. On the one hand, the reason for this is that the bedding planes are sealed by scarp fronts at many places, by this water flow could become local, on the other hand, because karren features probably already existed following the retreat of ice and the water flowing down the bedding planes passed through them and could get into the epikarst. Karren formation can be various also because the degree of the inclination of adjacent bedding plane surfaces can be extremely different too (see below).

Because of this, the karren formation of bedding planes separating from each other is individual and has karren characteristic features that are only characteristic for it. By this, the number, the density and the kind of karren features of a certain bedding plane differ from the others. However, the bedding planes may have certain characteristics (for example similar inclination and expansion) because of which their karren formation can be more or less similar.

On glaciokarst, four levels of karren phenomena can be differentiated: karren feature, karren assemblage, karren cell, and unit of karren formation.

4.2.1 Glacial Features and Surfaces Bearing Karren

Glacial erosional features are bordered by slopes with various inclinations. These slopes can be put into slope classes with small dip (0°–20°), medium dip (21°–40°), large dip (41°–60°, even 70°) and extremely large dip (71°–90°). Although the karren features of the slopes of various slope classes can also show partial similarity, the karren features of the slopes with various inclination are different (see below). Thus, glacial features show different karren formations if they are bordered by slopes with different inclinations (karren features will be different, but their density may be different too). Therefore, the karren formation of glacier valleys and that of their trough parts as well as that of roche moutonnées is heterogeneous since their slopes have different inclinations. The karren formation of troughs is heterogeneous because their side slopes are steeper than their floor. Heterogeneous karren formation will occur on their floor too if stepped surfaces developed there since they are constituted by bedding planes and heads of bed with various steepness. As the upstream and downstream side slopes of roche moutonnées are bordered by slopes with different steepness (Bennett and Glasser 2009, Fig. 3.4), the karren formation of these features is heterogeneous too.

The inclination of bedding planes depends on the dip of beds (however, the inclination of heads of bed is almost perpendicular to the dip of bedding planes). The bedding planes of beds with small inclination (smaller than 20°), the heads of beds with an inclination of 70°–90°, the slopes of giant dolines (paleodolines), the upstream slopes of roche moutonnées and the roofs of erratic blocks (if they have roofs) constitute slopes with small inclination or they are bordered by such slopes. However, as we will see below, the slopes belonging to the slope class with the small inclination can be put into two subclasses regarding their karren features: the slope subclass with a very small inclination (0°–10°) and the subclass with the less small inclination (11°–20°). The bedding planes of beds with a dip of 21°–40° and the heads of beds with a dip of 50°–69° constitute slopes with a medium dip. The slopes of giant dolines and the upstream slopes of roche moutonnées may be bordered by slopes of such inclination. It has to be noted that beds with a small or medium dip can be areas exempt from compression, but mainly the nappe structures of areas being affected by compression too. The bedding planes of beds with a dip of a 41°–60° (70°), the heads of beds of the beds with an inclination of 20°–49° constitute slopes with large inclination or some parts of the side slopes of paleodolines, the side slopes of rock basins, the side slopes of some glacier valley parts, some transverse steps, the downstream (leeward) slopes of roche moutonnées are bordered by such slopes. Finally, the bedding planes of beds with a large inclination (larger than 71°), the slopes with heads of the beds of beds with small dip (smaller than 19°), the side slopes of glacier valleys, the transverse steps and the side slopes of rock basins constitute slopes with an extremely large dip, or the downstream (leeward) slopes of roche moutonnées and the side slopes of erratic blocks are bordered by such slopes.

4.2.2 The Conditions of Karren Formation on Glaciokarst

The formation of karren is affected by the presence or lack of the cover and the soil, the characteristics of the surfaces created by glacial erosion and snow water.

Even on glaciokarst, karren are not only present on bare surfaces, although the main characteristic feature of this karst type is given by karren that developed on these surfaces. When karren formation does not occur on a bare surface, the cover can be soil, but also moraine (with soil or without soil), frost weathering material or material of mass movements (regarding its distribution, moraine is dominant among the latter ones). While soil stimulates karren formation (see Chap. 2), debris, if it is built up of limestone, hinders dissolution on the bedrock (Williams 1966, and see Chap. 2).

According to our opinion, under soil, mainly grikes (Kluftkarren) and kamenitzas develop. Mainly, kamenitzas can occur in such an environment (on bare surfaces between soil patches) where formation took place under soil, but it denuded later. (Soil denudation can be triggered by climate change, but by human activity too.) The diameter of kamenitzas that developed under soil is large as compared to their depth, the side slopes are of small dip, their margin has a wavy pattern, while the diameter of the kamenitzas of bare surfaces is small as compared to their depth, their side slopes are steep and is separated from the floor, and their margin is arcuate.

Some karren features (kamenitzas, grike, pitkarren) are formed at the margin of a soil patch (karst is partly covered with soil). In case of various kinds of karren features (rinnenkarren, grike, kamenitza) it occurs that soil and vegetation are formed in the feature that had developed on the bare surface. Because of this, the whole or some parts of these karren features deepen more intensively. Subsoil karren also occur in the lower part of the periglacial zone (in the so-called dissolutional subzone), but they become predominant in the temperate fluvial zone. Because of soil cover, we only have knowledge of their presence in case of the denudation of soil cover.

We get knowledge of karren developing under moraine if the moraine is denuded. Karren formation under moraine can take place if the thickness of the moraine with limestone material is small or if the moraine thins out (Fig. 2.15) since under moraine with thick limestone material, the bedrock does not become dissolved as the water flowing through the limestone debris becomes saturated before it reaches the bedrock (Williams 1966; Trudgill 1972, 1985). According to Jones (1965), dissolution also takes place under moraine if there is water with enough acidity which is enabled by CO_2 of biological origin. According to Williams (1966), under limestone cover, grikes can only develop on the bedrock if the thickness of bedrock is less than 45.72 cm. According to Sweeting (1973), no dissolution occurs on the bedrock under a calcareous cover with a thickness of more than 2 m. According to Trudgill (1972, 1975, 1985), the dissolution of the bedrock can take place under such a cover where the pH value is between 4 and 7, the $CaCO_3$ content is below 1%, but the bedrock does not dissolve under such a cover where the pH value is between 7 and 9 and calcareous content is larger than 10%.

Karren formation can happen under moraine on a glacially scoured surface. In this case, karren features and glacial striae occur together too (Kunaver 2009a). Karren formation can also happen in a way that the glacially scoured surface becomes smooth by dissolution. It is probable that water movement of percolation nature is dominant under the moraine and therefore karren features of seepage origin (see below) can develop. Karren features of flow origin are less likely to develop under moraine since the flow is impeded under moraine as a matter of course in addition to this the already developed karren features of seepage origin also hinder water movement.

The karren formation of bare rock surfaces can take place by the effect of flowing water and percolating water too if the water is unsaturated (Veress 2010). What karren feature is formed on a given bare rock surface basically depends on two circumstances: on the rock structure and the inclination of the bearing slope (in case of some karren features, the dip direction expansion of the slope can be determinant too). It can also be experienced that a certain karren feature type occurs on a surface situated in a certain slope angle dip interval though the capaciousness of an interval can be different in case of various karren feature types.

Karren (rillenkarren, rinnenkarren, meanderkarren, trittkarren, wandkarren), the longer axis of which coincides with the dip direction of the slope are of flow origin. The features developing during the process are presented in Fig. 4.3. Features of flow origin are called hydraulic forms by White (1988), and hydrodynamically controlled forms by Ford and Williams (2007), while features of seepage origin are called etched forms by White (1988) and fracture controlled forms by Ford and Williams (2007).

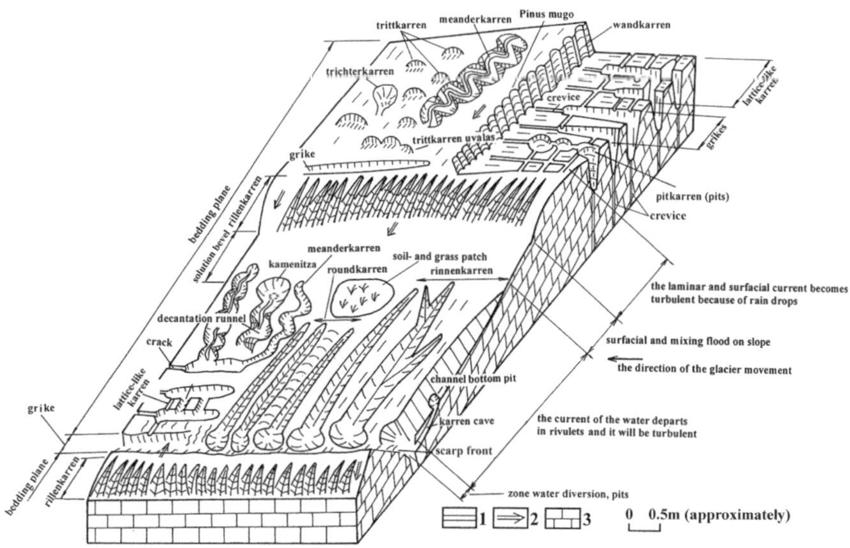

Fig. 4.3 Main karren forms on bedding planes truncated by glacier (Veress 2010, modified). 1. Crack, 2. dip direction of the surface, 3. limestone

Karren features of flow origin are mainly formed on surfaces with bedding planes having a smaller or larger expansion, the expansion of which can be some square metres or several hundred square metres too (maybe even some 1000 m^2) mainly if the flow is turbulent (Glew-Ford 1980; Trudgill 1985).

Karren features of flow origin develop on surfaces where water flow takes place in a continuous, undisturbed and widespread way. Therefore, the greater the chance for their development is and the larger some karren features are, the more expanded their bearing slope is in slope direction (in the direction of water flow) and the flatter and smoother surface is constituted by the slope. The expansion of a karren feature of flow origin is determined by the expansion of unchanged water flow. No flow develops if the expansion of the slope is small or if the characteristic features of the rocks of the slope hinder flow, but they stimulate percolation. A place that impedes water flow, but promotes percolation in the rock can be the fracture of the rock, the bedding plane (if it intersects the surface), skeletal remnant, calcite fill, debris, the already developed karren feature, the non-karstic rock that intercalated into the limestone (for example siliceous limestone, silica, sandstone).

Among surface-forming factors, glacial erosion is mostly prone to create expanded, smoothed surfaces and thus, surfaces which have undisturbed water flow, and mainly such surfaces that are separated from each other and which have various inclinations. The former surface favours the formation of extensive karren features, while the latter (with various inclination) favours the formation of various karren features. On flat and bare surfaces with similar size and inclination, the degree of karren formation and karren assemblages are considerably similar.

No smoothed surfaces are necessary for the development of karren features of seepage origin (grikes, kamenitza, pitkarren, schichtfugenkarren). Thus, they also develop at those places where the ice did not create extensive smooth surfaces or where the rock is disrupted with fractures and faults horizontally. In this case, the size and number of karren features mainly depend on fracture density, but the expansion of the bearing terrain also affects the size of karren features of seepage origin. Thus, the length of grikes is determined by the expansion of the bearing surface. However, in this case, the great expansion does not have to coincide with the dip direction. It is common that smooth surfaces with significant expansion are such bedding planes where fault lines with strike direction run. In this case, both karren features of flow origin and those of seepage origin develop adjacent to each other in slope direction below each other on the same terrain part.

On glaciokarst, mainly meltwater enables dissolution that causes the development of karren features. With the increase of altitude, the amount of snow and the duration of melting (and thus dissolution) increases too. The degree of snow accumulation and the local diversity of the duration of dissolution are caused by the exposure, the steepness of the slope, the presence or lack of depressions. The amount of meltwater and the duration of dissolution also affect the fact what karren features develop in a particular place and what their size and density are. There are karren features typically created by snow meltwater. Thus, features are trittkarren (Haserodt 1965), but the development of wandkarren, schichtfugenkarren and rinnenkarren is mainly caused by meltwater too (Veress 2010).

4.2.3 *Karren Features*

Karren features can be classified according to the coveredness of the bearing surface, as well as to their size, morphology (and genetics).

4.2.3.1 Classification of Karren Features According to Size

Bögli (1960, 1976) distinguished karren features that developed on bare surfaces, on surfaces half-covered with soil and on surfaces covered with soil. According to size, there are nanokarren, microkarren, mezokarren and megakarren (Ginés 2009; Grimes 2012). The width and depth of nanokarren is smaller than 1 mm, those of mikrokarren is below 1 cm. These sizes can extend from some centimetres to some metres in case of mezokarren and from some metres to several ten metres in case of megakarren. On glaciokarst, mainly mezokarren and microkarren are widespread, but megakarren may occur too. For example, such karren can be mentioned from the glaciokarst of Patagonia, where the amount of rainfall is of extreme degree (Veress et al. 2006; Maire et al. 2009). On the marble surfaces of Diego de Almagro Island megakarren, meanderkarren (Fig. 4.4a), wandkarren (Fig. 4.4b), giant grikes (Fig. 4.4c), and kamenitzas (Fig. 4.4d) are common.

On glaciokarst, common microkarren are microgrikes, microspitzkarren and microtrittkarren (Veress and Zentai 2004). Mezokarren have an even larger diversity which will be presented below. The distribution of microkarren is patchy, they mainly occur on terrains that became exempt from ice a short time ago (Veress and Zentai 2004). Mezokarren are more widespread than microkarren.

4.2.3.2 Classification of Karren Features According to Shape and Development

Karren Features of Flow Origin

Rillenkarren are constituted by features that are of great density, some centimetres wide and deep and some tens of centimetres long features with slope direction, but they wedge out in slope direction (Fig. 4.5). They develop at the upper margin of the slope under sheet water, where turbulent water flow develops to the effect of raindrops (Glew and Ford 1980).

Rillenkarren are cosmopolite karren which occur on almost every bare slope, their development is mainly patchy (the expansion of the patches rarely exceeds 1 m^2 though the size of the patches can be very diverse in some places). "Ausgleichsfläche" usually develops below the zone of rillenkarren which is a flat, small-inclined surface smoothed by dissolution where the intensity of dissolution decreases (Bögli 1960). Corrosion terraces (Kunaver 2009b), and flachkarren

Fig. 4.4 Megakarren on glaciokarst (Diego de Almagro, Chile, photo by Veress). **a** Meanderkarren, **b** wandkarren, **c** giant grike, **d** kamenitza

Fig. 4.5 Rillenkarren (Totes Gebirge, photo by Veress). 1. Destroying pit, 2. rillenkarren, 3. grike

(Veress 2010) are larger, but locally expanded, horizontal and flat, hardly dissected surfaces (Fig. 4.6). The latter bear different karren features (nischenkarren, kamenitza, meanderkarren).

Rinnenkarren (channels or runnels) are grooves with slope direction (Figs. 4.6, 4.7 and 4.8), the width and depth of which can spread from some centimetres to 1–2 m, their length can be some metres or several ten metres (Ford and Williams 2007; Veress 2009a). They have no surface drainage as they are connected to pits and grikes. They are formed under rivulets which develop by the splitting of sheet

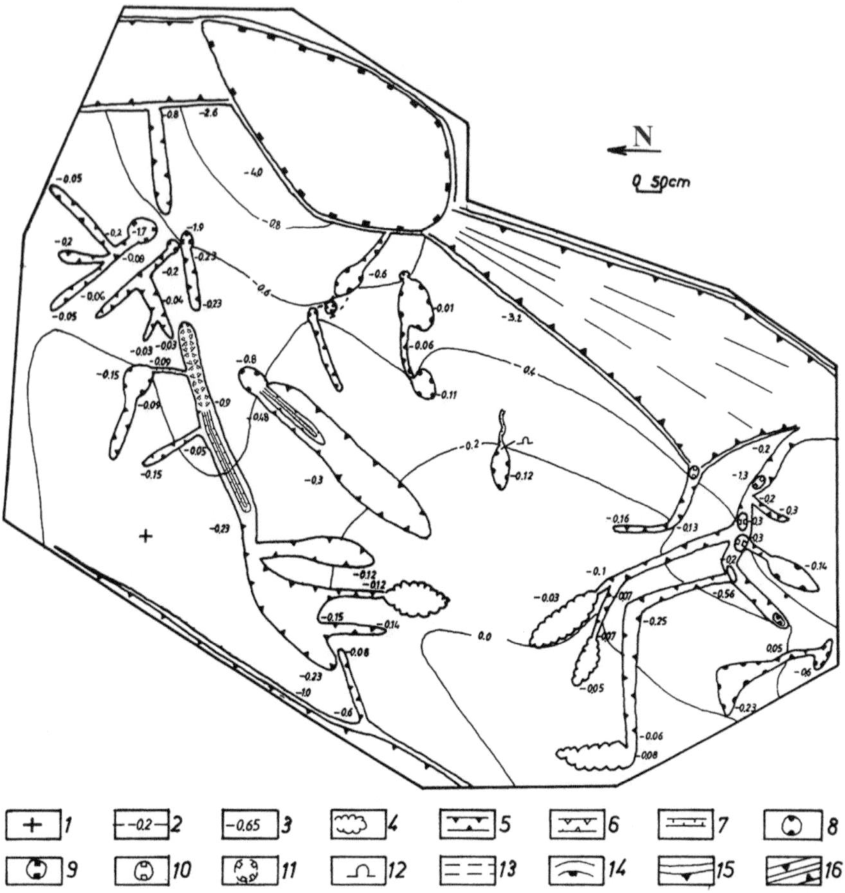

Fig. 4.6 Channels on a flachkarren (Totes Gebirge, Veress et al. 1995). 1. The place of theodolite, 2. contour line in local system, 3. the depth of the form (m), 4. nischenkarren, 5. type I channel or channel in general, 6. type II channel, 7. type III channel, 8. kamenitza, 9. pit in general, channel end pit, 10. channel-bottom pit, 11. wreck pit at the bottom of a channel, 12. entrance of spring karren cave, 13. through karren cave, 14. shaft, 15. grike 16. griant grike

water (Trudgill 1985; Ford and Williams 2007). On the bare slopes of glaciokarst, they occur in a large (up to several 100 m^2), adjoining expansion (Fig. 4.2). Their varieties are rundkarren, decantation channels, type A and type B channels, rinnenkarren systems, type I, II and III channels (Ford and Williams 2007; Veress 2010; Veress et al. 2015) . The ridges between the channels of rundkarren became rounded by dissolution (Fig. 4.8). Rounded ridges prove that dissolution took place under the soil cover. Soil cover is temporary (Haserodt 1965; Veress 2010) or these features were always under soil and thus, they developed under the soil, but they exposed because of soil denudation (Bögli 1976). Decantation runnels are unique, solitary features that originate from point-like water supply places (for example

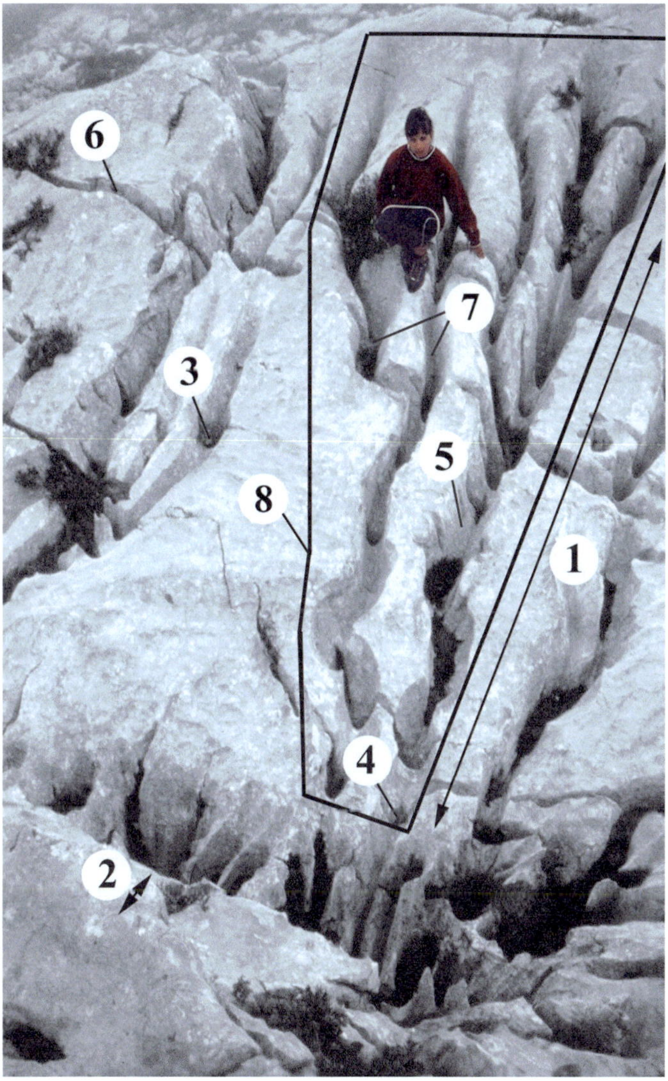

Fig. 4.7 The karren features of a part with bedding plane (Totes Gebirge, photo by Veress). 1. Bedding plane, 2. scarp front, 3. channel-bottom pit, 4. channel end pit, 5. main channel, 6. grike, 7. tributary channel, 8. channel system

kamenitza) (Fig. 4.9, Ford and Williams 2007). Type A channels are rinnenkarren with V-shaped cross section and a small size and have a small specific catchment area (the catchment area calculated to the 1-m length of the channel, which can be calculated from the quotient of the total catchment area of the channels and of the channel length) which have no tributary channels. Type B channels are karren features with a U-shaped cross section, they are larger and their specific catchment

Fig. 4.8 Semi-exhumated rounded rundkarren (Totes Gebirge, at Vd. Lahngang Lake, photo by Veress)

area is larger, which have tributary channels. Rinnenkarren systems are channel systems having type B main channels larger than the previous ones and having several type A and/or type B tributary channels (Veress et al. 2015, Fig. 4.7). The majority of rinnenkarren is given by type A and type B channels and rinnenkarren systems. According to their size, three channels can be distinguished: larger channels, being several tens of centimetres wide and deep (type I), smaller channels, the width of which is smaller than ten centimetres (type II) and even smaller channels which are some centimetres wide (type III) (Veress 2009a, Fig. 4.10).

Meanderkarren are channels with asymmetric cross section (Veress and Tóth 2004; Veress 2009b). The asymmetric cross section goes with the meandering ground plan of the feature only in the case of looping meanders (Fig. 4.11a). In case of other meanderkarren types the channel has a straight margin. The steep (often overhanging) and gentle slopes causing asymmetry alternate on the channel: the steep side is followed by a gentle one, then by a steep side again on the same side of the channel. Such karrenmeander types are the remnant meander (Fig. 4.11c), the developing meander (Fig. 4.11b) and the perishing meander (Veress and Tóth 2004; Veress 2009b). The asymmetric cross section is caused by the fact that the channel line of the water running down on the present or the former floor of karren meander was meandering or is meandering. The side of the feature will be steeper, where dissolution is more intensive. This occurs in case of faster water movement thus, where the channel line gets closer to the side wall during its meandering

Fig. 4.9 Kamenitza and decantation channel (photo by Veress). 1. Kamenitza, 2. decantation runnel, 3. channel with an opened up karren cave on its floor

(Veress 2009b). [According to Dubljanszkij (1987), faster flow maintains the concentration difference between the boundary layer and the flowing water and thus, it enables the transport of Ca ion.] The spread of karren meander is unique and they do not develop in a continuous way.

Wandkarren, (Fig. 4.12) are parallel channels with half-cylindrical or trapezoid cross section (Veress 2009c, 2010). They develop under the water that is flowing down the steep slope: wandkarren which are situated densely, are of smaller size,

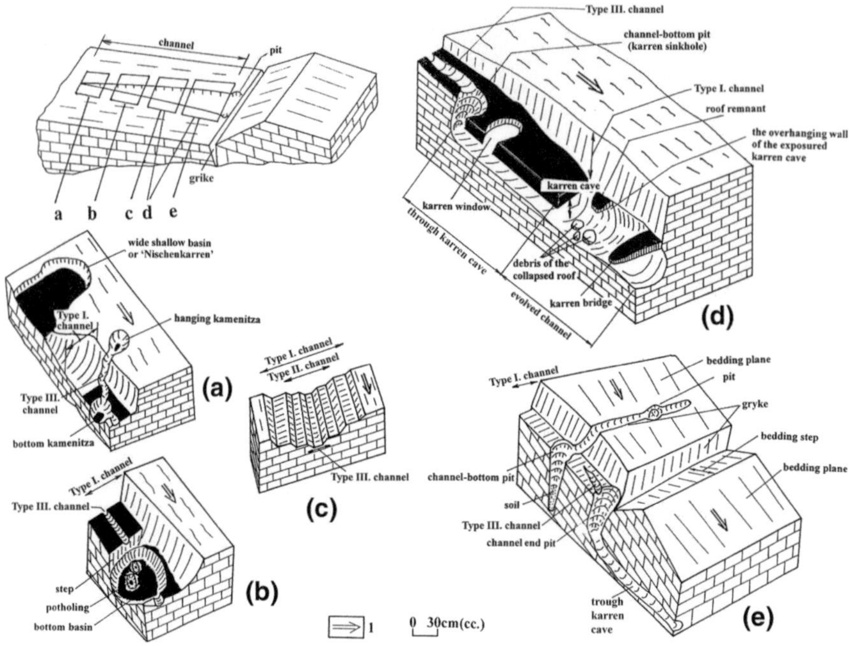

Fig. 4.10 The morphological complexes of different parts of a channel (Veress 2009a, modified). **a**, **b** Upper parts of the channel, **c**, **d** middle parts of the channel, and **e** lower part of the channel 1. line of dip of surface

having a half-pit cross section and being rillenkarren-like develop under sheet water, while those occurring less densely, having a larger size and trapezoid cross section and being rinnenkarren-like are formed under a rivulet (Veress 2009c, 2010). While their width and depth are between some centimetres and some tens of centimetres, their length is some metres or several ten metres. Wandkarren are not closed and not areic features. They are common karren features of steep slopes and develop in high densities and continuously. Their expansion is extremely variable at a given place. The area of the slope covered with wandkarren can be some m^2 or several hundred m^2.

Where the continuity of the rock is interrupted, for example at places where the surfaces being cut by ice are intersected by bedding planes (Fig. 3.34), small-sized passages with a diameter of some centimetres develop on the floors of rinnenkarren which drain the water of channels underground. These features are the karren channel swallets which continue in passages or in swallet karren caves (in karren caves) being some metres long (Veress 2010, Fig. 4.10). The percolating water also has a role in their development. Karren channel swallets convey the water flowing into the rinnenkarren into the karst. This water results in the development of karren channel caves first in a seepage way then in a flowing way. Karren channel swallets and swallet karren caves are unique and rarely occurring karren features.

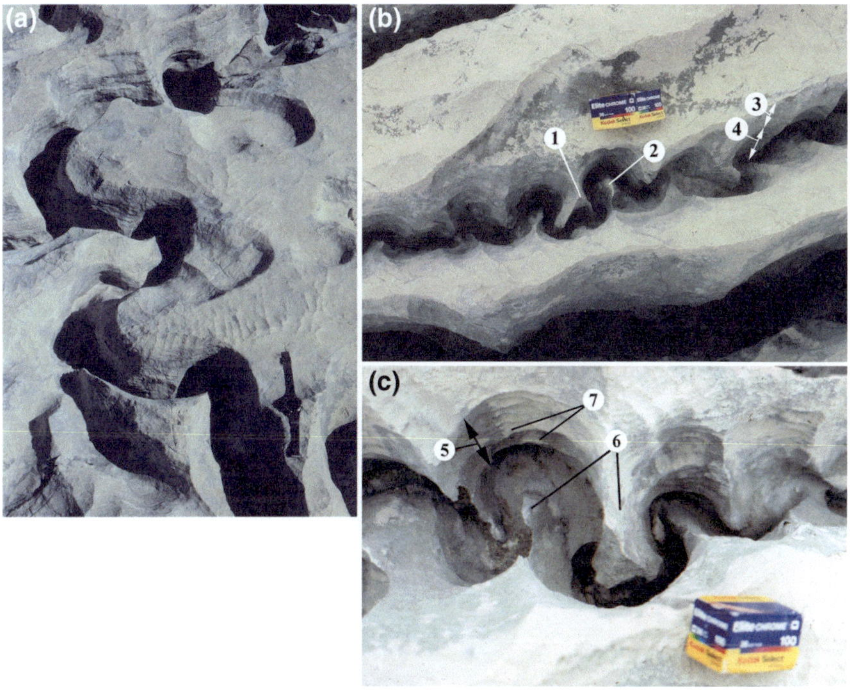

Fig. 4.11 Meanderkarren: loop meander (**a**) and developing meander (**b**), remnant meander (**c**) (photo by Veress). 1. Composite skirt, 2. partly detached skirt (karren "inselberg"), 3. symmetrical meanderkarren part (without skirt and overhanging wall), 4. asymmetrical meanderkarren part (with skirt and overhanging wall), 5. overhanging wall, 6. gently dip wall (skirt), 7. notch

Trittkarren are circular features with a diameter of some tens of centimetres (or maybe some metres), which are built up of two parts: the steeper riser with an arcuate pattern in plan view and the tread with a smaller inclination and near-circle ground plan (Bögli 1951; Veress 2009d, 2010).

The trittkarren of the increasingly steeper slopes have an increasingly higher riser, but their tread will be of increasingly smaller expansion (Fig. 4.13, Balogh 1998). The varieties of this karren feature are the typical trittkarren (Fig. 4.13b), the nischenkarren (Fig. 4.13a), the trittkarren uvala (Fig. 4.13b) and the trichterkarren (Haserodt 1965; Veress 2009d). The typical trittkarren has a tread and a riser. The common tread of trittkarren uvalas is surrounded by a riser built up of several arches. The wide tread of nischenkarren with an expansion of 1–2 m is surrounded by a wavy and arcuate riser with a small height (1–2 cm). The trichterkarren has a well-developed riser, but it does not have a tread. Trittkarren occur in patchy expansion only on a surface with bedding planes (the area of the patches is not large, only some m^2), however, typical trittkarren occur there in groups, in large density.

Fig. 4.12 Wandkarren (valley side under Tragl peak, Totes Gebirge, photo by Veress). 1. Wandkarren, 2. schichtfugenkarren

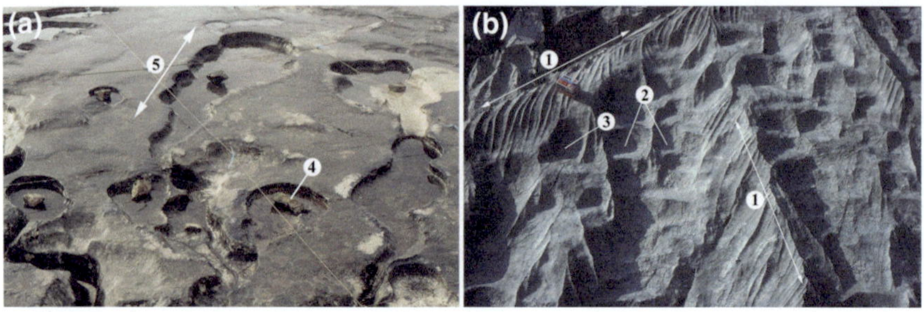

Fig. 4.13 Nischenkarren on a slope with a smaller dip (**a**), typical trittkarren on a slope with a larger dip (**b**) (photo by Veress). 1. Rillenkarren, 2. typical trittkarren, 3. trittkarren uvala, 4. typical trittkarren with a small riser, 5. Nischenkarren

Karren Features of Seepage Origin

Grikes (Fig. 4.5) are features with a width of some tens of centimetres, their depth is 1–2 dm and 1–2 m, their length is several 10 m, their side wall is vertical and their margin is straight. Typical grikes occur in groups, the distance between the depressions is 1–2 dm, and the depressions are parallel to each other. The

Fig. 4.14 Lattice-like karren (Totes Gebirge, photo by Veress)

expansion of terrains with grikes can also exceed several 100 m^2. One of their varieties is the lattice-like karren when the grikes develop in two directions being perpendicular to each other (Fig. 4.14). Grikes develop to the dissolution effect of the water percolating along fractures (Sweeting 1966; Goldie 2009), but they can also develop along vertical bedding planes. The directions of the grikes of bedding planes mainly coincide with the strike direction of bedding planes (Veress 2010, Fig. 4.7). Probably because the widening of fractures with strike direction is more intensive than that of fractures with dip direction. The reason for this is that the fractures with strike direction cross water flow and thus, they can get water, however fractures with dip direction do not cross water flow, water flows passing them thus, they get less water. Giant grikes which can be regarded as megakarren have several varieties according to their development, and therefore, this feature is presented in a separate chapter.

Kamenitzas are basin-like karren with a circular ground plan (Fig. 4.9). Kamenitzas with a diameter of several metres and even several ten metres occur for example on the Diego de Almagro Island (Veress et al. 2006). On the floor of large kamenitzas, other karren features with smaller size may develop too, such as karren channel swallet, rinnenkarren and trittkarren (Veress 1995; Veress et al. 2006). The density of kamenitzas is not large, they mostly occur alone, one by one with other karren features together.

Pitkarren are vertical tubes with a diameter of some centimetres and some tens of centimetres. Their vertical size can be some decimetres or some metres. They occur in groups (Fig. 4.15) or alone on the floor of other karren features (for example

Fig. 4.15 Pitkarren (Totes Gebirge, photo by Veress)

rinnenkarren) (Fig. 4.7). Their occurrence in groups is not common. The expansion of terrains with pitkarren is only of some m^2.

Schichtfugenkarren are features with a height of some centimetres or some tens of centimetres (Figs. 3.23, 3.29 and 4.12), which develop by dissolution along bedding planes. Their longitudinal expansion can be several 10 m. They most often occur with wandkarren together. Based on the number of bedding planes outcropping on the rock walls of valley sides, several schichtfugenkarren can also develop above each other. However, they can occur on shaft walls and grike walls too. On grike walls they were formed following soil development if the grikes are not inherited. Or they developed at the soil surface or under the soil surface where the water percolating laterally reaches the bedding planes of the rock.

Among the characteristic karren features of glaciokarst can also be mentioned the relict landforms such as arches (Veress 2010), relict ridges or tail-dune karren buttes on the leeside of rock blocks (Veress et al. 2006, Fig. 2.12) and karren tables (Bögli 1961). Arches are relict features above windows that develop with the coalescence of adjacent karren features. Tail-dune karren buttes are formed at those places, where the surface does not dissolve at the leeside of rock blocks (Veress et al. 2006; Maire et al. 2009). Karren tables develop at places where the surface dissolves everywhere around the rock blocks except under them (for example moraine) (Bögli 1961).

4.2.3.3 Karren Features of Glacial Erosional Surfaces and of Glacial Erosional Features

The development environment of karren features is given by the inclination and expansion of the bearing slopes. In Chap. 3, it has already been presented that the expansion of surfaces with bedding planes that developed by glacial erosion (and those of the scarp fronts too) depends on the inclination of the beds and on bed thickness. (Thus, also in the case of beds with the same inclination, their expansion in dip direction can be extremely different.) Since the development of some karren features also depends on the expansion of the slope (rinnenkarren, wandkarren), in case of the same inclination they can only develop if the bearing slope is wide enough in the dip direction. In the following, the development environment of different karren features are presented. The karren feature assemblage of the surfaces and features that are developed by glacial erosion can be described by this.

Rillenkarren occur on slopes with an inclination of about 10°–65°. Accordingly, they are mainly specific to both on bedding planes and on heads of bed, but they can also develop on roche moutonnées. Since they are also formed on slopes with an extremely small width, they can be present not only on glacial erosional features, but also on moraine blocks, on the blocks of falls and on the slopes of karren features (for example on the riser of trittkarren).

According to measurements carried out on slopes with bedding planes (Veress et al. 2015), type B channels occur on slopes with small inclination, smaller than 45°. Their density decreases with the increase of the inclination of the bearing slope. With the decrease of slope inclination, type B channels will be more complex and their total length increases too. In case of an inclination smaller than 20°–25° rinnenkarren systems become significant (Veress et al. 2015). The density of type A channels increases with the growth of slope inclination until 60° and their density exceeds the density of type B channels at an inclination of 30° (Veress et al. 2015). However, they do not occur on slopes with an inclination larger than 60° at all. Rinnenkarren mainly develop on the bedding planes of troughs (Fig. 4.2). Among them, type B channels develop on the bedding planes of beds with a smaller dip if the expansion of these surfaces is several metres in dip direction (5–10 m or even larger). Rinnenkarren systems develop in dip direction on bedding plane slopes with a width of several 10 m and with a small dip. Type A channels can also develop on bedding planes with a width of some metres and small expansion situated on beds with a larger dip. The larger dip of the beds is and the thicker the beds are, the bigger the chance is that type A channels develop on scarp fronts (on heads of bed) too. Rinnenkarren systems can also develop on the gentle side slopes of giant dolines, but they can also be formed on the upstream slopes of roche moutonnées (however, type B channels can be present too). Type B channels and rinnenkarren systems develop on such surfaces where thick beds were cut by glacial erosion (Fig. 3.34) since between the bedding planes there may be relatively wide surfaces where water flow is continuous too.

Meanderkarren develop on slopes with a dip smaller than 10° and thus, they occur on the bedding planes of almost horizontal beds. However, they can develop on unstratified rock (for example on the marble karst of the Diego de Almagro Island), where glacial erosion did not create stepped terrains, but it formed plain surfaces of small inclination.

Trittkarren are widespread on very smooth surfaces with an inclination of 0°–40°. Their type and morphology changes with the increasing dip. Especially on a horizontal slope (the dip of which is smaller than 5°–10°), if it is widespread (several metres), nischenkarren develop. With an increasing dip of the bearing terrain typical trittkarren develop, the riser of which increases with the increasing inclination of the bearing slope. The frequency of trittkarren groups decreases on slopes with a larger inclination. Trittkarren occur on the small-inclined bedding planes of troughs, on Flachkarren, on corrosion terraces and on the upstream sides of roche moutonnées.

On slopes with an inclination larger than 70°, wandkarren are formed. According to our measurements, they can occur in large densities even on slopes with a dip of 85°. Wandkarren are widespread on transverse steps, on longitudinal scarps, on the side slopes of rock basins, on the side walls with the head of bed of shaft walls, on moraine blocks, but mainly on the side slopes of glacier valleys. Wandkarren often develop on heads of the bed, which are surrounded by Schichtfugenkarren mainly if the beds are thick (2–4 m).

Grikes mainly develop on horizontal surfaces. Lattice-like karren are exclusively characteristic of slopes with an inclination smaller than 10°. With the increase of slope inclination, their density decreases, but grikes even occur on slopes with an inclination of 31°. Grikes being formed along fractures develop on bedding plane surfaces of beds with small dip and being not so thick, while those grikes which are formed along bedding planes develop on the surfaces of hogbacks constituted by beds with a large inclination (close to 90°). Therefore, grikes are specific of the floor of troughs where the beds are horizontal or of small dip (on the bedding planes of stepped surfaces) or they are almost vertical (on the heads of bed of hogbacks).

Solitary pits or those forming groups develop on slopes with a small dip (at most 30°). First of all, they develop on bedding planes with a small dip mainly on horizontal bedding planes of troughs.

Kamenitzas occur on the small-inclined slopes of troughs (mainly on slopes with an inclination of 0°–10°, at most 38°). Thus, they can be found on the bedding planes of beds with a small dip, at the roof level of roche moutonnées, or rarely on the downstream slopes of roche moutonnées (where the surface is less uneven). However, they can also appear on the flat terrains between roche moutonnées like on the Diego de Almagro Island (Veress et al. 2006).

The density and the frequency of occurrence of karren features of seepage origin decreases with the increase of slope inclination. However, with the exception of lattice-like karren, they increase on slopes with an inclination of 21°–30° as compared to slopes with a dip of 11°–20°. A probable cause of the local increase of density is that the fractures of the rock (their maturity and density) may modify the role of slope inclination in their development.

Schichtfugenkarren develop on slopes with an inclination of 70°–85°, where the dip direction of the beds is opposite the dip direction of the steep (but not vertical) slope that developed by glacial erosion. Mainly, if the beds are thick and wandkarren are aligned on the heads of bed. Such places are the slopes of troughs, rock basins and transverse steps.

We have already mentioned that among surface landforms being created by ice, glacier valleys, stepped surfaces and roche moutonnées show karren formation heterogeneity. The most different karren formation (karren features are different) can be experienced between the side slopes of troughs and their floor. On their steep-side slopes, wandkarren and schichtfugenkarren are characteristic. The karren features of the stepped surfaces of more gentle side slopes and of valley floors are significantly different from the karren features of steeper side slopes. Thus, rinnenkarren, meanderkarren and kamenitzas are predominant. Though the floor of the valley (especially that of the trough) is in itself complex: stepped surfaces, steps, rock basins, roche moutonnées, moraine (mountains falls) blocks and dolines may alternate. Therefore, individual erosional features (stepped surfaces, roche moutonnées) show karren heterogeneity in themselves too.

On the bedding planes of stepped surfaces, a variety of karren features (Figs. 4.2 and 4.16) can be present, karren assemblages according to environment are widespread, karren cells and karren formation units are common (see below). On the contrary, few karren feature types develop on scarp fronts (wandkarren and schichtfugenkarren), their size is small, their density is only locally large, their distribution is not continuous. On scarp fronts, karren do not constitute karren cells.

Since roche moutonnées are positive features, their karren formation is unique and specific. Karren features often enclose the top of roche moutonnées and if they are of flow origin, they are divided into branches radially mainly from the soil or vegetation patches being formed at the roof level (Veress 2012a). This structure proves the dissolution effect of soil and vegetation patches on bare surfaces. The karren formation of the upstream slopes of roche moutonnées (which is a plain slope with small inclination) resembles the karren formation of slopes with bedding planes to a great extent. At these places, features of flow origin such as rinnenkarren are common (Fig. 4.17a). However, the size and/or density of the features are smaller than that of the surfaces with bedding planes. On the lee side (the downstream slope) which is short and often dissected, few karren feature types occur (kamenitza and grike) and their density and size are small (Fig. 4.17b).

The karren formation of other glacial erosional features or features transformed by glacial erosion is less significant, it is discontinuous and usually, only some karren features are present. The side slopes of giant dolines are often covered by debris thus, either no karren developed on them at all or if they did, they become covered. If karren develop on the side slopes of giant dolines, they are mainly of flow origin. Thus, rinnenkarren develop which converge towards the direction of the doline floor (Fig. 4.18). The karren features of the blocks of moraines or of mountain failures occur sporadically, their karren formation is not continuous, it is dispersed (Fig. 4.19).

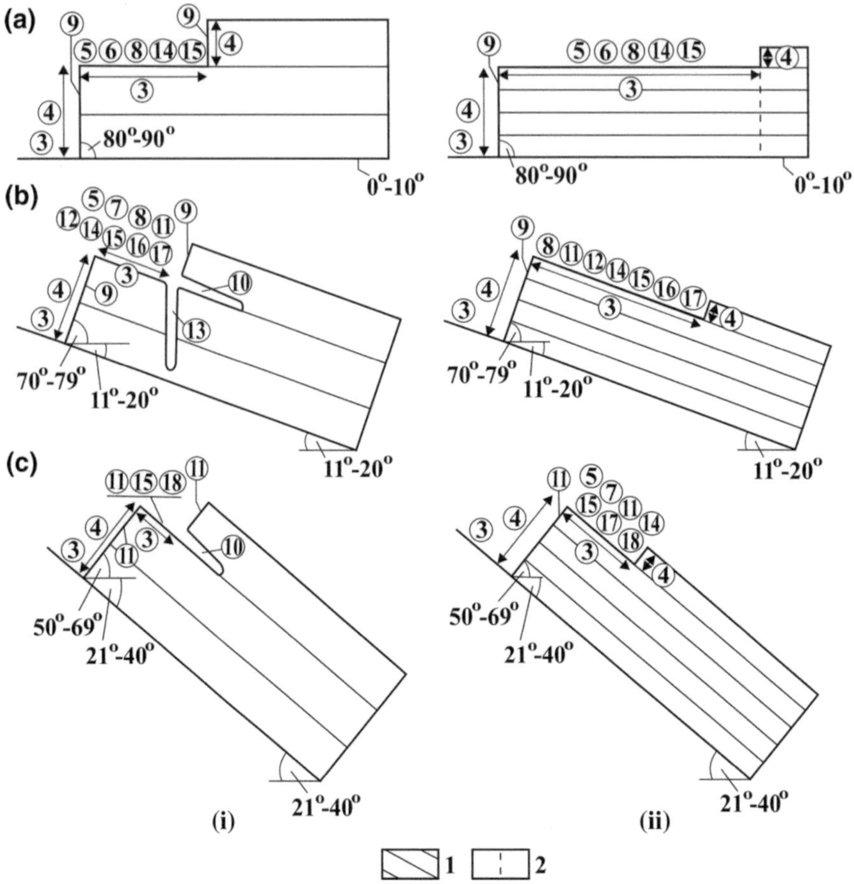

Fig. 4.16 Varieties and karren assemblages of stepped surfaces. **i** Bedding planes and scarp fronts that developed in case of large bed thickness and very small (**a**), less small (**b**) and medium (**c**) bed inclination, **ii** bedding planes and scarp fronts that developed in case of small bed thickness and very small (**a**), less small (**b**) and medium (**c**) bed inclination. 1. Bed, 2. fracture, 3. bedding plane, 4. scarp front, 5. grike, 6. nischenkarren, 7. typical trittkarren, 8. karrenmeander, 9. wandkarren, 10. Schichtfugenkarren, 11. rillenkarren, 12. „Ausgleichsfläche". 13. giant grike, 14. kamenitza, 15. rinnenkarren system, 16. type B channel, 17. type A channel

4.2.4 *Karren Assemblages*

On karren terrains, systems built up of various karren features are called karren assemblages. Bögli (1976) was the first to describe karren assemblages. Karren can constitute functional karren assemblages and karren assemblages according to the development environment. In case of the former ones, those features constitute the karren assemblage where water flows (or can flow) from one feature into the other. In case of karren assemblages of development environment, those features create the karren assemblage which occurs together on slopes with a similar inclination.

Fig. 4.17 Karren formation of a roche moutonnée (valley under Hetzkogel, Totes Gebirge, photo by Veress). **a** karren features of the upstream side, **b** karren features of the lee side; in photo A (top left) the upstream side of a roche moutonnée, (bottom right) the lee side of another roche moutonnée

Fig. 4.18 Karren features of the side slope of a giant doline: rinnenkarren systems are predominant (Totes Gebirge, photo by Veress)

Fig. 4.19 Wandkarren of a moraine block (near Simonyi Hütte in the valley of Hallstatt glacier, Dachstein, photo by Veress)

Within functional karren assemblages, Veress (2010) differentiated grike-rinnenkarren assemblages, rinnenkarren-grike assemblages, rinnenkarren-pit assemblages, rinnenkarren-giant grike assemblages, wandkarren-schichtfugenkarren assemblages, pit-grike assemblages and grike-schichtfugenkarren assemblages. In the case of grike-rinnenkarren assemblage, first rinnenkarren were formed, then grikes developed since grikes intersect the majority of rinnenkarren. However, in the case of rinnenkarren-grike assemblage, first grikes developed since rinnenkarren terminate at grikes and they do not continue over them. In the case of rinnenkarren-giant grike assemblage, giant grikes may occur inside the same terrain with bedding planes or at the foot of the step bordering the bedding plane. If the giant grike is situated inside a given bedding plane, there is often a small step (karren step) of 1–2 dm at the margin of the grike which falls in the dip direction of the surface that is inclined in the direction of the joining rinnenkarren. Its sizes (either the vertical or horizontal sizes) are significantly smaller than the sizes of the steps of stepped surfaces. The pit-grike assemblage exclusively occurs at the foot of stepped surfaces.

According to the development environment, on glacial erosional features and on their slopes the following karren assemblages can be differentiated (Fig. 4.16).

Among surfaces with a small dip (0°–20°) some surfaces with a dip of 0°–10° and with a relatively larger expansion are Flachkarren where nischenkarren, kamenitzas, meanderkarren and grikes can constitute karren assemblages on plain surfaces with a small dip. Such karren assemblages occur on those parts of troughs, where the dip of the bed is smaller than 10°. The karren formation of these surfaces can often be of two kinds. In the case of one kind, meanderkarren, various trittkarren (mainly nischenkarren) and rinnenkarren being wide as compared to their depth are predominant. This variety develops if water percolation is restrained on the bedding plane, there are no fractures (or their frequency is low), the precipitations of the former meltwater may create a crust on the surface of the bedding plane. In the case of the other variety, fracture density is large in the rock and in this case mainly grikes and lattice-like karren develop.

On surfaces with an inclination of 11°–20° and with an expansion of several metres or even several tens of metres in slope direction, rillenkarren, rinnenkarren systems, typical trittkarren, meanderkarren, kamenitzas, pitkarren and grikes can constitute the karren assemblages. These karren assemblages are widespread on the terrains with bedding planes of troughs, where the dip of the bed is 11°–20° and on the side slopes of paleodolines and roche moutonnées with such an inclination.

On surfaces with an inclination of 21°–40° and with an expansion of several metres or even several tens of metres in slope direction, rillenkarren, type B channels or maybe rinnenkarren systems, type A channels, grikes, trittkarren and kamenitzas can constitute the karren assemblages. (All types of the above-mentioned karren features are very rarely present together.) These karren assemblages occur on those terrains with bedding planes of troughs where the dip of the bed is 21°–40° and on those side slopes of paleodolines and roche moutonnées where the surface inclination is 21°–40°.

On surfaces with an inclination of 41°–60° (70°) and with an expansion of some metres, rillenkarren and type A channels can constitute the karren assemblages. These surfaces are the terrains with bedding planes of troughs where the dip of the bed is 41°–60° and those heads of the bed where the dip of the bed is 30°–49°, but also the slopes of those paleodolines and roche moutonnées where such an inclination occur.

On surfaces with a dip of 71°–90°, wandkarren and schichtfugenkarren can constitute the karren assemblages. Such slopes can constitute the side slopes of glacier valleys, the side slopes of steps and rock basins, the shaft walls and the walls of giant grikes.

Karren features are the most diverse on slopes with an inclination of 11°–20° and 21°–40°. Its significance is expressed by the fact that the existence, growth and coalescence of many karren features favour the denudation of limestone beds. Rinnenkarren systems (mainly on slopes with an inclination of 11°–20°), type B channels and grikes occur on these slopes. Thus, on slopes belonging to the above-mentioned categories, there is a possibility for the formation of grike-rinnenkarren, rinnenkarren-grike and rinnenkarren-giant grike assemblages. Mainly, these karren assemblages and grike-schichtfugenkarren assemblages (or the karren constituting them) contribute to the denudation of limestone.

4.2.5 Karren Cells

A karren cell is a morphologically closed, areic karren terrain of a surface with bedding planes (Veress 2010). Karstic water drainage takes place from its area, usually through pits. On a surface with bedding planes, several karren cells can be present which are separated from each other hydrologically and their karren features can be the same, but different too. On the area of karren cells, most often rinnenkarren, grikes and pitkarren occur (rinnenkarren-pitkarren and rinnenkarren-grike assemblage).

4.2.6 Karren Formation Units

By a karren formation unit, we mean the karren surface of a slope (Fig. 4.20) where rinnenkarren and grikes are predominant, but several other karren features may occur too (kamenitzas, rillenkarren, meanderkarren, trittkarren, pitkarren). However, not only karren features are present on the karren formation unit, but also giant grikes, schachtdolines and shafts too. On glaciokarst, they mostly developed on the bedding planes of stepped surfaces (Fig. 4.20). Their expansion coincides with the expansion of the bearing surface with bedding planes. They have a large expansion if the given bedding plane has a large expansion too. This is valid if small-inclined terrains built up of thin beds were destroyed by glacial erosion (see Chap. 3). The karren formation unit is dissected by karren cells of different maturities and expansions and karren assemblages may be repeated in their area where several water drainage places can exist.

4.2.7 The Degree of Karren Formation of Glaciokarstic Surfaces

The degree of karren formation of a terrain can be determined by the density and specific width of karren features occurring there (Veress 2010). The specific width is the average width of the measured karren features on a given length (along a profile, the selection of which takes places accidentally) calculated to 1 m, which can be determined if we divide the total width of the karren features occurring along the profile by the length of the profile.

In order to determine specific width data, the necessary profiles were created mainly on the surfaces with bedding planes of the glaciokarstic areas of the Alps (in the pine tree zone, in the Pinus mugo zone and on bare surface) (Veress 2010). The data represent the karren formation of surfaces with bedding planes. Both in the pine tree zone and the Pinus mugo zone, there were profiles with soil and vegetation patches and there were profiles which lacked them. However, this only reflects the

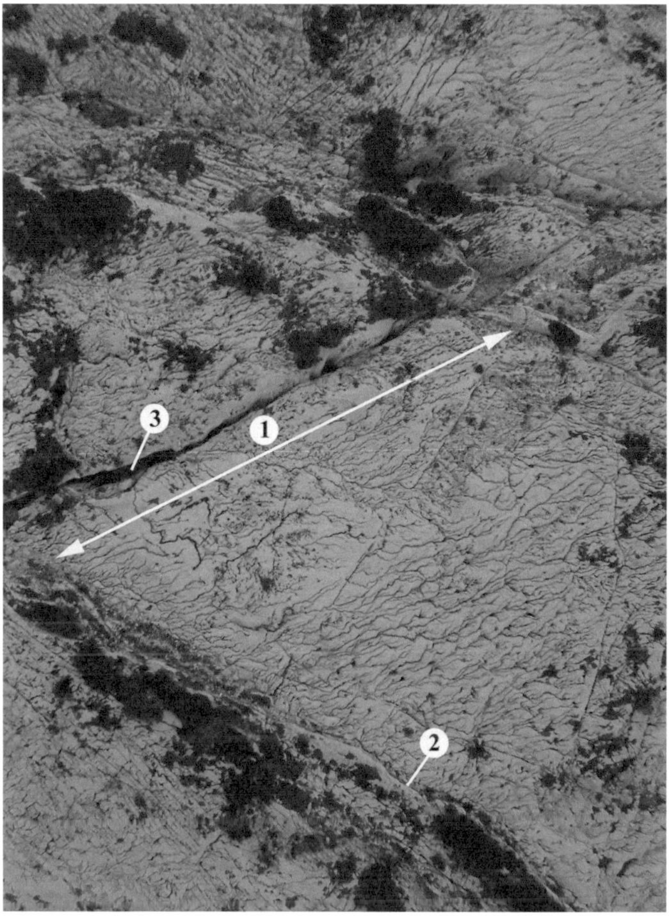

Fig. 4.20 A karren formation unit dominated by rinnenkarren systems (valley under Tragl peak, Totes Gebirge, aerial photograph, photo by Kalmár). 1. Bedding plane, 2. scarp front, 3. giant grike

momentary state at the measurement places. Since the initiation of karren formation, vegetation may have developed on the current bare slope (but the vegetation patch could be destroyed), while the current bedding plane with vegetation patches could also be bare at the beginning of karren formation.

However, in order to determine the actual (complete) karren formation characteristic features, the data of the terrain parts covered with soil or moraine should also be taken into account. But the data of the karren features of covered terrain parts are unknown. Thus, the data on the degree of karren formation described below only contain the data of bare surfaces. This does not contain the data of karren features occurring on all bare surfaces either. No measurements were carried out in case of small karren features (rillenkarren), those hard to access (wandkarren, schichtfugenkarren) and in case of karren features which occur very rarely (karren cave). However, because of inheritances, the values of dissolution (represented by specific widths) could not only develop under the present environmental circumstances.

According to Veress (2010), the density of karren features decreases by the increase of altitude. In the Alps, the average density along profile is 0.41 piece/m on the bare surfaces of the pine tree zone, 0.37 piece/m on the bare surfaces of the dwarf pine tree zone (Pinus mugo) and 0.26 piece/m in the vegetation free zone (partly the area of alpine meadow). If we consider the specific widths, it can be established that two kinds of karren features are predominant: rinnenkarren and grikes (Fig. 4.21, Veress 2010). The specific width of rinnenkarren is larger in the Pinus mugo zone, while the value of grikes is larger in the pine tree zone (Fig. 4.22). Therefore, grikes (or karren formation represented by them) are more dominant in the pine tree zone. Here, grikes developed by dissolution under the soil (disregarding those which were inherited from glacial and under soil cover they only continued their development) then they were exposed at the surface following the denudation of the soil. This is also proved by the fact that soil and vegetation are always present on the floor of the grikes of this zone. In the pine tree zone, the development of rinnenkarren was also possible preceding soil formation

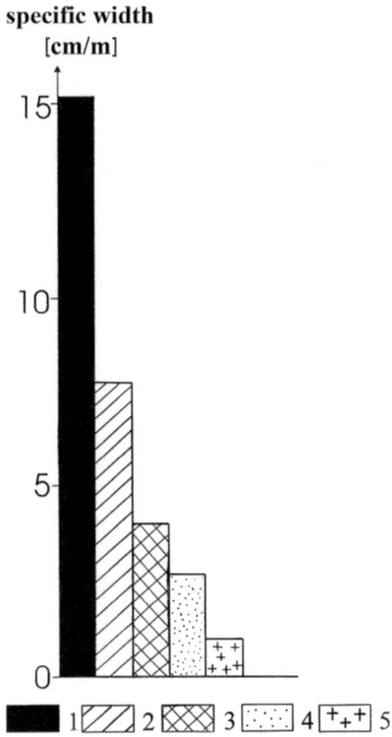

Fig. 4.21 The degree of specific width which refers to different karren forms by using data of profiles of valley of Triglav Lakes, Totes Gebirge, and Dachstein (Veress 2010, modified). 1. Rinnenkarren, 2. grike, 3. pitkarren, 4. trittkarren, 5. kamenitzas

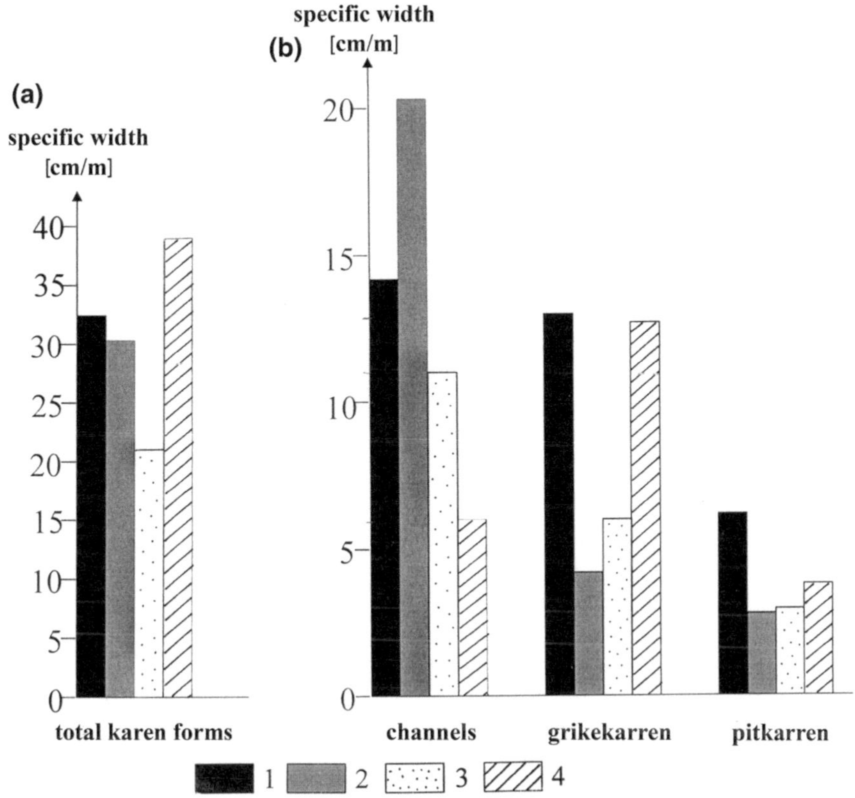

Fig. 4.22 Specific width with total value (**a**) and values belonging to various kinds of karren forms (**b**) in various plant belts (Veress 2010, modified). 1. Pine belt, 2. *Pinus mugo* belt, 3. bare belt (1–2–3 valley of Triglav Lakes, Totes Gebirge, Dachstein), and 4. *Pinus mugo* belt (Asiago Plateau)

(rinnenkarren that developed at this time are the present rundkarren). However, they could also develop on bare surface uncovered with soil or when bare surfaces developed because of soil denudation (which can be caused by human activity too). (From the glacial, only those rinnenkarren could be inherited which existed during the disappearance of ice cover. Those that had developed earlier were destroyed during glacial erosion because of their small depth.) In the Pinus mugo zone, the larger specific width of rinnenkarren is possible because here bare surfaces permanently survived after the retreat of ice where there were favourable conditions of water flow. At the same time, bare surfaces can also get CO_2 of soil origin from soil and vegetation patches, which can increase the intensity of dissolution (Veress 2012b).

4.3 Giant Grikes

The depth and width (mainly the depth) of giant grikes can exceed several metres, while their length can exceed even several hundred m (Fig. 4.23). They are often inherited to the cover too (Fig. 4.24). Giant grikes are common features on other karst areas, mainly on tropical karst (Grimes 2012; Ford and Williams 2007). However, they do not often occur on glaciokarst and in contrast with tropical karsts where they occur in groups, they are solitary forms. They are widespread both in the glacial and periglacial zone of glaciokarst. Ford (1984) mentions these features as solution corridors. On Nahanni karst (Canada), since they occur in large numbers, labyrinth karst develops (Brook and Ford 1978). They are described in a separate chapter as their development differs from grikes.

Grike walls are often divergent downwards. Various mezokarren features can occur on the walls (rillenkarren, wandkarren Fig. 4.4c). Those with a smaller (some tens of centimetres) width can be dissected by pits (Fig. 4.7) and shafts (Fig. 4.25). In case of grikes with pits, the pits are coalesced to a various extents, rinnenkarren are connected to them from the surrounding terrain with bedding planes. Giant grikes can develop on the floor of cirques (Fig. 4.24a) or on the floor of troughs (Fig. 4.24b), in karstic depression (Fig. 4.24c) at the scarp fronts of stepped surfaces (Fig. 4.26), but on bedding planes (Figs. 4.2 and 4.25) too. The longer ones can intersect a series of bedding planes and scarp fronts (Fig. 4.24b).

According to their position, they can be longitudinal and transverse. The longitudinal ones are parallel to the direction of the trough (Figs. 2.14 and 4.25) or to the dip direction of the bedding planes (Fig. 4.25), while the transverse ones are

Fig. 4.23 Giant grikes (Totes Gebirge, **a**, **b**, photo by Veress)

Fig. 4.24 Giant grikes with various morphological environment inherited over the superficial deposit (photo by Veress). **a** In a cirque (cirque under Triglav), **b** in a trough (Todorov valley, Durmitor), **c** in a giant doline (Totes Gebirge). 1. Anti-dip step (roche moutonnée), 2. covered karst of a cirque, 3. debris cone, 4. giant grike (covered with debris in photos **a**, **b**), 5. shafts, 6. suffosion dolines

perpendicular to the direction of the valley (Fig. 4.24b) and to the dip direction of the bed. Giant grikes are mostly straight in plan view, but those located along scarp front can be winding or saw-toothed. According to their cross section, there are symmetric and asymmetric giant grikes. In the case of the latter ones, which are mainly characteristic of those situated at scarp fronts, the side with the head of bed is longer than the slope on the side from the bedding plane. According to Ford (1984), giant grikes developed by subglacial dissolution. According to development, their main varieties are the following:

- Coalescing giant grike: it develops with the dissolution of the dividing walls of adjacent grikes or pits.
- Connecting giant grike: grikes connecting in dip direction create a larger feature (Fig. 4.26). Such giant grikes are formed in the area of bedding planes. Their development is enabled by the confluencing water of grikes.
- Polygenetic giant grike: the grike is wide, often partially filled with debris (Fig. 4.27). On the floor covered with debris, shafts and subsidence dolines may occur, on the scarp fronts there may be opened up (partially destroyed) cavity remnants. The grikes of this type probably developed by the coalescence of pits or maybe of shafts and then they widened by the frost weathering denudation of

Fig. 4.25 Giant grike with a shaft (valley under Tragl peak, Totes Gebirge, photo by Veress)

the scarp front. The debris is partially transported into the karst in which the fact also has a role that the channels (rinnenkarren) of the slopes with bedding planes transport water to the grike.

- Corridor on the dividing wall: corridors formed on the dividing wall between dolines may also belong to the group of giant grikes (Fig. 3.19). Their development is also complex and takes places to the effect of several factors (see Chap. 3.1.2).
- Meltwater giant grike: grikes belonging to this variety develop on the floor of troughs. Their development under present circumstances is not possible since they do not receive enough water for this. Therefore, they were formed to the effect of meltwater under the ice (Ford 1984).

4.4 Shafts

Transitional features can be put among surface and subsurface features of glaciokarst. Their common characteristic feature is that they have a sharp margin, they do not have a catchment area and they are often in a perched position as compared to their environment (Figs. 4.25 and 4.28). They have a vertical position and extend

Fig. 4.26 Giant grike that developed by the confluence of the water of grikes (Julian Alps, valley of Triglav Lakes, photo by Veress)

deep into the vadose zone. They are specific of the glacial and periglacial zone, but they may occur in the fluvial zone of the temperate belt too, and also on other karst types. The development of the shafts of the karst is explained by the dissolution effect of the sheet water which takes place on fault planes. This sheet water is characterized by slow saturation and solution capacity affecting large depths (White 1988).

Both their diameters and depths are extremely varied. Their width ranges between some metres and several ten metres, while their depth can be several 10 m

Fig. 4.27 Giant grike at scarp front (Totes Gebirge, photo by Veress). 1. Subsidence doline

or several 100 m. It is very difficult to draw a sharp line between schachtdolines and shafts, but while the depth and diameter of schachtdolines are almost the same, in case of shafts the depth can be hundred times larger than the diameter. Schachtdolines and shafts can occur together too (Fig. 4.29).

Their cross section can be elongated (Fig. 4.30) and circular (Fig. 4.31). According to their longitudinal section, they are simple (there is only one shaft) and complex (they are separated into part-shafts). Their sidewalls may be dissected by giant grikes (Fig. 4.25), by schichtfugenkarren (Fig. 4.31) and by wandkarren

Fig. 4.28 Shaft (valley under Tragl peak, Totes Gebirge, photo by Veress)

Fig. 4.29 Shaft (1) and schachtdoline (2) (valley under Tragl peak, photo by Veress)

Fig. 4.30 Shaft row (Totes Gebirge, photo by Veress)

(Fig. 4.32). They occur both in cirques (Fig. 4.24a) and troughs (Fig. 4.25), individually or in groups. In case of a grouped occurrence, they are often arranged in rows. (Fig. 4.30). Especially, these latter occur with giant grikes together (they are the local widenings of giant grikes). They often occur at the stem of the scarp fronts of stepped surfaces creating shaft rows too. They may occur on elevations (Ford 1979) indicating their subglacial origin.

Shafts develop by dissolution among fractures. According to this, they are not only arranged into rows, but the direction of the row, the direction of the elongatedness of each shaft coincides with the direction of that fault plane along which

Fig. 4.31 Shaft with a circular cross section with schichtfugenkarren on its walls (Totes Gebirge, photo by Veress)

they developed. Shafts were mainly formed in subglacial environment and are of glacial age. Shaft development is favoured by the water film developing on shaft walls which becomes saturated slowly and thus, it preserves its dissolution capacity in large depths too (White 1988). The development of slow water percolation with low discharge must have had favourable conditions in the subglacial environment, when the karst water level subsided below the ice.

However, it is probable that the shafts, the width of which hardly differs from the width of the bearing grike (Fig. 4.30), are of postglacial age and they developed in the periglacial zone by snowmelt. Shafts of glacial origin can develop by the dissolution effect of the meltwater of ice (the meltwater can exert its effect under the glacier or in the foreground of ice) (Corbel 1957; Ford 1984; Ford and Williams 1989; Stepišnik et al. 2010), or by the erosion of the meltwater of ice (Ford 1979) and by the glacial erosional destruction of the rock above the cavities (Ford 1977; Kunaver 1983, 1996; Klimchouk et al. 2006), or by the postglacial mass movement of the rock above the cavities (Klimchouk et al. 2006). In their formation and further development, snow may also have an effect. The meltwater of the snow accumulating on the floor of the shafts causes permanent dissolution (Veress et al. 1996). Although such shafts are of glacial age, their postglacial growth survives. Postglacial shaft development is favoured by fractures that developed during tensional tectonical regime, where dissolution can penetrate into the depth below the surface (Ford and Williams 2007). In postglacial shaft development, the dissolution by condensing water of the air that circulates towards the surface also has a role (Klimchouk 1995).

4.5 Karstic Depressions

Karstic depressions are closed, areic features, they are the most typical features of karst areas. On glaciokarst, depressions being older than the given glacier can be of four types as compared to the turnover of glacier ice: ice receiving (the ice flowing into the depression does not leave it), ice storing (in this case the ice developing in them stays there), conveying (glacier ice enters the depression then leaves it), overflowing (ice is formed in the depression, but it flows out of it). The more elevated and the smaller the depression is, the greater the chance for the overflowing of the ice is. The lower position it has and the larger the depression is, the larger the chance is to receive ice. In case of large glaciers and in the case of depressions with a higher elevation the phenomenon of through-flow develops.

The researchers' opinion is different on whether depressions survive glacial erosion. According to Smart (2004), superficial karst features of karst are completely destroyed by denudation or infilling. However, Kunaver (1983) states that even small-sized features (for example schachtdolines) survive glacial erosion, though truncated. Ford (1996) emphasizes the infilling of karstic depressions. We think that in case of warm-based glaciers, where ice is also present in the inner part of the feature they are not denuded since glacial erosion affects their inner part too (they deepen), only those are destroyed which have a small size thus, the ice cannot enter them. (In case of cold-based glaciers, karst features survive in any case in lack of glacial erosion.) Such features are karren, but probably giant grikes, schachtdolines and shafts too. The surface between karren (but also in the case of the other above-mentioned features) is destroyed by glacial erosion except for their floor (Fig. 4.33). However, if meltwater has a solution capacity, even karren can survive

Fig. 4.32 Wandkarren on shaft wall (valley under Tragl peak, photo by Veress)

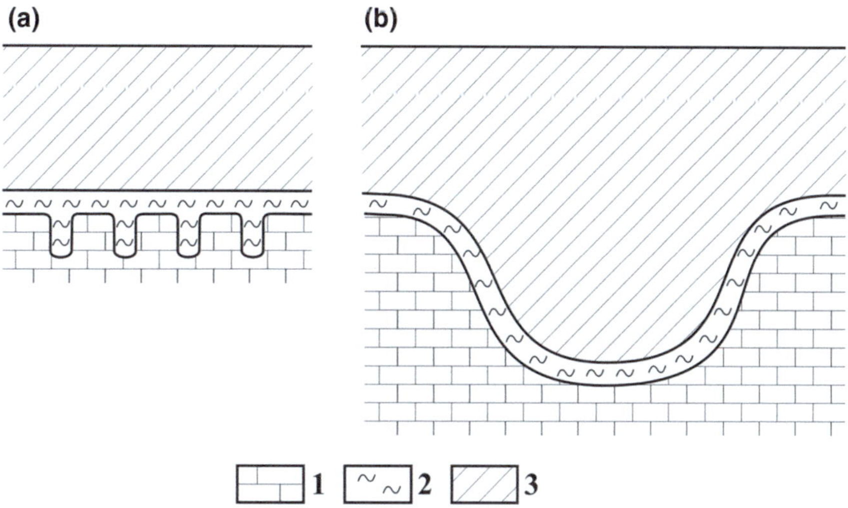

Fig. 4.33 Relation between the nature of glacial erosion and the size of the karren feature. **a** Only the surface between the features is destroyed since the ice is not able to penetrate into the feature, **b** the inner part of the feature is destroyed too as ice is also present there. 1. Limestone, 2. meltwater, 3. ice

and redevelop during dissolution (see Chap. 3). Larger sized karst features can only be truncated. This occurs where although there is ice in the depression, it does not move and thus, there is no glacial erosion (see Chap. 3). At places where the degree of glacial erosion is the same inside and outside the depression, the depth of the depression will stay the same.

The degree of truncation is also affected by the degree of denudation outside the depression since there is a relationship between the superficial elevation difference of glaciokarst and its environment and the intensity of glacial erosion. It is smaller where the former is smaller too (Djurovič et al. 2010). However, elevation difference is also small on plateaus, the inclination of the glacier surface too thus, glacial erosion will be small or smaller too. It is extremely small where the ice cap is smaller and does not reach the margin of the plateau. Thus, on a karst with a small dip and a small height as compared to its environment, even the environment of depressions is less denuded. Therefore, the lack of depressions of preglacial or interglacial age on glaciokarst can probably be explained by the fact that they did not develop at all. However, also depressions affect the destruction of ice. According to Stepišnik et al. (2010), karstic depressions reduce glacial erosion.

Depressions can be categorized according to their type (doline, uvala, ponor, polje), and development age (preglacial, interglacial, postglacial). They can be described by their size, morphology and morphological environment. The following outline is based on their types.

4.5.1 Dolines and Uvalas

The density of dolines and uvalas is smaller than that of karren. They mainly occur on surfaces with a small inclination. Independent of their type, the proportion of the area with dolines is the largest on surfaces with a dip of 2°–7° in the Mecsek Mountains (Hungary) (Lippmann et al. 2008). In the Miroč Mountains (Telbisz et al. 2007) 85% of the surfaces with a dip of less than 12° are areas with dolines, while only 23% of the surfaces with a larger inclination bear them. Because of the above-mentioned facts, these features mostly occur on the floor of cirques and of long troughs with a small dip. They are absent on the non-areic valley floors with a large dip of cliff glaciers and on slopes bordering plateaus.

Dolines can be of solution origin, collapse origin, caprock origin and subsidence origin (Waltham and Fookes 2003; Williams 2004, Fig. 4.34). Solution dolines develop by dissolution taking place on the surface of bedrock or in the epikarstic zone, while collapse dolines are formed with the cave-in of cavities (Jakucs 1977; Williams 2004; Ford and Williams 2007). On glaciokarst, the number of collapse dolines and caprock dolines is small. Collapse dolines are mentioned from the Canadian Rocky Mountains (Ford 1979), the Biokovo Mountains (Telbisz et al. 2005) and Hochschwab (Plan and Decker 2006). It has to be noted that some dolines described in the literature are not collapsed dolines but schachtdolines.

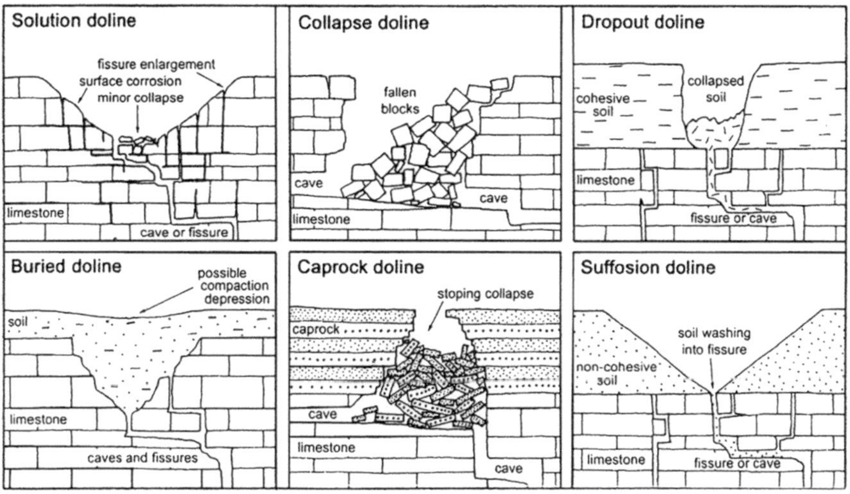

Fig. 4.34 Doline types (after Waltham and Fookes 2003, modified)

Thus, Miotke (1968) mentions collapse dolines from the Picos de Europa. However, in the figure presented by him, it can be seen that the feature termed as collapse doline is a schachtdoline.

4.5.1.1 Solution Dolines and Uvalas

These features are easily distinguishable from subsidence dolines since they subside into the karstic rock and thus, their side slopes are built up of limestone. However, the karstic rock does not always crop out on their slopes either because they are covered by soil or they became covered by sediment subsequently. For their development, the following conditions are necessary:

- At the time of their formation karst water level should not be at the level of the surface or close to the surface.
- In case of features of preglacial and interglacial age, a bare or at the most soil-covered karstic surface should exist preceding glacial erosion.
- Water storage is of long duration in the epikarst as water drainage is slow in the vadose zone (Williams 1983; Ford and Williams 2007).
- The surface of the water stored in the epikarst should form a depression (Williams 1983, 2008; Ford and Williams 2007).

Provided that any of the above-mentioned conditions are not fulfilled, no solution doline can develop.

Giant Dolines and Uvalas

On glaciokarst, three varieties of solution depressions (doline and uvala) can be distinguished: giant doline, small-sized solution doline and uvala and schachtdoline (Veress 2017a). These dolines mainly occur in the periglacial belt, but giant dolines can also be found in the fluvial erosional zone of the temperate belt with postglacial temperate solution dolines together. Large-sized karstic depressions are also termed as mega dolines (Auly 2008) and Großdoline (Mix and Küfmann 2012). An exact size limit between giant solution dolines and small-sized solution dolines is hard to be determined. However, while the diameter of small-sized solution dolines can range from some metres to 10–15 m (exceptionally it can reach 30–35 m), the diameter of giant dolines can be between 40 and 50 m and several 1000 m. An example of this latter size is the Štirovača depression from Croatia, its length is 8.2 km and its width is 1.2 km (Bočič et al. 2012) or another example is the basin bearing Lake Medicine (Canada), the length of which is 6 km and its width is 2 km (Ford 1979). This latter feature is of mixed origin (Ford 1979). The two varieties can be more clearly differentiated if we consider the shape (the quotient of the diameter and the depth) or the average slope angle (which can be calculated from the tangent of the quotient of the depth and the half-diameter). In case of giant dolines, taking the longitudinal diameter into consideration, based on Kunaver's (1983) and Veress's (2016a) data (n = 28), the shape spreads between the values of 8.25 and 26.28 (average: 15.22), while in case of small-sized solution dolines (without the data of schachtdolines, if we considered them, the average shape would be even smaller) this value is between 2.00 and 5.00 (n = 16, average: 3.13). In case of giant dolines, the average slope angle is between 4.4° and 13.63° (their average is 8.24°), in case of small-sized solution dolines this value is between 21.8° and 45° (their average is 34.25°). (The data of small-sized solution dolines were received by measuring dolines of Hochschwab and Durmitor.) According to their size, two further varieties of giant dolines can be distinguished. The diameter of smaller giant dolines can spread from 40–50 to 100 m (perhaps some 100 m), while the diameter of the larger can range from several 100 m to several km.

The large size of giant dolines is probably due to several factors which are the following:

- They are of old development and thus, they could increase to a large degree by dissolution of long duration, karstification could happen again in their area.
- Dissolution exerted by meltwater could also contribute to their growth similarly in case of poljes (see below). For this, there was a chance in case of depressions which were located at the margin of glaciers.
- Repetitive and intensive glacial erosion. There was a greater chance for the latter in the area of ice caps.

Large-sized karstic depressions, being 1000 m or even larger are called kontas in the Dinarides (Kunaver 1983; Stepišnik et al. 2010; Žebre and Stepišnik 2015a, b), and jou in the Picos de Europa (Miotke 1968). The characteristic features of kontas according to Kunaver (1983), based on examples from the Kanin Mountains: they are corroded basins with a diameter of 80–680 m, with slopes covered by debris and with a floor partly covered by moraine where ponors occur too. They are located on old valley floors, along fault lines. In their development, in addition to karstification, glacial and periglacial effects played a role too, so they are polygenetic (Kunaver 1983). Kontas occur in the area of former ice caps (Žebre and Stepišnik 2015a), but also in areas exempt from former ice cover, though these features became filled with ice (Žebre and Stepišnik 2015b). According to Kunaver (1983) and Fu and Harbor (2011), kontas are overdeepened cirques. However, this is disproved by the fact that kontas do not occur in valley heads (Žebre and Stepišnik 2015a). Žebre and Stepišnik (2015a) emphasize their complex origin. Thus, snow erosion played a role in their development since they were local snow accumulation places (Kunaver 1983; Hughes et al. 2006) and periglacial processes also contributed to their formation (Žebre and Stepišnik 2015a).

Turloughs probably belong to this group too. These are closed features with a large expansion in the British Islands, they occur in the area of the former ice sheet (Williams 1970; Sweeting 1973; Coxon 1987), there are katavotra and intermittent lakes on their floor. According to Coxon (1986), they are karstic features which were not destroyed by glacial erosion. According to Gunn (2006), they are features with complex genetics, dissolution, karstic water drainage and glacial erosion also played a role in their development. Glacial erosion removed older sediment from their inner part (Sweeting 1973). The appearance of intermittent lakes is explained by the increased amount of inward flowing water, by water drainage of small degree and by the elevated position of karst water level (Sweeting 1973). Depressions with large diameter (their diameter can even reach half a mile) are also mentioned from the Pennines (Clayton 1966; Moisley 1955).

Giant dolines most often occur in glacier valleys, but they can also be found on plateaus subject to glacial erosion (Bočič et al. 2012; Bognar and Faivre 2006). The characteristic features of giant dolines are the following:

- Their diameter is several times larger than their depth (Fig. 4.35).
- They can occur in cirques and troughs (Fig. 4.36), but also on plateaus without glacier valleys (Fig. 4.37), and on ridges too for example on the Snježnik-Guslica ridge (Žebre and Stepišnik 2015b) or on the ridge-like plateau of Hochschwab.
- The morphology of their inner part is varied, their features can be of karstic and non-karstic origin. Their karstic features are small-sized solution dolines (Figs. 4.35 and 4.37), subsidence dolines (Figs. 4.35, 4.36, 4.38 and 4.39), karren (Fig. 4.18) and ponors. The floor of the largest can also be dissected by inner giant dolines (Žebre and Stepišnik 2015a). With the growth of the diameter, the diversity of the features of non-karstic origin increases too. Thus, roche moutonnées, rock basins, hogbacks, hummocky moraines, debris cones (Figs. 4.40 and 4.41), hummocks of mountain failures, alluvial cones (Figs. 4.38 and 4.39), valleys and channels (Fig. 4.42) can occur inside them. According to Žebre and Stepišnik (2015a), outwash plains occur in their area too.

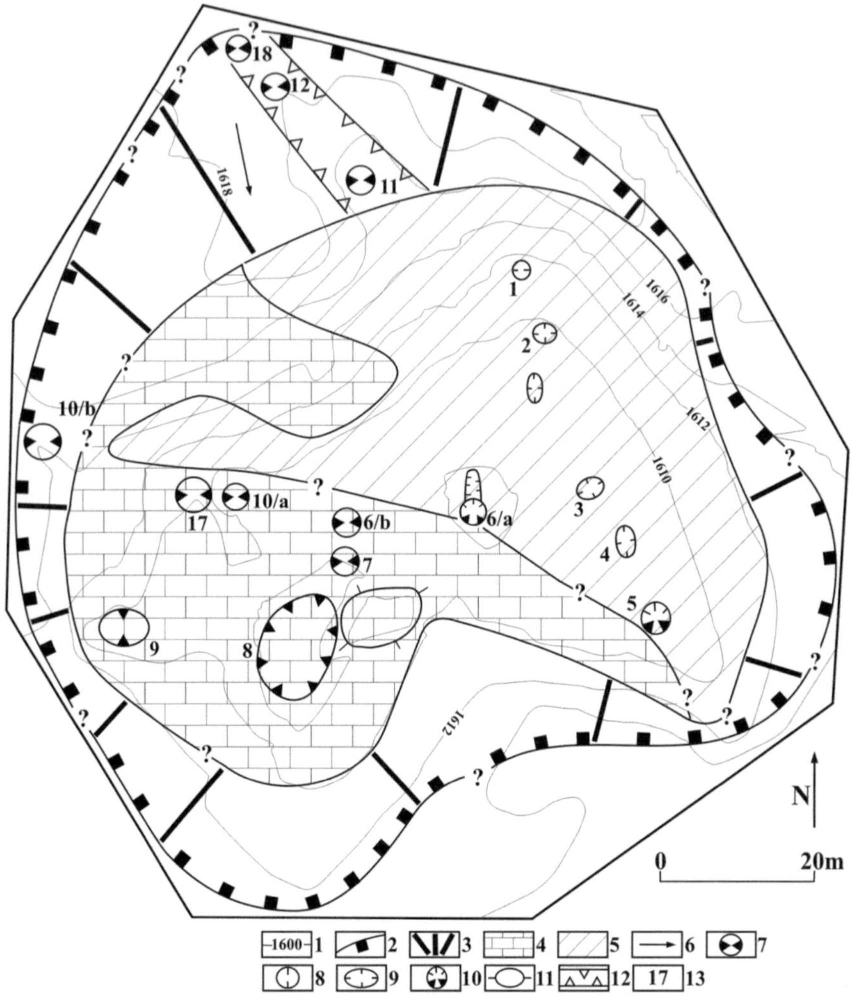

Fig. 4.35 Map of the paleodoline near Tauplitz alm (Veress 2016a, modified). 1. Contour line, 2. paleodoline, 3. uncovered side slope of paleodoline, 4. uncovered floor of paleodoline, 5. covered floor of paleodoline, 6. dip direction of layer, 7. solution doline, 8. subsidence doline, 9. elongated subsidence doline, 10. half-doline, 11. roche moutonnée, 12. solution giant grike, 13. identification code of doline

- According to the ground plan, they can be dolines (Figs. 4.36 and 4.37) and uvalas (Fig. 2.9). Several varieties of uvalas can be distinguished which can be the following:
  - Constituting dolines of the uvala constitute a grouped pattern. Former dolines could coalesce by karstification or glacial erosion.
  - Constituting dolines of the uvala create a row as a part of a trough.

Fig. 4.36 Interglacial giant solution doline (Totes Gebirge, photo by Veress). 1. Interglacial giant doline, 2. floor remnant of glacier trough, 3. suffosion doline

Fig. 4.37 Giant doline from Asiago plateau (photo by Veress). 1. Solution doline, 2. roche moutonnée

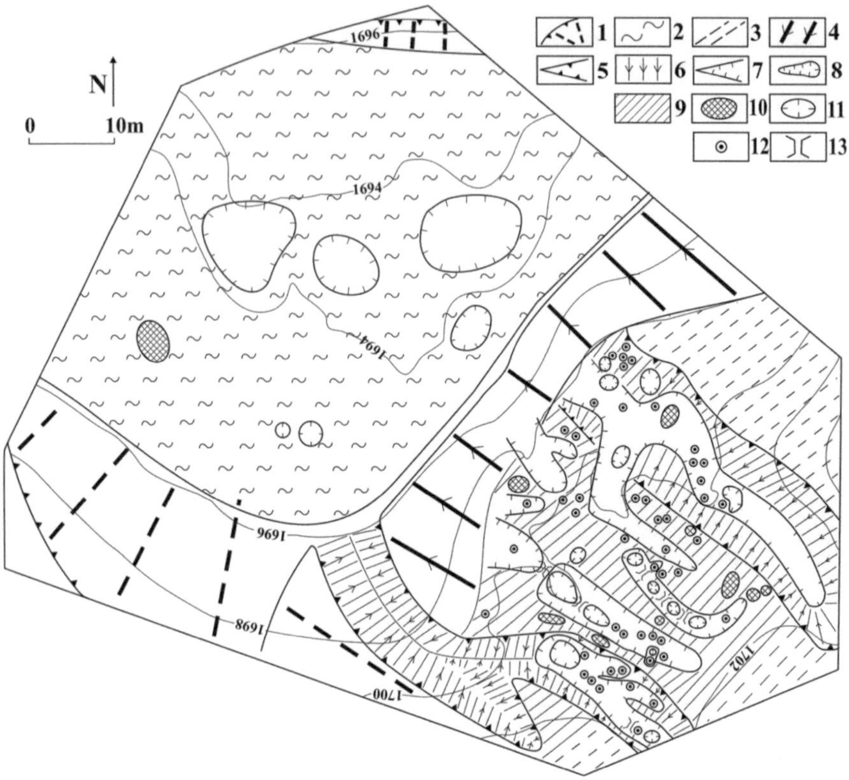

Fig. 4.38 Karstification on alluvial fan (Dachstein) (Veress 2016a, modified). 1. Paleodoline and its side slope, 2. floor of paleodoline, 3. surface of alluvial fan, 4. front of alluvial fan, 5. ravine, 6. side slope of ravine, 7. gully, 8. blind suffosion gully, 9. ridge between ravines, 10. elevation, 11. suffosion doline with diameter more than 2 m, 12. suffosion doline with 0.5–2.0 m diameter, 13. col

  - The floor of uvalas is dissected by giant dolines. These latter could also coalesce by karstification or glacial erosion. Inner dolines are separated by thresholds or plain floor remnants from each other. Inner dolines can also be dissected by small-sized solution dolines and subsidence dolines.

- As compared to the glacier valley, the dolines or uvalas can be valley dolines, valley floor dolines, cirque dolines, dolines transformed into cirques (cirque of karstic origin) and dolines outside the valley.
  - Valley dolines expand on the whole valley thus, the valley margin has an arcuate pattern. The doline and the slope of the valley do not separate from each other. The dolines are complex: larger dolines or uvalas have inner dolines or uvalas (Fig. 2.9).
  - Valley floor dolines deepen into the floor of the trough, by this, the remnants of valley floors could survive in different width at these features

Fig. 4.39 Subsidence dolines in a paleouvala (Dachstein, photo by Veress)

(Veress 2012a, Fig. 4.36). Valley floor dolines are of smaller diameter than valley dolines. As compared to the diameter of the latter, their depth is smaller too than in the case of valley dolines. Their slope is separated from the slope of the bearing valley (the slope of the doline is less steep than that of the valley). Valley floor dolines do not coalesce. The margin and side slope of the bearing valley are not arcuate.

- The dolines of the cirques can be of various size. Both the dolines and the cirques which developed by the transformation of dolines can have straight or arcuate margins (Fig. 2.1). While the margin of the smaller is arcuate, there occur features with arcuate, straight and irregular margin among the larger. The margin of arcuate depressions with a larger diameter can be many times arcuate. Cirques could develop in the valley head of a river valley (see Chap. 2.4). They could develop by the transformation of a preglacial karstic depression (Fels 1929; Smart 1986), where glacial erosion, snow erosion and mass movements might have had a role, while cirque dolines could develop with the karstification of the cirque (Barrére 1964; Djurovič et al. 2010). However, closed cirques can develop on glaciokarst too by glacial erosion only (Ford and Williams 2007). Cirques being developed from preglacial dolines (Fells 1929) are believed to be mixed features that continue their formation in the glacials and interglacials (Ford 1979).

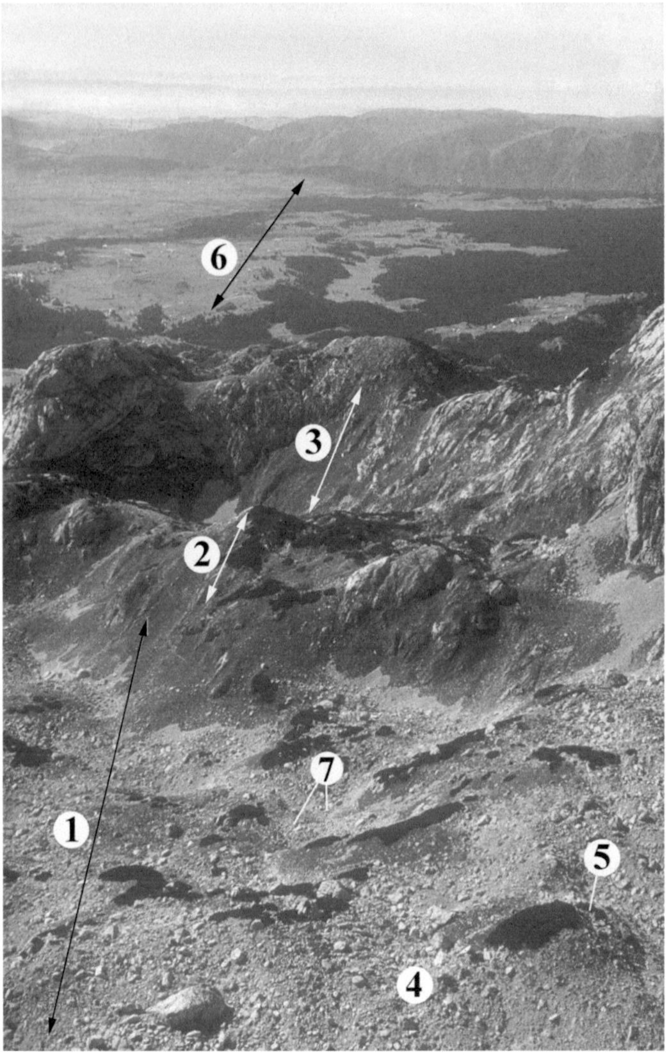

Fig. 4.40 Cirque dissected by glacial erosion (Valoviti do, Durmitor Mts, photo by Veress). 1. Closed cirque developed from a preglacial doline (Valoviti do), 2. ridge, 3. preglacial giant doline, 4. moraines, 5. roche moutonnée, 6. lower surface around the mountains covered with morainic deposit, 7. suffosion doline

- Their floor can be exclusively bare karst, soil-covered karst and concealed karst. It may happen that uncovered and covered parts alternate in their area (Fig. 4.35). If there is a cover on their floor, the following karst types can occur in their area:

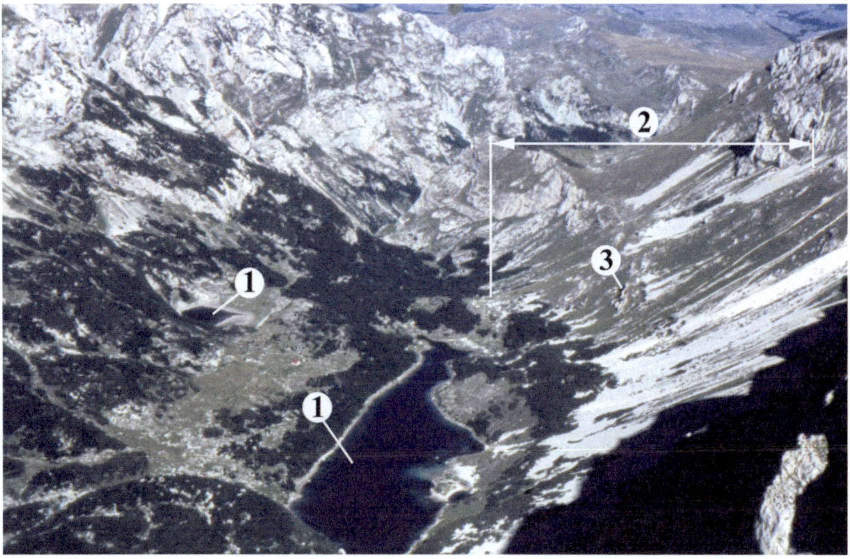

Fig. 4.41 Cirque formed from giant doline (uvala) with rock basin (Skryke, Durmitor Mountains, photo by Veress). 1. Rock basin, 2. debris slope, 3. margin of partial doline covered with debris

- allogenic crypto karst and concealed karst with ponors and subsidence dolines,
- concealed karst and bare karst with subsidence dolines, karren, small-sized solution dolines and schachtdolines,
- bare karst and soil-covered karst with karren, small-sized solution dolines and schachtdolines,
- bare karst, soil-covered karst and concealed karst with karren, small-sized solution dolines, schachtdolines and subsidence dolines,
- allogenic karst, concealed karst and bare karst, in this case, all mentioned features occur in the depression.

- Mixed, polygenetic features which were transformed genetically by glacial erosion (Ford 1979; Plan and Decker 2006; Annys et al. 2014).

The morphology of their shape can refer to the impact that affected them. According to their shape, their development can be the following:

- The depression is a doline, according to this, it is of single-arcuate margin, there are giant dolines on its floor. In this case, a two-phase karstification took place which were probably interrupted by glacial erosion.
- The depression is a doline, according to this, its margin is single-arcuate, in its inner part there are large-sized solution dolines and features of glacial erosion (roche moutonnée, rock basin) and/or moraine. In these features karstification and glacial effects were repeated several times.

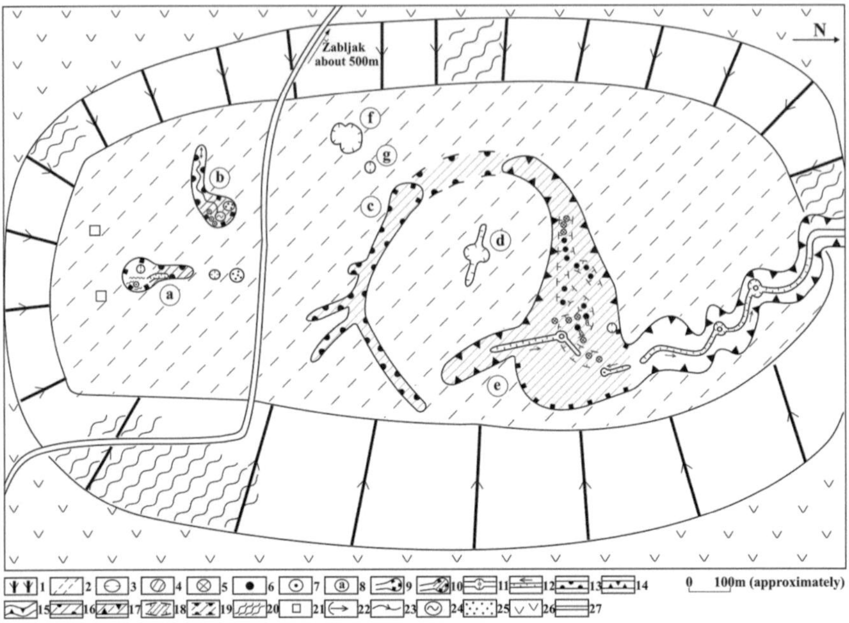

Fig. 4.42 Morphological map of a giant depression near Žabljak (sketch, not surveyed) (Veress 2016a, modified). 1. Side slope of paleodoline (giant solution doline), 2. infilled floor of paleodoline, 3. suffosion doline (diameter more than 1–2 m), 4. infilled suffosion doline, 5. suffosion doline (diameter less than 1–2 m), 6. suffosion doline with debris on floor (diameter less than 1–2 m), 7. dropout doline or opening in cover (diameter less than 1–2 m), 8. doline code, 9. ponor-like doline (covered karst ponor), 10. infilled covered karst ponor, 11. shallow solution doline on channel floor, 12. slope direction of gully and its floor, 13. ravine, 14. valley, 15. valley with incised meanders, 16. infilled ravine, 17. infilled valley, 18. buried gully, 19. buried ravine, 20. limestone outcrop, 21. limestone block (moraine), 22. artificial water inflow, 23. Watercourse, 24. lake, 25. bog in doline, 26. karst terrain bordering paleodoline, 27. road, a–c, covered karst ponor with creek (on the floor of types "a" and "b" there are subsidence dolines of different types), (d) suffusion doline with creek, (e) valley with a channel (gully) on its floor, there are a lot of small suffosion dolines on the valley floor, (f) coalesced suffosion doline which is separated into constituting dolines, (g) larger suffosion doline

- The depression is an uvala, according to this, its margin is many times arcuate, karstic depressions, glacial features and a moraine occur in its inner part. This feature was formed with the coalescence of former dolines (during karstification, but mainly in case of those uvalas where constituting dolines constitute a row, in their coalescence the role of glacial erosion cannot be excluded either), then glacial erosion and karstification could be repeated several times.

The coveredness and feature assemblages is different in case of valley dolines, valley floor dolines and cirques.

- Coveredness is more widespread in valley dolines and cirques. The cover partly originates from the frost weathering debris covering the slopes and partly from the ground moraine and terminal moraine covering the floor. On the contrary, in troughs where valley floor dolines exist, ground moraines have a smaller expansion and thickness.
- In valley dolines (and in valleys) stepped surfaces are absent (or if they exist, they are buried). Thus, karren features are different and less widespread than in valley floor dolines and in the valleys bearing them. This can probably be traced back to the fact that karren formation could not take place or it took place only to a limited extent on the relatively steep slopes of giant dolines (which developed preceding ice formation) (grikes could not develop). However, it is also possible that glaciers had a weaker ability to remove bed units on the steep slopes (plucking is replaced by abrasion).

Exhuming depressions are called depressions of superficial deposit (DSD) by Veress (2009e, 2012a, 2016a) if their sediments were or are transported into the karst. Such an exhumed depression is mentioned from the Canadian Rocky Mountains by Ford (1979). The depression with a diameter of 6 × 2 km became filled during Wisconsinan Glaciation. Since then the fill has been transported into the passages of the karst. Exhumation (the loss of the cover) can happen in the following ways:

- During its dissolution, limestone debris gets into solution and is transported into the karst (Trudgill 1985).
- The cover is transported into the karst through ponors and subsidence dolines by mechanical erosion (rainwater, intermittent and permanent streams) (Veress 2009e; Bočič et al. 2012; Žebre et al. 2016).

According to the degree of exhumation, giant dolines (DSDs) can be the following (Veress 2016b):

- A giant doline becomes exhumed to a small extent (Fig. 4.43) if there are few gullies and subsidence dolines without passages on its floor in addition to this, the passages of the bedrock are confined to the epikarst and thus, their sediment receiving capacity is small.
- A giant doline becomes exhumed to a medium degree if the cavities of the karst are larger, for example there are shafts (Fig. 4.24a), the number of subsidence dolines is larger and at least some of them have passages.
- A giant doline becomes exhumed to a great extent if on the floor there are subsidence dolines or ponors (with blind valley) having gullies (Figs. 4.44 and 4.45). Mostly, if there is a well-developed cave system in the karst.

Djurovič et al. (2010) typified large-sized depressions considering karstic and non-karstic processes taking place inside them. The following depression types were distinguished:

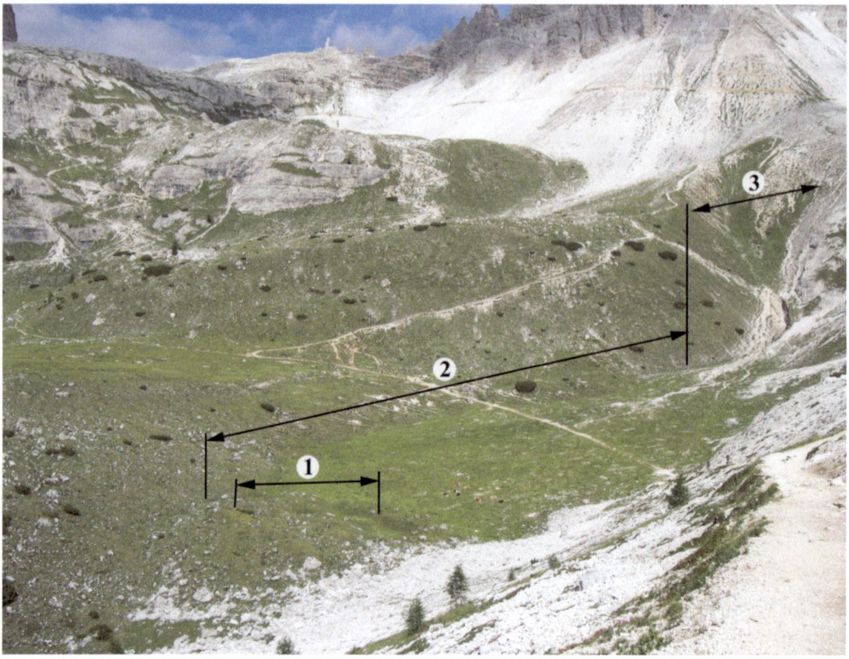

Fig. 4.43 Depression of superficial deposit in the Grava Lunga cirque (Dolomites, photo by Veress). 1. Doline zone of depression where material transport takes place into the karst, 2. depression, 3. ravine in morainic deposit

- Depressions of glacial uvala type have the most elevated position. Inside them, cryogenic-nivation processes operate. A karstic process is manifested in global (areal) dissolution. A lot of debris is generated, karst features are represented by karren, the density of which is small.
- Depressions of glacial-karst uvala type are of lower elevation (or temperature is higher inside them). In their area, intermittent streams occur which are fed by snow patches. Their length is short and they are drained into the karst. Intermittent lakes can also be formed in their karst features. Karstification prevails in their area: ponors, sinkoles (subsidence dolines?), karren occur on their floor.
- Depressions of glaciofluvial uvala type are of the lowest elevation. Their streams which can also be permanent create blind valleys. Karstic processes are determinant, sinkholes, ponors, permanent lakes may occur in their area.

It is indisputable that in giant depressions of higher elevation where vegetation is absent, snow erosion and frost weathering are powerful. The intensity of the above-mentioned processes decreases in giant depressions of lower elevation. (In the latter, with the growingly larger expansion of soil and vegetation, dissolution which can be traced back to biogenic CO_2 is increasingly significant and the role of

Fig. 4.44 DSD of exhuming paleodepression (Hochschwab, photo by Veress). 1. Ponor, 2. lake, 3. infilled ponor, 4. end of blind valley, 5. probable former level of fill, 6. valley, 7. gully, 8. allogenic karstic part of depression floor, 9. concealed karst part of depression floor, 10. margin of DSD (paleodoline)

snow erosion and frost weathering decreases increasingly in their formation.) However, the type of the developing features (ponor or subsidence doline) is not elevation dependent (and thus, the degree of the transportation of the cover into the karst either), but it depends on the expansion, thickness, grain size and composition of the superficial deposit. However, frost weathering has an effect to the degree of inward sediment transport in the depressions: continuously less material of such origin arrives at depressions of increasingly lower elevation.

Depending on the fact whether inward sediment transport is smaller or larger than sediment transport from the depression into the karst, giant depressions can be exhuming (Figs. 4.43, 4.44 and 4.45) or infilling (Fig. 4.46) and infilled (Fig. 4.47). These processes do not only happen at present, but they also took place during the Pleistocene. Cave fills (Bočič et al. 2012; Kiernan et al. 2001) refer to the fact that accumulation and exhumation alternated and were repeated in various depressions. On karst, the degree of exhumation depends on the degree and intensity of cover denudation and on the possibility of sediment transport into the karst (on the floor, it depends on the number of ponors and subsidence dolines and on their sediment transport capacity), on the sediment receiving capacity of the karst (on the density, size, morphology and pattern of cavities).

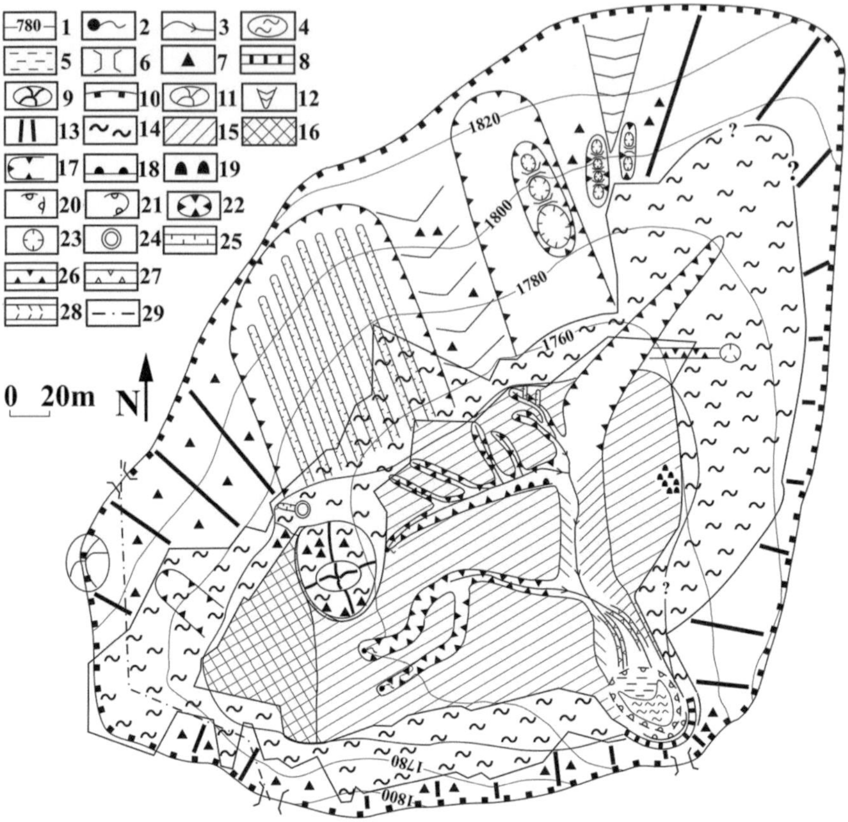

Fig. 4.45 Geomorphological map of the DSD of Hochschwab (Veress 2016a, modified). 1. Contour line, 2. spring (permanent and intermittent), 3. stream, 4. lake, 5. waterlogged area, 6. col, 7. exposed limestone, 8. cliff wall, 9. roche moutonnée or karstic residual feature, 10. edge of the DSD, 11. mound on the margin of the depression, 12. limestone ridge, 13. denuded side slope of the depression, 14. accumulational slide slope of depression, 15. allogenic karstic part of the depression floor, 16. concealed karstic part of the depression floor, 17. nivation and mass movement features of the side slope, 18. slide, 19. thufur field, 20. infilled ponor, 21. active ponor, 22. recent uvala, 23. small-sized solution doline, 24. shaft, 25. rain furrow, 26. gully, 27. channel on valley floor, 28. valley, 29. tourist track

On glaciokarst, sediment transport and thus, its arrival at the cavities of the karst is promoted by the increase of the amount of meltwater that happens to the effect of warming at the end of the glacial. (However, stream load can also be significant in the interglacial too because of the increased amount of precipitation.)

According to Woodward et al. (2008), the amount of water flowing down the surface (and thus, the degree of sediment transport) is also different in the glacials and interglacials since ice and the developed moraine-blocked percolation in the glacial. However, as the percolation of meltwater also takes place below the ice, only a part of this could leave the glacier. However, the amount of escaping

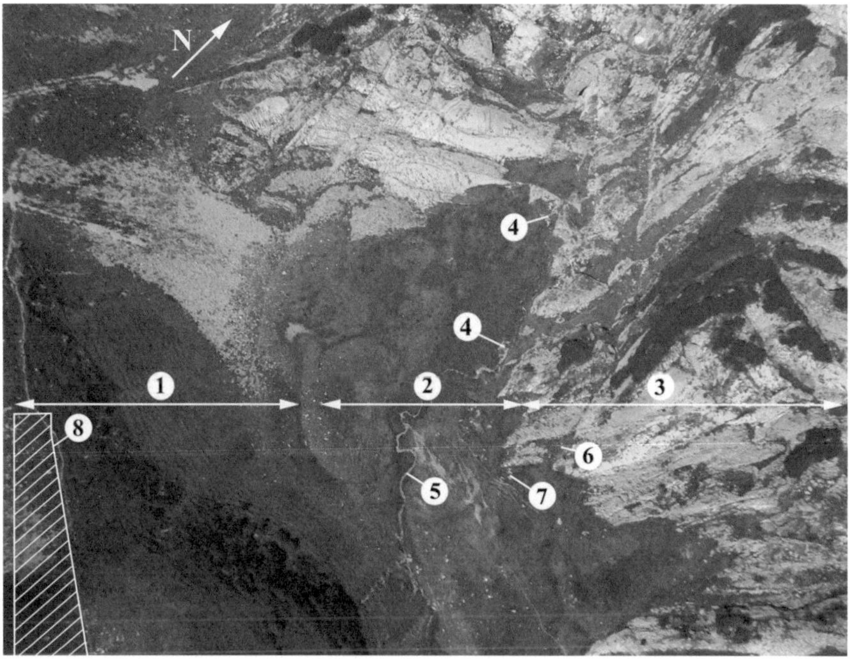

Fig. 4.46 Aerial photograph taken under Wildgössl (cirque formed from giant doline, Totes Gebirge, photo by Veress). 1. Side of giant solution doline with debris fans, 2. floor of giant solution doline with superficial deposit, 3. slope of giant solution doline eroded glacially, 4. covered karst ponor, 5. gully, 6. covered karst terrain intruding into the glacially eroded surface, 7. a subsidence doline in the zone of thin superficial deposit next to rock boundary, 8. zone of springs

meltwater might have depended on several factors too; thus, on the maturity of the epikarst and on the inclination of the bearing terrain.

However, the increased amount of sediment can contribute to the plugging of ponors, partly in a way that caves become upfilled and infilled and partly in a way that the entrance of ponors becomes plugged. While preceding the plugging of the ponor, the exhumation of karstic depressions is characteristic, following plugging, accumulation is specific. The increased amount of water can contribute to the rise of the karst water level which further increases the upfilling of the caves (Bočič et al. 2012). At the same time, the decline of the karst water level contributes to the further transportation of cave sediment. Fine-grained sediment contributes to the plugging of drainage passages (if they are underdeveloped), while coarser-grained sediment contributes to the upfilling of the caves (if sediment gets into the cave).

Surface accumulation is caused by the advance of glaciers at the start of cold phases when the amount of meltwater decreases. In the glacials, accumulation spreads to caves too (Kiernan et al. 2001). However, floods also cause accumulation, the water of which originates from the lakes plugged by moraine if karstic water drainage ceases from the lakes (Stepišnik et al. 2010). Further, transportation

Fig. 4.47 Infilled, smaller sized (truncated?) giant doline (northern foreground of Durmitor, photo by Veress)

and reworking of the sediment is slowed down both on the surface and in the caves (thus, on the surface, exhumation, the clearing of caves and thus, the reception of newer sediment is slowed down) by the cementation of the sediment which takes place in the interglacials (Hughes et al. 2006; Adamson et al. 2014; Žebre and Stepišnik 2015a), by vertical water drainage under the glaciers (Žebre and Stepišnik 2015a) and by ground ice. Ground ice may have a lower position in the glacials and a higher position in the interglacials.

Small-Sized Solution Dolines and Schachtdolines

Small-sized solution dolines (Fig. 4.48) have steep sides. As it has already been mentioned, their width or depth is mostly some metres, those with a diameter larger than 10 m rarely occur. Because of their specific shape, these features are also called trichterdolines (Fink 1973; Plan 2005).

These features can occur in giant dolines, in troughs, in cirques or on surfaces and plateaus with slopes being scoured by glacial erosion. According to Bauer (1962), small-sized solution dolines are of postglacial age, which is also proved by the fact that they often occur in giant dolines (uvalas). Their small size can be explained by their young age. Their characteristic features are the following:

Fig. 4.48 Small-sized solution dolines (Maglič Mountains, Montenegro, photo by Veress)

- Their density is small, they mainly occur individually, their occurrence in groups is rare.
- They are vertically well developed: their depth is hardly smaller than their upper (marginal) diameter, their floor diameter is very small, and so they are funnel-like.
- Their side slopes are steep.
- Rarely, but it can be observed that despite their small size, they can form uvalas. It is probable that they do not coalesce by their horizontal growth, but they already coalesce during their development. They can also be complex. In this case, the floor of the long shaped, but narrow depression is dissected into constituting dolines (Fig. 4.49).

Fig. 4.49 Complex feature: small-sized solution dolines in a larger depression (Hochschwab, photo by Veress)

- The dolines deepen into the karstic rock. Therefore, can be differentiated from subsidence dolines. On their side of slopes, there is limestone in smaller or larger patches (In this case, it crops out from below the soil), on their floor there is soil and rock debris.
- If they occur on a steep slope, they are elongated in slope direction and have an asymmetric cross section. Their slope falling in the dip direction of the bearing slope is smaller and steeper, while the opposite is more gentle and longer.
- They occur above the tree line (mainly in the lower, dissolutional subzone of the periglacial zone). The shape of dolines below the tree line, although their sizes are smaller than those of giant depressions, is characterized by gentle slopes and a large diameter as compared to depth (Fig. 4.50) thus, the latter are temperate dolines (Ford and Williams 2007).

Schachtdolines are dissolutional features with vertical or almost vertical sides and a plain floor (Figs. 4.51 and 4.52), with frost weathering debris and with snow often surviving during summer months on their floor. Above the tree line, they are mostly characteristic of the upper part of the periglacial zone (in the subzone of mass movement and frost weathering). They are also of small size: according to Kunaver (1983), on the Kanin plateau, their average diameter is 7.1 m (the largest diameter does not exceed 20 m), while their average depth is 5.9 m. A significant characteristic of the steep side slope is that it is bare and exempt from vegetation. (In contrast with this, the side slope of small-sized dolines is not bare or it is partly bare.) The lack of vegetation can not only be explained by low temperature

Fig. 4.50 Temperate solution doline with subsoil karren on its floor (Asiago plateau, Italy, photo by Veress)

Fig. 4.51 Schachtdolines with twin-like development (Totes Gebirge, photo by Veress)

resulting from high altitude (since Pinus mugo can occur in their environment too), but also by the fact that the snow fill of long duration impedes the formation and development of vegetation on their slope. Their diameter exceeds their depth to a less degree than in case of small-sized solution dolines. They can occur in groups too. In this case, they can even be uvala-like. The dividing walls between them can get holed by corrosional, frost weathering processes (arches may develop) or they can collapse. Their density is changing, it can be large too. Bauer (1962) mentions

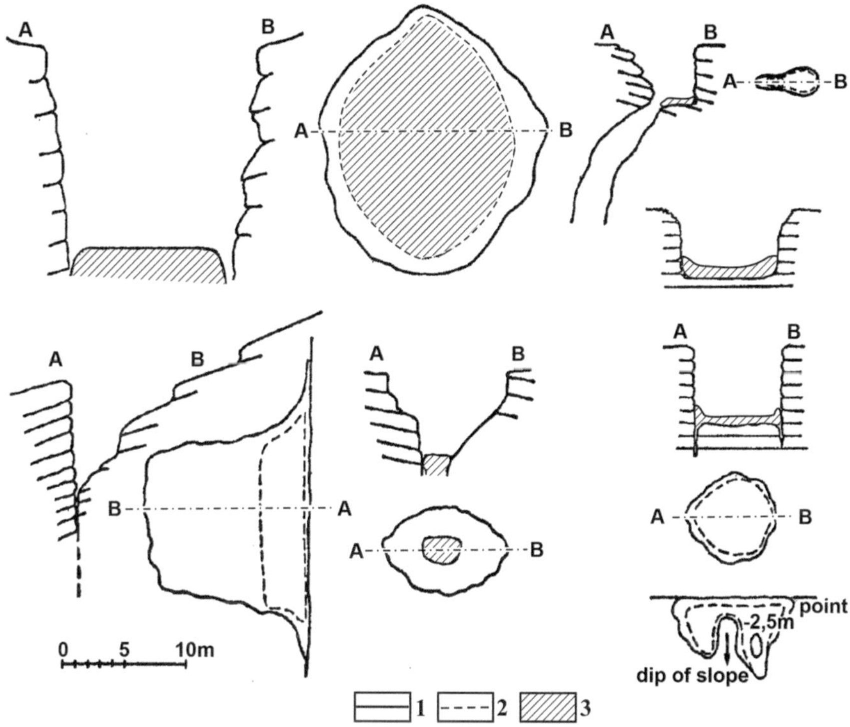

Fig. 4.52 Schachtdolines from Kanin Mountains (Kunaver 1983, modified). 1. Bedding plane, 2. fracture, 3. snow

900 schachtdolines from an area of 1 km^2 from the Alps. They have a varied morphological environment: they occur on lower combe-ridges, on small-inclined slopes of glacier valleys, but they are less specific in giant dolines. According to Kunaver (1983), they can be found over 1600 m on the Kanin plateau.

Schachtdolines often have a mixed occurrence: small-sized solution dolines or if superficial deposit is present, subsidence dolines may occur in their environment (Fig. 4.53). According to Kunaver (1983), about half of the schachtdolines of Kanin are elongated in the direction of those fractures along which they developed. However, they can also be elongated in the dip direction of the bearing slope or in the dip direction of the beds. Schachtdolines being elongated in slope direction or in the dip direction of the bed are asymmetric. Their side slopes dipping opposite slope direction and opposite dip direction of the bed is steeper and shorter. Their floor does not always terminate blindly in the bedrock, but they continue in a grike or passage (Fig. 4.52). Ford (1984) describes cenote-like schachtdolines from the Canadian Rocky Mountains too. Intermittent lakes were formed in these features when an ice plug developed below them in the drainage passage.

Fig. 4.53 Schachtdolines (1), small-sized solution doline (2), subsidence doline (3) (Maglič-Mountains, photo by Veress)

According to Kunaver (1961, 1983) and Ford (1979), schachtdolines are post-glacial features. However, according to Kunaver (1983, 2009a), fossil schachtdolines having been truncated by glacial erosion may occur too, which can refer to the fact that the conditions of their formation were also present in the interglacial(s). They develop during the local dissolution of the meltwater of snow accumulations (Kunaver 1983; Ford 1979). According to Stepišnik et al. (2010), they are formed by the meltwater of glaciers.

The cause of the small diameter of both the small-sized solution dolines and the schachtdolines is that they do not increase laterally which can be explained by the lack of horizontal dissolution. The fine-grained cover intensifies horizontal water movement. The coarser-grained the cover, the more vertical the water movement (Veress 2016a). Since there is no fine-grained sediment on the floor of these features, there is at most soil (it is probable that soil appeared in the late phase of their development when there were favourable climatic conditions inside them), snow water gets rapidly into the karst without lateral percolation, promoting downward dissolution. Vertical maturity and snow accumulation strengthen each other: the deeper the doline, the more snow accumulation can take place in it, however percolating snow water maintains further deepening. These features can develop in the periglacial zone because the survival of snow is favoured by high altitude and snow accumulation (snow drifts) is favoured by the lack of trees. However, the permanent survival of snow cannot be only favoured by higher altitude, but by local circumstances (wind conditions, exposure, the morphology of the environment) too. Therefore, the pattern of their distribution can be varied and influenced by local conditions.

It is probable that a schachtdoline develops if complete snow fill survives for a long time even in summer in the deepening depressions. By this, the doline deepens in its whole expansion below the snow and thus, a plain-floored feature develops with steep sides. A small-sized solution doline develops if the snow fill in the depression decreases for a long duration in summer. By this, the expansion of snow will be smaller thus, dissolution is focused to a smaller area on the doline floor, causing the formation of a tunnel-like shape with narrowing, gentle slopes.

The Comparison of Giant Dolines with Small-Sized Solution Dolines and Schachtdolines

If we compare the solution giant dolines, the small-sized solution dolines and the schachtdolines of glaciokarst, then they show differences in the following characteristics:

- Giant dolines (and uvalas) are of great size, (their width can exceed 50–100 m and their depth can exceed several 10 m), while the size of small solution dolines and schachtdolines is small (their depth and width are some metres).
- The shape of giant dolines is large, while that of small-sized solution dolines (but of schachtdolines too) is small.
- As it has already been mentioned, the inclination of the side slopes of giant dolines is small, while the dip of small-sized solution dolines and schachtdolines is large.

Consequently, giant depressions are horizontally developed, while small-sized solution dolines and schachtdolines are vertically developed. Although glacial erosion contributed to the horizontal maturity of giant depressions, they could mainly achieve this characteristic by horizontal dissolution. Regarding their characteristic features, they are similar to the dolines of the temperate belt, which are characterized by large diameter as compared to the depth and also by gentle side slopes (Ford and Williams 2007). Therefore, giant depressions (dolines, uvalas) did not develop in a glaciokarstic environment, but in another one which favoured their horizontal growth of karstic origin (lack or smaller amount of snowmelt, soil or fine-grained cover on the floor). On glaciokarst, such circumstances can be expected in case of a climate warmer than the present one, which could exist when the glaciokarst was of a lower elevation than today. Thus, they can be regarded as paleokarstic features.

Their shape was further modified by glacial erosion. Thus, the inclination of their side slopes and their depth could decrease during glacial erosion and accumulation. However, their further growth and transformation were also contributed by periglacial processes and snow erosion (Žebre and Stepišnik 2015b).

The differences in the shape and the size of giant solution dolines as compared to small-sized solution dolines and schachtdolines can be explained by their different development age. The preglacial or interglacial development of giant dolines is proved by the following things:

- There may be a moraine on their floor (Kunaver 1983).
- Valley floor dolines often with a diameter of several hundred metres could not reach their large size during postglacial dissolution of short duration.
- Inside them, there may be small-sized solution dolines with steep sides and of postglacial age, subsidence dolines that developed in moraine and roche moutonnées referring to glacial erosion (Fig. 4.37) and hogbacks.
- In these karstic depressions and in caves situated below them, fluvioglacial sediments may occur (Bočič et al. 2012), which indicates that they had a sediment receiving capacity in the last glacial.
- The direction of the elongatedness of karstic depressions is identical with the direction of the movement of the ice sheet on the Burren glaciokarst in Ireland (Drew 2004). The concordance of the orientation of depressions with the direction of ice movement refers to glacial erosion which proves that they had already existed in the last glacial.
- They can form cirques and troughs (Figs. 2.1, 2.9, 4.40 and 4.41).

According to their development age, valley giant dolines and uvalas can be put into two groups:

- Those constituting troughs or cirques and there are further karstic features inside them, mixed features and thus, they underwent several karstifications and glaciations. Thus, they are probably of preglacial age or they developed in any of the older interglacials.
- Those which are not complex and they were formed on the floor of glacier valleys, they were probably developed in the last interglacial.

4.5.1.2 Subsidence Dolines

Their Varieties and Characteristic Features on Glaciokarst

Subsidence dolines develop on covered (concealed) karst. These dolines are formed either in their complete expansion or predominantly in the cover of karst (Fig. 4.34). Their varieties are dropout dolines, suffosion dolines and compaction depressions (Waltham and Fookes 2003; Williams 2004; Waltham et al. 2005) Dropout dolines develop on cohesive (on clay, clayey sediment) rock (Waltham and Fookes 2003) by collapse. According to this, their side slopes are steep (nearly vertical) surfaces. According to Waltham et al. (2005), they develop by the collapse of the cavity that was formed in the cover. However, dolines of collapse origin can also develop on less cohesive rock if collapse already occurs on the bedrock which is inherited to the cover (Veress 2016a). They can also be formed if there is a well-soluble rock under the non-cohesive cover. Ford (1979) describes such dropout dolines from the Canadian Rocky Mountains which were formed by the collapse of the moraine, since the gypsum cavities of the bedrock collapsed. The development of suffosion dolines mainly takes place by suffosion: since the non-cohesive grains of the cover are accumulated in the passages of the bedrock by meteoric water percolating through the cover. Thus, the side slope that was formed in the loose superficial deposit is not vertical though it can be of extremely diverse steepness. A compaction depression (compaction doline) develops by the compaction of the grains of the cover which is the fill of an older, filled doline. Its side slopes are of very small inclination. Compaction dolines should not be necessarily differentiated from suffosion dolines: on the one hand, compaction takes place in the cover in the case of suffusion dolines too, on the other hand, the process is preceded by material transport from the cover here too, and third not only these dolines can develop in the fill of older karstic depressions.

Ground water or karst water are important factors in the development of subsidence dolines (Xu and Zhao 1988). Because of the decrease of the water level in the cover, the support of the cover decreases (this favours the development of dropout dolines) and the possibility for rainwater percolating through the cover increases, which promotes the chance for suffosion doline formation. The fall of the water level can be both of natural and artificial origin (Beck and Sinclair 1986; Waltham and Smart 1988). On glaciokarst, the formation of subsidence dolines related to ground water level decrease can take place during the retreat of glaciers. In this case, a lot of meltwater is generated which creates groundwater in the cover. However, later during vertical water drainage, the water level of groundwater that developed in this way decreases.

On glaciokarst, surfaces of superficial deposit with smaller or larger expansion (thus, covered karst) are widespread, where subsidence dolines can develop. Covered karst and thus, subsidence dolines can be present in the periglacial zone, but in the fluvial erosional zone of the temperate belt too. Subsidence dolines can occur everywhere where superficial deposit exists: thus, in cirques (Fig. 4.24a), on the floor of troughs (Fig. 4.24b) and in giant depressions. They can also occur in giant depressions which are outside a glacier valley (Figs. 4.54 and 4.55) and in those which are situated on the floor of troughs (Fig. 4.56). Subsidence dolines can be found on alluvial cones (Fig. 4.38) at the margin of the debris mounds of mountain failures (Figs. 4.57 and 4.58a), in the gaps of hummocky moraines (Fig. 4.59) in gullies and creeks (Fig. 4.42). However, according to Ford and Williams (2007), subsidence dolines can develop on limestone pavements covered with frost weathering debris. However, it can be experienced that the frequency of suffosion dolines is larger in giant dolines and cirques, while dropout dolines are more characteristic on the floor of troughs exempt from giant dolines. The size and morphology of the suffosion dolines of various giant dolines show considerable similarity.

The cover, in which subsidence dolines develop on glaciokarst, can be moraine, glaciofluvial sediment, materials of mountain failure, frost weathering, and rock avalanches or purely fluvial sediment. On the British Islands, suffosion dolines that developed on terrains with moraines are termed shake-holes (Sweeting 1973). According to Sweeting (1973), the side slopes of these features are unstable, dissected by slides and their shape changes. They are elongated in slope direction and have an asymmetric cross section. According to Sweeting (1973), their size depends on the width of joints, on the thickness of the till, on the intensity of water drainage

Fig. 4.54 Giant doline with suffosion dolines on its floor that constitute an arcuate row (Rax, Austria, photo by Veress)

into the karst and on the inclination of the surrounding surface. According to Sweeting (1973), their average depth is 3 m and their diameter is 8–11 m.

On glaciokarst, subsidence dolines have specific characteristic features. These are the following (Veress 2016b, 2017b):

- Although subsidence dolines also have a smaller size on karst of other types than the dolines of other doline types (for example solution dolines or collapse dolines), except the subsidence dolines of evaporite covered karst, on glaciokarst, the small size is predominant (Sweeting 1973; Waltham et al. 2005, Figs. 4.59, 4.60 and 4.61). Their width and depth are some metres, but their depth does not often reach 1 m either. Thus, in the area of Mlječni do (Durmitor) 5 out of 9 suffosion dolines have a depth smaller than 1 m, but neither of them exceeds a depth of 2 m. As another example can be mentioned the suffosion dolines being formed on the alluvial cone of a giant uvala in Dachstein (Figs. 4.38 and 4.39). Here, only 14 out of 64 suffosion dolines have a diameter larger than 2 m (but neither of them have a diameter exceeding 5 m). Although small-sized dolines (with a diameter of 1–2 m) occur on other covered karsts too, there may be dolines with a larger diameter and depth in a significant proportion too. Thus, on Pádis karst (Romania) in addition to the above-mentioned smaller ones, there occur dolines the size of which exceeds 60 m along their longitudinal axis, but subsidence dolines with a diameter of 30–50 m and with a depth of 5–8 m often occur here too. The possible causes of small size are the following (Veress 2016a):
  - Their development could only start after the retreat of the ice. This can be some thousand years in the periglacial zone. The dolines of non-glaciokarst

Fig. 4.55 Subsidence dolines (**1**) developed in a giant doline (**2**) (under Triglav, Julian Alps, photo by Veress)

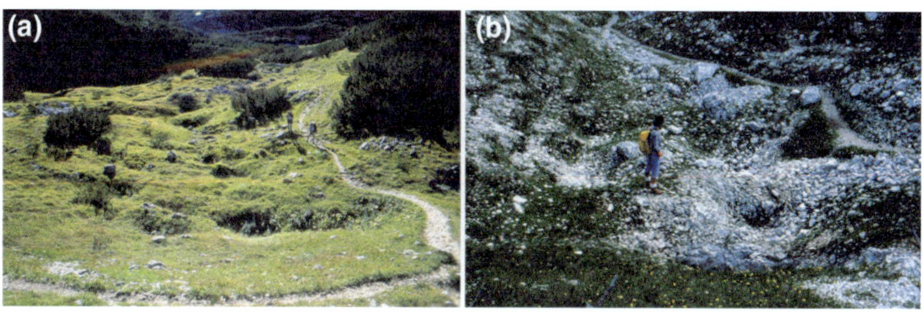

Fig. 4.56 Suffosion dolines (**a**, **b**) of giant dolines on the trough floor (Totes Gebirge, photo by Veress)

Fig. 4.57 Suffosion dolines that developed on the blocks of mountain failure in a giant doline (near Lake Vd. Lahngang, Totes Gebirge, photo by Veress)

can also be of young development age, but the possible development age (potential development age) is not limited by the disappearance of ice cover. Therefore, while the age can be diverse in case of the former, the age of the latter is always young.

- In the bedrock, cavity size is small since because of glacial erosion, the epikarst zone under the actual surface is young. (The large caves of the karst are situated far from the majority of subsidence dolines in a large depth.) There are few and small cavities which are only able to receive a small amount of cover, the cavities become filled and plugged fast and easily. Because of this, even if a cavity with a more significant size exists deeper, the cover is not able to be transported there.
- The cover is mostly coarse-grained. Because of this, suffosional material transport is limited. Coarser material (rock debris) is accumulated even in the initial section of the cavities. This filters the later inward transported sediment, the fine-grained material gets stuck on the debris. Thus, the sediment is not able to be transported deeper.
- The cover is often thin. Therefore, only suffosion dolines with a small depth can be formed in it. However, dropout dolines have a small depth not only because of the above-mentioned cause, but also because as a matter of course only small-sized cavities can develop in the cover with a small thickness.

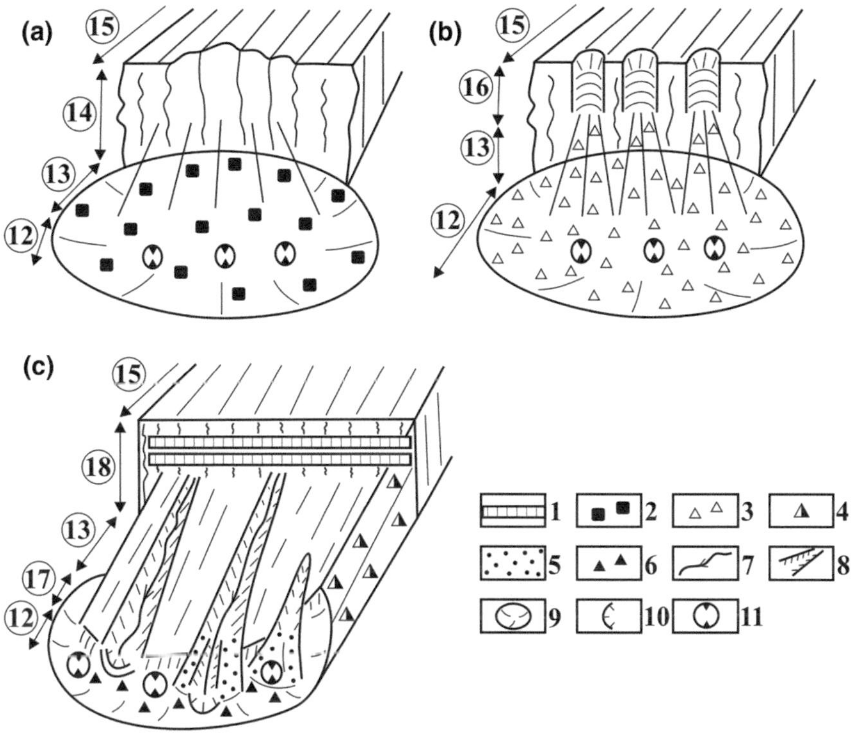

Fig. 4.58 Karstification of paleodolines partially covered on the material of mountain failures (**a**), on debris fans of rock avalanches (**b**) and on debris slopes (**c**) (Veress 2016a, modified). 1. Siliceous layer, 2. collapsed blocks, 3. frost-shattered debris with limestone, 4. frost-shattered debris partly of limestone, partly of siliceous material, 5. reworked debris, 6. morainic deposit, 7. watercourse, 8. ravine, valley, 9. paleodoline, 10. ponor, 11. subsidence doline, 12. outer part of cover, where the cover is thin, and therefore karstification takes place in the doline, 13. heap of mountain failure, zone of debris fans, buried section of paleodoline (no karstification) 14. side of glacial valley with mountain failure, 15. plateau, aréte, 16. slope of glacial trough dissected with cliff niches and rock avalanches, 17. debris reworked by watercourses (alluvial fan), 18. slope of glacial trough with siliceous intercalations

- The material of the cover is mainly limestone. Because of this, if the cover is extremely thick, the meteoric water percolating through it arrives at the bedrock saturated. Under such a cover there is no cavity development in the bedrock at present. By this, on the bedrock, there are only cavities that had been inherited earlier (from the period preceding covering).
- In case of a thinner cover with limestone debris, the bedrock can be dissolved, but only to a small degree. During percolation through the cover, the water becomes partly saturated thus, its solution capacity is small on the bedrock and thus, dissolution is of a small degree too.
- In areas bearing subsidence dolines, following the retreat of ice, ground ice is present in one part of the year. Because of this, water percolation and

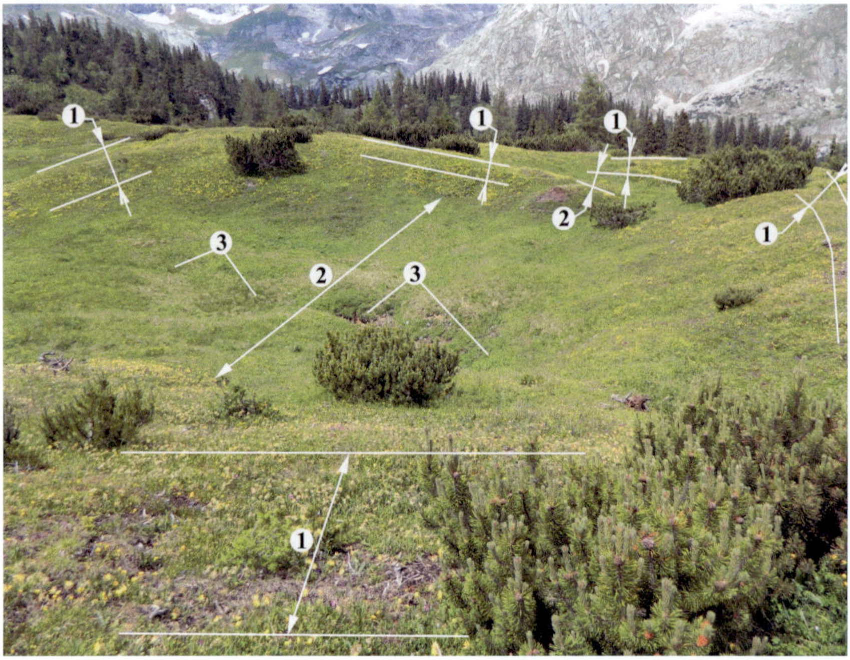

Fig. 4.59 Karstification of terrains enclosed by morainic hills (cirque of a glacial valley in the Hochschwab, photo by Veress). 1. Morainic hill, 2. covered karst terrain enclosed by morainic hills, 3. subsidence doline

material transport through the cover was and is only possible only during a short period of the year.

- Because of their large density, their catchment area is small or does not exist at all thus, they get a small amount of water and suffosion is of a small degree too.

- The distribution of dolines is diverse. There occur terrains exempt from dolines, but there are such small-expanding doline patches where their density is extremely high. Thus, on the already mentioned alluvial cone of Dachstein giant depression, the density is 0.031 doline/m^2. Although larger subsidence doline densities also occur on covered karst of mountain of medium-height (thus, on the floor of a karstic depression of the already mentioned Pádis, the density can be 0.021 doline/m^2 too), small doline density is more characteristic on non-glaciokarst. Thus, in the Bakony Mountains (Hungary) subsidence doline density is 15 doline/km^2 (Veress 1982). The density can be so high on glaciokarst that small mounds (with a circular or elongated ground plan) remain from the former flat terrain between the dolines (Fig. 4.62). The cause of the large density, which is mainly characteristic of the subsidence dolines of giant depressions, is that the meteoric water having fallen to the area of giant

Fig. 4.60 Typical small-sized suffosion dolines (Tauplitz alm, Totes Gebirge, photo by Veress)

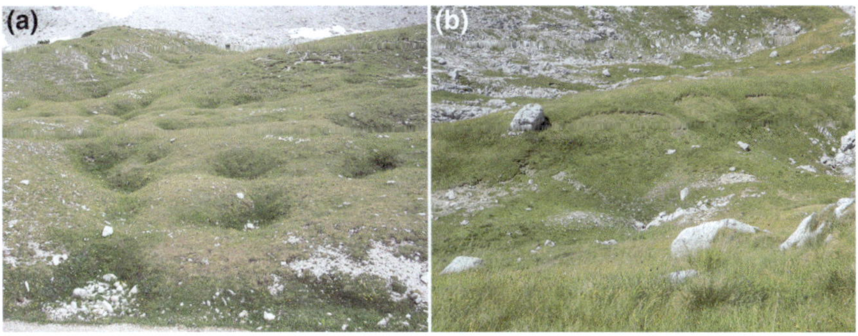

Fig. 4.61 Suffosion (**a**) dolines (Dolomites) and dropout (**b**) dolines (Durmitor) that developed in large densities (photo by Veress)

depressions can get into the karst only at this place. In addition to this, snow accumulates in the depressions from their environment thus, the size of snow packs exceeds the area or the catchment area of giant dolines. (Snow blown by the wind can originate from outside the catchment area.) Therefore, a relatively large amount of water can get into the giant depressions and into their superficial deposit which promotes the chance for the development of subsidence dolines

Fig. 4.62 Mounds survived from the original terrain because of large doline density (Totes Gebirge, photo by Veress)

on their floor. Great doline density may be not only specific of giant depressions. Therefore, it is probable that the large density can be traced back to the large density of karren features (which can also occur outside giant dolines) too since these latter may also be capable of receiving sediment (Fig. 4.63). Clayton (1966) described this way of development. The density of subsidence dolines is extremely high if there is halite under the moraine (Ford and Williams 2007). It can also be observed that the frequency of suffosion dolines is larger than that of dropout dolines. This can probably be explained by the grain size of the cover. Dropout dolines are formed on fine-grained, clayey cover, the occurrence of which is less frequent on glaciokarst than that of coarse-grained cover.

- Their development is young, they were formed after the retreat of glaciers, following the development of the moraine. Therefore, the higher their elevation, the younger they are. It is probable that if they are absent on terrains of higher elevation it cannot always be traced back to the lack of development conditions, but it can also be explained by the fact that the short duration of deglaciation has not made their formation possible.
- Their formation is rapid. Thus, in the cirque under Triglav there was ice 50 years ago (Gams 2002). As they could only be formed following deglaciation, those occurring here are younger than 50 years old.

As regards their development age and rate of development, they show similarities with the formation and development of subsidence dolines that were formed in

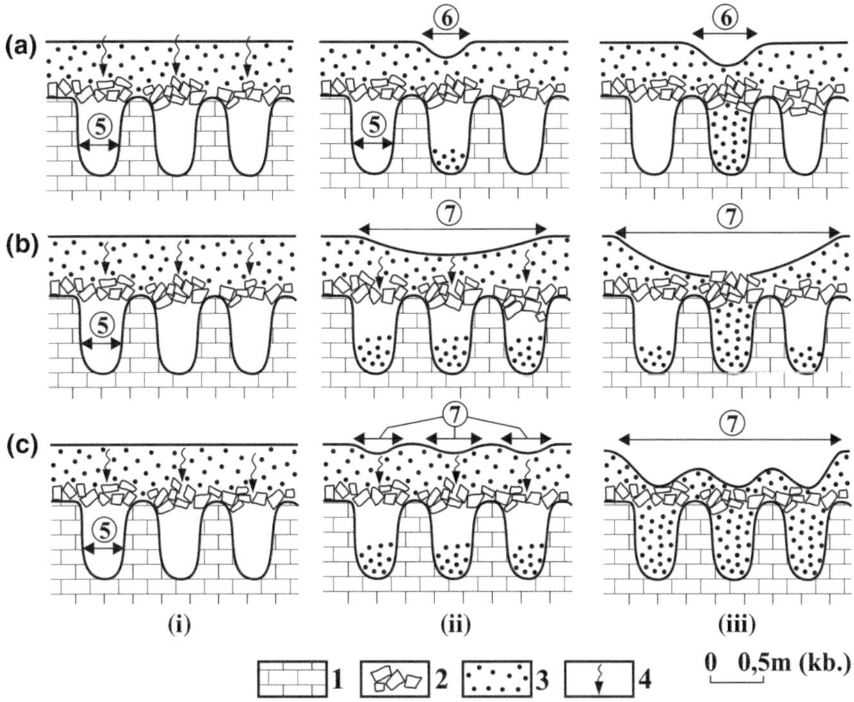

Fig. 4.63 Suffosion dolines formed on cover above grikes (Veress 2016a, modified). 1. Limestone, 2. coarser grained cover, 3. finer grained cover, 4. water infiltration and debris redeposition, 5. grike, 6. small doline with a diameter smaller than 1–2 m, 7. larger doline with some metres' diameter, **a** small scale doline above grike, **b** larger doline formed above several grikes, **c** development of larger doline with different rates of redeposition of debris into the grike, the floor of which will be uneven, **i** initial stage, **ii–iii** doline formation

extreme development environment, but are not of glaciokarstic origin. Thus, in the valley of the river Flint (USA) 312 dolines were formed within 48 h in 1994 during a tropical storm (Hyatt and Jacobs 1996). A gypsum bedrock or water extraction also results in a rapid development on non-glaciokarst. Thus, above a gypsum bedrock in the valley of Elbro River (Spain) a subsidence rate of 64.5 mm/year was also measured on a doline floor (Soriano and Simon 2001), while following a water extraction of 8 years 1032 dolines developed near the town of Liupunshui (China) (Waltham and Smart 1988). However, a rapid formation can be experienced under less extreme circumstances too. On loess, on the covered karst of Montgomery County (Tennessee state), the age of dolines was estimated to be 65 years (Kemmerly and Towe 1978).

Their Morphology

The characteristics of their morphology are the following:

- Among the subsidence dolines of glaciokarst, the varieties of dropout dolines can be well-distinguished morphologically which are the following (Veress 2016a). Dropout doline with soil developed in soil (Fig. 4.64a). This variety represents the initial phase of doline development. It also occurs on other covered karst. The floor of the dropout doline with a plain floor is flat and wide as compared to its depth. While its depth is of some tens of centimetres, its diameter can be several metres. The margins of the dolines of this doline variety are straight and in this case, the neighbouring margins enclose an angle with each other or they are separated into smaller, arcuate sections. They are the specific dropout doline variety of glaciokarst (Figs. 4.61b and 4.64b). The floor of the dropout doline with boulders is covered by boulders (Fig. 4.64c). The depth of dropout dolines with boulders is relatively large (0.5–1 m) as compared to their diameter (1–2 m). Although the side slopes of asymmetric dropout dolines are not vertical, their slopes intersect the soil and the cover too. The downstream side slope is shorter than the upstream (Fig. 4.64d). It is not a common doline variety, it only occurs on glaciokarst. The suffosion dolines of glaciokarst are less likely to constitute morphological varieties. However, suffosional blind-ended gully can be mentioned (Veress 2016a), which is a closed, channel-like feature (Fig. 4.38). It is a non-widespread, rare doline variety.
- Subsidence dolines can occur individually or in groups. On glaciokarst, when they occur in groups, they can form a straight row (Fig. 4.65), an arcuate row (Fig. 4.54) and an irregular group (Figs. 4.55 and 4.61). Subsidence dolines outside giant dolines are grouped in a straight row, while those situated in dolines can constitute a straight row (Fig. 4.35), an arcuate row (Fig. 4.54), or an irregular group (Fig. 4.61a).
- Subsidence dolines have passages which can be in the cover (passage in fill or non-karstic pipe) and in the bedrock too, or exclusively in the karstic rock if there is no cover on the floor. A floor without a cover is not characteristic of the dolines of glaciokarst and thus, the passages can be formed in the cover too (Fig. 4.66).
- Subsidence dolines can be simple and complex (Figs. 4.67 and 4.68). In the latter, in addition to passages, inner dolines, scars of mass movements, erosional features (gullies, creeks), debris cones and alluvial cones may occur. The subsidence dolines of glaciokarst belong to the simple variety. However, according to Ford (1984), on the floor of the subsidence dolines of glaciokarst there may be inner depressions of collapse origin too. According to him, these inner depressions are of postglacial age, but the bearing features are older than the last glaciation. On the side slopes of the subsidence dolines of glaciokarst, at most smaller scars of mass movements can occur. These are soil creeps, gelisolifluctional mass movements and rarely soil falls.

Fig. 4.64 Dropout doline varieties: **a** dropout doline with soil (Durmitor), **b** dropout doline with plain floor (Durmitor), **c** dropout doline with boulders (Durmitor), **d** asymmetric dropout doline (Hochschwab, photo by Veress)

Fig. 4.65 Subsidence dolines formed on till along fracture (crack) (Durmitor, photo by Veress). 1. Active subsidence doline, 2. inactive subsidence doline, 3. rills produced by runoff from rainfall or snowmelt, 4. Sušica canyon

Fig. 4.66 Passage with a diameter of some cm (Tauplitz alm, Totes Gebirge, photo by Veress). 1. Embryonic suffosion doline, 2. passage in fill

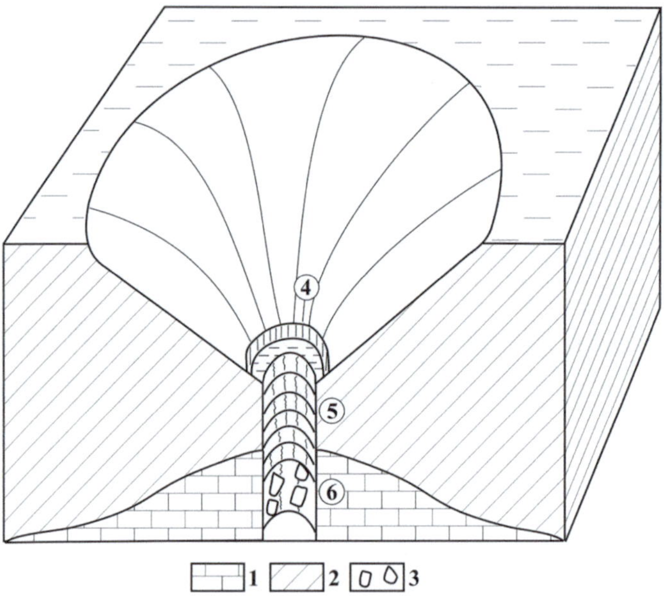

Fig. 4.67 Elements of a simple suffosion doline (Veress 2016a, modified). 1. Limestone, 2. superficial deposit, 3. collapsed material, 4. interior subsidence doline, 5. passage, 6. shaft

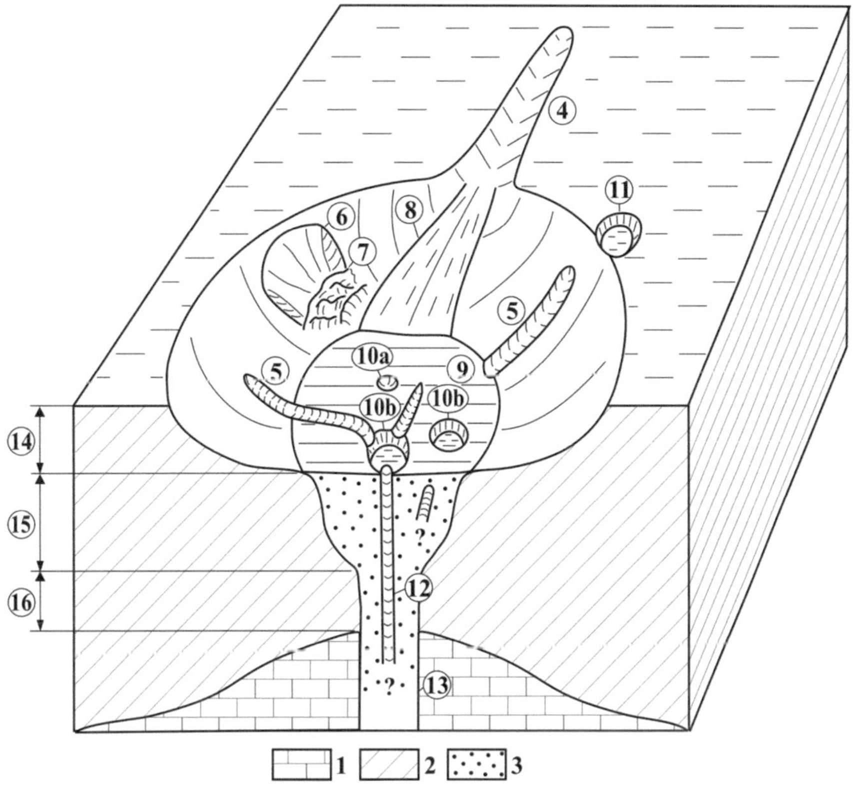

Fig. 4.68 Elements and morphology of a composite suffosion doline (Veress 2016a, modified). 1. Limestone, 2. superficial deposit, 3. bearing surface, 4. doline fill, 5. ravine, 6. gully, 7. slide, 8. slide heap, 9. alluvial fan, 10. flat floor, 11.a. interior suffosion doline, 11.b. interior dropout doline, 12. subsidence doline in marginal position, 13. passage in fill, 14. chimney, shaft, 15. upper part of doline without fill: with the erosion of slopes the doline was broadening and its slopes were getting gentler, 16. infilled lower part of the doline, 17. infilled non-karstic pipes in the doline

- The floor of subsidence dolines is often flat because of their filling up. This morphological characteristic is not specific to the dolines of glaciokarst. A plain floor occurs in case of dropout dolines because of their development by collapse. The floor of suffosion dolines is usually not plain, but funnel-shaped. This is caused by continuous material transport from the doline into the karst.
- Rain furrows, gullies, and creeks are often connected to dolines. On glaciokarst, the dolines of glaciokarst are less likely characterized by these features. Rain furrows occur more often (Fig. 4.65), while gullies (Fig. 4.69) are rarer. The smaller frequency of gullies and creeks can be caused by the lawn, the longer duration of frozenness of the soil, the smaller frequency of more intensive rainfalls and by the fact that surface runoffs are short because the dolines are situated close to each other.

Fig. 4.69 Gully (marked b) of one of the subsidence dolines of the giant depression that can be seen in Fig. 4.42 (Durmitor, photo by Veress)

- Subsidence dolines can be circular, elongated or grike-like (in this latter case, the opposite margins are parallel). Subsidence dolines with an elongated ground plan are formed along fractures or on slopes with a larger inclination. On glaciokarst, elongated suffosion dolines can develop along fractures (Fig. 4.65), but they can be elongated in the direction being identical with the dip direction of the slope too. Although suffosion dolines that developed on surfaces with slopes are not always elongated (their diameter may be larger in the strike direction of the slope), they can be asymmetric (Fig. 4.70). In this case, the side dipping oppositely the inclination of the slope is steeper and shorter. The dropout dolines of glaciokarst do not often have an arcuate ground plan, but they have a square, rectangle one.
- Subsidence dolines are mostly formed in fine-grained sediment and rarely develop in coarse-grained sediment. A reversed situation is specific of glaciokarst: they are more often formed in coarse-grained cover (Figs. 4.56b and 4.61a).
- The catchment area of subsidence dolines is small and usually it is not an individual catchment area which could be well separated from other dolines. On glaciokarst, this is true to a larger extent and the dolines often lack a catchment area completely, they only receive water from the meteoric water falling in their area.

Fig. 4.70 Asymmetric doline (Totes Gebirge, photo by Veress)

- The side slopes of subsidence dolines can be convex, concave, straight and normal slopes (they consist of convex and concave parts). The side slopes of the dolines of glaciokarst are mainly short and straight.
- The diameter of the suffosion dolines of glaciokarst is smaller as compared to their depth than that of dropout dolines. These latter have a small depth and their diameter is large compared to this.
- Intermittent or permanent lakes often develop in subsidence dolines. In the lack of detailed investigations, it cannot be estimated either whether lakes are formed more rarely or more often in the dolines of glaciokarst as compared to other covered karsts. At the same time, one of the inducing factors of lake development is only specific of the subsidence dolines of glaciokarst: in a permafrost environment, an ice plug can develop in the drainage passages which results in the back swelling of water in the passage (Ford 1984).

Their Evolution

The development age of subsidence dolines may be identical with the age of the sediment receiving feature being formed on the bedrock or it can be younger than this. In the former case, the doline development is syngenetic, while in the latter it is postgenetic (Veress 2016a). On glaciokarst, where the superficial deposit is mainly of limestone material, the water percolating through this can become saturated. Syngenetic, subsidence dolines can develop in case of limestone cover if the percolating water still has a dissolution capacity on the bedrock and therefore, it is able to create sediment receiving passages there. However, if it does not have a dissolution capacity, subsidence dolines can only develop, which are thus, postgenetic if sediment-receiving features had already been formed on the bedrock. In case of a postgenetic development, the sediment receiving features of the bedrock could be formed by meltwater (under the ice or in the foreground of the ice), but they could also develop preceding ice formation.

If the water arrives at the bedrock in a saturated state because of large cover thickness, the already existing drainage passages do not increase, karstic passages may become filled and there is a large chance for the cessation of doline development later. (Moreover, as the grain size of the cover increases, it is increasingly difficult to be transported by suffosion and the sediment receiving passages of the bedrock should be increasingly wide to be able to receive the grains.) If the water arriving at the bedrock still has a solution capacity, the doline development still takes place because of the increase of the drainage passage. A chance of the subsequent increase of the bedrock passages increases even if surface water arrives at the karstic passages through the drainage passage of the postgenetically developed doline. Since it does not percolate through the cover, this water does not become saturated, because of this getting into the passages of the bedrock, it widens them by dissolution. The non-percolating away surface water not only becomes saturated, but it can also transport CO_2 surplus. This is proved by the fact that a relation can be determined between the cave flood wave (which takes place because of the streams flowing inward through ponors) and the increase of the CO_2 content of cave air in the Hungarian Baradla cave and Béke cave (Stieber and Leél-Őssy 2016).

Suffosion dolines can be formed above the karren features of the bedrock (Clayton 1966), above its shafts and giant grikes (Ford and Williams 2007). However, the larger the features that developed on the bedrock, the larger the chance that instead of suffosion (or together with it), gravitational rock fall and the subsidence of the cover play a role in the development of dolines. Suffosion dolines can develop in the following ways (Veress 2016a).

- The cover is transported into the karren mainly by suffosion. In this case, small-sized suffosion dolines develop either in a syngenetic or postgenetic way too (Fig. 4.63). Dolines developing in this way have a small size and a large density.
- The superficial deposit gets into the shafts by the fall of debris parts, perhaps by subsidence or collapse. The dolines are postgenetic and relatively large sized. The material of the cover can get into the shafts directly or indirectly. In case of a direct development, the debris parts of the cover fall into the shaft that has no sediment fill (Figs. 4.24a and 4.55). An indirect development takes place if the upper part of the shaft is filled. In this case, during depression formation, the material of the cover can get into the shaft in several ways. Thus, the finer

grained superficial deposit is accumulated in the gaps of the debris filling the shaft and is transported to a greater depth between the gaps (Fig. 4.71a). It can happen in a way that the boulders move downwards because of the melting of snow and ice located between them, the parts of the cover fall into the space developing in this way or the cover subsides into it (Fig. 4.71b). It can also take place in a way that the debris gets into a lower position in the shaft because of the melting of the ice or snow plug which also causes the parts of the cover to fall into it or the cover subsides into the developing space (Fig. 4.71c). Finally, rock fall or subsidence can be replaced by the collapse of the cover if the material of the shaft accumulates at a lower part of the shaft (Fig. 4.71d) or if the upper part of the shaft becomes wide (Fig. 4.71e).

- The superficial deposit is transported into giant grikes by debris fall, by collapse, by subsidence, but even by suffosion too. Doline rows develop (Fig. 4.72a), but if the grike is wide, the dolines may be irregularly arranged (Fig. 4.72b). With the dissolution along the bedding planes of the beds between hogbacks, grikes develop, above which dolines with a larger diameter are formed if the cover subsides uniformly (Fig. 4.72c). If the subsidence is of smaller expansion, doline rows develop which consist of elongated dolines (Fig. 4.72d–e). In the cases presented in figures c–e, the development of suffosion dolines is syngenetic. However, dolines with a circular ground plan develop in the cover above buried roche moutonnées, even if grikes or pits are present in the rock (Figs. 4.72f and 4.73).
- Finally, small-sized dolines are formed also in the case if the limestone becomes detrital with dissolution. Below the debris parts, the bedrock is dissolved locally and its physical weathering continues. The debris subsides into the depression that is formed by dissolution. This doline variety is a transition between solution dolines and covered karst dolines (Veress 2016a).

Dropout dolines can develop by the collapse of the cavity of the cover (Waltham et al. 2005) or by the collapse of the cavity of the bedrock (Veress 2016a). Dolines with a plain floor are formed by the collapse of the cavity of the cover, while dolines with boulders and asymmetric dolines develop by the collapse of the cavity of the bedrock (Veress 2016a).

Since suffosion dolines develop mainly on fine-grained cover by suffusion, there is a greater opportunity for their occurrence in the lower part of the periglacial zone and on terrains below the tree line, where fine-grained sediment (clayey moraine, fluvioglacial sediment) is more widespread. As the fine-grained cover can also be transported into karren features by suffusion thus, at the above-mentioned sites, the size of suffosion dolines is small, their density is high, and they often occur with dropout dolines together. Because of clayey cover that is soluble to a less degree or is not soluble at all, syngenetic subsidence dolines are more likely to occur here.

In the upper part of the periglacial zone, where the cover is coarser grained (and thicker) suffosion dolines mainly develop by rock fall and collapse. The size of suffosion dolines is larger, their density is smaller and their development is predominantly postgenetic since calcareous cover is more widespread.

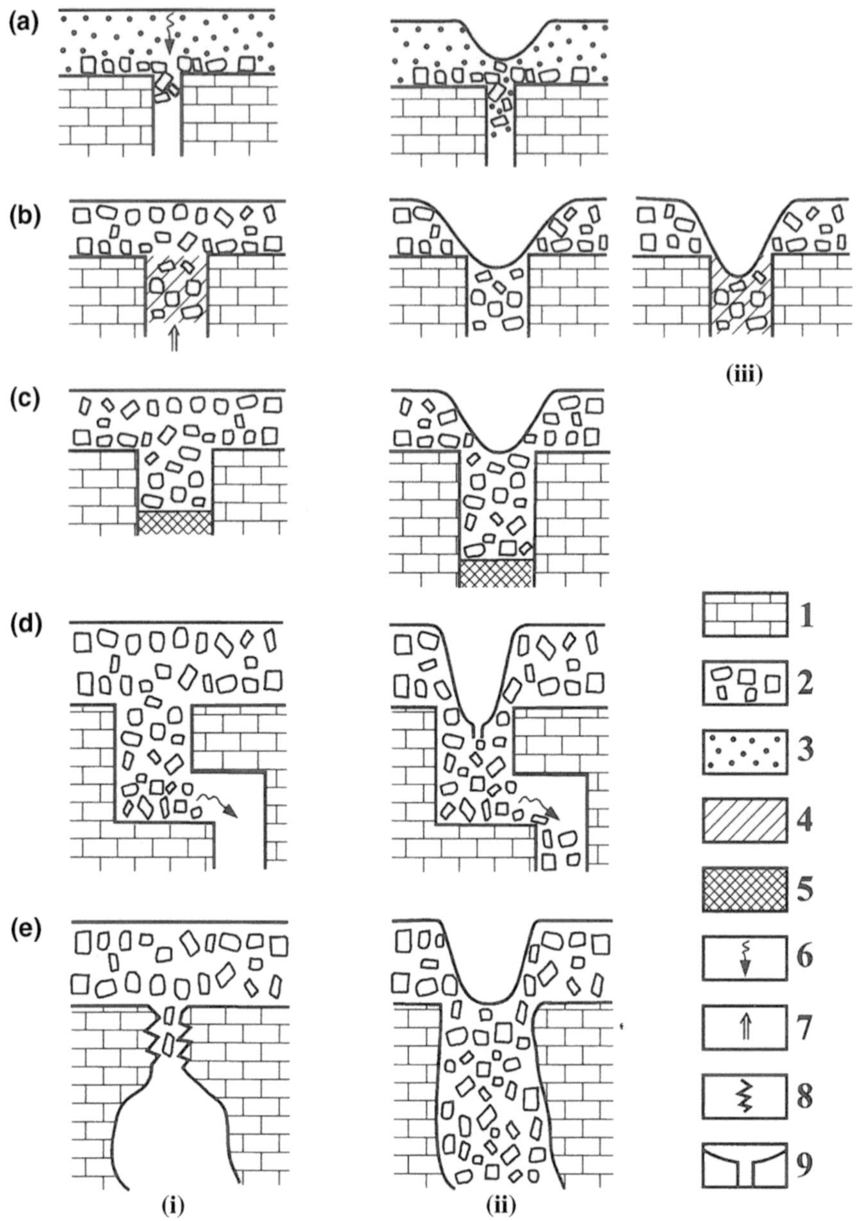

Fig. 4.71 Modes of postgenetic doline formation above grikes and shafts of the mountain karst. (Veress 2016a, modified). **a** the frost-shattered debris is transported deep into the shafts, **b** the ground ice between the blocks melts and the blocks are displaced downwards in the shaft, **c** under the blocks the top part of the snow and ice fill melt and the blocks above the fill are displaced downwards, **d** the collapsed material is redeposited from the vertical shaft section into the bottom of the shaft, **e** the top part of the shaft is broadened by solution and the collapsed material falls into the shaft, 1. limestone, 2. collapsed and morainic material, 3. frost-shattered debris, 4. ground ice, 5. snow, 6. water infiltration, 7. upward air current, 8. surface formed by collapse, 9. doline, **i** initial stage, **ii**–**iii** mature stage

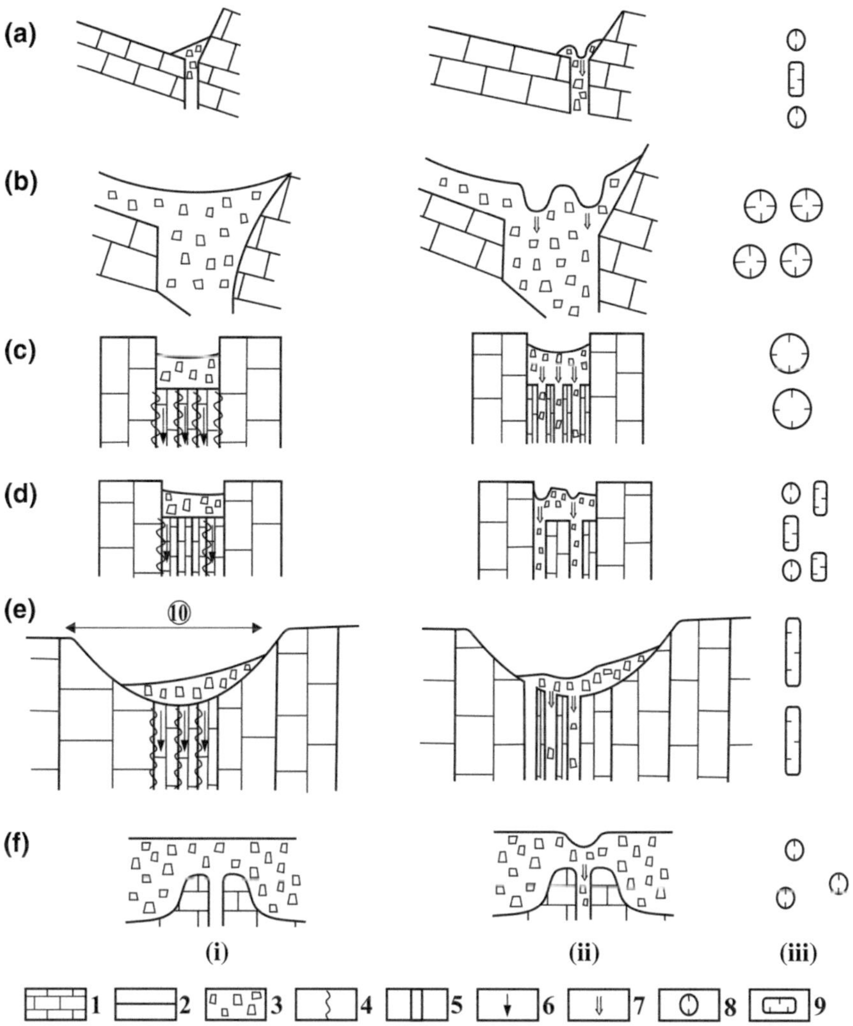
(a)
(b)
(c)
(d)
(e)
⑩
(f)
(i)
(ii)
(iii)
1
2
3
4
5
6
7
8
9

◂**Fig. 4.72** Formation of subsidence dolines on terrains affected by glacial erosion (Veress 2016a, modified). In cross section: **i** initial stage, **ii** mature stage, **iii** mature stage in plan view, 1. limestone, 2. bedding plane, 3. morainic deposit, 4. solution along bedding plane, 5. solution grike along bedding plane, 6. water infiltration, 7. debris fall, 8. subsidence doline of circular groundplan, 9. subsidence doline of elongated groundplan, **a** doline row of dolines with circular and elongated groundplan develop in the debris accumulating above the solution grike at the front of the scarp front, **b** doline row of large dolines with circular groundplan develop in the debris above giant grike at the scarp front, **c** doline row of large dolines with circular groundplan develop in the debris covering the more denuded terrain of thin beds between hogbacks, **d** doline row of dolines with circular and elongated groundplan develops in the debris above thin beds between hogbacks, **e** doline row of dolines with elongated groundplan develops in the debris of a rock basin formed above the thin beds of terrain with hogbacks, **f** dolines with circular groundplan develop in the debris above pits (giant grike) formed on roches moutonnées. Remark: In the parts **a**, **b** of the figure, glacial scarp fronts are shown on terrain with strata of 10–20° dip, while in the c, d and e parts it is shown that on the terrain built of subvertical strata (the profile in c and d is perpendicular to the strikes of strata, while in e it is in the direction of strike) the ice created hogbacks with depressions between them. In the parts **c–e** actual solution takes place along the bedding planes

Fig. 4.73 Subsidence dolines that developed on covered hogbacks and in their spaces (Todorov valley, near Sedlo, Durmitor, photo by Veress)

4.5.2 Ponors

The characteristic features of the ponors of glaciokarst mainly manifest in the fact that they were mostly formed in karstic depressions or in glacial erosional depressions. Because of this, the size of the catchment area of the ponor or maybe its shape too is determined by the size and shape of the bearing depression. The significance of the ponors is given by the fact that they have a determinant role in the turnover of the depressions of glaciokarst. It is also another characteristic feature that as compared to other karsts, there are several other features functioning as ponors on glaciokarsts (see below). By a ponor, we mean a drainage depression formed at the rock boundary (allogenic karst) (Jakucs 1977; Gams 1994; Ford and Williams 1989, 2007) which develops on the floor of the valley, blind valley that was formed on a non-karstic terrain and that terminates at rock boundary. However, covered karst ponors (Veress 2016a) and the features functioning as ponors also belong to these features. Covered karst ponors occur inside covered karst at the boundary of the impermeable and permeable rock. Therefore, limestone does not crop out at the ponor (Veress 2016a). Features functioning as ponors mostly occur on bare karst and thus, they did not develop at rock boundary. Ponor development and the functioning of some features as ponors is favoured by the increased amount of meltwater which mainly takes place during the retreat of glaciers (Ford 1996).

Drainage features which do not develop at rock boundary thus, functioning as ponors may be the following:

- Meltwater flows into the shafts and giant grikes that are subglacial or situated in front of the ice (Fig. 4.74a).
- There are passages under the lakes into which the water of the lake flows. Such passages were described by Bauer and Zötl (1972) from Lakes Elm and Steyree (Totes Gebirge).
- The water overflowing from the lake of depressions is drained in karren features (for example the water of the lake under Wildgössl peak in Totes Gebirge flows into a karren feature).

The water of sloping surfaces flows into shafts and giant grikes.

The water flowing from the permafrost terrain that developed in the foreground of the glacier flows into karstic passages.

- However, those subsidence dolines also function as ponors, which receive water from their environment often through the gullies leading to them.

Ponors can be of preglacial, glacial, interglacial and postglacial age. From the Kanin Mountains, Kunaver (1983) mentions such ponors which are filled by till. According to him, this proves the development of these features in the last interglacial. Ponors of glacial age could develop under the glacier (Fig. 4.74a, Corbel 1957; Ford 1979, 1984; Smart 1986, 2004; Žebre and Stepišnik 2015a) or at the margin of the glacier (ice sheet) (Žebre and Stepišnik 2015a). Neither varieties develop necessarily at rock boundary. However, glacial erosion and the meltwater

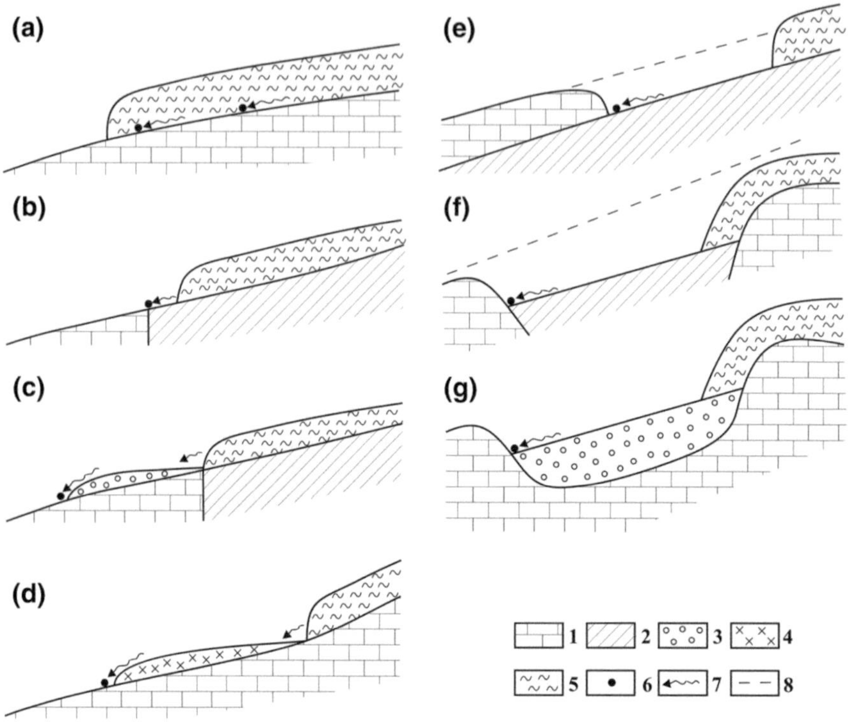

Fig. 4.74 Cases of ponor development on glaciokarst. 1. Limestone, 2. non-karstic consolidated rock, 3. clayey moraine of non-karstic material, 4. clayey moraine of karstic material, 5. glacier, 6. ponor, 7. meltwater or surface stream, 8. former surface of the karstic rock (**e**) and non-karstic rock (**f**) destroyed by glacial erosion, **a** subglacial ponor development, **b** meltwater leaving the glacier creates a ponor at the termination of the non-karstic rock, **c**, **d** ponor develops at the termination of the moraine deposited to the karstic rock, **e** the non-karstic rock outcropping from below the limestone destroyed by glacial erosion constitutes a rock boundary where a ponor develops, **f** the non-karstic rock is exposed in a rock basin and a ponor develops at its margin, **g** a ponor develops at the margin of the fill in a karstic depression transformed by glacial erosion

of the glacier may often create a rock boundary and thus, a typical ponor too. These ponors may already develop when the glacier still exists. Ponors of glacial age especially develop in case of retreating glaciers (Ford 1996), when a lot of meltwater is generated, but they can also develop after the disappearance of glaciers (ponors of postglacial age). Ponors of an ice marginal position can develop to the effect of the meltwater leaving the glacier or at the termination of the glacier (Auly 2008) or farther from that (Kiernan et al. 2001; Žebre et al. 2016). Ponors situated increasingly farther have a growingly larger chance for a development at rock boundary (Fig. 4.74c–d). Since the meltwater can only get farther from the glacier if the surface is covered because if it is uncovered, then the meltwater already percolates away close to the ice (diffuse drainage). It is probable that some ponors, but mainly the features functioning as ponors become inactive with the

Fig. 4.75 Ponor in the giant depressions presented in Figs. 4.44 and 4.45 (Hochschwab, Austria, photo by Veress). 1. Lake and marshy area which were developed by the infilling of an older ponor, 2. present water drainage sites, 3. rock boundary, 4. side slope of depression

disappearance of the glacier or they fill up. The rock boundaries connecting to the glaciers and thus, ponors can develop during accumulation, at the termination of the moraine (Ford 1996) (Fig. 4.74c–d) and they can be formed during glacial erosion (Fig. 4.74e–f). In case of a development related to glacial erosion, in the first phase, the non-karstic rock crops out during the advance of the glacier, then in the second phase during retreat, the meltwater creates ponors at rock boundary. The accumulation can also take place in the area of a depression (Fig. 4.74g). In this case, the fill creates a rock boundary at the steep side slope of the depression. Such a situation can be seen in Fig. 4.75. The fill can be purely till, when the ice filled the depression at least partly (Fig. 4.74g), it can also be fluvioglacial, when the ice terminated outside the depression. It is also possible that ponor development is not directly related to glacial geomorphic activity. Ponors often develop in such karstic depressions having been transformed by glacial erosion which became later partially filled with sediment of fluvial origin. Ponors can develop at the termination of alluvial cones that developed during the process (Fig. 4.46). Ponors that were formed here are of postglacial age.

Ponor points can be shifted. In this case during the denudation of the cover, newer ponors can be formed in a lower position (Goeppert et al. 2011) or during accumulation in a higher position (Veress 2016a; Žebre et al. 2016).

4.5.3 Poljes

Classifying poljes, Gams (1974, 1977, 1978) differentiated border polje, overflow polje, piedmont polje, peripheral polje and polje in the piezometric. Ford and Williams (2007) distinguished three genetical types of poljes: structural polje, border polje and baselevel polje. Lehmann (1959) differentiated Hachfächenepolje, Talpolje and Semipolje. Talpoljes developed in older valleys. The term small polje is used for depressions of a small area (with a diameter of about 100 m), but also for depressions with the characteristic features of a polje (Goeppert et al. 2011). Poljes can bear several karst features thus, karren, dolines (solution dolines and subsidence dolines), shafts and blind valleys with ponors.

In the formation and development of poljes, the smooth becoming of the polje floor by dissolution under permeable superficial deposit and the dissolution taking place at the termination of the cover, at the margin of the polje, resulting in the widening of the polje are emphasized (Gams 1978). According to Gams (1978), marginal corrosion takes place if the cover is impermeable, while marginal and subsoil corrosion take place if the cover is semi-permeable and suballuvial corrosion occurs at the foot of the slope if the cover is permeable. Glaciers may affect the development of poljes (Gams 1994).

Piedmont poljes deepened and became smooth under glacial and fluvioglacial sediment to the effect of the abundant amount of meltwater (Sweeting 1973; Gams 1978). It is probable that the evolution of piedmont poljes already began before the formation of glaciers. Their development becomes intensive when the advancing glaciers approach and penetrate into the polje (Žebre et al. 2016). However, non-piedmont type poljes—dependent on their position—could also have been affected by glacial effects (glacial erosion and the accumulation of stream load of glacial origin).

As compared to glaciers, poljes can be of various positions. Thus, the polje could be completely covered by ice which is proved by the till occurring in poljes (Auly 2008) or the glacier penetrating into the area of the polje (Žebre et al. 2016) covered a part of it. In both cases, the ice and the polje can form a common system and they can mutually affect each other's development. This is increasingly effective, the closer they are to each other at the beginning of the glacial. The orientation of the glacier is determined by the polje (if it already exists), then following this the glacier guides the way and degree of polje development. The meltwater of the glacier contributes to the formation of the polje or to its further growth. The margin of the polje, the area of which was formed or increased by the glacier, shifts away from the ice margin then this latter advances into the polje part that was formed by it. If the glacier advances further to the area of the polje, as it has already been mentioned, it can completely cover the polje or it can even exceed it (Fig. 4.76, Auly 2008). In case of ice caps, the ice may not only cover the polje, but its surroundings too (Adamson et al. 2014). In case of a glacier filling the polje, the polje is located at the end of the glacier valley, if it went beyond it, then the polje has a transitory position on the valley floor (Auly 2008). It can be generally worded

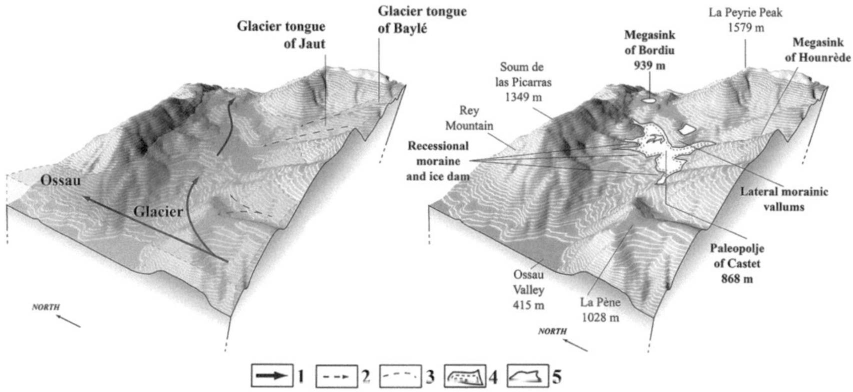

Fig. 4.76 Polje on the floor of a trough in the Pyrenees (after Auly 2008, modified). 1. Main flow direction of the glacier, 2. small local glaciers, 3. supposed limit of glacier, 4. hanging paleopolje, 5. megasink

that independently of the polje type, the higher altitude the polje has and the smaller its expansion is, the greater the chance is that it was filled by ice. The lower its altitude and the larger its expansion, the smaller the possibility for this is. In case of a complete covering, glacial erosion is predominant (though accumulation exists in this case too). In case of a partial covering, glacial erosion, accumulation, fluvial reworking, dissolution, karstic drainage and a relating fluvial erosion exert an effect.

The glacier if it did not enter the polje, could terminate at its margin or farther from it. If the glacier is of marginal position, accumulation and dissolution operate in the polje. If there is a ponor or there are subsidence dolines in the polje, karstic drainage as well as fluvial erosion take place and thus, the cover is transported into the karst.

If the margin of the polje is farther from the glacier end, glacial erosion does not shape the polje, but the polje receives stream load of ice origin (Adamson et al. 2014; Žebre et al. 2016). The farther the polje from the glacier end is, the more fluvioglacial the accumulation material is. The fluvioglacial sediment can be replaced by fluvial sediment and colluvial sediment.

The glaciers primarily control the development of poljes by their meltwater. If the amount of meltwater increases, but the sediment of the polje floor is coarse-grained, percolation intensifies. Subsidence dolines may develop. The karstic floor can also get dissolved if the calcareous content of the cover is low. Karst water level can rise, which results in the transformation of polje hydrology. Because of the rise of karst water level, sediment transportation into the karst can decrease, the intensity and nature of dissolution can change (vadose dissolution may be replaced by phreatic dissolution). If the cover is fine-grained, since the impermeability of the floor increases, the inclination to ponor development increases. The amount of meltwater can increase in the polje if the glacier is in the polje or at its margin and retreats. (However, the amount of water can increase in the polje independently of glaciers in the interglacials.)

The decrease in the amount of meltwater can occur in the poljes during the advance of glaciers, in the glacial(s). However, it can also take place during the retreat of glaciers thus, during the interglacial (or at its beginning) too, if the glacier gets far from the polje during its retreat since in this case the meltwater gets into the karst outside the polje.

During the advance of a glacier, a coarse-grained sediment is accumulated in the polje, which favours the above-mentioned seepage. Glacier retreat increases the possibility that a fine-grained (fluvioglacial) sediment accumulates in the polje. Thus, the impermeable nature of the floor and the inclination to ponor development increase too.

The sediment of the polje may be the following:

- Fill of till.
- Glaciofluvial sediment (Adamson et al. 2014; Žebre and Stepišnik 2015a), the grain size of which is extremely diverse and the roundness of the grains may also be very different depending on the distance of transport. Farther from the former glacier end, the grain size of the cover becomes finer, its stratification increases and its thickness decreases (Zebre and Stepisnik 2015a). These sediments may be interrupted by non-reworked moraine wedgings in (buried moraine ridges) (Žebre et al. 2016).
- Lacustrine sediment (Farias et al. 1996; Menkovič et al. 2004; Žebre et al. 2016) which may originate from lakes that may develop in ponors of the polje margin, but inside the polje too. Lacustrine sediment is formed if the karstic drainage system becomes plugged (Sweeting 1973), karstic cavities become filled, karst water level reaches the polje floor or a clayey sediment arrives at the depressions of the polje.
- Debris originating from the non-glacigenic denudation of slopes.

All these sediments contribute to the fact that the polje floor is transformed into allogenic karst, buried karst or concealed karst. During the repeating sediment transport inward and outward, the expansion and nature of covered karst changes (concealed karst turns into allogenic karst).

The features of the polje that can be associated with glaciers are the following:

- Outwash plains may develop (Adamson et al. 2014; Žebre and Stepišnik 2015a; Žebre et al. 2016).
- Plain floor which dips from the direction of the former glacier into the direction of the ponor.
- The floor may be dissected by positive features, by moraine ridges (Žebre and Stepisnik 2015a; Žebre et al. 2016) or by ground moraine that is dissected by hummocky moraines. (The place of moraine ridges determines the termination of the glacier in the polje.)
- At the margin of the polje, alluvial cones (Žebre et al. 2016) may be arranged which are of glaciofluvial or fluvial origin, but these may be of flood origin too. Floods develop from proglacial lakes (Stepišnik et al. 2010).

- Because of the expansion of the moraine or the fluvioglacial sediment, a ponor develops at the polje margin.
- In relation to glaciers, intermittent or permanent lakes may be formed in the poljes. Thus, a lake developed in the Gomance polje in the glacial which retreated in the interglacial (Žebre et al. 2016). The development of intermittent lakes can be contributed by the under-development of the passage, its plugging, the increase of the amount of meltwater, the formation of glacier floods and by the rise of karst water level.

The development of piedmont poljes is controlled by the amount and type of the sediment transported into the polje and the water arriving there. During the glacials, when a lot of coarse-grained stream load arrives, especially if the glacier reaches the polje or at the end of the glacial when there is a lot of meltwater, accumulation takes place in the polje (Adamson et al. 2014; Žebre and Stepišnik 2015b). The karstification of the polje is determined by the inward transported sediment. The cover will be impermeable if the inward transported material is clay or clayey, if the non-karstic rock that is intercalated into the limestone is denuded and transported and if the deposited material becomes cemented. If the cover is impermeable, a ponor develops, while if it is permeable, subsidence dolines develop. In case of the presence of both, the karstic assemblages will be mixed too. In case of a fine-grained sediment transport (this depends on the grain size of the sediment created by the glacier, on the intensity of transport by water, on the distance of transportation) the impermeable nature of the sediments of the floor also increases. In the beginning, accumulation favours ponor development (karstic drainage), then further accumulation may also result in the plugging (burial) of the ponor of the polje, allogenic karst turns into buried karst. If the proportion of non-karstic rock is high in the inward transported sediment, at the end of the glacial when the amount of meltwater increases and the water cannot be drained in the plugged ponors, but it percolates on the polje floor, dissolution increases on the floor and at the margin of the polje (the polje grows by dissolution).

4.6 Karstic Relict Landforms

These are elevations with various sizes and shapes. Since such features can develop on a surface of lower altitude or on tropical karsts, they do not frequently occur on glaciokarst. As an example the Pyrenees can be mentioned where the glacier ice truncated the elevations (Auly 2008). Bayari et al. (2003) described tropical cockpit karst being transformed by glacial erosion from the Aladağlar Mountains (Turkey). Ford (1996) mentions limestone hills being transformed by the ice from Canada. Increasingly farther from the margin of the former glacier, the elevations are truncated to an increasingly larger degree (Auly 2008). The shape of elevations can also be modified by the denudation of their side slopes (Fig. 4.77).

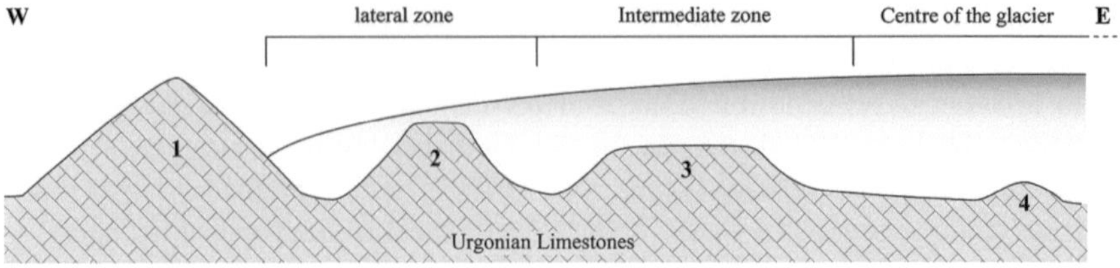

Fig. 4.77 Karstic mounds being truncated to different degrees from the Pyrenees (after Auly 2008, modified). 1. Karstic mound (mogote type), 2. karstic mound truncated to a smaller degree, 3. karstic mound truncated to a larger degree, 4. karstic mound that lost both its height and expansion to a large degree

4.7 Lakes

Although lakes are hydrological features, they have to be mentioned since their existence is strongly connected to glaciokarst. The lakes of glaciokarst may be permanent and intermittent lakes. Mainly, the lakes belonging to the former group can be of various size and they may often be related to the glacial erosional depressions too, on the floor of which the non-karstic rock is exposed. Permanent lakes may also have karstic drainage (Bauer and Zötl 1972).

Lakes that can be related to the karstic features of glaciokarst may develop in giant dolines, in poljes, in subsidence dolines and in ponors. The reason for their development may be the impermeable sediment depositing in karstic depressions, the plugging of passages with sediment or with ice, the rise of karst water level to their floor. They can also be formed if the karstic passages are underdeveloped and the water supply increases to a great extent.

4.8 Conclusions

On glaciokarst, with the exception of tropical karst features, all surface karst features are present thus, karren, giant grikes, shafts, various dolines, ponors, and poljes. However, the occurrence and frequency of various features depend on in what climatic environment the glaciokarst occurs. At the same time, it can be established that karren dominate on glaciokarst independently of the environment. They are widespread and occur in great diversity and they are landscape features of glaciokarst. Karren are postglacial features, but they can be of glacial age too. The latter are mostly those which existed during the disappearance of ice because of the continuousness of karren formation under ice. As the time of the disappearance of ice cover can be different even within a glacier, the absolute age of inherited karren can be different too. From karren of glacial age mainly grikes and schichtfugenkarren could survive (though they continued their development during the Holocene). Their four levels can be differentiated: karren features, karren assemblages, karren cells and karren formation units. Their dominance can be attributed to minutely dissected glacial erosional terrains, where stepped surfaces (limestone pavement) are predominant. They are constituted by a series of surfaces which have various expansion and inclination and which are mainly bare (or partly bare) and more or less plain. The dissected nature is intensified by other glacial erosional slopes and features (the slopes of glacier valleys, roche moutonnées, rock basins, transverse steps) and by karstic depressions. Karren, mainly of flow origin could probably develop on these bare, glacial erosional surfaces of the periglacial zone. On glaciokarst, karren constitute patches of various expansions and therefore, the distribution of karren is mosaical. The pattern and size of these patches and the fact what karren and karren assemblages occur in a place depend on the dip direction of surfaces with bedding planes and the surfaces with heads of bed, on the degree of inclination and on their expansion. The dip direction depends on the relation between ice motion and the spatial position of the beds, the degree of the dip and the expansion of the surface that developed by glacial erosion depend on the thickness and the dip of the beds and on the sublacial karren formation (see Chap. 3).

The appearance of a given karren feature and karren assemblage can be determined by the dip and expansion of the slope and the structure of the constituting rock (for example the presence or lack of joints) or the formation of a karren feature of the karren assemblage can be generated by the existence of another one. Karren features of flow origin are characteristic of surfaces with bedding planes which are of small dip and and larger expansion as well as of the side slopes of roche moutonnées and paleodolines where water flow is undisturbed. Wandkarren are exceptions which develop on slopes with a large dip (in case of a surface being large enough). Thus, wandkarren occur on the side slopes of glacier valleys, on steps, on the wall of rock basins and on the heads of bed of stepped surfaces (scarp fronts). Karren features of seepage origin are formed on slopes with a smaller dip (exceptions are schichtfugenkarren) or even on slopes with a smaller expansion as small dip favours percolation, but the expansion of the slope does not influence percolation or its degree. Specific features of heterogenous karren formation are troughs, stepped surfaces and roche moutonnées. On slopes with different inclination (and thus, on a glacial erosional feature), various karren features (karren assemblages) may occur.

On slopes with a small dip, with the increase of the slope angle (until 60°–70°) the number of the types of karren features constituting karren assemblages and their size decrease and karren of flow origin will be increasingly predominant at the expense of karren of seepage origin. On slopes with a large dip, (from 60° to 70°) if the slope angle increases, the richness of the types of karren assemblages that have a small amount of features as a matter of course decreases, and karren of flow origin (wandkarren) will be predominant at the expense of karren of seepage origin (exceptions are schichtfugenkarren). It can be established that the greatest diversity of karren features occurs on slopes with a dip angle of 11°–20° mainly if the expansion of the bearing slope in dip direction is 10 m or greater.

Giant grikes could be formed by meltwater (these are of glacial age) or by the coalescence of grikes and by the coalescence of pits, grikes and shafts. The latter are of postglacial age.

Shafts rarely occur in groups in spite of this they are characteristic features of glaciokarst. They were mainly developed by meltwater and thus, they are of glacial age, but they can further develop on a surface exempt from ice. Shafts with a small width are of postglacial age. The depressions of glaciokarst are solution dolines and uvalas, subsidence dolines, ponors and poljes.

Former karstification causing the development of giant dolines ceased, nowadays this process is manifold and operates at several points inside them. As a result of this, glacial features (roche moutonnées), periglacial features (debris cones) and karstic features (karren, giant grikes, small-sized solution dolines, subsidence dolines and ponors) may occur inside them. According to their position, giant dolines may be valley dolines (preglacial giant dolines and uvalas), valley floor dolines, cirque dolines, depressions that turned into cirques, dolines and uvalas outside the valley. Valley dolines determine the shape of the trough, they are preglacial, mixed features. Valley floor dolines are situated on the floor of the trough and they are of interglacial age. Karstic depressions that were transformed into cirques are also of preglacial or interglacial age. Karstification causing transformation from a cirque into a cirque doline may be of Holocene age or older than this if the cirque already developed in a former glacial. Giant depressions are horizontally developed: their diameter is several times larger than their depth, the dip of their side slope is relatively small. These characteristic features of giant depressions, but the large size also prove that they did not develop in a glaciokarstic environment. Since their sediments can be transported into the karst, giant depressions can be exhumed.

Small-sized solution dolines and schachtdolines are vertically developed: as compared to their diameter, their depth is large, their side slopes are steep. These features are formed during local dissolution. This is promoted by snow accumulation, which is supported by the increase of altitude and local circumstances promoting accumulation and the fact that since no fine-grained sediment exists inside them, drainage from their area is vertical. If they are not fossil, they are postglacial features that developed in glaciokarstic environment. Their shape and thus, the thing that recent dolines are transformed into small-sized solution dolines or schachtdolines depends on the duration of complete snow fill. If this time is long, a schachtdoline develops, if this duration is short, a small-sized solution doline develops.

The subsidence dolines of glaciokarst are of small size and if they occur in groups, their density is high. The small size can be explained by the small size of karstic passages, by the small thickness and large grain size of the cover. While high density can be attributed to the large density of sediment receiving karstic features (karren). As compared to the subsidence dolines of other karst types, their morphology is simple (they mostly lack inner dolines, mass movements and gullies). Subsidence dolines can be syngenetic or postgenetic. In case of the former, the sediment receiving feature of the bedrock and the depression of the cover develop simultaneously, while regarding the latter, the sediment receiving passages of the bedrock developed earlier (in any one of the glacials or interglacials). The karstic passages of postgenetic dolines do not always become filled with inwashed cover since the water flowing in through the passages that developed in the cover can maintain their subsequent growth. Suffosion dolines can develop not only by suffusion, but by debris fall if the superficial deposit covers a larger sized shaft or giant grike. These suffusion dolines are transitions to dropout dolines. Suffosion dolines that developed by suffosion are formed in areas covered by fine-grained cover. Such terrains occur at the lower part of the periglacial zone and below the tree line. On the upper part of the periglacial zone, their development takes place by rock fall.

The ponors of glaciokarst can develop at drainage karst features (shafts, karren) or at rock boundary. The latter are typical ponors which could be formed at rock boundaries of glacial erosional origin or accumulation origin. However, they can also occur at the rock boundary that developed during the accumulation of depressions. Typical ponors that developed at rock boundary could be formed in the glacial and they could also maintain their active nature at present too. However, their development can also be postglacial. In the above-mentioned cases, the formation of the rock boundary (and thus, the ponor too) is directly caused by the geomorphic activity of ice. The role of ice is indirect if the rock boundary is formed subsequently by other processes (fluvial transport) in a depression.

On glaciokarst, the development of poljes is determined by glaciers to a large degree since the glacier directly contributes to the formation and development of poljes by erosion and by its meltwater. However, poljes can also affect the expansion and position of glaciers. The closer the glacier to the polje is, the closer this relation is. In case of glaciers being close to poljes or situated inside them, glacial erosion and accumulation directly caused by the glacier are significant, in case of a glacier being farther from the polje, accumulation is predominant in the polje, in which fluvial transport has an increasing role. In case of permeable sediments of glacial origin, the polje deepens and widens by dissolution, the sediment in its area can only be transported into the karst through subsidence dolines to a limited extent. In case of an impermeable sediment, the sediments of the polje are transported into the karst through ponors, the plugging of the ponor promotes the widening of the polje. The above-mentioned developmental ways of the polje depend on the size (type) of the glacier, on the amount and grain size of the transported material and on the fact whether the glacier advances or retreats. Glaciers create the conditions (the transportation of till to the area of the polje directly or by meltwater and because of this the formation of a polje) which result in the transportation of the debris produced by them into the karst.

References

Adamson KR, Woodward JC, Hughes PD (2014) Glaciers and rivers: Pleistocene uncoupling in a Mediterranean mountain karst. Quat Sci Rew 94:28–43

Annysa K, Frankla A, Spalević V, Čurović M, Borotac D, Nyssena J (2014) Geomorphology of the Durmitor Mountains and surrounding plateau Jezerska Površ (Montenegro). J Maps. http://dx.doi.org/10.1080/17445647.2014.909338

Auly T (2008) Quelques morphologies de rapport karst/glaciaire dans les Pyrénées (France). In: Tyc A, Stefaniak K (eds) Karst and Cryokarst, University of Silesia Faculty of Earth Sciences, University of Wrocaw Zoological Institute, Sosnowiec-Wroclaw, Pologne, pp 129–154

Balogh Z (1998) Saroknyomkarrok vizsgálata az ausztriai Totes-Gebirgében. Karsztfejlődés II:149–168

Barrére P (1964) Le relief karstique dans l'ouest des Pyrénées centrales. Rev. Belge Géogr. Ed. Soc. Roy. Belge Géogr. Special Publ., Karst et Climats Froids 88(1–2):9–62

Bauer F (1962) Nacheiszeitliche Karstalpen. In: Proceedings of 2nd international congress speleology, Bari-Lecce-Salerno 1958.1, pp 299–329

Bauer F, Zötl J (1972) Karst of Austria. In: Herak M, Stringfield VT (eds) Karst, important Karst of the Northern Hemisphera. Elsevier, Amsterdam, London, New York, pp 225–265

Bayari, S, Zreda, M, Ciner, A, Nazik, L, Törk, K, Ozyurt, N, Klimchouk A, Sarikaya A (2003) The extent of Pleistocene ice cap, glacial deposits and glaciokarst in the Aladaglar Massif: Central Taurids Range, Southern Turkey. XVI INqUA Congress, Paper #55360, XVI INqUA Congress, Reno Nevada USA, 23–30 July 2003, Abstracts, 144 p

Beck BF, Sinclair WC (1986) Sinkholes in Florida: an introduction. Florida Sinkhole Research Institute Report 85-86-4, 16 p

Bennett MR, Glasser NF (2009) Glacial geology: ice sheets and landforms. Wiley, Chichester, UK, p 385

Bočič N, Faivre S, Kovacic M, Horvatincic N (2012) Cave development under the influence of Pleistocene glaciation in the Dinarides—an example from Štirovača Ice Cave (Velebit Mt., Croatia). Zeitschrift für Geomorphologie 56(4):409–433

Bögli A (1951) Probleme der Karrenbildung. Geographica Helvetica 6:181–204

Bögli A (1960) Kalklösung und Karrenbildung. Zeitsch. f. Geomorph. N. E. 2:4–21

Bögli A (1961) Karrentische, ein Beitrug zur Karstmorphologie. Zeits. f. Geomorphologie 5:185–193

Bögli A (1964) Le Schichttreppenkarst. Un exemple de complexe glaciokarstique. Revue Belge de Geographie 1(2):64–82

Bögli A (1976) Die wichtigsten Karrenformen der Kalkalpen. Karst processes and relevant landforms. International Speleological Union, Commission on Karst denudation. Department of Geography, Philosophical Faculty, University of Ljubljana. Ljubljana, pp 141–149

Bögli A (1980) Karst hydrology and physical speleology. Springer, Berlin, p 284

Bognar A, Faivre S (2006) Geomorphological traces of the younger Pleistocene glaciation in the central part of the Velebit Mt. Hrvatski geografski glasnik 68(2):19–30

Brook GA, Ford DC (1978) The origin of labyrinth and tower karst and the climatic conditions necessary for their development. Nature 275:493–496

Clayton KM (1966) The origin of the landforms of the Malham area. Field Stud 2:359–384

Corbel J (1957) Les karsts du Nord – Ouest de l' Europe et de quelques Regions de Comparaison: Etude sur le Rôle du Climat dans l' Erosion des Calcaires – Institut des Etudes Bhodaniennes de l' Université de Lyon, Mémoires et Documents 12

Coxon CE (1986) A study of the hydrology and geomorphology of turloughs. Unpublished Ph.D. thesis, University of Dublin

Coxon CE (1987) The spatial distribution of turloughs. Irish Geogr 20:11–23

Djurović P, Petrović AS, Simić S (2010) The overall impact of Pleistocene glaciation on morphological diversity of uvalas at Durmitor and Žijovo. Serbian Geogr Soci 90:17–34. https://doi.org/10.2298/gsgd1001017d

Doughty PS (1968) Joint densities and their relation to the Great Scar Limestone. Proc Yorks Geol Soc 36(4):479–512
Drew D (2004) Burren Glaciokarst, Ireland. In: Gunn J (ed) Encyclopedia of caves and karst science. Fitzroy Dearborn, New York, pp 169–171
Dubljanszkij JV (1987) Teoreticseszkoje modelirovanije dinamiki formirovanija gidrotermokarsztovüh polosztyej – Metodi i izucssenyija geologicseszkih javlenyij, Novoszibirszk, pp 97–111
Farias P, Monserrat J, Martínez J (1996) Nuevos datos sobre la estratigrafía del relleno cuaternario de la depresión de Comeya (Picos de Europa, Asturias). Geogaceta 20(5):1116–1119
Fels E (1929) Das Problem der Karbildung in den Ostalpen. Petermanns Mitt., Ergänzungsh 202:1–84
Fink MH (1973) Multilingual glossary of karst and speleological terminology. ISU, Subcommission on Karstterminology Project, p 53
Ford DC (1977) Genetic classification of solutional cave systems. In: Proceedings of the 7th international congress of speleology, Sheffield, pp 189–192
Ford DC (1979) A review of alpine karst in the Southern Rocky Mountains of Canada. Bull Nat Speleol Soci 41:53–65
Ford DC (1984) Karst groundwater activity and landform genesis in modern permafrost regions of Canada. In: LaFleur RG (ed) Groundwater as a geomorphic agent. Allen&Unwin, London, pp 340–350
Ford DC (1996) Karst in a cold climate: effects of glaciation and permafrost conditions upon the karst landfrom systems of Canada. In: McCann SB, Ford DC (eds) Geomorphology Sans frontieres. John, Chichester, pp 153–179
Ford DC, Williams PW (1989) Karst geomorphology and hydrology. Unwin Hyman, London, p 601
Ford DC, Williams PW (2007) Karst hydrogeology and geomorphology. John, Chichester, p 562
Fu P, Harbor J (2011) Glacial erosion. In: Singh VP, Singh P, Haritashya UK (eds) Encyclopedia of snow, ice and glaciers. Springer, Dordrecht, pp 332–341
Gams I (1974) Kras. Zgodovinski, naravoslovni in geografski oris (A historical, natural-scientific and geographical outline). Slovenska matica, Ljubljana, 358 p
Gams I (1977) Towards the terminology of the polje. In: Proceedings of the international speleology. Congress Sheffield, England, pp 201–203
Gams I (1978) The polje: the problem of definition. Zeits. für Geomorphology 22:170–181
Gams I (1994) Types of contact karst. Geografia Fisica e Dinamica Quateraria 17:37–46
Gams I (2002) Changes of the Triglav Glacier in 1945-94 period int he light of climatic indicators. http://ai.ijs.si/mezi/personal/triglav/
Ginés Á (2009) Karrenfield landscapes and karren landforms. In: Ginés Á, Knez M, Slabe T, Dreybrodt W (eds) Karst rock features. Karren sculpturing. Zalozba ZRC. Institut za raziskovanje krasa ZRC SAZU, Postojna-Ljubljana, Carsologica 9, pp 13–24
Glew JR, Ford DC (1980) Simulation study of the development of rillenkarren. Earth Surf Process 5:25–36
Goeppert N, Goldscheider N, Scholz H (2011) Karst geomorphology of carbanatic conglomerates in the Folded Molasse zone of the Northern Alps (Austria/Germany). Geomorphology 130:289–298
Goldie HS (2009) Kluftkarren or grikes as fundamental karstic phenomena. In: Ginés Á, Knez M, Slabe T, Dreybrodt W (eds) Karst rock features, Karren sculpturing Zalozba ZRC. Institut za raziskovanje krasa ZRC SAZU, Postojna-Ljubljana, Carsologica 9, pp 89–102
Grimes KG (2012) Surface Karst features of the Judbarra/Gregory National Park, Northern Territory, Australia. Helectite 41:15–36
Gunn J (2006) Turloughs and tiankengs: distinctive doline forms. Speleogenesis Evol Karst Aquifers 4(1):1–4
Haserodt K (1965) Untersuchungen zur Hohen- und Altersgliederung der Karstformen in den nördlichen Kalkalpen. Münchner Geogr. H. 27

Hughes PD, Woodward JC, Gibbard PL, Macklin MG, Gilmour MA, Smith GR (2006) The glacial history of the Pindus Mountains. Greece J Geol 114:413–434
Hyatt JA, Jacobs PM (1996) Distribution and morphology of sinkholes triggered by flooding Tropical Storm Alberto at Albany, Georgia, USA. Geomorphology 17:305–316
Jakucs L (1977) Morphogenetics of karst regions. Adam Hilgar, Bristol, p 284
Jones RJ (1965) Aspects of the biological weathering of limestone pavement. Proc Geol Ass 76:421–433
Kemmerly RR, Towe SK (1978) Karst depressions in a time context. Earth Surf Proc Land 35:355–361
Kiernan K, Lauritzen S-E, Duhig N (2001) Glaciation and cave sediment aggradation around the margins of the Mt. Field Plateau Tasmania. Aust J Earth Sci 48:251–263
Klimchouk A (1995) Karst morphogenesis in the Epikarstic Zone. Cave Karst Sci 21(2):45–50
Klimchouk A, Bayari S, Nazik L, Tork K (2006) Glacial destruction of cave systems in high mountains, with a special reference to the Aladaglar Massif, Central Taurus, Turkey. Acta Carsologica 35(1):7–21
Kunaver J (1961) Visokogorski Kras v Vzhodnih Julijskh in Kamniskih Alpah. Geografski Vestnik 33:95–135
Kunaver J (1983) Geomorphology of the Kanin Mountains with special regard to the glaciokarst. Geografski zbornik XXII 1:201–343
Kunaver J (1996) Kotlič—a special depression of Subnival Alpine Karst. In: Proceedings of the 6th international congress of speleology, Academia Praha, pp 223–230
Kunaver J (2009a) The nature of limestone pavements in the central part of the southern Kanin plateau (Kaninski podi) Western Julian Alps. In: Ginés A, Knez M, Slabe T, Dreybrodt W (eds) Karst rock features. Karren Sculpturing Zalozba ZRC. Institut za raziskovanje krasa ZRC SAZU, Postojna, Ljubljana, Carsologica, 9, pp 299–312
Kunaver J (2009b) Corrosion terraces, a megaausgleichsfläche on a specific landforms of bare glaciokarst. In: Ginés A, Knez M, Slabe T, Dreybrodt W (eds) Karst rock features. Karren Sculpturing Zalozba ZRC. Institut za raziskovanje krasa ZRC SAZU, Postojna, Ljubljana, Carsologica, 9, pp 161–168
Lauritzen S-E (1986) Kvithola at Fauske, northern Norway: an example of ice-contact speleologenesis. Norks Geografisk Tidesskrift 66:153–161
Lehmann H (1959) Studien über Poljen in den Venezianischen Voralpen und im Hochapennin. Erdkunde 13:258–289
Lepirica A (2008) Geomorphological characteristics of the massif Prenj. Acta Carsologica 37 (2-3):307–329
Lippmann L, Kiss K, Móga J (2008) Az Abaliget-Orfűi karszt karsztos felszínformáinak vizsgálata térinformatikai módszerekkel (Investigation of the karstic phenomenon near Orfű and Abaliget by GIS methods). Karsztfejlődés XIII:151–166 (in Hungarian)
Maire R, Jaillet S, Hoblea F (2009) Karren in Patagonia, a natural laboratory for hydrogeolian dissolution. In: Ginés A, Knez M, Slabe T, Dreybrodt W (eds) Karst rock features. Karren Sculpturing Zalozba ZRC. Institut za raziskovanje krasa ZRC SAZU, Postojna, Ljubljana, Carsologica, 9, pp 329–348
Menkovic LJ, Markovic M, Cupkovic T, Pavlovic R, Trivic B, Banjac N (2004) Glacial morphology of Serbia, with comments ont he Pleistocene Glaciation of Monte Negro, Macedonia and Albania. In: Ehlers J, Gibbard PL (eds) Quaternary glaciations: extent and chronology. Part 1: Europe Elsevier BV, pp 379–384
Miotke FD (1968) Karstmorphologische Studien in der glazialuberformten Hohenstufe der Picos de Europe' Nordspanien. Jahrbuch der geographischen Gesellschaft zu Hannover Sonderhef, vol 4, 161 p
Mix C, Küfmann K (2012) Genezis and controlling factors of dolines in subalpin karst, Northern Calcareous Alps (Plateau Zahmer Kaiser, Austria). Zeits. f. Geomorph. 56(2):141–163
Moisley HA (1955) Some karstic features in the Malham Tarn district. Ann Rep Counc Promot Field Stud 1953–1954:33–42

Monroe WH (1970) A glossary of Karst terminology. Geol Surv Water-Supply Paper 1899-K:1–26

Plan L (2005) Karstwasserschutz und andere umweltrelevante fragestellungen im bereichrax, Schneeberg und Hochswab. Projeketbericht

Plan L, Decker K (2006) Quantitative karst morphology of the Hochschwab plateau Eastern Alps, Austria. Zeits. f. Geomorph. N. F. 147:29–54

Rose L, Vincent P (1983) Some aspects of the morphometry of grikes: a micture model approach. In: Paterson K, Sweeting MM (eds) New direction in karst, proceedings of the Anglo-French Karst symposium, Geo Books, Norwich, pp 497–514

Sauro U (2009) Glaciokarst landforms of the lower Adige and Sarca valleys. In: Ginés Á, Knez M, Slabe T, Dreybrodt W (eds) Karst rock features, Karren sculpturing Zalozba ZRC. Institut za raziskovanje krasa ZRC SAZU, Postojna-Ljubljana, Carsologica 9, pp 323–328

Smart PI (1986) Origin and development of glacio-karst closed depressions in the Picos de Europa. Spain. Zeits. f. Geomorph. 32(4):423–443

Smart C (2004) Glacierized and glaciated karst. In: Gunn J (ed) Encyclopedia of caves and karst science. Fitzroy-Dearborn, New York, London, pp 389–390

Soriano MA, Simón JL (2001) Subsidence rates of alluvial dolines in the central Ebro basin, Northeastern Spain. In: Beck BF, Herring JG (eds) Geotechnical and environmental applications of Karst geology and hydrology. Balkema, Lisse, pp 47–52

Stepišnik U, Ferk M, Kodelja B, Medenjak G, Mihevc A, Natek K, Žebre M (2010) Glaciokarst of western Orjen, Montenegro. Cave Karst Sci 36(1):21–28

Stieber J, Leél-Őssy S (2016) Vízszintváltozások és hatásaik a Béke- és a Baradla-barlangokban (Water level changes and their effects on the Béke and Baradla Caves) Karsztfejlődés XXI:175–185 (in Hungarian)

Sweeting MM (1966) The weathering of limestones (with particular to the Carboniferus limestones of Northern England). In: Dury GH (ed) essays in geomorphology. Heinemann, London, pp 177–210

Sweeting MM (1973) Karst landforms. Columbia University Press, New York, p 362

Telbisz T, Dragašice H, Nagy B (2005) A horvátországi Biokovo-hegység karsztmorfológiai jellemzése terepi megfigyelések és digitális domborzatelemzés alapján (A karstmorphological analysis of the Croatian Biokovo Mountains based on field observations and digital relief analysis) Karsztfejlődés X:229–243 (in Hungarian)

Telbisz T, Mari L, Kohán B, Jelena C (2007) A szerbiai Miroč-hegység töbreinek térinformatikai GPS-es terepi vizsgálata (a spatial informatics GPS study of the dolines of the Serbian Miroč Mountains). Karsztfejlődés XII:71–90 (in Hungarian)

Trudgill ST (1972) The influence of drifts and soils on limestone weathering in N.W. Claire, Ireland. Proc Univ Bristol Speleol Soc 13:113–118

Trudgill ST (1975) Measurement of erosional weight-loss of rock tables. Br Geomorphol Res Group, Tech Bull 17:13–19

Trudgill ST (1985) Limestone geomorphology. Longman, New York, p 196 p

Veress M (1982) Adatok a Hárskúti-fennsík morfogenetikájához (Data to the morphogenetics of the Hárskút plateau). Karszt és Barlang II:71–82 (in Hungarian)

Veress M (1995) Karros folyamatok és formák rendszerezésének szempontjai Totes Gebirge-i példák alapján (aspects of systematization of karren processes and features based on examples from Totes Gebirge) Karsztfejlődés I:7–40 (in Hungarian)

Veress M (2009a) Rinnenkarren. In: Ginés Á, Knez M, Slabe T, Dreybrodt W (eds) Karst rock features, Karren Sculpturing Zalozba ZRC. Institut za raziskovanje krasa ZRC SAZU, Postojna-Ljubljana, Carsologica 9, pp 151–159

Veress M (2009b) Meanderkarren. In: Ginés Á, Knez M, Slabe T, Dreybrodt W (eds) Karst rock features, Karren Sculpturing Zalozba ZRC. Institut za raziskovanje krasa ZRC SAZU, Postojna-Ljubljana. Carsologica 9, pp 223–235

Veress M (2009c) Wandkarren. In: Ginés Á, Knez M, Slabe T, Dreybrodt W (eds) Karst rock features, Karren Sculpturing Zalozba ZRC. Institut za raziskovanje krasa ZRC SAZU, Postojna-Ljubljana. Carsologica 9, pp 237–248

Veress M (2009d) Trittkarren. In: Ginés Á, Knez M, Slabe T, Dreybrodt W (eds) Karst rock features, Karren Sculpturing Zalozba ZRC. Institut za raziskovanje krasa ZRC SAZU, Postojna-Ljubljana. Carsologica 9, pp 151–159

Veress M (2009e) Investigation of covered karst form development using geophysical measurements. Zeitschrift für Geomorph 53(4):469–486

Veress M (2010) Karst environments—Karren formation in high mountains. Springer, Dordrecht, Heidelberg, London, New York, p 230

Veress M (2012a) Glacial erosion and Karst evolution (Karren formation on the surfaces formed by glaciers). In: Veress B, Szigethy J (eds) Horizons in earth science research 8. Nova, New York, pp 1–94

Veress M (2012b) Morphology and solution relationships of three Karren slopes in different environments (Totes Gebirge, Eastern Alps). Zeitschrift für Geomorphologie 56(Suppl 2): 47–62

Veress M (2016a) Covered karst. Springer, 536 p. https://doi.org/10.1007/978-94-017-7518-2

Veress M (2016b) Postglacial evolution of paleodepressions in glaciokarst areas of the Alps and Dinarides. Zeitschrift für Geomorph 60(4):343–358

Veress M (2017a) Solution doline development on glaciokarst in alpine and Dinaric areas. Earth Sci Rev 173:31–48

Veress M (2017b) Subsidence dolines of glaciokarst. In: Veress B, Szigethy J (eds) Horizons in earth science research 16. Nova, New York, pp 101–120

Veress M, Tóth G (2004) Types of meandering karren. Zeitschrift für Geomorph 48(1):53–77

Veress M, Zentai Z (2004) Karros lejtőfejlődés a Triglav északi előterében (Karren slope development in the northern foreground of Triglav) Karsztfejlődés IX:177–196 (in Hungarian)

Veress N, Nacsa T, Széles GY, Dombi L (1995) Néhány totesi karros forma domborzatrajzi ábrázolása (map representation of some karren features from Totes Gebirge) Karsztfejlődés I:31–40 (in Hungarian)

Veress M, Zentai Z, Horváth ET (1996) Egy magashegységi karsztterület vertikális karsztformáinak vizsgálata (Totes Gebrige) (the study of vertical karst features of a high mountain karst area, Totes Gebirge) A BDTF Tud. Közl. X. Természettudományok 5:141–157 (in Hungarian)

Veress M, Szunyogh G, Tóth G, Zentai Z, Czöpek I (2006) The effect of the wind on karren formation on the Island of Diego de Almagro (Chile). Zeitschrift für Geomorphologie 50:425–445

Veress M, Samu S, Mitre Z (2015) The effect of slope angle on the development of type a and type b channels of rinnenkarren with field and laboratory measurements. Geomorphology 228:60–70

Vincent P (2009) Limestone pavements in the British Isles. In: Ginés A, Knez M, Slabe T, Dreybrodt W (eds) Karst rock features. Karren Sculpturing Zalozba ZRC. Institut za raziskovanje krasa ZRC SAZU, Postojna, Ljubljana, Carsologica, 9, pp 267–274

Zebre M, Stepišnik U (2015a) Glaciokarst landforms and processes of the southern Dinaric Alps. Earth Surf Process Land. https://doi.org/10.1002/esp.3731

Žebre M, Stepišnik U (2015b) Glaciokarst geomorphology of the Northern Dinaric Alps: Snežnik (Slovenia) and Gorski Kotar (Croatia). J Maps. https://doi.org/10.1080/17445647.2015.1095133

Žebre M, Stepišnik U, Colucci RR, Forte E, Monegato G (2016) Evolution of a karst polje influenced by glaciation: the Gomance piedmont polje (northern Dinaric Alps). Geomorphology 257:143–154

Waltham AC, Fookes PG (2003) Engineering classification of karst ground conditions. Quart J Eng Geol Hydrogeol 36:101–118

Waltham AC, Smart PL (1988) Civil engineering difficulties in the karst of China. Quart J Eng Geol 21:2–6

Waltham T, Bell F, Culshaw M (2005) Sinkholes and subsidence. Springer, Berlin, Heidelberg, 382 p

White WB (1988) Geomorphology and hydrology of Karst terrains. Oxford University Press, New York, Oxford, 464 p

Williams PW (1966) Limestone pavements: with special reference to Western Ireland. Trans Inst Br Greogr 40:155–172

Williams PW (1970) Limestone geomorphology in Ireland. In: Stephens N, Glasscock RE (eds) Irish geographical studies. Queens University, Belfast, pp 105–124

Williams PW (1983) The role of the subcutaneous zone in karst hydrology. J Hydrol 61:45–67

Williams PW (2004) Dolines. In: Gunn J (ed) Encyclopedia of caves and karst science. Fitzroy Dearborn, New York, London, pp 304–310

Williams PW (2008) The role of the epikarst in karst and cave hydrogeology: a review. Int J Speleol 37(1):1–10

Woodward JC, Hamlin RHB, Macklin MG, Hughes PD, Lewin J (2008) Glacial activity and catchment dynamics in norteast Greece: long-term river behaviour and the slackwater sediment record for the last glacial to interglacial transition. Geomorphology 101:44–67

Xu W, Zhao G (1988) Mechanism and prevention of karst collapse near mine areas in China. Environ Geol Water Sci 12:37–42

Chapter 5
Characteristics and Genesis of Subsurface Features in Glaciokarst Terrains

Tamás Telbisz

Abstract Glaciokarst terrains are rich not only in specific landforms, but in subsurface forms as well. Long, complex cave systems are widespread in glaciokarst terrains, and the deepest caves are almost all found in glaciokarsts. On the other hand, as for the volume of cave chambers and passage dimensions, glaciokarst caves are not among the largest ones. One of the most important questions about glaciokarst speleogenesis is whether subglacial cave development exists at all, and if so, how effective it is. Other important issues are the age of glaciokarst caves and the karst hydrology of glaciokarst terrains. Characteristic features of alpine caves are vadose shafts and (sub)horizontal passage levels. The two main variations of passage profiles are the tubular phreatic and the canyon-like vadose cross-sections, moreover, the combination of the previous two also exists, it is the so-called keyhole profile. Among small-scale cave features, paragenetic shapes and scallops are presented in this chapter. Characteristic glaciokarst cave sediments are coarse debris, which are mainly the results of extreme high discharges, fine-grained varved carbonates, which are deposited due to back-flooding conditions, and speleothems, which grow mostly during warm periods, but if some special conditions are satisfied, they may grow even below actually glacier-covered terrains due to the so-called "common-ion effect". Further on, cryogenic cave calcites are also formed in glaciokarst caves, but their amount is insignificant. As for the karst hydrology, extreme fluctuations are characteristic to glaciokarsts, meaning both high seasonal changes and relatively high daily changes according to melt cycles. Using U-series and cosmogenic nuclide methodology to date speleothems and detrital cave sediments, it is now evident that the majority of glaciokarst caves are polygenetic in origin, surviving one or more glacial periods. Preglacial caves (i.e. caves evolving since at least the Pliocene) are common in the Alps. On the other hand, there are approved postglacial caves as well, which are related to drumlins or isostatic fissures. Finally, subglacial speleogenesis is also proved to be possible, though it has a low rate. Ice-contact cave development takes place when a connected aquifer is formed in the glacier ice and in the neighbouring karstic rock mass.

M. Veress et al., *Glaciokarsts*, Springer Geography,
https://doi.org/10.1007/978-3-319-97292-3_5

Keywords Subglacial speleogenesis · Ice-contact speleogenesis
Preglacial · Postglacial · Speleothems · Varved sediments · Phreatic
Vadose · U-series dating · Cosmogenic nuclides · Common-ion effect
Paragenesis

5.1 Introduction

Most of the glaciokarst terrains are rich in subsurface forms, too. Speleogenesis is influenced by a number of factors. The first type of factors, notably lithology and tectonic settings are independent of climatic conditions and glaciations, therefore, they are called *passive factors* (Skoglund and Lauritzen 2011). On the other hand, hydrology and geochemistry are tightly related to climatic settings and glaciations, thus they are called *active factors* (Skoglund and Lauritzen 2011). Due to the active factors, speleogenesis in glaciokarst terrains has some specific characteristics. The evolution of caves in glaciokarst settings can be considered a kind of competition between karstic and other (glacial or periglacial) processes, such as frost shattering. "Cold conditions, seasonally abundant water, rapid runoff, steep gradients strongly sustain other geomorphic processes, most of which can be viewed as competing with karst development" as Smart (2004) summarized it. The impact of glaciation on karstification and consequently on speleogenesis may be deranging, destructive, inhibitive, preservative or stimulative (Ford and Williams 2007).

Speleogenesis is possible if dissolution operates faster than erosion (Palmer 2003), which, in turn, is at a very fast pace in glacial and periglacial environments. The inception phase of carbonate cave development requires generally 10^4–10^5 years, it is the time necessary for the slow dissolutional enlargement of the initial openings to become wide enough to throughput relatively high discharge of water. This threshold time is called *breakthrough time* (Palmer 1991, 2003). Most frequently, cave development leads to branchwork cave pattern, but in certain cases, maze type caves are formed. This latter type is characteristic, if the hydraulic gradient is steep and the cave is occasionally flooded, or if the subsurface flow paths are short, or if the water recharge is uniform, or if waters of different chemistry are mixed (Palmer 1991, 2003). Cave passage horizons are often developed, they are related to erosion base levels, but the relation may be quite complicated in certain cases and influenced by a variety of factors (Ford 1971; Palmer 1991, 2003; Audra 1994). Caves near the surface are subject to destruction, partly due to collapses, partly due to direct surface erosion (Ford 1983a; Audra et al. 2002; Palmer 2003). All of these general speleogenetic processes can be more or less influenced by glaciations as it is discussed in the followings.

Physical and scientific exploration of caves in glaciokarst terrains has been presented in Chap. 1. Scientific questions related to speleogenesis in glaciokarst environments have emerged in parallel with explorations and expeditions. Many of these questions have been answered by subsequent karst research, but still, there are certain details in general or in local that remained unanswered. This chapter aims at presenting our knowledge about speleogenesis in glaciokarst settings by going through the following questions:

- What are the sizes typical of subsurface glaciokarst features?
- Is speleogenesis possible below glaciers, and if so, how effective is it?
- What is the age of caves found in terrains glaciated during some periods of Quaternary, are they of preglacial, interglacial, subglacial or deglacial origin?
- What is the relationship of speleothem growth and climate? Is active speleothem growth possible in caves under actual glacier cover?
- How partly of fully glaciated karsts work from a hydrological point of view?

5.2 Morphology and Sediments of Glaciokarst Caves

5.2.1 Dimensions and Morphology

A basic fact is that caves are abundant in glaciokarst terrains. In case of alpine mountains, it was essentially unambiguous from the beginning of karst research, but as for the karst development of glacier-covered and permafrost terrains, they remained unexplored for a long time, and it was not obvious whether significant caves exist in these environments or not, and if so, how high the intensity of karst processes is. Due to the later exploration of these hardly accessible glaciokarsts, it became clear that karst processes and cave development are active even in arctic glaciokarsts (Brook and Ford 1980; Ford 1983a; Salvigsen and Elgersma 1985; Lauritzen 2006).

The caves of alpine glaciokarsts have *variegated dimensions*. As for their length, there are really remarkable long systems, such as Hölloch (203 km) and Siebenhengste (157 km) in the Swiss Alps, the Schönberg-Höhlensystem (143 km) in Totes Gebirge (Eastern Alps, Austria), the Sistema del Mortillano (139 km) in the Cantabrian Mountains (Spain), or the Gouffre de la Pierre Saint Martin (80 km) in the Pyrenees (France). In the karst terrains of the United Kingdom that were covered formerly by continental ice sheet, there are also considerable systems, though they are somewhat shorter than the previously mentioned examples. Notably, the Three Counties System (86 km) at the edge of Yorkshire or the Ogof Draenen (70 km) in South Wales. In more arctic terrains, there are no such long caves, but still, there are caves longer than 20 km, for instance, in Norway (Tjoarvekrajgge 25 km) and in Canada (Castleguard Cave 20 km) as well. Given the above data, it is evident that if only length is considered, glaciokarst caves are well developed and are among the longest cave systems in the world. Nevertheless, it is not true for the 3D size of passages and rooms. The largest cave room in glaciokarst terrains is the Salle de la Verna found in the Gouffre de la Pierre Saint Martin, but even this large feature is only 13th in the list of the largest underground chambers by volume, whereas mostly tropical and subtropical caves are found in the top twelve (www.caverbob.com). Nonetheless, aside from the records, it is stated that glaciokarst caves in general have smaller passage diameters, especially in case of arctic terrains (Lauritzen and Skoglund 2013).

One of the most remarkable features of alpine caves is that they often consist of *vadose shafts* (Fig. 5.1) and roughly *horizontal passage levels* (Smart 2004; Audra 2004; Szabó 2008, 2009; Plan et al. 2009). Naturally, this morphology is only partly related to glaciations because it is also the result of large topographic relief, tectonic uplift and deep incision (explained later in details). The deepest caves on the Earth are almost exclusively found in glaciokarst terrains, namely, in the Western Caucasus, where the Arabika Massif hosts the four deepest caves: Veryovkina (−2204 m), Krubera (−2197 m), Sarma (−1830 m), and Illuzia-Snezhnaja-Mezhonnogo (−1853 m). The Lamprechtsofen-Vogelschacht (−1632 m) is found in the Eastern Alps (Austria), whereas the Gouffre Mirolda (−1626 m) and the Réseau Jean Bernard (−1602 m) are situated in the French Alps (www.caverbob.com). Shafts may have extreme vertical dimensions, but their diameters are relatively small. The deepest vertical drop with 603 m pitch is situated in Vrtiglavica Cave (Kanin Massif, Slovenia). Shafts are generally created by water flowing from the surface, thus they are called *invasion shafts* (Ford 1983b). Horizontal or subhorizontal passages commonly have elliptical or tubular cross-sections and they are considered epiphreatic or phreatic in origin, i.e. they are formed at or below the karst water table (Ford 1983b; Plan et al. 2009; Bočić et al.

Fig. 5.1 Characteristic shaft in Gortani Cave (Canin Plateau, Italy, photo by Egri)

Fig. 5.2 Phreatic passage in Gortani Cave (Canin Plateau, Italy). Note that the profile is influenced by layer dip and by a vertical fissure as well (photo by Egri)

2012, Fig. 5.2). While horizontal cave passages usually do not show significant vertical amplitudes in the Alps (Audra et al. 2007), there are remarkable vertical excursions in the geometry of caves in the Canadian Rocky Mountains. These latter features are called phreatic loops (Ford 1983b). Another characteristic of glaciokarst caves is that maze sections are often found either in alpine (White 1979; Plan et al. 2009) or in arctic settings (Skoglund et al. 2010; Farrant and Simms 2011; Lauritzen and Skoglund 2013). A special layout of Norwegian karsts is the so-called *stripe karst* that means a narrow band of karstifiable rocks on the surface that is bordered by non-karstic rocks at both sides. It is a result of metamorphic rock structure typical in the Scandinavian Mountains (Horn 1937; Lauritzen and Skoglund 2013). In stripe karst settings, underground passages are usually constrained by a relatively thin karstifiable bed, commonly marble, thus caves are in fact 2D features, but usually not horizontal. Stripe karsts contain only short phreatic passages, occasionally crossed by vadose branches (Faulkner 2006).

Passage shapes are determined by several factors. First, by the hydrological regime, because the shape of the completely water-filled *phreatic* passages and the shape of the only partly and periodically water-filled *vadose* passages are completely different. In the first case, the cross-section is tubular or rounded, while in the second case, the passages are vertically elongated with a *subsurface canyon-like* morphology. As a result of uplift relative to base level, phreatic passages gradually become vadose, therefore, their shapes are transformed and characteristic keyhole-profile may evolve (Chevalier 1944; Ford and Williams 2007; Bočić et al. 2012). Naturally, these cross-sections are not exclusive to glaciokarsts, however, it is true that the hydrologic regime, i.e. the phreatic or vadose formation of passages quite frequently changed in the geological history of glaciokarsts due to glacial cycles. Nonetheless, changes in karst water level can be caused by other factors as well, namely by tectonic uplift or subsidence, or by other climatic changes, and by sea level oscillations. Passage cross-sections are also influenced by general geologic factors, such as bedding planes (see e.g. Ford 1983b), tectonic fissures, joints, fault planes (see e.g. Bodenhamer 2007; Farrant and Simms 2011) or weakness zones due to metamorphosis (see e.g. Lauritzen 1984) similarly to other, non-glaciokarst caves.

A characteristic impact of glacial erosion is the *truncation* of cave segments near the surface. Unroofed caves (called *Ruinenhöhlen* by Lechner in Audra et al. 2002) are transitional forms between surface and subsurface features. They are very common phenomena in the Alps (Audra et al. 2002; Plan et al. 2009), but also in other places like in the Aladağlar Mountains (Turkey), where Klimchouk et al. (2006) discriminated several subtypes of unroofed caves. There are shafts, whose upper parts were completely destroyed by ice including a certain thickness of bedrock that he calls *decapitated shafts* (Fig. 5.3). In some other cases, only the epikarst segment is cleared away, these forms are *epikarst shafts*. Finally, there are *unwalled shafts* (Fig. 5.4), whose wall was partly destroyed by valley entrenchment, they are geologically short-lived phenomena. It is observed that fluvial incision opens rather the (sub)horizontal passages, whereas vertical shafts are principally opened by glacial erosion (Klimchouk et al. 2006).

Due to *glacial erosion*, cave entrances often get to anomalistic position, i.e. the entrance is found hanging high at the valley side (Fig. 5.5), without either spring or sink functionality, if only the present topographic and hydrologic situation is taken into consideration (Oxaal 1914; Ford 1983b; Lauritzen 1984, 1986). Cave entrances can be partially destroyed by direct glacial scouring, but frost shattering is also an active agent. It worked efficiently in the glacial periods, but it is still active in periglacial conditions (Ford 1983b; Murphy et al. 2008, 2015). If destructive effects are considered at a larger scale, then it must be taken into account that valley entrenchment may demolish certain preglacial passages, others may get to different hydrologic positions, and in general, the karst aquifer may become disordered, and formerly contiguous caves may turn to be dissected (Ford 1983a; Ford and Williams 2007).

Among the small-scale dissolutional cave features, *paragenetic forms* are mentioned first (Skoglund and Lauritzen 2011; Farrant and Smart 2011; Bočić et al. 2012). They are created when a passage is filled with sediments, and water finds a way between the sediment and the cave wall, and so dissolves the limestone profile. In many cases, the ceiling is the target of the dissolution. Paragenesis is usually possible under phreatic conditions. Paragenetic features include ceiling half tubes (Fig. 5.6), ceiling meanders and bypasses (Farrant and Smart 2011; Plan et al. 2009). It is important to note that paragenetic features are not restricted to

Fig. 5.3 Shaft entrance at a tectonic fissure and the glacially eroded surface (Tennengebirge, Austria, photo by Egri)

glaciokarst caves, they are present in different climates as well, but they are typical in glaciokarst areas, because the filling of passages by sediments is a frequent event during glacial–interglacial transitions (Farrant and Smart 2011). Other common small scale cave features are the *scallops* (Fig. 5.7), which are created by dissolution due to flowing water in cave passages (Ford 1983b). Again, they are not confined to glaciokarst caves, but they frequently occur in them, and in the lack of other formations (e.g. speleothems), they are often the dominant small-scale features of glaciokarst caves. In order to highlight their significance, it is noted that former flow velocities and directions can be calculated from scallops morphology, therefore former hydrologic conditions can be reconstructed. Notably, the hydraulic gradient can be calculated based on the roughness of scallops (Lauritzen 1986). By the examination of several Norwegian caves, Lauritzen and Skoglund (2013) concluded that relatively low flow velocities created the scallops of the studied caves, and in certain cases, they found that flow directions were just the opposite of what would seem logical based on the present-day topography-related directions. In the area of Yorkshire Dales, Murphy et al. (2015) observed particular *scoops*, which could be also flow generated features. However, as these forms are larger than normal scallops, their development can be explained by faster flow, namely by the water flowing between the glacier ice and the rock wall.

Fig. 5.4 Unwalled shaft at the Canin Plateau (Italy, photo by Telbisz)

In permafrost areas, most of the subsurface water is in a frozen state that means the interruption of karstification. Notwithstanding, there exist some karst terrains in permafrost areas, too. However, their development took place in different climatic conditions (Ford and Williams 2007). Among permafrost terrains, Svalbard has been studied by Lauritzen (2006), who described some caverns as well. These caverns are small and fully plugged with ice or ice-cemented debris in some meters from their entrances, but even these small segments contain some scallops and paragenetic features.

Fig. 5.5 Entrance to Eisriesenwelt Ice Cave (Austria) situated in a hanging position (photo by Egri)

Fig. 5.6 Ceiling half tube in Gortani Cave (Canin Plateau, Italy, photo by Egri)

Fig. 5.7 Phreatic passage with scallops (Lamprechtsofen, Austria, photo by Egri)

5.2.2 Characteristic Glaciokarst Cave Sediments and Depositions

The most peculiar sediment deposited in glaciokarst caves is *ice* (Fig. 5.8). However, the formation of ice is not restricted to glaciokarst caves at all. There are a number of ice caves in temperate climate terrains as well, such as the Dobšinská Ice Cave in the Carpathians (Slovakia) or the Scarisoara Ice Cave in the Apuseni Mountains (Romania), which are found in medium mountains in continental climate. Obviously, the best-known ice caves (e.g. the Eisriesenwelt or the Dachstein Ice Cave, both in the eastern Alps, Austria) are found in glaciokarst terrains, since these are the environments where the present climate is cold enough for the accumulation of ice in the caves. On the contrary, the previously mentioned ice caves in temperate climate medium mountains could be formed only due to specific topographic settings. Turning to actually glacierized karsts, the presence of ice can be even more substantial. The expedition into Castleguard Cave led by Derek Ford explored the cave starting from the spring mouth by going upwards, and when they reached the upstream end, they were the first to discover an ice plug in a cave that completely filled the passage and that was created by the penetration of glacier ice into the cave (Ford 1983b). Moreover, in the Alps and other high karst mountains, caves, especially shafts, which are partly open to the surface are often permanently filled with snow and ice. Depth of such an ice or snow plug may reach several hundred meters. Smart (2004) mentions, for instance that cavers descended down to 200 m deep in a shaft between the ice plug and the rock wall in the Julian Alps

Fig. 5.8 Ice formation in Eiskogelhöhle Cave (Tennengebirge, Austria), which is the highest show cave in Europe (photo by Egri)

(Slovenia), and Plan et al. (2009) described a 250 m deep ice plug from Totes Gebirge (Austria). The ice plugs may be permanent or periodic, occasionally blocking the infiltration, but when the ice is partly decayed, water may flow down, and even paragenetic features may be created this way (see Murphy et al. 2015).

Coarse debris accumulations are commonly found in glaciokarst caves, they are typical of the vadose sections (Audra et al. 2002; Bočić et al. 2012). Source of the coarse debris may be directly the frost-shattered rock, but flowing water may also bring glacial debris and morainic material into the cave (Audra et al. 2002). This latter option requires streams with enough discharge and energy. The sediments detached from glacial debris, transported and reworked by flowing water are called fluvioglacial deposits (Farrant et al. 2014; Bočić et al. 2012). Fluvioglacial deposits are usually rounded or sub-rounded, and the clast material depends on the source, which may be karstic in case of limestone dominated mountains or non-karstic if there are other bedrocks within the glacier domain. Size of the clasts ranges from gravels to boulders and fluvioglacial sediments are usually poorly sorted (Plan et al. 2009; Bočić et al. 2012). These deposits may be porous, but occasionally they may completely fill shafts and less frequently horizontal passages as well (Plan et al. 2009; Bočić et al. 2012). If the bottom of the ice plug melts in a shaft, then the bottom debris gradually becomes cemented, and so-called *cemented diamict* is created (Klimchouk et al. 2004).

Beside the accumulation of coarse debris, *fine-grained sediments* are also frequently deposited in glaciokarst caves (Fig. 5.9). They are called *varved carbonate sediments*, i.e. they are finely laminated and contain carbonate material (Maire 1990; Audra et al. 2002; Plan et al. 2009). The carbonate content often consists of fine calcite flakes, and the deposits occasionally contain high amount of quartz grains, too (Audra et al. 2002, 2004). The typical grain size is silt. Commonly, there are alternating light and dark laminae according to the high seasonal changes of discharge (Audra et al. 2002). The thickness of fine-grained sediments often reaches several meters, and they may effectively clog cave passages. Given the grain size of that material, it is obvious that they are deposited from suspension (Riviere 1977 in: Audra et al. 2002). Typically, there are black-floodings in the caves that causes the deposition of these varved sediments (Audra et al. 2002; Plan et al. 2009; Farrant and Smart 2011; Bočić et al. 2012).

Both the fine and coarse-grained sediments may result in the *clogging* of certain cave passages either near the entrance or deeper within the cave. The deeper parts are mainly plugged by fine-grained sediments, whereas the sections near the entrance are typically blocked by coarse sediments. The entrance of a cave can be directly dammed by glacial till as well (Waltham et al. 1997; Skoglund et al. 2010; Bočić et al. 2012). Sedimentary plugs may be flushed by extreme high discharges, which are characteristic to the end of glacial phases due to the sudden increase of meltwaters or they may occur during interglacial periods in general.

Fig. 5.9 Laminated, fine-grained sediments in Gortani Cave (Canin Plateau, Italy, photo by Egri)

Commonly, the best known and most peculiar cave sediments are *speleothems*. They are relatively rare phenomena in glaciokarst caves with respect to other climatic environments, but they do exist. According to classical theories of speleothem growth, they can form only in caves below soil-covered terrains (Ford and Williams 2007). Thus, even if they occur in some glaciokarst caves, their formation is linked to pre-, inter-, or postglacial warm periods (Audra et al. 2002). Given this general view, it was surprising new information in the 1970s that in the Castleguard Cave, below an actually glacier-covered terrain, only some 150 m from the ice plug at the end of the cave, small speleothems were found by Ford and his companions, and a bit farther away from the ice plug, even larger speleothems were observed, which showed traces of active growth (Ford 1983b). The mechanism of speleothem growth is presented in the following subchapter. The presence of speleothems may provide information about the activity erosion processes as well. In several caves of the Alps, speleothems are well preserved in seemingly active passages. However, the presence of apparently intact speleothems proves that both the mechanical erosion and the chemical aggressivity of waters is at a low level in the given passages since the formation of the speleothems (Audra 2004).

Cryogenic cave calcites (CCC) are rare phenomena, created when the carbonate content of the water precipitates during freezing (Clark and Lauriol 1992; Žák et al. 2004). They accumulate at the bottom of caves, and they consist of relatively small size (1–10 mm), and varied shapes of calcite crystals. The amount of cryogenic carbonates is negligible with respect to other deposits (Spötl and Mangini 2007a, b), but they can be used as climate markers. They are formed either in glacial or in periglacial environments, when liquid water is occasionally present in the karst system but the temperature is usually below zero. They bear useful information for climate chronology because they can be numerically dated (Žák et al. 2009, 2012).

5.3 Subsurface Processes in Glaciokarst Terrains

5.3.1 Age and Formation of Speleothems

As it was already mentioned, speleothem growth was previously linked to warmer (interglacial, preglacial, postglacial or occasionally to interstadial) periods, when the surface was covered with soil (Ford 1976a, b; Harmon et al. 1977; Atkinson et al. 1978). In some cases, it was even stated that active speleothem growth is possible only below the treeline (Lauritzen and Gascoyne 1980). Thus, it was quite unexpected when Ford (1983b) and Gascoyne et al. (1983) described subglacial speleothem formation. However, this new discovery only modified the earlier views, as it is still true that the main speleothem growth periods can be linked to warmer climates. This fact is also supported by the increasing number of speleothem age data (Atkinson 1983; Gascoyne et al. 1983). Flowstones were observed even on Svalbard, their age proved to be interstadial (Lauritzen 2006). On the other hand, it must be emphasized that hiatuses in speleothem growth are caused not only by cold, permafrost periods, but also by the modification of infiltration paths, or by aridity (Lauritzen and Mylroie 2000). The potential of subglacial speleothem formation in the Alps was studied by Spötl and Mangini (2007a, b), who concluded that while the temperature is above freezing point in the cave and the passage is not completely filled with meltwater, speleothem growth is possible. Notably, the MIS 3 was unambiguously colder than the Holocene, however, the rate of speleothem growth was similar during interstadials and stadials. According to his analysis, speleothem growth begins as soon as the glacier above the cave becomes warm-based and promptly interrupts whenever the cave temperature decreases below freezing point. Further on, he observed that there are no unequivocal macroscopic differences in stalagmites and flowstones formed during the warmer Holocene or during the cooler MIS 3 and MIS 7.

Given the above observations, scientists began to seek for chemical explanations for the formation of subglacial speleothems. Dreybrodt (1982) explained calcite precipitation by the temperature effect. According to his theory, water infiltrating from the surface at a temperature near zero, is capable to dissolve a certain amount of $CaCO_3$. When the solution enters the cave and water temperature increases by some °C, the solution becomes supersaturated with respect to $CaCO_3$ and precipitation begins within the cave. He stated that this mechanism can explain the generation of speleothems below glacier-covered terrains and also below surfaces, which are bare due to other reasons. Atkinson (1983) examined Dreybrodt's theory and three other mechanisms as well, and he directly measured the geochemistry of karst waters in Castleguard Cave. One of the possible mechanisms was that the partial pressure of CO_2 is increased in the infiltrating water due to some non-biogenic reasons. However, this possibility could be excluded based on the field measurements. The temperature effect described by Dreybrodt (1982) is also insignificant according to Atkinson's measurements, though it is responsible for some percents of the calcite precipitation. The precipitation caused by the evaporation of the infiltrating water is also minimal, again responsible to few percents only. Finally, the essential process behind calcite precipitation in Castleguard Cave is the so-called *common-ion effect* (Atkinson 1983). The infiltrating water reacts with sulfuric acid originating from the oxidation of pyrite that makes the dissolution of carbonate minerals (calcite and dolomite) possible even at low partial pressure of CO_2. If the solution is already saturated with respect to calcite, and further gypsum or dolomite is dissolved, then precipitation of calcite begins. Thus, in case of appropriate rock composition, this mechanism can explain speleothem formation even below glaciated terrains, until the temperature is above 0 °C in the cave, and the passage is not fully inundated. Based on the analysis of alpine speleothems, Spötl and Mangini (2007a, b) observed the same effect. In his Austrian study area, marble was dissolved by sulphide-oxidation, and sulphates originated from gneiss layers situated above the marble. Besides, the mechanism of dissolution-precipitation, he also studied the stable isotope content of speleothems. He observed that speleothems formed by sulphide-oxidation have lower $\delta^{13}O$ values, i.e. they are closer to the isotope ratio of meltwaters. The relatively high $\delta^{13}C$ values of subglacial speleothems are even more remarkable, that suggests their formation without soil. Using stable isotope and age data of speleothems together, it is possible to determine when the glaciers above the cave were warm-based (Spötl 2007). According to the above facts, speleothems sensibly react to temperature changes around 0 °C and to hydrological changes caused by the glacier, thus they are appropriate means to study changes in the mass balance of alpine glaciers (Luetscher et al. 2011).

Speleothem growth data can be also used to infer valley entrenchment rates of the surrounding terrain. The basic principle behind this calculation is that speleothem growth can start only if the given passage is above the local karst water table. Hence, the level of the karst water table is often related to the base level of the valley bottom, thus speleothem age data also provides a minimum age for a given level. This principle was used for Norwegian, Canadian and alpine study areas by Lauritzen and Gascoyne (1980), Ford et al. (1981), and Häuselmann et al. (2008), respectively.

5.3.2 Hydrological Characteristics of Glaciokarsts

In case of remote, rarely populated areas, the exploration of the karst hydrological system is important mainly from a scientific point of view, but in the densely inhabited surroundings of high karst terrains, notably in many parts of the Alps, the hydrologic knowledge of glaciokarsts is of utmost importance from the viewpoint of water management. As an example, Vienna (Austria) gets 60% of its water from alpine karst aquifers (Smart 2004; Plan and Decker 2006).

One of the most remarkable characteristics of alpine type glaciokarst hydrology is the high fluctuation of discharge. The first type of fluctuation is due to seasonality, peak discharges are usually the results of the rainfall combined meltwater periods, i.e. they occur generally in summer (Skoglund and Lauritzen 2010). The second type of fluctuation means daily oscillations according to daily changes of icemelt. Naturally, based on the local hydrologic properties of the aquifer, there may be a certain lag between the highest melting time and the spring discharge. Peak discharges often cause flood conditions, occasionally completely filling certain passages, thus a significant proportion of glaciokarst caves can be safely visited only during winter. Among others, summer floods in Castleguard Cave also caused serious situations in the early period of explorations (Ford 1983b).

Based on hydrologic measurements, Smart (1983) classified waters in glaciokarsts into two classes. The first type includes *supraglacial meltwaters*, which cause high daily fluctuations in discharge. In addition, these waters are able to dissolve a certain amount of limestone, and through large sink points, they can transport coarse debris into the cave. The second includes the *subglacial meltwaters*, which are the results of high pressure below warm-based glaciers. They infiltrate into the karst principally through smaller fissures, and occasionally, they may lead to speleothem formation if conditions of the above-described mechanism are satisfied. Both types of waters are active mostly in summer, but in exceptional cases, a small quantity of liquid water may be generated from the glacier even in winter. Smart (1983) identified three conditions for the active functioning of karst hydrology. First, supraglacial meltwater must reach the bottom of the ice—it requires joints and fissures in the glacier ice. Second, it is important that there is no impervious glacial drift below the glacier that could hinder the penetration of water into the bedrock. Third, the hydraulic gradient must be high enough to make water flowing.

The aforementioned varved deposits can be explained by the extreme high discharges in the summer period (Maire 1990; Audra et al. 2002, 2004). Due to the high amount of meltwater flowing into the passages, the karst water level may increase by several hundreds of meters (Ford 1983b). The fine-grained, partly carbonate sediments are transported by the flood to different parts of the cave, and when the water is drawn back, then the silty material gradually settles out from the suspension. In some parts of the cave, temporary lakes may be formed from the remaining flood water. As floods are typical in glaciokarst settings, these fine-grained sediments can be the markers of glacial effects (Maire 1990).

By geochemical analysis, Ek (1964) and Ford (1971) ascertained that the dissolution capacity of subglacial waters is extremely low with respect to supraglacial meltwaters. Low dissolution capacity was also demonstrated by Skoglund and Skoglund and Lauritzen (2013) in Norwegian karst water samples, where they found total hardness values of 15–36 $CaCO_3$ equivalent mg/l.

5.3.3 *Speleogenesis in Glaciokarst Terrains*

As for the speleogenesis of glaciated karsts, there existed quite different views simultaneously for a long time. Based on the observation that caves and related karst forms are so abundant in the Alps, Maire (1978) suggested that speleogenesis should have been very active during glacial periods. The basic reason behind this activity could be the high amount of water, which flowed into the karst during seasonal snowmelts contributing to the development of cave passages. Corbel (1957), arguing by the fact that there is a high density of caves in the arctic glaciokarst terrains near the margins of the former continental ice sheet, also hypothesized intense karstification during glacial periods, that he explained by high corrosion rates. His theory was later unambiguously refuted by the measurements of Ek (1964) and Ford (1971). Jennings (1971 in Cooper and Mylroie 2015) supposed that in glaciated karsts, all caves are relatively young, i.e. postglacial, because glaciers necessarily destroyed preglacial (or interglacial) forms. Although glacier erosion is really significant, and in some places more than 100 m thick material was eroded by the continental ice sheet (Mangerud et al. 2011; Braun 1989), but the thickness of erosion is highly dependent on topography, ice thickness, ice movement and thermal conditions at the glacier base, and there are areas, where glacial erosion remained only minimal. Later, measurements refuted Jennings's opinion, too. According to Horn (1935), caves are generally small in Scandinavia, so he concluded that glaciokarst caves are generally young and there are certainly no preglacial caves in Norway. However, the small size of caves can be also explained by the low rates and by the fact that cave development was interrupted several times during the Pleistocene (Lauritzen 1986).

Until the 1970s, karst researchers had only indirect ways to imply the age of glaciokarst caves (Waltham 1974 in Cooper and Mylroie 2015), and a more precise estimation of the age of glaciokarst speleogenesis became available only after quantitative dating methods were developed. Speleothem dating has been already discussed above, but additionally, since the 2000s, the age of other cave sediments became also measurable with the help of cosmogenic nuclides. The cosmogenic ^{10}Be and ^{26}Al nuclides occur in quartz grains, and they are often washed into caves. But as they get to the cave, they are shielded from further cosmogenic radiation, and the amount of ^{10}Be and ^{26}Al starts to decay by radioactivity. As the half-times of ^{10}Be and ^{26}Al are different, the ratio of the two decays can be used to infer the date, when the sediment got into the cave (Häuselmann 2007). Notwithstanding, like speleothem age data, cosmogenic nuclides also provide a minimum age for the cave formation, that may be significantly younger than the true age of the cave. A particular advantage of the cosmogenic nuclide methodology, that even several million years old burial ages can be determined, while the upper limit of the U-series dating method, which is generally applied to speleothems, is only around 750 ka (Häuselmann and Granger 2005).

When scientists started to use quantitative speleothem dating methods, the main goal was to determine the eldest possible formations. These results quickly evidenced that either in arctic glaciokarsts (Harmon et al. 1975, 1977; Ford 1976a, b; Atkinson et al. 1978, 1987; Gascoyne et al. 1981; Gascoyne and Ford 1984; Lauritzen and Gascoyne 1980; Lauritzen 1983, 1984), or in alpine glaciokarsts (Audra and Quinif 1997; Spötl et al. 2002a, b; Spötl and Mangini 2007a, b; Holzkämper et al. 2005; Audra et al. 2007; Häuselmann 2007; Luetscher et al. 2011), some speleothems are older than the last glacial period, i.e. they survived at least one glaciation, which means, of course, only a minimum age for the formation of the caves. Later on, especially due to cosmogenic nuclide studies, de facto Pliocene age of certain caves has been also proved, for instance to Siebenhengste cave in Switzerland (Häuselmann et al. 2008). Thus, at present, the most widely accepted view is that the majority of caves are older than the last glacial, and in many cases, they are even preglacial. It is generally valid for the glaciokarsts in the Alps (Audra et al. 2007, Fig. 5.10), the British Isles (Farrant et al. 2014), Norway (Skoglund and Lauritzen 2010) and North America (Atkinson 1983; Burger 2004) as well.

Naturally, there are exceptions, too, i.e. there are caves, whose postglacial age were reliably demonstrated (Cooper 2014; Lauritzen 1986). The postglacial age cannot be attested purely by speleothem data, since the lack of old speleothems or even the absolute lack of speleothems is not a strict evidence for the young age of a cave. Instead, other morphological markers were used to prove the postglacial age. An important observation is that young fissures are often generated in rocks due to postglacial isostatic rebound. These fissures are sometimes enlarged by karst processes. If the fissure is postglacial by genetics, then the cave formed from it must be

Fig. 5.10 Wide, possibly old (preglacial) passage with ice accumulation in Eiskogelhöhle Cave (Tennengebirge, Austria, photo by Egri)

also postglacial in age. Some examples for such features are described by Cooper (2014) in the Appalachian Mountains, and by Faulkner (2006) in Scotland. Similarly, the walls of alpine type glacier troughs are often weakened after glacier withdrawal and stress-release sheet fractures may be created parallel with the valley side. Such forms are presented by Lauritzen (1986), Skoglund and Lauritzen (2010) in Norway, and by Murphy et al. (2015) in Britain. In some other places, caves are related to subglacially created drumlins. For example, if a drumlin blocks a former cave entrance, then a new bypass branch may develop. Or runoff water may create new stream sinks along the edge of drumlins, and thus new cave passages may evolve connected to these sink points. As drumlins were formed below the continental ice sheet, drumlin-related cave passages are necessarily postglacial (Cooper and Mylroie 2015).

In principle, the age of cave sediments (speleothems, coarse debris, fine-grained deposits) is not enough to determine the exact age or the formation period of a given cave passage. Speleogenesis can be related to warm (i.e. preglacial, interglacial or Holocene), cold (i.e. subglacial) or transitional (i.e. deglacial or proglacial) periods. Audra et al. (2007) concluded that in the Alps, the main periods of speleogenesis and of speleothem growth were the interglacials. On the contrary, glacials can be characterized principally by destruction and by filling of passages by sediments. Real speleogenesis, i.e. the formation of new passages is of minor importance during glacial periods. New passages and shafts can be formed only where there are holes in the glaciers (Audra 2004), or where the non-karstic cover layers are eroded by the glaciers and karst bedrocks come to light constituting the new surface. This latter possibility is called the *indirect role* of glaciers by Audra (2004). Ford (1983c), after carefully examining glaciokarsts of the Canadian Rocky Mountains, stated that none of the essential parts of Castleguard Karst can be considered truly preglacial, instead, glacial and karstic processes jointly formed the landscape. Burger (2004) identified three phases of glaciokarst processes based on his research in Colorado. First, during glacial periods, calcite deposition and roof breakdown are characteristic, second, in periglacial conditions, active stream erosion and filling of passages are typical due to high energy meltwaters transporting loads of sediment into the cave, and third, interglacials are characterized by roof collapse, calcite precipitation and sediment deposition. Similar phases have been outlined by Murphy et al. (2008) using Gaping Gill Cave (Yorkshire, UK) as an example. Block breakdowns mostly occur during interglacials, but discharges are not high enough to dissolve and transport the large blocks. Glacial periods are characterized by intense frost shattering, while the transportation and the distribution of the shattered material within the cave is related to deglaciation phases.

The idea of speleogenesis related to high discharges caused by meltwaters during deglaciations is not new, it was already suggested by the Norwegian Oxaal in 1914. When glaciers retreated, meltwaters created huge lakes, which were occasionally dammed by ice. The waters of these lakes had higher dissolution capacity than waters of subglacial origin, thus, where the water of these ice-dammed lakes got into the karst, speleogenesis could be significant due to the high discharge and the relatively high dissolution capacity. This process was influenced by the remarkable fluctuations of the water level of these lakes. Faulkner (2006) and Murphy et al. (2015) stated that this factor could play an important role in the development of several caves in the British Isles and Norway. Faulkner (2006) and Murphy et al. (2015) also suggested that extreme discharge events, such as jökulhlaups could also contribute to the formation of caves and gorges.

The idea of *subglacial speleogenesis* emerged also early in the history of glaciokarst research (Horn 1935, 1947). The observation of Werenskiold (1953) that there is no permafrost below glaciers wider than 400 m provided a further argument in favour of subglacial cave development. However, the demonstration that subglacial waters are usually saturated, and at very low concentrations became a strong argument against the hypothesis of subglacial speleogenesis (Ek 1964; Ford 1971). Lauritzen and Skoglund (2013) and Skoglund et al. (2010) calculated by computer simulation that subglacial wall retreat is cca. 50 times slower than wall retreat during interglacials. Moreover, the breakthrough time of cave inception is 26 times longer in glacials than in interglacials. Nevertheless, cave development is not completely interrupted during glacials, only during permafrost periods. In the Norwegian karst areas studied by Lauritzen and Skoglund, there were large enough glaciers during a significant part of the Quaternary, when permafrost was not present below the glaciers, therefore, cave development was possible. In case of non-karstic rocks, the erosion of subglacial waters creates *Nye-channels*, which mean relatively small streambeds incised into the bedrock. In case of karstic bedrock, these Nye-channels are specifically transformed and may be occasionally created as small, near surface cave passages (Lauritzen 1986). However, the dissolution caused by supraglacial waters is much more significant. These waters enter the karst at ice margins, so Lauritzen and Skoglund (2013) called their theory *ice-contact speleogenesis*. Supraglacial waters reach the karst bedrock either directly from the ice surface, or from lower parts of the glacier ice due to crevasses. The different types and sizes of crevasses in the glacier form a glacial aquifer, which is, in turn, joint to the karst aquifer formed within the bedrock. Whenever the subsurface flow is directed towards the rock, speleogenesis becomes more intensive, and due to diffuse recharge conditions, the cave evolves to a maze type (cf. Palmer 2003), which is a frequent phenomenon in Norwegian karsts. This process could be manifested by computer simulation as well (Skoglund et al. 2010). Scallops also demonstrate the existence of a joint aquifer between the ice and the rock mass. Low flow velocities calculated based on the scallops imply low hydraulic gradients, which is feasible only if the glacial and karstic aquifers were connected. Moreover, flow directions opposed to the present-day topographic gradients can be most easily explained if the glacier surface is also taken into account (Lauritzen 1986). Depending on the ice thickness of the glacier, even the same passages could have opposite flow directions in different glacial periods, that sounds paradoxical at first. The key is the ice thickness, which determines the hydraulic gradient. This occasionally alternating flow direction is called *see-saw effect* by Farrant and Simms (2011) and by Farrant et al. (2014).

The style of cave development is largely determined by the karst water level, as it determines whether phreatic or vadose processes form a certain passage. On the other hand, hydraulic gradient is also highly important. Based on these factors, alpine type glaciokarsts can be grouped into two types (Ford and Williams 2007). The first is the *Canadian type alpine glaciokarst*, where glaciers are present both in the higher zones and in the lower zones of the mountains, in the valleys, occasionally spreading to the piedmont area as well. This arrangement results that not only the water recharge zone but also the spring zone of the karst is affected. The valleys filled with glaciers and partly by till block the outflowing springs and usually cause black-flooding in the karst aquifer. Moreover, melting of glaciers provides an extra water recharge into the karst system. Consequently, a remarkable karst water level increase can be generally attributed to glacial periods, that significantly influences speleogenesis as well (Ford 1983b; Audra 2004). Further on, it is noted that if the karst aquifer is dammed at the output side, it may also cause the development of maze sections (Palmer 2003; Lauritzen and Skoglund 2013). The Canadian type is characteristic in the Alps, the Scandinavian Mountains, and in the Canadian Rocky Mountains from where it got its name (Ford and Williams 2007). The second group is the *Pyrenean type alpine glaciokarst*, where glaciers are restricted to the higher zones, while the lower parts of the valleys remain unglaciated. In this type, the outflow is not hindered, thus glacial periods do not cause significant water level increase. This group includes obviously the Pyrenees, the southern parts of the Alps, the Taurus, the Caucasus and the US part of the Rocky Mountains (Ford and Williams 2007). Nevertheless, Audra et al. (2007) remarked that the characteristic cave horizons present in many large alpine systems cannot be interpreted as parts of a former, contiguous regional karst water level because water level fluctuations are influenced more locally by individual valley glaciers.

Glacial erosion is one of the most effective forms of erosion, so alpine valley entrenchments were remarkable during glacial periods, therefore, erosion base levels and, consequently, karst springs were significantly lowered after deglaciation (Ford and Williams 2007; Skoglund and Lauritzen 2010). Speleogenesis usually follows these relatively abrupt base level drops by a certain time lag.

Finally, it is necessary to mention another significant factor influencing base level, that is not related at all to glacial processes. However, it affected many glaciokarst terrains at the southern side of the Alps. It is observed in many places that caves were adapted to a much lower karst water table and consequently to a much lower erosion base level. The reason for this is the so-called *Messinian Salinity Crisis*, during which the Mediterranean Sea nearly completely desiccated between 6 and 5.3 Ma ago. This event lowered base levels in some places by more than 1000 m, so it must be taken into consideration when cave horizons are interpreted, but should not be confused with "normally" low interglacial base levels (Audra 2004).

5.4 Conclusions

Glaciokarst terrains are often rich in caves. In many cases, the exploration of these caves was a real challenge to cavers and posed several scientific questions to karst researchers. Among glaciokarst caves, there are many compounds, very long systems, but passage volumes are generally not so large. On the other hand, vertical dimensions are remarkable, and most of the deepest caves on Earth are found in glaciokarst terrains (in the Caucasus, the Alps, and the Pyrenees).

Besides, the thorough morphological description of caves, mainly the cave sediments (speleothems, coarse debris, fine-grained deposits) helped scientists to decipher the age of speleogenesis, that, in turn, provided the key to understand the mechanisms of cave development. Due to modern, quantitative dating methodology, namely U-series dating and cosmogenic nuclide techniques, it is now unequivocal that caves of the different glaciokarst terrains have a wide range of age, there are postglacial caves as well, but most of them are significantly older.

Field observations, geochemical analysis, compound models and computer simulations all contributed to demonstrate that speleothem growth and even speleogenesis is possible in subglacial environments if liquid water exists in the karst system. However, it allows only limited growth and enlargement, whereas speleothems and cave passages increase dominantly during warmer (preglacial, interglacial or Holocene) periods. As for the relative significance of glacial or deglacial periods, opinions are divided so far, and it is feasible that there is not a global answer, but situations are locally different.

In general, it is stated that glaciokarst terrains were repeatedly affected by Quaternary climate fluctuations, and as the majority of caves are older than the last glacial period, their development is polygenetic in the meaning of Ford and Williams (2007). During cave evolution, the effects of several phases have been stacked together (Skoglund and Lauritzen 2010), but these phases were not equal neither in duration nor in their speleogenetic significance. Taken into consideration permafrost periods as well, it is clear that cave development was intermittent in most glaciokarst areas (Skoglund et al. 2010). The image is even more complicated in tectonically active areas, where the effects of climate change and tectonic uplift, notably karst water level changes, are intermingled (Bočić et al. 2012). This is why Häuselmann et al. (2008) emphasized that the interpretation of paleoclimate or paleogeography from cave information is not straightforward!

References

Atkinson TC, Harmon RS, Smart PL, Waltham AC (1978) Palaoclimatic and geomorphic implications of ^{230}Th/^{234}U dates on speleothems from Britain. Nature 272:24–28

Atkinson TC, Lawson TJ, Smart PL, Harmon RS, Hess JW (1987) New data on speleothem deposition and palaeoclimate in Britain over the last forty thousand years. J Quat Sci 1:67–72

Atkinson TC (1983) Growth mechanisms of speleothems in Castleguard Cave, Columbia Icefields, Alberta Canada. Arctic Alp Res 15(4):523–536

Audra P, Bini A, Gabrovšek F, Häuselmann P, Hobléa F, Jeannin PY, Kunaver J, Monbaron M, Šušteršič F, Tognini P, Trimmel H, Wildberger A (2007) Cave and karst evolution in the Alps and their relation to paleoclimate and paleotopography. Acta Carsologica 36(1):53–68

Audra P, Quinif Y (1997) Une cavité de haute-montagne originale: la grotte Téophile (Alpe d'Huez, France) Rôle des paléoclimats pléistocènes dans la spéléogenèse. Spéléochronos 8:23–32

Audra P (1994) Alpine karst speleogenesis: case studies from France (Vercors, Chartreuse, Ile de Crémieu) and Austria (Tennengebirge). Cave Karst Sci 21(3):75–80

Audra P (2004) An overview of the current research carried out in the French Western Alps karsts. Acta Carsologica 33(1):25–44

Audra P, Quinif Y, Rochette P (2002) The genesis of the Tennengebirge karst and caves (Salzburg, Austria). J Cave Karst Stud 64(3):153–164

Audra P, Mocochain L, Camus H, Gilli É, Clauzon G, Bigot J (2004) The effect of the Messinian Deep Stage on karst development around the Mediterranean Sea. Examples from Southern France. Geodinamica Acta 17(6):27–38

Bočić N, Faivre S, Kovačić M, Horvatinčić N (2012) Cave development under the influence of Pleistocene glaciation in the Dinarides—an example from Štirovača Ice Cave (Velebit Mt., Croatia). Zeitschrift für Geomorphologie 56(4):409–433

Bodenhamer HG (2007) Preglacial development of caves at structural duplexes on the Lewis Thrust, Glacier National Park, Montana. J Cave and Karst Stud 69(3):326–341

Braun DD (1989) Glacial and periglacial erosion of the Appalachians. Geomorphology 2(1–3):233–256

Brook GA, Ford DC (1980) Hydrology of the Nahanni Karst, northern Canada, and the importance of extreme summer storms. J Hydrol 46(1–2):103–121
Burger PA (2004) Glacially-influenced sediment cycles in the Lime creek karst, Eagle County, Colorado. In Studies of cave sediments, Springer, pp 107–122
Chevalier P (1944) Distinctions morphologiques entre deux types d'érosion souterraine. Revue de géographie alpine 32(3):475–486
Clark ID, Lauriol B (1992) Kinetic enrichment of stable isotopes in cryogenic calcites. Chem Geol Isot. Geosci Sect 102:217–228
Cooper M (2014) Verification of post-glacial speleogenesis and the origins of epigene maze caves in New York. MSc dissertation, Mississippi State University
Cooper MP, Mylroie JE (2015) Glaciation and Speleogenesis. Springer
Corbel J (1957) Les karsts du Nord-Ouest de l'Europe et de quelques régions de comparaison: étude sur le rôle du climat dans l'érosion des calcaires. Revue de Géographie de Lyon 12
Dreybrodt W (1982) A possible mechanism for growth of calcite speleothems without participation of biogenic carbon dioxide. Earth Planet Sci Lett 58(2):293–299
Ek C (1964) Note sur les eaux de fonte des glaciers de la Haute Maurienne (Savoie, France): leur action sur les carbonates. Revue belge de Géographie 88(1–2):127–156
Farrant AR, Simms MJ (2011) Ogof Draenen: speleogenesis of a hydrological see-saw from the karst of South Wales. Cave Karst Sci 38(1):31–52
Farrant AR, Smart PL (2011) Role of sediment in speleogenesis; sedimentation and paragenesis. Geomorphology 134(1):79–93
Farrant AR, Smith CJ, Noble SR, Simms MJ, Richards DA (2014) Speleogenetic evidence from Ogof Draenen for a pre-Devensian glaciation in the Brecon Beacons, South Wales UK. J Quat Sci 29(8):815–826
Faulkner T (2006) The impact of the deglaciation of central Scandinavia on karst caves and the implications for Craven's limestone landscape. In: Re-thinking Craven's limestone landscape, p 4
Ford DC (1971) Alpine Karst in the Mt. Castleguard-Columbia icefield area, Canadian rocky mountains. Arctic Alp Res 15(4):239–252
Ford DC (1976a) Evidences of multiple glaciation in South Nahanni National Park, Mackenzie Mountains, Northwest Territories. Can J Earth Sci 13(10):1433–1445
Ford DC (1976b) Evidence of multiple glaciation in South Nahanni National Park, Mackenzie Mountains, Northwest Territories. Can J Earth Sci 13:1433–1445
Ford DC (1983a) Effects of glaciations upon karst aquifers in Canada. J Hydrol 61(1–3):149–158
Ford DC (1983b) The physiography and speleogenesis of castleguard cave, Columbia icefields, Alberta Canada. Arct Alp Res 15(4):437–450
Ford DC (1983c) Alpine karst systems at crowsnest pass, Alberta-British Columbia Canada. J Hydrol 61(1):187–192
Ford DC, Williams P (2007) Karst hydrogeology and geomorphology. John Wiley and Sons Ltd, West Sussex, England
Ford DC, Schwarcz HP, Drake JJ, Gascoyne M, Harmon RS, Latham AG (1981) On the age of the extant relief in the southern Rocky Mountains of Canada. Arct Alp Res 13(1):1–10
Gascoyne M, Ford DC (1984) Uranium series dating of speleothems. II: results from the Yorkshire Dales and implications for cave development and quaternary climates. Cave Sci 11(2):65–85
Gascoyne M, Currant AP, Lord TC (1981) Ipswichian fauna of Victoria Cave and the marine palaeoclimatic record. Nature 294(5842):652–654
Gascoyne M, Latham AG, Harmon RS, Ford DC (1983) The antiquity of Castleguard Cave, Columbia Icefields, Alberta Canada. Arc Alp Res 15(4):463–470
Harmon RS, Ford DC, Schwarz HP (1977) Interglacial chronology of the Rocky and MacKenzie Mountains based upon $^{230}Th/^{234}U$ dating of calcite speleothems. Can J Earth Sci 14:2543–2552
Harmon RS, Thompson P, Schwarcz HP, Ford DC (1975) Uranium-series dating of speleothems. Nat Speleol Soc Bull 37:21–33

Häuselmann P, Granger DE (2005) Dating of caves by cosmogenic nucleides: method, possibilities, and the Siebenhengste example (Switzerland). Acta Carsologica 34(1):43–50
Häuselmann P, Lauritzen SE, Jeannin PY, Monbaron M (2008) Glacier advances during the last 400 ka as evidenced in St. Beatus Caves (BE, Switzerland). Quat Int 189(1):173–189
Häuselmann P (2007) How to date nothing with cosmogenic nuclides. Acta Carsologica 36(1): 93–100
Holzkämper S, Spötl C, Mangini A (2005) High-precision constraints on timing of Alpine warm periods during the middle to late Pleistocene using speleothem growth periods. Earth Planet Sci Lett 236(3):751–764
Horn G (1935) Űber die Bildung von Karsthöhlen unter einem Gletcher. Nor Geogr Tidsskr 5:494–498
Horn G (1937) Über einige Karsthöhlen in Norwegen. Mitteilungen für Höhlen und Karstforschung 1–15
Horn G (1947) Karsthuler i Nordland. Norg Geol Unders 165:1–177
Klimchouk A, Bayari S, Nazik L, Törk K (2006) Glacial destruction of cave systems in high mountains, with a special reference to the Aladaglar massif, Central Taurus Turkey. Acta Carsologica 35(2):111–121
Klimchouk A, Nazik L, Bayari S, Tork K, Kasjan Y (2004) Kuzgun Cave and its Context: the first super-deep cave in the Aladaglar Massif, Turkey. www.researchgate.net
Lauritzen SE (1983) Arctic and alpine karst symposium. Program and field guide. Dept. of Chemistry, University of Oslo. 89 p
Lauritzen SE (1984) Evidence of subglacial karstification in Glomdal, Svartisen, Norway. Nor Geogr Tidsskr 38:169–170
Lauritzen SE (1986) Kvithola at Fauske; Northern Norway: an example of ice- contact speleogenesis. Nor Geol Tidsskr 66:153–161
Lauritzen SE, Gascoyne M (1980) The first radiometric dating of Norwegian stalagmites–Evidence of pre-Weichselian karst caves. Nor Geogr Tidsskr 34:77–82
Lauritzen SE (2006) Caves and speleogenesis at Blomstrandsøya, Kongsfjord W. Spitsbergen. Int J Speleol 35(1):37–58
Lauritzen SE, Mylroie JE (2000) Results of a speleothem U/Th dating reconnaissance from the Helderberg plateau New York. J Cave Karst Stud 62(1):20–26
Lauritzen SE, Skoglund RØ (2013) Glacier ice-contact speleogenesis. Treatise Geomorphol 6:363–396
Luetscher M, Hoffmann DL, Frisia S, Spötl C (2011) Holocene glacier history from alpine speleothems, Milchbach cave Switzerland. Earth Planet Sci Lett 302(1):95–106
Maire R (1978) Les karsts sous-glaciaires et leurs relations avec le karst profond. Revue de géographie alpine 66(2):139–148
Maire R (1990) La haute montagne calcaire: karsts, cavités, remplissages, paléoclimats, Quaternaire. Karstologia-Mémoires 3:1–731
Mangerud J, Gyllencreutz R, Lohne Ö, Svendsen JI (2011) Glacial history of Norway. In: Ehlers J, Gibbard PL, Hughes PD (eds) Quaternary glaciations-extent and chronology: a closer look. Elsevier
Murphy PJ, Faulkner TL, Lord TC, Thorp JA (2015) The caves of Giggleswick Scar-examples of deglacial speleogenesis? Cave Karst Science 42(1):42–53
Murphy P, Westerman AR, Clark R, Booth A, Parr A (2008) Enhancing understanding of breakdown and collapse in the Yorkshire Dales using ground penetrating radar on cave sediments. Eng Geol 99(3):160–168
Oxaal L (1914) Kalkstenshuler i Ranen. Norg Geol Unders 69:1–47
Palmer AN (2003) Speleogenesis in carbonate rocks. Speleogenesis Evol Karst Aquifers 1(1):2–11
Palmer AN (1991) Origin and morphology of limestone caves. Geol Soc Am Bull 103(1):1–21
Plan L, Decker K (2006) Quantitative karst morphology of the Hochschwab plateau, Eastern Alps, Austria. Zeitschrift Für Geomorphologie Supplementband 147:29–54

Plan L, Filipponi M, Behm M, Seebacher R, Jeutter P (2009) Constraints on alpine speleogenesis from cave morphology—a case study from the eastern Totes Gebirge (Northern Calcareous Alps, Austria). Geomorphology 106(1):118–129

Salvigsen O, Elgersma A (1985) Large-scale karst features and open taliks at Vardeborgsletta, outer Isfjorden. Svalbard, Polar Res 3(2):145–153

Skoglund RØ, Lauritzen SE (2010) Morphology and speleogenesis of Okshola (Fauske, northern Norway): example of a multi-stage network cave in a glacial landscape. Norw J Geol 90:123–137

Skoglund RØ, Lauritzen SE, Gabrovšek F (2010) The impact of glacier ice-contact and subglacial hydrochemistry on evolution of maze caves: a modelling approach. J Hydrol 388(1):157–172

Skoglund RØ, Lauritzen SE (2011) Subglacial maze origin in low-dip marble stripe karst: examples from Norway. J Cave Karst Stud 73(1):31–43

Skoglund RØ, Lauritzen SE (2013) Characterisation of a post-glacial invasion aquifer: the Grønli-Seter karst system, northern Norway. Norwegian J Geol 93(1):61–73

Smart CC (2004) Glacierized and glaciated karst. Gunn J (edt) Encyclopedia of caves and karst science New York. Fitzroy Dearborn, NY, pp 804–809

Smart CC (1983) The hydrology of the castleguard karst, columbia icefields, Alberta Canada. Arctic Alp Res 15(4):471–486

Spötl C, Mangini A (2007a) Speleothems and paleoglaciers. Earth Planet Sci Lett 254(3):323–331

Spötl C, Mangini A (2007b) Speleothems and paleoglaciers. Earth Planet Sci Lett 254(3):323–331

Spötl C, Mangini A, Frank N, Eichstädter R, Burns SJ (2002a) Start of the last interglacial period at 135 ka: evidence from a high Alpine speleothem. Geology 30(9):815–818

Spötl C, Unterwurzacher M, Mangini A, Longstaffe FJ (2002b) Carbonate speleothems in the dry, inneralpine Vinschgau valley, northernmost Italy: witnesses of changes in climate and hydrology since the last glacial maximum. J Sediment Res 72(6):793–808

Szabó L (2008) Barlangfejlődés a Canin-fennsík mélyén. Karsztfejlődés 13:247–267

Szabó L (2009) Cave area of the Canin-Plateau—a naturally geodiverse land in the middle of Europe. Acta Climatologica et Chorologica 42–43:143–150

Waltham AC, Simms MJ, Farrant AR, Goldie HS (1997) Karst and caves of Great Britain. Geol Conserv Rev 12, 358 p

Werenskiold W (1953) The extent of frozen ground under the sea bottom and glacier beds. J Glaciol 2:197–200

White WB (1979) Karst landforms in the Wasatch and Uinta Mountains, Utah. NSS Bulletin 41:80–88

Žák K, Hercman H, Orvošová M, Jačková I (2009) Cryogenic cave carbonates from the Cold Wind Cave, Nízke Tatry Mountains, Slovakia: Extending the age range of cryogenic cave carbonate formation to the Saalian. Int J Speleol 38(2):139–152

Žák K, Richter DK, Filippi M, Živor R, Deininger M, Mangini A, Scholz D (2012) Cryogenic cave carbonate–a new tool for estimation of the Last Glacial permafrost depth of the Central Europe. Clim Past Discuss 8(3):2145–2185

Žák K, Urban J, Cílek V, Hercman H (2004) Cryogenic cave calcite from several Central European caves: age, carbon and oxygen isotopes and a genetic model. Chem Geol 206(1):119–136

Chapter 6
Karstic Pattern of Glaciokarst

Márton Veress

Abstract The recent karstic patterns of glaciokarst are presented. Karstic pattern is described taking karst types and karst-type assemblages into consideration. The recent karstification and pattern of the areas of former cirque glaciers, valley glaciers, ice caps, ice sheets and piedmont glaciers are presented. The recent karstic pattern and development of the karst of glaciokarst types (plateau glaciokarst, semi-plateau glaciokarst, mountain glaciokarst, rock fortress, ridge glaciokarst, complex glaciokarst and island glaciokarst) are analysed and described.

Keywords Karstic pattern · Karst type · Karst type assemblage
Cirque · Trough · Karst water level

6.1 Introduction

Karstic pattern is essentially determined by karst type. Thus, on the temperate karst, karstic pattern is given by the pattern of dolines, on tropical karst, it is determined by karstic inselbergs. Glaciokarst also has its own karstic pattern which is primarily determined by glacial erosion.

On glaciokarst, Veress (2012) described surface types created by glaciers and the recent karstification of these types. He distinguished surfaces with roche moutonnées, flat surfaces, surfaces with dolines and uvalas, surfaces slightly dissected by troughs, troughs with rock basins and glacial troughs with giant dolines and uvalas, then he described their karstification and turnover. The covered karstic pattern of glaciokarst and the karstification associated with it was presented in his later work (Veress 2016a). In this chapter, the karstic pattern of glaciokarst is presented taking the factors influencing karstic pattern into account. Karstic pattern depends on the geomorphological belts of glaciokarst, on glacier types and patterns, on the type of the features created by the glaciers (e.g. valley or combe-ridge) and on karst types that were formed on glaciokarst (mainly in the area of glacial features).

The zones of glaciokarst are determined by the distance of the zone from the equator as well as its altitude (see Chap. 2). Typical glaciokarst develops in the

M. Veress et al., *Glaciokarsts*, Springer Geography,
https://doi.org/10.1007/978-3-319-97292-3_6

periglacial zone. In the fluvial erosional zone of the temperate belt of glaciokarst, since it is covered by soil and vegetation, glaciokarst is less characteristic, in addition to this, karstic effects of the temperate belt may be present too. Therefore, the glaciokarst is pronounced if the periglacial zone is wide and well-developed. This is characteristic of high karst plateaus. The width and the change of the width of the periglacial zone significantly affect the expansion and landscape of glaciokarst and karstification pattern.

The pattern of karstification also depends on glacier types. If cold-based glaciers were formed on the karst, no glacial erosion occurred. Preglacial karst features were modified at most in the foreground of retreating glaciers by meltwater and accumulation. Postglacial karstification takes place on surfaces which were not modified by glacial erosion. Features developing to the joint effect of ice and karstification or those being of subglacial origin did not form under cold-based glaciers. There is a lack of stepped surfaces, karren formation is more limited and different as compared to such a karst where warm-based glaciers occurred. Because of smaller erosion, there are only a few moraines or there are no moraines at all. Superficial deposit originates from periglacial processes. Covered karst and allogenic karst are probably only locally developed on such glaciokarst.

In case of warm-based glaciers, the glacial erosional surfaces of ice caps (ice sheets), valley glaciers and cirque glaciers are different. The erosion of ice cap or ice sheet results in less dissected, expanded and smoothed surfaces, while valley glaciers and cirque glaciers create more dissected surfaces where glacial erosional features are elongated (valley glaciers), patchy and island-like (cirque glaciers). In various karst areas, these differences give a completely different basis and conditions for recent karstification, and thus for its pattern.

Karstic pattern depends on the intensity of glaciation too. Glaciokarst is simple if one glaciation occurred in its area and it is complex if many glaciations took place. Repeating glaciation overrode the already existing glacial features and terrains. The intensity of younger and older glacials, as compared to each other, affect the landscape. If it was smaller, there may be terrains affected by glaciation once, twice or three times next to each other. It can be observed that the intensity of glacials decreased towards the latest glacial in karst areas (Djurovič 2009; Adamson et al. 2014; Van Husen 2000). This resulted in the decrease of the area of glacial erosion and the surfaces exempt from younger glacial erosion, but affected by older glacial erosion became covered with moraine and fluvioglacial sediments. Mainly transformed glacial features contribute to the transformation of karstification pattern as compared to the former state. In case of repeating glaciation, complex features develop, when there are smaller features (also of glacial origin or glacially transformed) in a larger karstic feature (or in glacial erosional features such as in cirques) or features being truncated to a various extent (for example shafts) develop next to each other. The cause of different truncation may be due to the fact that the given feature was only affected by glacial erosion once (it was not covered with ice during later glaciation) or the feature being less destroyed is younger (it was affected once by glaciation which was younger but more intensive) than the one being destroyed to a larger degree.

The karstification and pattern of features formed by the glacier will be different. In the area of glacial erosional features (mainly if accumulation took place too), karstification is of significant intensity, varied and it is relatively widespread even in glacier valleys too. On relict landforms (combe-ridge), karstification is of smaller intensity, less varied and its width is limited by the width of the bearing feature.

The presence, distribution or lack of karst types and karst-type assemblages are determinant in the formation of the karstic pattern of glaciokarst. According to Gvozdetskiy (1965), karst can be bare karst, soil-covered karst (karst is covered by soil), covered karst and buried karst. Covered karst can be cryptokarst when the non-karstic cover is impermeable and concealed karst when the non-karstic cover is permeable (Veress 2016a, b). On glaciokarst, concealed karst should be interpreted in a wider sense since the cover can be of non-karstic, but karstic material too. The calcareous content of the cover may be extremely diverse—it can range from 0 to 100%. It is 0% if it originates from non-karstic rock or it is low if such rock is mixed with karstic debris. It is higher if the karstic rock contains non-karstic intercalations or pollutants. It can reach 100% if it develops during the physical weathering of such a karstic rock which does not contain non-karstic intercalations or pollutants. (If the cover is only of partly limestone material, its calcareous content decreases during dissolution.) The glaciokarst owning a cover with calcareous content has to be regarded as concealed karst since (postgenetic) subsidence dolines may occur on its surface. However, subsidence dolines can only develop if solution features had already existed on the cover before they became covered or maybe if the cover is extremely thin (see Chap. 4). On glaciokarst, concealed karst also has the characteristic feature that the cover is dissolved, whereby it can become thinner without mechanical transport too.

On cryptokarst, karstification is represented by caprock dolines, while on concealed karst by subsidence dolines. Glacial cryptokarst can be mentioned, for example from the gypsum karst of Canada (Ford 1996), but it is a less widespread karst type on glaciokarst. On buried karst, the karstic bedrock is covered by impermeable, non-karstic cover to such a thickness that karstification neither occurs in the bedrock nor on the surface.

Veress (2016a) distinguished four types of cryptokarst: allogenic cryptokarst, autogenic cryptokarst, transitional cryptokarst and the cryptokarst developing from buried karst. Allogenic cryptokarst is equal to allogenic karst (see below). Local karstification can take place inside autogenic cryptokarst above well-karstifying rocks (halite, gypsum) or paleokarstic cavities. It can also take place on transversal cryptokarst at the thinning out of the cover. On cryptokarst developing from buried karst, karstification takes place at the floor of inheriting valleys. On glaciokarst, concealed karst is widespread. From cryptokarst types, allogenic cryptokarst (allogenic karst) and cryptokarst that develops from buried karst during karst development occur. Buried karst is rather only buried-like karst on glaciokarst since caves may occur in the area of this type too as the caves of the surrounding areas may also spread over the area of buried karst during their development. From a hydrological point of view, autogenic karst and allogenic karst are differentiated (Jakucs 1977; Ford and Williams 2007). Autogenic karst is an uncovered karst (bare karst or soil-covered karst), the allogenic karst receives water from the bordering surface covered with impermeable cover (or from the area of non-karstic rock outcrop), which favours ponor development at the interface of the karstic and non-karstic rock.

Due to the formation of ground ice, the permeable cover with limestone debris becomes impermeable temporarily. If it takes place on concealed karst, such a karst is called allogenic-like karst.

On glaciokarst, various karst types can occur next to each other. Although this characteristic feature can also be observed on karsts belonging to other types, on glaciokarst the change of karst types is extremely specific because of rapid climate change, the lack or diverse expansion and thickness of the superficial deposit (moraine or frost weathering debris). Since karst types directly affect the character of karstification, on glaciokarst, mostly, various karst types are responsible for the diversity of karstification.

Different karst types can constitute stripes and patches which can frequently alternate and be repeated as compared to each other or even be transformed into each other within a short time.

Bare karst, concealed karst and soil-covered karst are widespread in the whole width of the area of glacial erosional features. However, concealed karst is often local, while the continuity of soil-covered karst is often interrupted by the patches of bare karst. Buried karst and allogenic karst have a patchy development.

Features occurring on bare karst are giant dolines and uvalas, shafts, giant grikes, schachtdolines, small-sized solution dolines, opened-up cavities, relics of caves, karren (mainly of flow origin) and stepped surfaces. Features of soil-covered karst are the above-mentioned features, but karren show differences as compared to those that developed on bare karst. Instead of karren of the flow origin, karren of seepage origin are determinant, but these latter were also formed under soil. Both schachtdolines and small-sized solution dolines can occur in a soil environment above the tree line, but they are absent below it. Subsidence dolines, giant grikes and shafts are specific of concealed karst. Buried karst has only small patches (in dolines) where no recent karst features exist, but such areas can be those of debris cones too. Ponors and blind valleys are widespread on allogenic karst.

Karst types may be connected so close to each other both in space and functionally (processes working in the area of one type affect the other too) that they constitute a system called karst-type assemblage (karst-type union). In this, the fact also has a role that the expansion of a karst type is small and because of the great intensity of processes they have a larger chance to spread from the area of a given karst type to the area of another one. The karst types of a karst type assemblage can affect each other first of all, in the form of material transports. Thus, for example the debris that was formed on bare karst is transported into the area of covered karst.

The karst types of karst-type assemblages can create larger and smaller patches. The karst types of assemblages can be situated next to each other or a karst type develops inside of another. The following karst type assemblages can be distinguished:

- bare karst and soil-covered karst type-assemblage,
- bare karst and concealed karst-type assemblage,
- bare karst and buried-like karst-type assemblage,
- bare karst and allogenic karst-type assemblage,
- soil-covered karst and concealed karst-type assemblage,
- soil-covered karst and buried-like karst-type assemblage,
- soil-covered karst and allogenic karst-type assemblage,
- concealed karst and allogenic karst-type assemblage,
- concealed karst and buried-like karst-type assemblage,
- bare karst, soil-covered karst and concealed karst-type assemblage,
- bare karst, soil-covered karst and buried-like karst-type assemblage,
- bare karst, soil-covered karst and allogenic karst-type assemblage,
- bare karst, concealed karst and allogenic karst type assemblage,
- bare karst, concealed karst and buried-like karst type assemblage,
- bare karst, buried-like karst and allogenic karst-type assemblage.

From karst types and karst-type assemblages, bare karst type, bare and soil-covered karst-type assemblage and bare, soil-covered and concealed karst-type assemblage are widespread on terrains subject to glacial erosion in the periglacial zone, while the soil-covered and concealed karst-type assemblage is widespread in the temperate fluvial morphological belt. On accumulation terrains, soil-covered karst, concealed karst and allogenic karst and soil-covered, concealed and allogenic karst assemblage are characteristics.

The joint development of karst types and karst-type assemblages in the area of a glacial erosional feature can be homogenous, patchy or heterogeneous (zonal). In case of a homogenous development, there is only one karst type or karst-type assemblage in the same glacial erosional feature (Fig. 6.1), while in case of a heterogeneous development, several karst types or karst-type assemblages develop

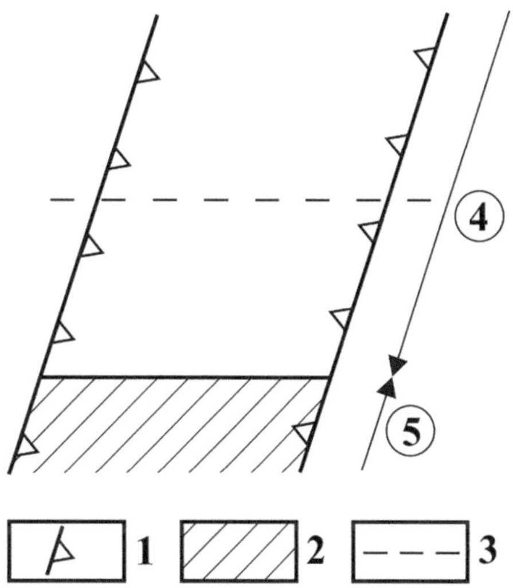

Fig. 6.1 Karstifying and non-karstifying part of a trough. 1. Trough, 2. karst water level close to the surface or permeable rock on the surface or thick moraine, 3. tree line, 4. karstifying valley part, 5. non-karstifying valley part

in the same glacial erosional feature (Fig. 6.2a). Patchy development is formed as a result of a local effect if the karst type differs from the karst type of its environment, for example, in a giant doline or on a cover accumulating at a cliff (Fig. 6.2b) as well as on relict features with small expansion (combe-ridge, horn, Fig. 6.3).

The number of karst types, the density and distribution of each type, the spread of karst-type assemblages and the spread of the karst types constituting the karst-type assemblage as compared to each other are affected by the following:

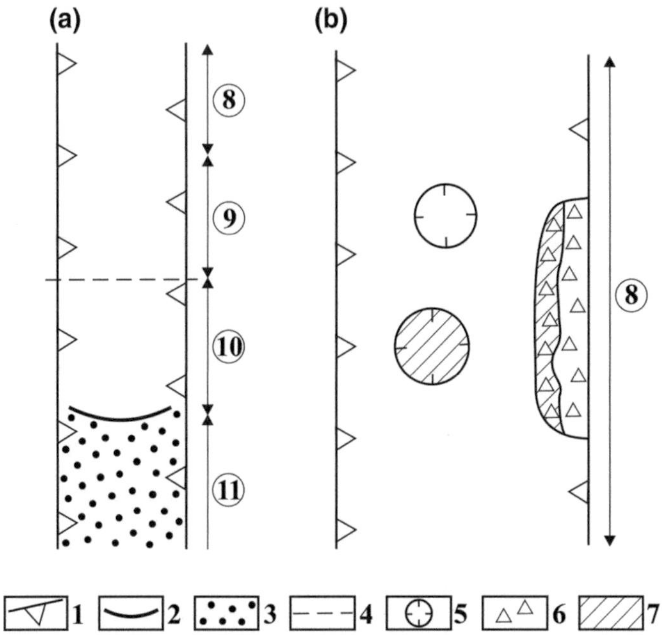

Fig. 6.2 Karst types and karst-type assemblages in a trough. 1. Trough, 2. terminal moraine, 3. fluvioglacial sediment, 4. tree line, 5. giant depression, 6. frost weathering cover, 7. concealed karst, 8. bare karst, 9. bare karst and concealed karst-type assemblages or bare karst and soil-covered karst-type assemblages, 10. soil-covered karst and concealed karst-type assemblage or soil-covered karst and allogenic karst-type assemblage, 11. concealed karst, soil-covered karst and allogenic karst-type assemblage or buried-like karst type, **a** trough of heterogeneous karst type, **b** trough of patchy karst type

Fig. 6.3 The karstification of a combe-ridge. 1. Combe-ridge, 2. there is no karstification, 3. there is karren formation, 4. karren, small-sized solution dolines, schachtdolines and in case of superficial deposit, suffosion dolines can occur too

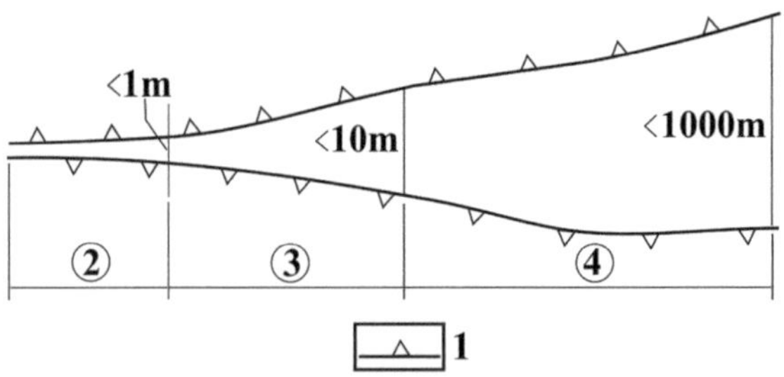

- It depends on preglacial karstification and on the developed karst features. In karstic depressions, karst types being different from their environment may be formed because of accumulation. For example, in a bare karst environment, concealed or allogenic karst develops. These interrupt the expansion of bare karst type. The density of concealed or allogenic karst depends on the density of karstic depressions, while their expansion depends on the size of the depressions.
- It also depends on glacier types: karst types with wide expansion are formed in the area of ice caps, while they are of striped, linear development in the troughs of valley glaciers. These karst types, if they are not bare, are separated by the narrow bare karst types and karst-type assemblages of combe-ridges (or at a lower altitude by soil-covered karst type or bare karst and soil-covered karst type).
- However, it also depends on the character of glacial erosion. According to the thickness and type of the developing cover, three zones can be distinguished. The erosional zone where moraine occurs in a limited expansion (the expansion and thickness of which increases towards the lower end of the glacier). The accumulation zone which is the most characteristic in the area of end moraines and lateral moraines. This is lined by a third zone which is the fluvioglacial zone. The sediments of this zone are not necessarily directly connected to the end moraine and are not necessarily developed continuously. According to the position of poljes (and other karstic depressions), the zone can be of three different kinds. The first kind is when the polje is directly connected to the trough (piedmont polje), the second is when it situated farther from the valley and the third kind is the combination of the above-mentioned cases. In the first case, there is a moraine in the polje, glacial erosion exerted its effect inside the polje too. In the second case, no glacial erosion took place in the polje and the fill of the polje is fluvioglacial sediment. Terrains exempt from glacial erosion occur between the poljes. The fluvioglacial zone can be formed in the foreground of glaciers exempt from poljes and also in the area of former ice sheet. Both the erosional zone and the fluvioglacial zone may be interrupted by tongue-like, coarse depositions of outlet glaciers at ice caps and at ice sheets. In the erosional zone, there is a greater chance for the development of bare karst (if the terrain is of higher altitude) or soil-covered karst (if the terrain is lower). In the accumulation zone, concealed karst, buried-like karst and allogenic karst are more likely to develop. Towards the lower end of glacial erosional features, especially if they are more expanded (for example troughs) the chance of the development of karst types related to superficial deposit increases. Fluvioglacial accumulations favour the formation of concealed karst type, allogenic karst type, buried-like karst type and concealed karst and allogenic karst type assemblages on terrains outside glaciokarst (in the fluvioglacial zone). In the Dinarides, the fluvioglacial zone dissected by poljes can constitute a separated zone where concealed karst type, allogenic karst type, buried-like karst type as well as soil-covered and concealed karst type assemblages occurred too.

- During periglacial processes, debris develops which also favours the formation of covered karst type (concealed karst). The cover created by periglacial processes can increase towards the upper end of glacial erosional features (thus, they can also develop at the upper sections of these features). However, the cover that was formed by these processes can locally develop anywhere, where the side slopes of glacial erosional features are constituted by cliffs with northern exposure built up of vegetation-free, well-stratified rocks.
- It depends on the climate predominant in the area of glacial feature (this is determined by altitude, the distance from the Equator and by exposure). The climate controls the appearance and expansion of bare karst type and soil-covered karst type, though the large inclination of the slopes (which mainly depends on the intensity of glacial erosion) can also contribute to the development of bare karst.
- It also depends on the work of meltwater since it affects the site, thickness and expansion of cover accumulation in the areas formerly covered and uncovered with ice too. Thus, meltwater significantly contributes to the development of concealed karst and allogenic karst as well as concealed karst and allogenic karst-type assemblages.
- It depends on the length of the bearing features since the longer they are, the more karst types and karst-type assemblages being determined by climate can develop. If a longer erosional feature was formed on a terrain with a smaller altitude-difference, the expansion of some karst types and karst-type assemblages can increase.
- The composition of the moraine depends on the lithological structure of the karst, however, this influences the development of buried-like karst, concealed karst and allogenic karst. The exposure of the non-karstic rock being intercalated into the karst favours the development of allogenic karst.
- It also depends on the intensity of subsequent glaciations. Intensity influences the length of glaciers. Because of the changes in glacier length, morainic terrains (mainly end morainic terrains) can be repeated. This contributes to the repetition of some karst types (concealed karst type, allogenic karst types) on the valley floor.

Although theoretically, every section of the glaciokarst can present larger and smaller karstification, first of all, this is focused on terrains subject to former glacial erosion. Glacial erosional depressions, troughs and cirques are karstified more intensively. Relict landforms of glacial erosion karstify to a less extent (or if they are karstified, bare karstification occurs there).

6.2 Karstic Pattern in Areas and Features of Glacial Erosion

The pattern will be presented in glacier valleys (troughs), in cirques, in karstic depressions on surfaces transformed by ice caps (ice sheets) and on combe-ridges.

6.2.1 Karstic Pattern of Glacier Valleys and Troughs

On glaciokarst dissected by glacier valleys, karstic pattern depends on the pattern of valleys and on the pattern of karst types and karst-type assemblages being formed in the valleys. The pattern and morphology of glacier valleys depend on the type of glaciers. The pattern and morphology of valleys are determined by various ice formation circumstances.

6.2.1.1 The Effect of Ice Development Circumstances to Valley Pattern and Morphology

Ice development may happen on uncovered or covered karst (see Chap. 2). On uncovered karst, places of ice development are karstic depressions (Smart 1986) and ridges creating lee (Smart 1986; Žebre and Stepisnik 2015a). The ice developing in karstic depressions can flow out in one direction, but in several directions too; according to this, several glacier snouts (troughs) can be formed in the environment of depressions. Their place and direction, and thus those of troughs are also determined by the dip direction and altitude circumstances of the preglacial karstic surface. The length of troughs does not only depend on the altitude of the bearing terrain, but it also depends on the amount of developing ice. This latter fact is affected by the distance from the seas, the intensity of glaciation and the size and morphology of the karstic depression (in case of steeper side slopes, snow accumulation and thus, ice development is more intensive). However, the density of troughs does not only depend on the number of outflowing glacier snouts, but it also depends on the number of karstic depressions. The margin of such troughs is straight, there may be giant dolines on their floor which developed in the interglacials.

On covered karst (cryptokarst developing from buried karst), epigenetic valleys develop. This may happen in two ways: karst water is situated below the valley floor which is called former, hanging karst water level environment or it is situated at the inheriting valley floor which is called former, non-hanging karst water level environment. If the karst water level is located below the inherited valley floor, karstification takes place on the valley floor (Jakucs 1977) and dolines develop. The dolines will be the places of snow accumulation. The ice leaving the dolines transforms the valley and its doline row (the epigenetic valley) into a trough. Such a valley is characterized by the following:

- the valley margin is separated into arcuate sections (Fig. 6.4),
- there is no separating, well-developed cirque,
- the valley has no tributary valleys or it has only some (Fig. 6.5).

The reason for this latter is that surficial waters percolate on the inherited valley floor. Consequently, there is a smaller chance of the development of regressional tributary valley. However, even if a tributary valley develops, it is not deep, and the stream does not cut through the cover and thus, the valley is destroyed by the denudation of the cover. At the same time, tributary valleys may be formed by the glaciers leaving the karstic depressions of the bare becoming terrain between the valleys. The junction angle of the so developed tributary troughs into the main trough depends to a less degree on the dip direction of the bearing slope. Therefore, the junction angle can be large too (Fig. 6.5c).

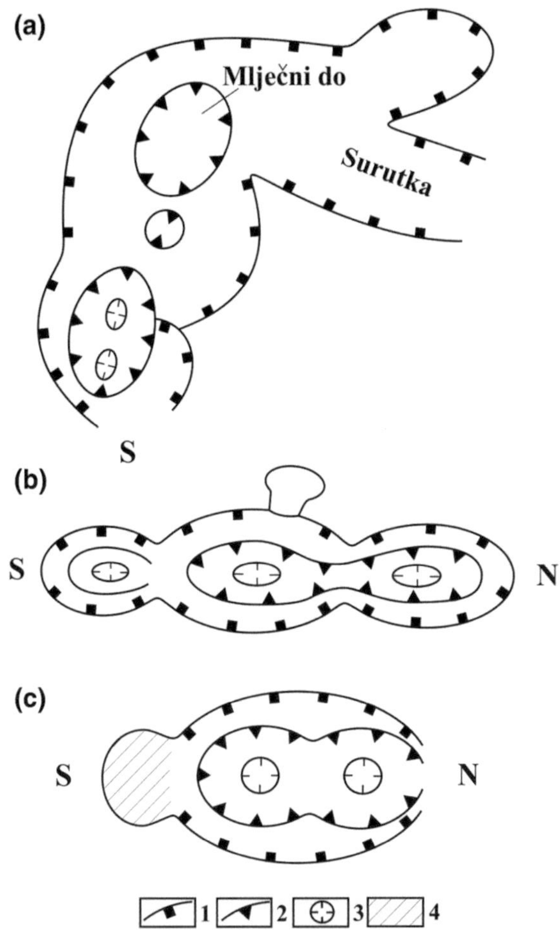

Fig. 6.4 Decomplex troughs with arcuate margins (ground plan sketches on some trough parts from Durmitor based on a map M 1:25,000). 1. Combe-ridge surrounding a glacier valley, 2. doline or uvala that follows the pattern of a combe-ridge, 3. valley floor doline, 4. glacier, **a** Mlječni-Urdeni do, **b** Lokvice, **c** Velika kalica

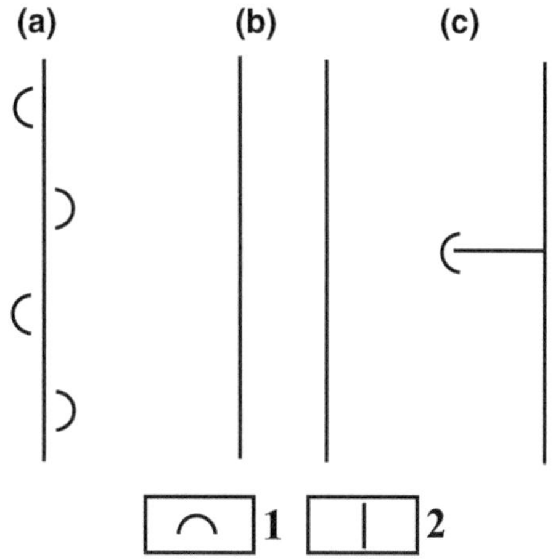

Fig. 6.5 Glacier valley patterns that developed on uncovered karst. 1. Cirque, 2. trough, **a** straight trough with cirques at its margin, **b** straight parallel troughs which developed along rows of depressions by glaciers or by outlet glaciers that left an ice cap, **c** tributary trough being perpendicular to the main trough

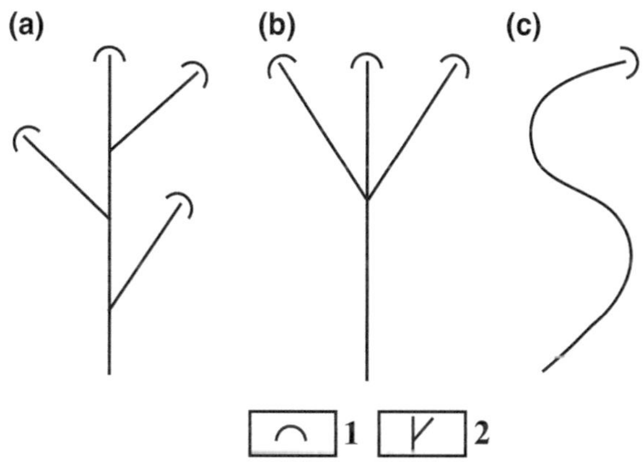

Fig. 6.6 Glacier valley patterns on karst where there is an impermeable cover (the glacier developed in an epigenetic valley). 1. Cirque, 2. trough, **a**, **b** bunchy valley network where the valleys have straight margin, the valley network is built up of main valleys and tributary valleys enclosing an acute angle with it, **c** meandering glacier valley

On covered (buried) karst, the karst water level can be situated at the surface of bedrock during valley inheritance (non-hanging karst water level environment). In this case, valley development is not interrupted by valley floor karstification (Hevesi 1980). Fluvial valley development may continue in the bedrock until the karst water level is situated at the valley floor. On the covered terrain, the already developed valley may have several tributary valleys. During glaciation, the sites of snow accumulation and ice development will be the valley heads of these valleys. Such trough network is characterized by large valley density since a number of tributary valleys developed (Fig. 6.6). The headwaters of streams bifurcate like a fork. According to this, the sections of troughs under the cirque often develop like twins, in pairs. The margin of troughs is non-arcuate, but dolines having been developed in the interglacials may occur on their floor. Valley floor dolines are formed when the floor of the valley that underwent glaciation gets above the karst water level because of the sinking of karst water level or the rising of the bearing terrain.

It can be seen that on bare karst, the density of troughs is determined by the density of karstic depressions and the density of glaciers branching out of them, while on cryptokarst developing from buried karst it is determined by the density of streams. On bare karst, there is a chance of the formation of branching out troughs, while on covered (buried) karst, bunchy, coalescing valleys develop in large density, where the karst water level was at the bedrock. On a covered (buried) karst, where the karst water level is below the bedrock, solitary valleys develop with arcuate margin and without tributary valley. On bare karst, the direction of troughs was determined by the dip direction of the preglacial karstic surface, while on covered (buried) karst, it was determined by the dip direction of the surface of the non-karstic rock. While in case of bare karst and covered, but non-hanging karst water level environment there are cirques, in case of hanging karst water level environment there are not any separating cirques.

6.2.1.2 The Patterns of Ice Development Environments

On a glaciokarst there may occur one or several types of ice development environments (Fig. 6.7). Terrains with manifold ice development environments may have extremely different shape and size which is determined by the site and size of covered and uncovered areas as well as by the position of karst water as compared to the surface of the bedrock. The karstification of glaciokarst is also affected by the position of present karst water where the karst water reaches the surface (present karst water level environment), the floor of glacier valleys is not karstifying now.

On a glaciokarst one, two or three environments may occur from various glacier development environments. In case of a circular development, the middle part of the mountains which if it is in an elevated position did not become covered. It was surrounded by the hanging and non-hanging karst water belt (Fig. 6.8a). Because of the elevated position of the middle part of the mountains, the karst water level is the deepest here as compared to the surface of the bedrock. It may happen that later the present karst water level environment also develops at the margin of the mountains (Fig. 6.9a). Bare karst may also evolve in a way that it was only bordered by the belt of the non-hanging karst water level (Figs. 6.8b and 6.9b) or by the belt of the former hanging karst water level (Figs. 6.8c and 6.9c).

It is possible that the whole area of the glaciokarst was covered by impermeable rock. In this case, both the belt of the former hanging karst water level environment and that of the former non-hanging karst water level environment were present in the area of the karst (Fig. 6.10a). In the case of this glaciokarst too, the present karst water level may be situated deeper than the valley floors or at the present valley floors (Fig. 6.11a). It may also be the case that there was only one development environment and the present karst water level is also situated below the valley floors (Fig. 6.11b) or at the valley floors (Fig. 6.11c).

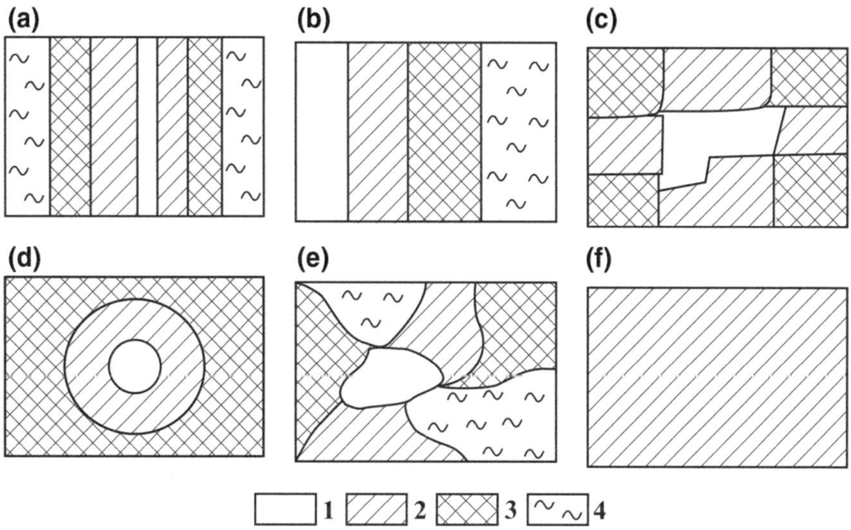

Fig. 6.7 Theoretical patterns of the parts of the glaciokarst which have different evolution. patterns: **a** symmetrically striped, **b** asymmetrically striped, **c** mosaical, **d** ringed, **e** irregular, **f** homogeneous, 1. the karst was bare preceding glaciation, 2. there was an impermeable cover overlying the karst preceding glaciation, but the karst water level was below the surface of the bedrock, 3. there was an impermeable cover overlying the karst preceding glaciation, but the former karst water level was at the surface of the bedrock, 4. the karst water level is at the valley bottoms of the karst at present

6.2.1.3 Feature Types of Glacier Valley

On glaciokarst, the following glacier valley feature types may occur:

- Trough which does not have a separating, individual cirque (Figs. 6.4, 6.5 and 6.12a). Former cirques were the dolines constituting the valley which became parts of the ablation area. Troughs are of arcuate margin (even many times) since they are constituted by dolines.

The denser the inherited valleys developed and the more complete karstification was on valley floors, the smaller the chance of the occurrence of tributary valleys. Consequently, glacier (mainly tributary glacier) valley density is small, the valleys are parallel to each other, if there is a tributary valley, it can be even perpendicular to the main valley.

The cirques of outlet glaciers branching out of ice caps or ice sheet are also absent. The troughs of outlet glaciers are parallel to each other or branch out radially from the area of the former ice cap (ice sheet).

- Cirque with a trough, two varieties can be distinguished. One of them develops on uncovered karst, while the other evolves on covered (buried) karst. On covered karst, river valleys develop if the karst water level is located at the

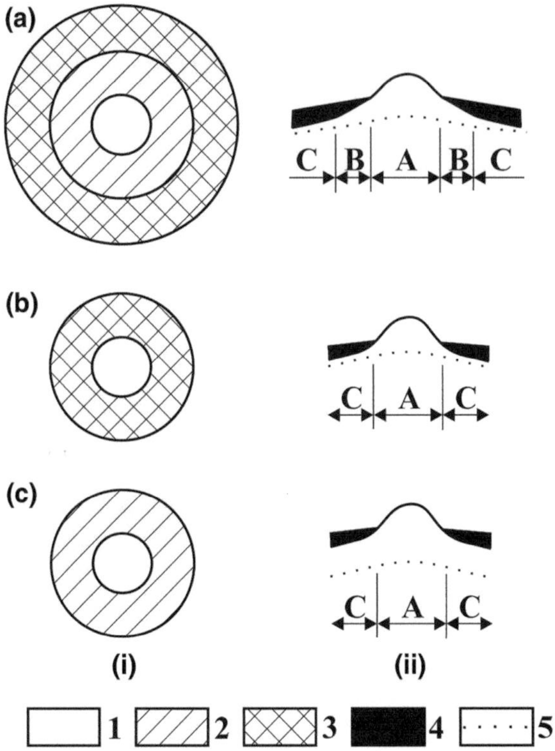

Fig. 6.8 The parts of glaciokarst with various evolution in case of a ringed development when the karst had both a bare karst part and a part where there was an impermeable cover overlying the karst preceding glaciation. **i** Top view, **ii** lateral view, **a** preceding glaciation bare karst was present, in case of a karst where there was an impermeable cover overlying the karst, the karst water level could be situated below the surface of the bedrock, but at the surface of the bedrock too, **b** preceding glaciation bare karst was present, but in case of a karst where there was an impermeable cover overlying the karst, the former karst water level was at the surface of the bedrock, **c** preceding glaciation bare karst was present, but in case of a karst where there was an impermeable cover overlying the karst, the former karst water level was lower than the surface of the bedrock, 1. the karst was bare preceding glaciation (*A* in the profile), 2. preceding glaciation the karst water level was lower than the surface of the bedrock (*B* in the profile), 3. preceding glaciation the karst water level was at the surface of the bedrock (*C* in the profile), 4. cover, 5. former karst water level

surface (no karstification takes place). In this case, as already mentioned in Sect. 6.2.1.1, tributary streams of the main stream develop by retreating, by which the water network and valley network will have a bunchy pattern. In the valley heads cirques, while in the valleys, troughs develop. The headwaters of the streams bifurcate like a fork, therefore, glacier valleys of one order also present a fork-like pattern. Valley density is large and the valley margin is straight (Fig. 6.6).

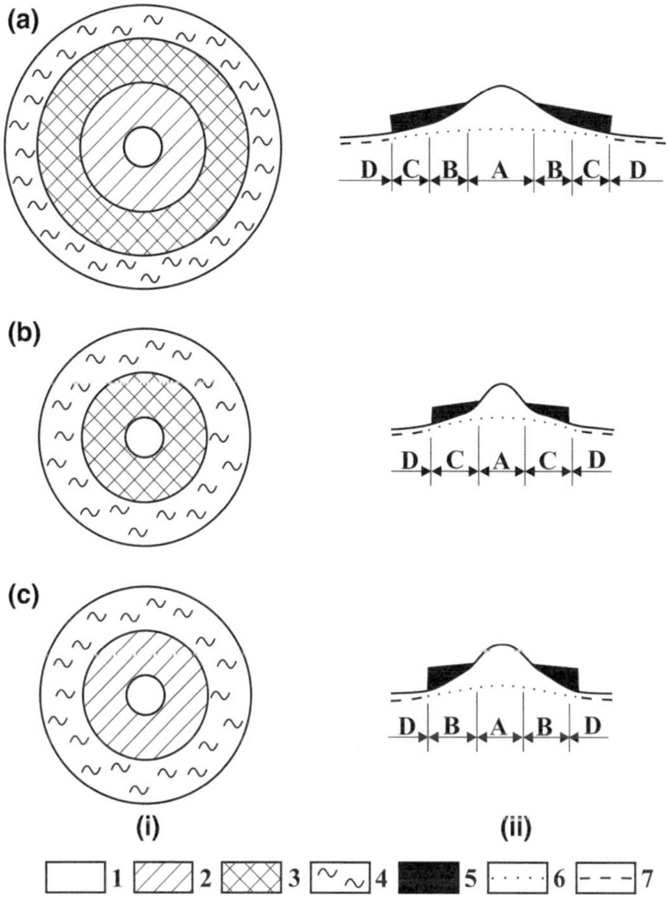

Fig. 6.9 The parts of glaciokarst with various evolution in case of a ringed development when the karst had both a bare karst part and a part where there was an impermeable cover overlying the karst preceding glaciation and the karst water level is at the floors of valleys at the margin of the mountain at present. **i** Top view, **ii** lateral view, **a** preceding glaciation bare karst was present, in case of a karst where there was an impermeable cover overlying the karst, the karst water level could be situated below the surface of the bedrock, but at the surface of the bedrock too, **b** preceding glaciation bare karst was present, but in case of a karst where there was an impermeable cover overlying the karst, the former karst water level was at the surface of the bedrock, **c** preceding glaciation bare karst was present, but in case of a karst where there was an impermeable cover overlying the karst the former karst water level was lower than the surface of the bedrock, 1. the karst was bare preceding glaciation (*A* in the profile), 2. preceding glaciation, there was an impermeable cover overlying the karst, but the karst water level was lower than the surface of the bedrock (at the valley floor) (*B* in the profile), 3. preceding glaciation, there was an impermeable cover overlying the karst, but the karst water level was at the surface bedrock (at the valley floor) (*C* in the profile), 4. the karst water level is at the valley floors (at the margin of the mountain) of the karst at present (*D* in the profile), 5. cover, 6. former karst water level, 7. present karst water level

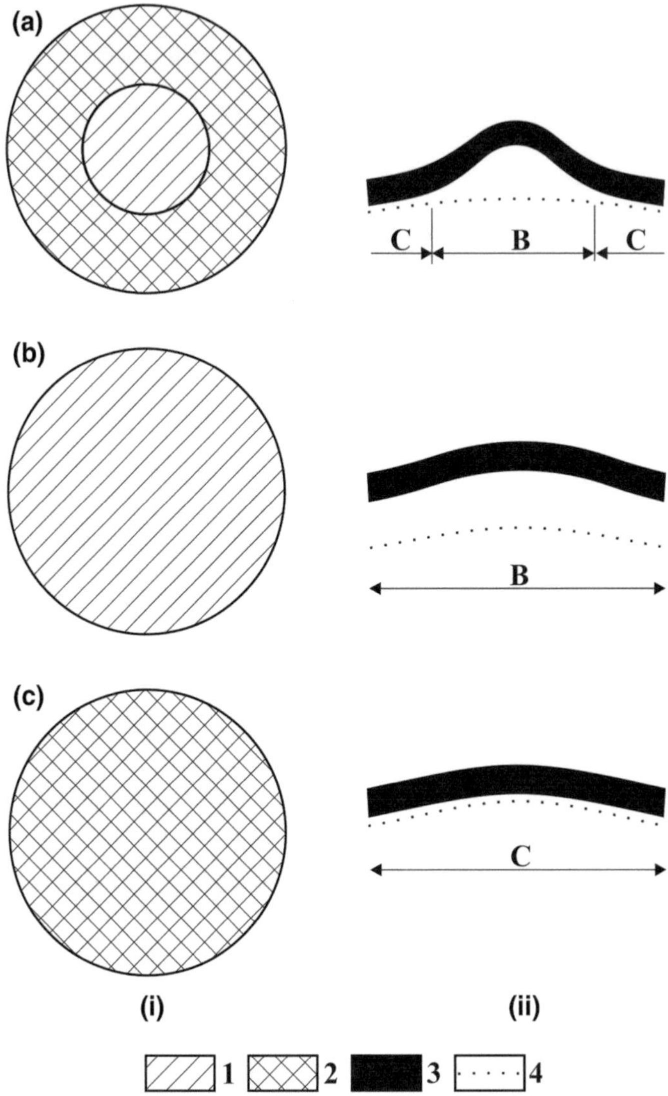

Fig. 6.10 The parts of glaciokarst with various evolution in case of a ringed development when there was an impermeable cover overlying the karst. **i** top view, **ii** lateral view, **a** preceding glaciation there was an impermeable cover overlying the karst, the former karst water level was at the surface of the bedrock and below the surface of the bedrock, **b** preceding glaciation, there was an impermeable cover overlying the karst and the former karst water level was located below the surface of the bedrock, **c** preceding glaciation, there was an impermeable cover overlying the karst, but the former karst water level was located at the surface of the bedrock, 1. preceding glaciation, the karst was covered (buried), but the former karst water level was of a lower position than the surface of the bedrock (*B* in the profile), 2. preceding glaciation, the karst was covered (buried), but the former karst water level was at the surface of the bedrock (*C* in the profile), 3. cover, 4. former karst water level

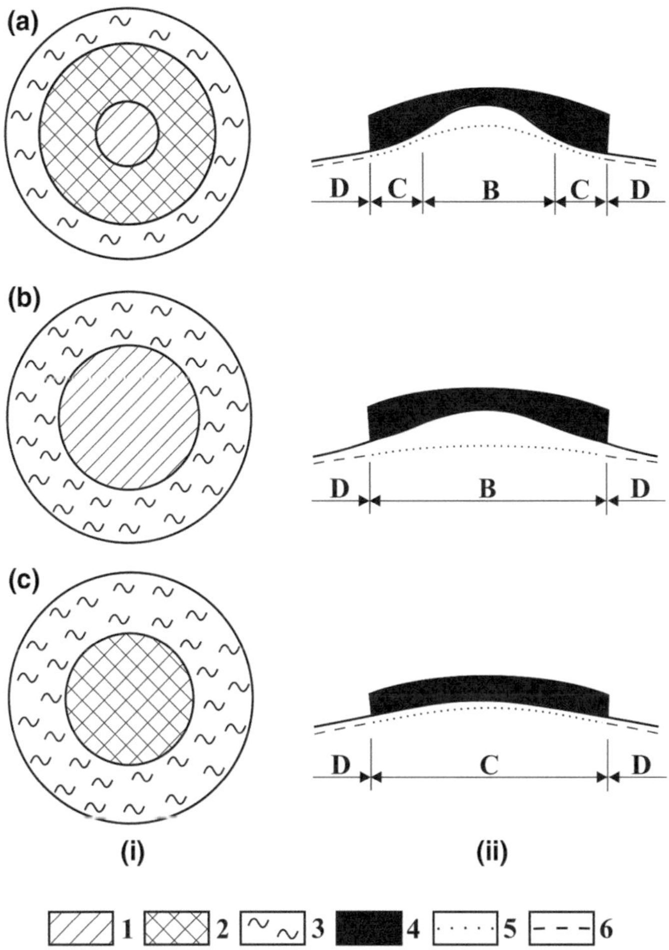

Fig. 6.11 The parts of glaciokarst with various evolution in case of a ringed development when there was an impermeable cover overlying the karst and the present karst water level is at the valley (at the margin of the mountain) bottom. **i** Top view, **ii** lateral view, **a** preceding glaciation, there was an impermeable cover overlying the karst, the former karst water level was located at the surface of the bedrock and below the surface of the bedrock, **b** preceding glaciation, there was an impermeable cover overlying the karst and the former karst water level was situated below the surface of the bedrock, **c** preceding glaciation, there was an impermeable cover overlying the karst, but the former karst water level was at the surface of the bedrock, 1. the former karst water level was of a position lower than the surface of the bedrock (*B* in the profile), 2. the former karst water level was situated at the surface of the bedrock (*C* in the profile), 3. the karst water level is at the valley bottoms of the karst at present (*D* in the profile), 4. cover, 5. former karst water level, 6. present karst water level

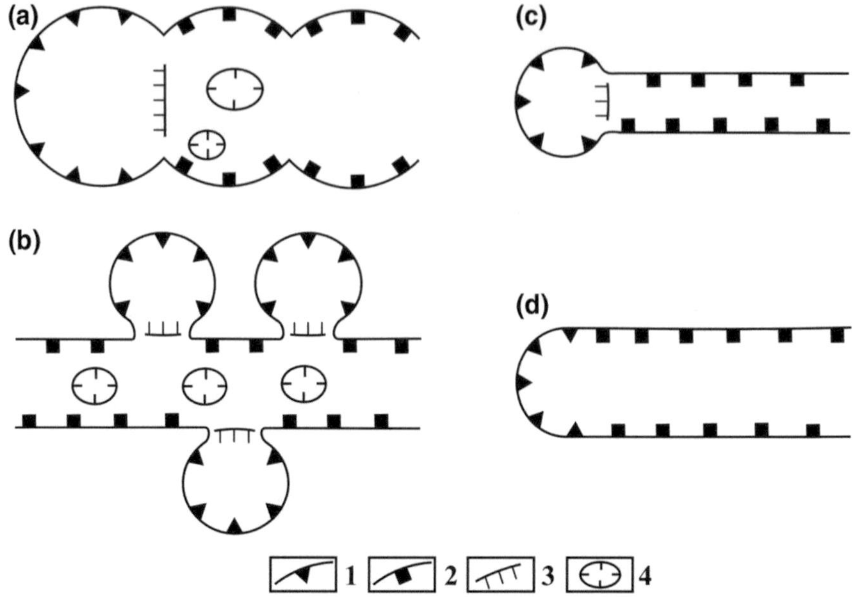

Fig. 6.12 Singly complex troughs with arcuate margin (**a**) and straight troughs (**b–d**). 1. Cirque, 2. trough, 3. anti-dip step of a cirque, 4. giant doline

A cirque with trough may develop on bare karst too. The snow pack is the karstic depression that developed on the uncovered terrain. The density and pattern of cirques are determined by preglacial karstic depressions. In this case, the following patterns may occur:

- The ice did not leave the karstic depression, the trough is missing, a terminal moraine may occur at the margin of the depression (Fig. 6.13a, c, see at the cirques).
- The glacier left the karstic depression and created a trough with a straight margin (Fig. 6.13g, h) or troughs with a straight margin (Fig. 6.13f) On the floor of the trough(s), if it (they) did not develop in the last glacial, there may be giant doline(s) (valley floor doline).
- The trough is bordered by a cirque from the side. The cirque may hang above the trough. Mostly a short tributary trough with a large gradient (maybe a hanging one) is connected to the main trough (Fig. 6.13d, e).
- The trough is lined by a series of cirques (Figs. 6.5a and 6.12b). In this case, the trough was probably created by the confluencing ice that left the cirques (karstic depressions). The main trough has a straight margin, however, its margin may be dissected by hanging cirques. (The Pirine doline in the Durmitor may have such a development.)

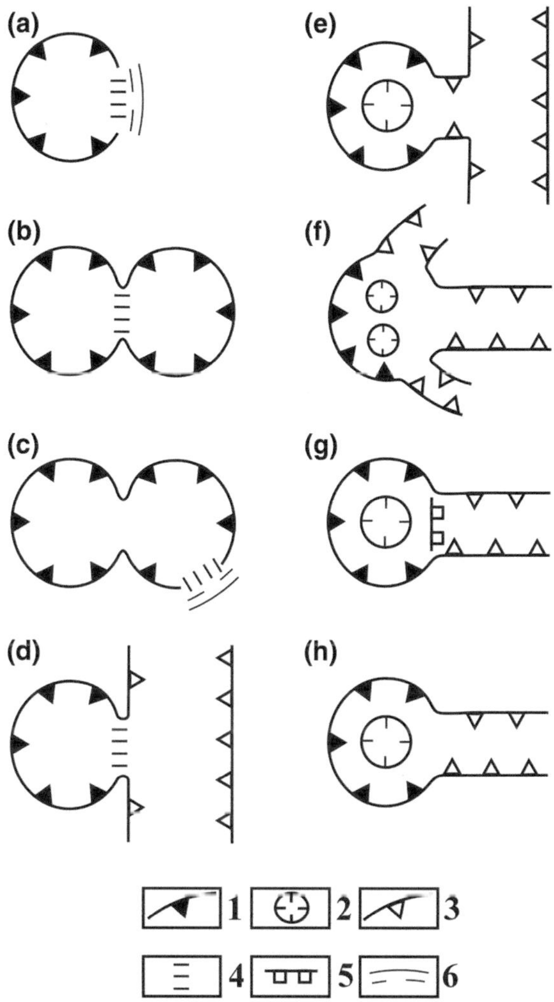

Fig. 6.13 Cirque varieties. 1. Cirque, 2. karstic depression, 3. trough, 4. threshold sealing the cirque, 5. transverse step, 6. terminal moraine, **a**–**c** cirques without troughs, **a** the cirque developed from a doline, the ice flowed over its margin, **b** the cirque developed from an uvala, **c** the cirque developed from an uvala, the ice flowed over its margin, **d**, **e** cirques connecting to a trough, **f**, **g** cirques with troughs (the cirques of figures g and h might have developed in karstic depressions, but in a valley head too), **f** several troughs are connected to a cirque, **g**, **h** a trough is connected to the cirque

6.2.1.4 The Karstic Pattern of the Trough

The trough can be karstified downwards in its total length as long as the valley deepens into a karstic rock or the valley floor is not covered by till in a large thickness and at sites where the karst water level has not been situated at the valley floor yet (Fig. 6.1). The karst features, but also the karst types have a striped, linear pattern because of linear, striped glacial erosion. This may be emphasized the giant dolines of troughs and the concealed karstic patches that developed in them (Fig. 6.14). An exception is first of all the bare karst which spreads to the combe-ridges at large altitudes (subzone of mass movement and frost weathering), and thus it may have a wider expansion.

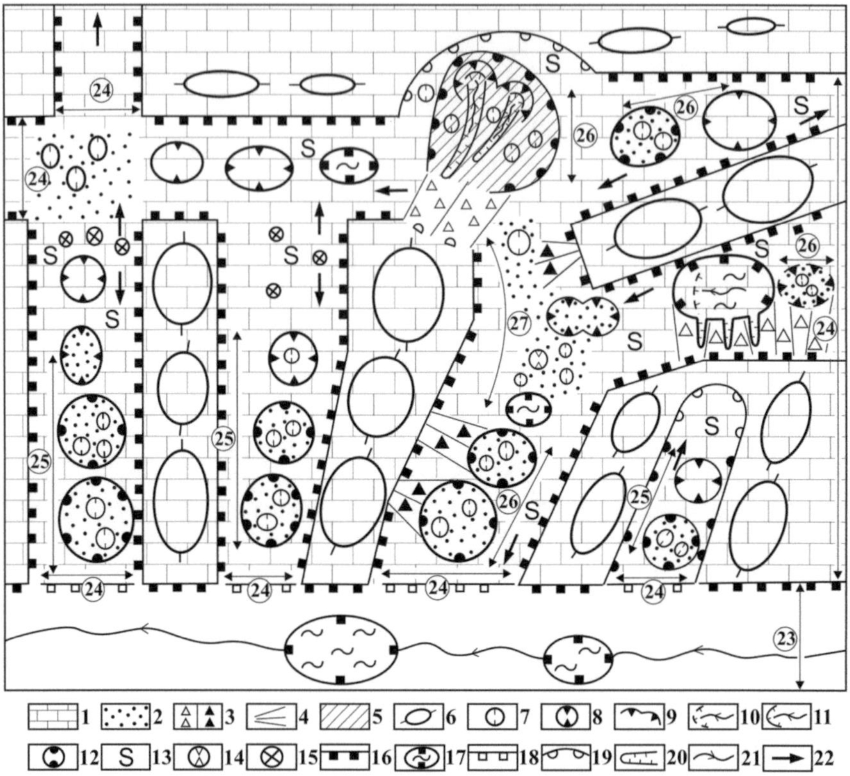

Fig. 6.14 Concealed karst patches aligned linearly on the floor of troughs that developed parallel to each other (Veress 2016a modified). 1. Limestone, 2. moraine, 3. frost-shattered debris of siliceous limestone, blocks and debris of mountain failures, 4. debris cone, 5. reworked, transformed rock debris (with high clay content), 6. horn, 7. subsidence doline, 8. paleodoline, 9. paleouvala, 10. subaqueous ponor, 11. ponor, 12. depression of superficial deposit, 13. schichttrippenkarst, 14. recent solution doline, 15. schachtdoline, 16. trough, 17. rock basin with lake, 18. transverse step, 19. cirque, 20. gully, ravine, 21. watercourse, 22. dip direction of surface, 23. main trough, 24. tributary trough, 25. striped-patchy karst (concealed karst), patches are aligned in the interglacial dolines of troughs (with uncovered karst between patches), 26. allogenic karst, 27. covered karst (concealed karst) terrain composed of moraine patch and local rock debris accumulation and interrupted by uncovered patches with solution dolines

The elongatedness of the areas covered by karst types and karst type assemblages coincides with valley directions. The karst types and karst-type assemblages of the valleys may also have a zonal nature (Fig. 6.2a). If there is a karstic surface at the termination of the trough too, the trough may be connected to a karstic depression (polje) (Fig. 6.15). The shape and expansion of karst-type belts may be modified by giant dolines and debris cones that developed at the stem of steep valley sides (Fig. 6.2). The karst types and karst-type assemblages may also be repeated in the trough, for example, if superficial deposit appears again on the valley floor.

Fig. 6.15 Non-zonal karst types of polje-trough. 1. Trough, 2. polje, 3. karstic depression, 4. debris cone, 5. bare karst, 6. concealed karst, 7. allogenic karst

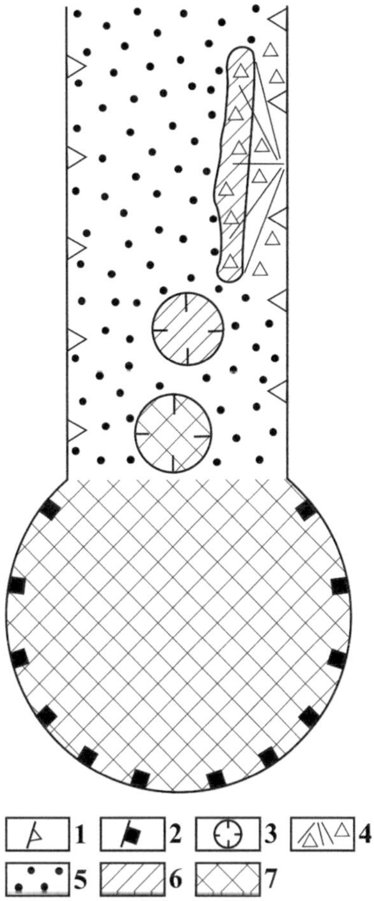

The distribution of the karst types and karst-type assemblages of the trough can be modified by the change of the intensity of glaciation. A glaciation of decreasing intensity results in the shortening of glaciers on glaciokarst too (Djurovič 2009) or the formation of an inner trough. Because of glacier shortening, the same karst type or karst-type assemblage can be repeated on the valley floor and as a result of an inner valley development, the karst types of the floor of the inner valley may be different from the karst types of the outer valley at the same place.

Trough of Homogenous Karst Type

The cause of its development is that the valley is short or there are factors hindering karstification on the lower part of the trough (Fig. 6.1).

The karst type of the trough is bare karst. It occurs in troughs of higher altitude, situated farther from the equator, being short and having a northern exposure. In the area of this type, karren stepped surfaces (limestone pavements) are widespread. Karren formation is predominant. On valley floors, mainly the exposed surfaces with bedding planes, while in the valley sides mostly the head of beds (wandkarren) are affected by karren formation. The pattern of karren depends on the relation between stratal dip and valley direction as compared to each other. If the inclination of beds is perpendicular to valley direction, the direction of karren of flow origin is also perpendicular to the direction of the valley. If the stratal dip is identical with the direction of the valley, the direction of the karren features of flow origin is identical with the direction of the valley (see Chap. 3).

The karren formation of stepped surfaces is interrupted and modified by roche moutonnées, karstic depressions, rock basins and steps. On roche moutonnées, karren of flow origin with bifurcating pattern develop, while in karstic depressions and in rock basins, a converging karren (mainly rinnenkarren) are formed.

Small-sized solution dolines and schachtdolines are widespread along fractures or at sites favouring snow accumulation (rock basins, giant dolines or the leeward slopes of elevations).

In the area of bare karst (in giant dolines, at the margin of debris cones), concealed karst may occur in patchy expansion (Fig. 6.2b).

Trough of Heterogeneous Karst Type

It is characteristic of long troughs with large gradient. With the increase of valley length more and more karst types and karst type assemblages may develop in these features. The expansion of karst types can be various within karst-type assemblages too. This depends on the altitude and morphology of the trough or the exposure of the valley. The belts are the following:

- Bare karst is on the uppermost part of the trough which has already been mentioned.
- Above the tree line, in the dissolutional subzone, the bare karst and soil-covered karst type assemblage, the bare karst and concealed karst-type assemblage as well as the bare karst, soil-covered karst and concealed karst-type assemblages are the most widespread, but soil-covered karst, concealed karst and allogenic karst-type assemblages also occur. Karren of bare surfaces are predominant, but subsoil karren occur too. On bare karst sections, shafts and schachtdolines can be found, while small-sized solution dolines occur both on soil karst and bare karst sections. Karren formation is greater and more intensive in the area of bare karst and soil-covered karst assemblage than on bare surfaces (Veress 2010). This can be traced back to the larger solution capacity of waters originating from surfaces with soil patches (Veress 2010). Concealed karsts are widespread inside giant dolines, but outside them too. In giant dolines allogenic karsts may occur too (Fig. 6.14).
- Below the tree line, until terminal moraines, bare karst and soil-covered karst-type assemblage, soil-covered and concealed karst-type assemblage as well as

soil-covered karst, concealed karst and allogenic karst-type assemblages occur. Karren formation is mainly represented by subsoil karren. Concealed karst can be found on valley floors without depressions, while allogenic karst type occurs in depressions. Shafts and schachtdolines are absent, but dolines of the temperate belt may occur.
- Under the terminal moraine, in the area of the fluvioglacial cover, soil-covered karst and conc ealed karst-type assemblage, soil-covered karst, concealed karst and allogenic karst-type assemblage may occur. In case of present fluvial erosion, the area covered by fluvioglacial sediment can develop into fluviokarst.

However, in troughs with an arcuate margin, stepped surfaces are absent or if there are any, they are covered (the possible reasons for this were mentioned in Chap. 4). Because of this, karren formation is less intensive, bare karren are subordinated and less varied. Troughs of both homogenous and heterogeneous karstification can be connected to depressions (to poljes), which are partly infilled by fluvioglacial sediment or till (for example the Gomance polje, Fig. 6.15). Concealed karst, buried-like karst and allogenic karst or concealed karst, buried-like karst and allogenic type karst assemblages may occur in them. Consequently, karren formation is less widespread, karren features of flow origin are subordinated and rather karren features of seepage origin are widespread.

6.2.2 *The Karstic Pattern of Karstic Depressions*

Both the morphology and karst types of troughs are affected by the fact whether there are any old karstic depressions (giant dolines and uvalas) in its area. If they are present, the morphology of the valley and the pattern of its karst types are affected by their number and size. The larger the karstic depressions, the greater the chance that an independent karstification and karstic pattern being different from the environment develops in their area. The larger their density and the karstification represented by them, the greater extent this karstification determines the karstification of the valley. However, the karstification of various karstic depressions, often situated next to each other can be basically different too. The difference may be caused by the karstic development of karstic depression, the glacial erosion and the accumulation. According to their position, as compared to the valley (Veress 2012, 2013, 2016a, Chap. 4) they may be valley dolines and valley floor dolines.

As compared to the features and karst types of the bearing valley, the features, karst types and karst type-assemblages of depressions may be the following:

- There are different karst features, karst types and karst-type assemblages in the karstic depression than on the bearing valley floor. It often occurs that in the karstic depression situated on the bare valley floor, there is soil-covered karst type or a soil-covered karst and concealed karst-type assemblage. More rarely, but a reverse situation also occurs: there is a soil-covered karst and concealed karst-type assemblage on the valley floor, while bare karst type can be found in the karstic depression occurring there.

- The same postglacial features occur as in their environment, but the size and density of the features is larger or smaller (for example, the density of subsidence dolines may be larger).
- There are not any postglacial features in the karstic depression.

In karstic depressions above the tree line, bare karst, bare karst and soil-covered karst-type assemblage as well as soil-covered karst and concealed karst-type assemblage as well as soil-covered, concealed karst and allogenic karst-type assemblage are widespread. Their postglacial karst features are karren, giant grikes, small-sized solution dolines and ponors. In karstic depressions below the tree line, soil-covered karst type, a soil-covered and concealed karst-type assemblage as well as a soil-covered karst, concealed karst and allogenic karst-type assemblage may occur. Their postglacial karst features are karren (mainly subsoil karren) subsidence dolines and ponors.

6.2.3 The Karstic Pattern of Cirques

Cirques may develop from karstic depressions (Fels 1929) from poljes (Goeppert et al. 2011) too. However, the former cirque can also develop into a karstic feature (Barrére 1964). In the former case, cirques may be separated into constituting depressions, they are large-sized (Miotke 1968; Smart 1986), and their margins may even be multiple arcuated. (In this case, the cirque was formed by the coalescence of karstic depressions.) During its development into a karstic depression, its diameter increases by dissolution, its floor is dissolved into a plain (Barrére 1964), but later karstic depressions can also be formed on the floor. On the surface planated in a karstic way, rock debris develops along fractures and bedding planes (Corbel 1957).

The karstic pattern of cirques and that of their troughs are determined by the following characteristic features of cirques:

- If the cirque is formed in a karstic depression, a trough can be connected to the cirque or it can be absent too. If the trough is missing, the cirque preserved the closed nature of the karstic depression. Terminal moraine may occur on longer–shorter sections, at the margin of closed cirques where the ice left the depression (Fig. 6.13a, c). In this case, the environment of the cirque can be exempt from glacial erosion or it may be a glacial erosional karst surface (see below). If there is a trough, one or several could develop. In this case, the number and size of troughs affect the karstic pattern in the environment of the cirque. Depending on altitude of the karst, the terrains between the troughs can also be glacial erosional or exempt from glacial erosion. If the trough of the cirque is a tributary trough, it is short and has a large gradient and hangs over the main trough (Fig. 6.13e).

- The cirques (former karstic depressions) can be located on glacial erosional terrain or on a terrain exempt from glacial erosion. In the latter case, the cirque constitutes an island-like glaciokarst pattern. Such features occur, for example, in the Velez Mountains (Žebre and Stepišnik 2015a). When it is on a glacial erosional terrain, the karst type or karst-type assemblages of the cirque can be the same or different from the karst type or karst-type assemblage of the bearing terrain, but its karstification varies from the former karstification of the depression. The karstification is overprinted. When the cirque (the karstic depression, respectively) is situated on a terrain exempt from glacial erosion, then its karst is separated from the karst (karstic features) of its environment, since during glacial erosion its size and shape were modified. A more significant fact is that the karstification was not interrupted in its environment, the features that already existed during glaciation are developing further without modification at present too. However, the karstic development of glacially transformed karstic depressions was interrupted and the whole feature as a karstic feature is not active. Inside it, non-karstic processes operate too. In their area karstification is mosaical. Karstification is of various nature regarding not only its area but also its nature (bare karst, covered karst) is heterogeneous, while it is rather homogeneous in its environment.
- The floor of cirques can be non-dissected or dissected. Features causing a dissected nature can be preglacial karstic features (giant dolines) or glacial erosional features (roche moutonnées, hummocky moraine), subglacial features (shafts, giant grikes), postglacial features (karren, schachtdolines, small-sized solution dolines, ponors). In the giant dolines of its floor, the thickness, composition and expansion of the cover may be different. As a result of this, various karst types may develop in the giant dolines.
- On the karst, cirques can occur individually or in groups. In the latter case, they can constitute an irregular group or row. The above-mentioned patterns of cirques modify the karstic pattern of glaciokarst.
- During glaciation of decreasing intensity, the ice of cirques can shrink uniformly or separately, by which the ice is focused in the inner karstic depressions during the process. The way of ice retreat affects the expansion and thickness change of the moraine within the cirque and thus, there postglacial karstification.

The postglacial karstification and the karst types of the cirque mainly depend on the elevation of the cirque and on the thickness of the cover that accumulated inside it. In case of a cover with large thickness, no karstification occurs in the cirque (buried-like karst). If there is impermeable cover or a thinner and permeable cover on the floor, allogenic and concealed karst develop. According to its elevation, it can be of a higher position (at the upper part of the periglacial belt) or of a lower position (at the lower part of the periglacial belt, but it can occur below the tree line too).

The area of the cirque can be separated into three morphological belts which are the following:

- The cirque that developed in a steep bedrock is a wall which is of wreathed, circular development and it can be dissected by rock shelters and gullies of debris avalanche.
- A debris cone accumulated at the foot of the cirque wall which is more or less interrupted and may be dissected by gullies and creeks, but it is also of wreathed development.
- The floor, which is patch-like, is circular, elliptical or irregular in plan view.

The side wall of the cirques with a higher elevation is bare karst. If no trough is connected to the cirque, its circular development is complete, but if a trough is connected to it, it is not complete. The side walls are rarely stepped. According to this, mainly wandkarren and schichtfugenkarren are the predominant karren features. However, beside karren, opened-up cavities may also occur.

The debris cone system of the cirque is buried-like karst.

On the floor of a cirque with a higher elevation, concealed karst type, bare karst and concealed karst-type assemblage as well as concealed karst and allogenic karst-type assemblage may occur. The expansion and shape of karst types and karst-type assemblages is guided by the expansion and shape of the cirque floor. Its subglacial and postglacial karst features are the shafts, schachtdolines, giant grikes, suffosion dolines, ponors and subordinately karren.

On the side wall of cirques with a lower position, bare karst and soil-covered karst-type assemblage is characteristic. According to this, in addition to bare karren features subsoil karren may also occur. In the zone of debris cone, the debris is less thick, therefore, concealed karst may be present here too. The karstic pattern of the floor is varied too. If the cirque creates only one depression, only one karst type is characteristic which may be soil-covered karst, concealed karst or allogenic karst. If the cirque is larger and the floor is dissected by karstic depressions, roche moutonnées, moraine hummocks, soil-covered karst, concealed karst and allogenic karst-type assemblage, soil-covered karst and concealed karst-type assemblage, soil-covered karst and allogenic karst-type assemblage as well as concealed karst and allogenic karst-type assemblage may occur in its area. Its postglacial karst features are similar to the karst features of the cirques with a higher position, but subsoil karren and concealed karst features may be more dominant.

6.2.4 General Characteristics of the Karstification of Ice Caps, Ice Sheets and Piedmont Glaciers

While ice caps can be of higher elevation, ice sheets predominantly developed on low terrains. Consequently, in the former area of ice caps, bare karst or such a karst assemblage may occur where bare karst is dominant, while in the former area of ice sheets, bare karst is absent or there are such karst assemblages where bare karst is subordinate (British Islands).

In case of these glacier types, areal erosion was predominant. As a result, stripped, less dissected erosional surfaces developed. Subglacial and postglacial features are widespread. If the features and feature assemblages have orientation, it was determined by the flow inside the ice and the movement direction of the ice sheet (Drew 2004) as well as tectonics and to a less extent by glacial landforms. The development of areal erosion can be modified in the following cases:

- In the area of ice sheets, ice flows developed (Davies 2014) which could create troughs.
- During the retreat of ice sheet, dead-ice developed. At these places, present karstification started later than outside the dead-ice. At the same time, subglacial karstification could survive permanently.
- Outlet glaciers left the area of ice cap. They could form troughs either outside the glacial erosional surface (the outlet glacier expanded beyond the maximum expansion of the ice cap) or on the glacial erosional surface (the outlet glacier left the area of the younger ice cap with smaller expansion). While in the former case the trough deepens into the terrain exempt from glacial erosion, in the latter case it deepens into the areal erosional terrain (in case of outlet glaciers being dense enough, combe-ridges may develop too).
- A cirque glacier developed in the area of the karstic depression situated in the area of former ice cap. This is indicated by the terminal moraine at the margin of the depression which is situated on a glacial erosional surface (Žebre and Stepišnik 2015a).

6.2.4.1 The Karstic Pattern of Ice Caps

The intensity of subsequent glacials was different. Thus, on Mount Orjen the intensity of MIS 12 glaciation was large, then it was followed by glaciations of smaller intensity in MIS 6 and an even smaller in MIS 5d-2 (Adamson et al. 2014). Therefore, the expansion of the ice caps of younger glacials decreased. Taking the spread of terminal moraines into account, four types of ice cap patterns can be distinguished.

In case of the first variety, the ice cap has one centre. There is only one terminal moraine system which follows the margin of the maximum expansion of the former ice cap. The shrinking of the ice cap was probably continuous, as a result only a basal moraine, but no terminal moraine developed in its area. The ice cap which was formed on Lovčen might have had such a development (Žebre and Stepišnik 2015a).

In case of the second variety, the ice cap has several centres (more exactly it developed into an ice cap having several centres). In this case, the terminal moraine system is double. The outer developed at the margin of the former ice cap with maximum expansion. Within this, there are several terminal moraine systems, which developed at the margin of former ice caps with smaller expansion (which were formed in younger glacial(s) of smaller intensity). At most, a basal moraine exists in the area of former smaller and younger ice caps. If the bearing area is of an altitude being high enough, there is a greater chance of the development of bare karst since the coveredness with moraines is of smaller degree. In the area between younger ice caps, moraine occurs in a larger number and in larger expansion (terminal moraine and basal moraine, moraine developed in the area of outlet glaciers), and fluvioglacial sediment may be widespread too. Outlet glaciers developed not only on the terrain surrounding the ice cap with maximum expansion, but also in the area between the inner ice caps. Such a pattern can be seen in Mount Orjen (Adamson et al. 2014). Here, smaller ice caps were probably separated into even smaller ice caps during MIS 5d-2 (Adamson et al. 2014), which is proved by a third terminal moraine system in Mount Orjen. Thus, in the area of MIS 12, outlet glaciers developed both in MIS 6 and in MIS 5d-2 glacials.

The third ice cap pattern develops when smaller ice caps survived in karstic depressions within the area of former and larger ice cap during the younger glacial of smaller intensity. In this case, terminal moraine was formed in karstic depressions and at their margin within the area surrounded by the terminal moraine system of the ice cap with maximum expansion. Such a development occurs in Lovčen Mountains (Žebre and Stepišnik 2015a).

The fourth variety of the ice cap is when the ice cap is one-sided and asymmetric (Fig. 6.16). Such a glacier development can be detected in Velez Mountains (Žebre and Stepišnik 2015a). The ice caps lean in one direction, from which outlet glaciers branch out. (In the area of the ice cap, development type 3 also occurs.) Cirques may occur at the upper termination of the ice cap.

In the area of ice caps, the types of karst types and karst type assemblages depend on the elevation of the karst, while the expansion of some karst types and karst type assemblages depend on the inclination of the karst surface and on the expansion of the former ice cap. Because of the plateau characteristic, the inclination of the surface is small and this favours the development of wide belts.

In the area of single-centred ice cap, the karst types have a ring-shaped development (Fig. 6.17). On the most surviving part of the ice cap, if the surface of the karst was of large elevation, bare karst developed, which is surrounded by the belt of bare karst and soil-covered karst-type assemblages and the belt of soil-covered karst and concealed karst-type assemblages. The distribution of karst features is not striped on glacial erosional terrains. Mainly the distribution of karren can be areal. The elevation differences of stepped surfaces are smaller than that of troughs. This depends on the inclination of the preglacial surface, on the resistance of the rock and on the degree of glacial erosion. The elevation of stepped surfaces may be nearly similar, but different too (Fig. 6.18). Mainly stepped surfaces determine the distribution of bare karren surfaces. If the karst has an altitude being large enough and the stepped surfaces have a similar elevation, the distribution of bare karren is also large.

The troughs of outlet glaciers can interrupt the karstic terrain exempt from glaciation (when the glaciers branched out of ice caps with maximum expansion) or

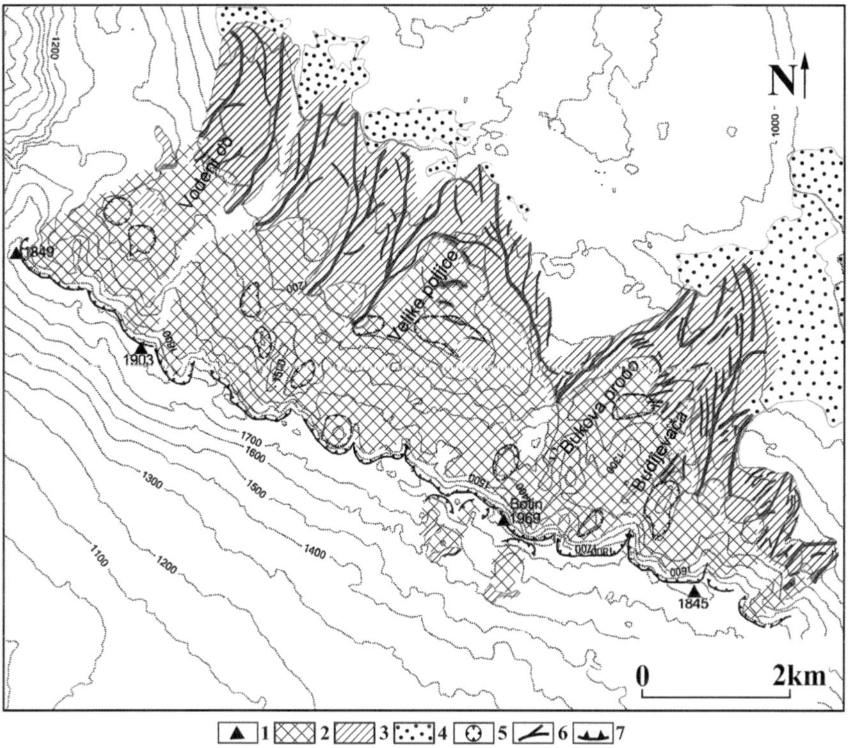

Fig. 6.16 Asymmetric ice cap (based on Žebre and Stepišnik's data 2015a). 1. Summit (metres), 2. ice cap, 3. outlet glacier, 4. glacial outwash, 5. glaciokarst depression, 6. moraine ridge, 7. cirque

the glacially transformed surface. In the latter case, glaciers branched out of the smaller ice caps of ice cap having several centres. At these sites, the pattern of karst types is changed to striped nature. If the valley is of higher elevation and shorter, bare karst type may be predominant. If it is of lower altitude, soil-covered karst as well as bare karst and soil-covered karst type assemblage may occur in its area. If there is a moraine on the valley floor, bare karst and concealed karst type assemblage as well as soil-covered karst, concealed karst type assemblage is characteristic. If outlet glaciers created thick depositions, allogenic karst, buried-like karst or maybe concealed karst may be present on the valley floor.

In case of multicentred ice cap, several smaller bare karst types or bare karst and soil-covered karst type assemblage may occur on the karst. In the former case, bare karst is surrounded by bare karst and soil-covered karst-type assemblage, while in the latter case by soil-covered karst and concealed karst-type assemblage. These can develop around former smaller ice caps zonally or only one of them or both of them in the area of former ice cap with maximum expansion (Fig. 6.19).

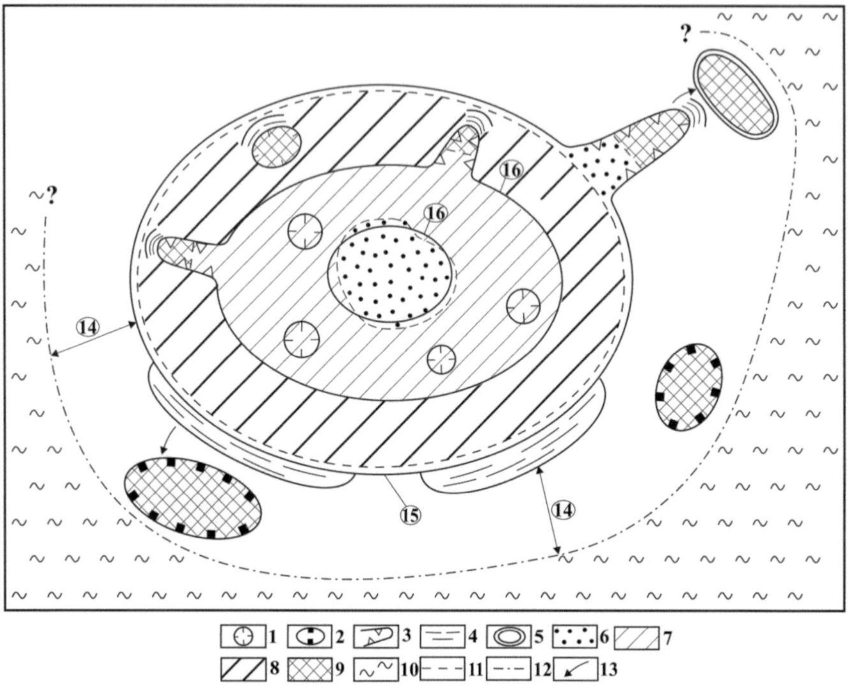

Fig. 6.17 Karstic zonality in the area of an ice cap. 1. Giant doline, 2. polje, 3. trough of an outlet glacier, 4. terminal moraine, 5. fluvioglacial sediment in non-karstic derpression, 6. bare karst, 7. bare karst and soil-covered karst-type assemblage, 8. soil-covered karst and concealed karst-type assemblage, 9. allogenic karst, 10. karst exempt from the effects of ice cap, 11. border of karst types, 12. border of fluvioglacial zone, 13. sediment transport by meltwater, 14. fluvioglacial zone, 15. maximum expansion of ice cap, 16. limit of the smaller expansion of ice cap

In the area of ice caps, preglacial karst features are represented by giant dolines and uvalas having been transformed to various extent. These are rarely arranged into rows (which is caused by tectonic preformation), but they create an irregular group. The karst type that developed in karstic depressions may be different from the karst type of the environment or can be the same with that. It may happen that karst type assemblages are formed in karstic depressions, while there is only one karst type in their environment. In case of different karst type, it often occurs that while there is soil-covered karst type in its environment, concealed karst or allogenic karst type develops in the depressions. In the area of ground moraine, subsidence dolines may be widespread, while on bare surfaces, karren, on terrains with higher elevation, shafts, schachtdolines and small-sized solution dolines may be widespread. Among the above-mentioned features, the distribution of karren is continuous. The pattern of other features is irregular, or as it has already been mentioned, they can be situated in rows if the fracture (fault) was a determinant factor in their development.

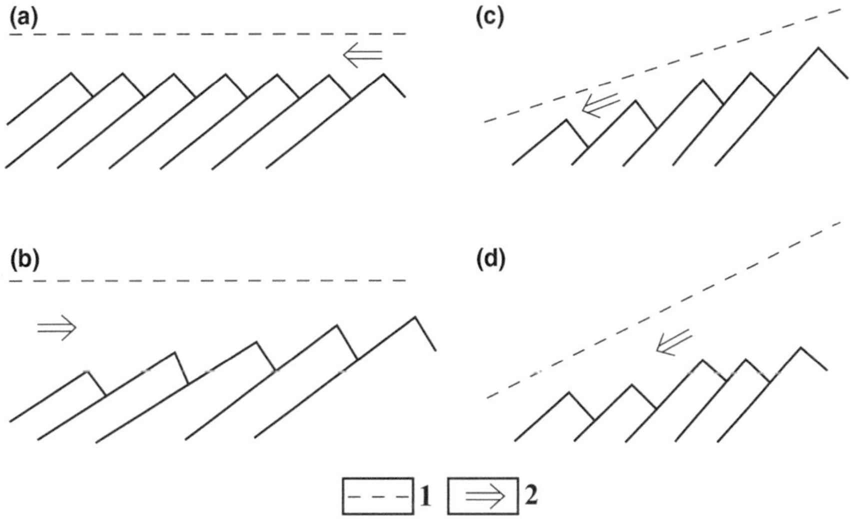

Fig. 6.18 Inclination of stepped surfaces on areal glacial erosional terrain. **a** The stepped surface is horizontal: the preglacial terrain was horizontal, the intensity of glacial erosion was the same, **b** the stepped surface is oblique: the preglacial surface was horizontal, but the intensity of glacial erosion was different, **c** the stepped surface is oblique: the preglacial surface was inclined, but the intensity of glacial erosion was similar, **d** the stepped surface is oblique: the preglacial terrain was oblique, but the intensity of glacial erosion was different, 1. original surface, 2. direction of ice movement

Both the single-centred and the multicentred ice cap (in case of multicentred, the younger, smaller ice caps too) are lined by the fluvioglacial zone.

In the area of the fluvioglacial zone, fluvial effects, the effects of former outlet glaciers and that of ground ice had a role. Mostly, the area of fluvial effects can be of significant expansion because of the large quantity of meltwater and the frozenness of the surface in the environment of current ice. The fluvial effect may manifest itself in superficial erosion and subsurface erosion as well as in accumulation. However, because of the cooling effect of ice caps, permafrost patches could also be formed (Fig. 6.20). Formerly, allogenic karst patches could develop at these places which can be represented by fossil ponors today. This zone can also be present in the area of the former ice cap where the fluvioglacial sediment is deposited on glacial erosional surfaces and may be dissected by features that were developed by outlet glaciers or it may be dissected by sediment accumulations. However, it can also be present on the karstic terrain situated outside the margin of the ice sheet that did not undergo glacial erosion. Although fluvioglacial sediment patches may occur on plain, hardly dissected surfaces, they mostly fill the depressions. The fluvioglacial sediment can fill non-karstic features, for example, valleys (Adamson et al. 2014) partly or completely, but the sediment can cover any karstic terrain part or karstic feature (Adamson et al. 2014; Žebre and Stepišnik 2015a) too. The sediments being accumulated by meltwater altered the type of

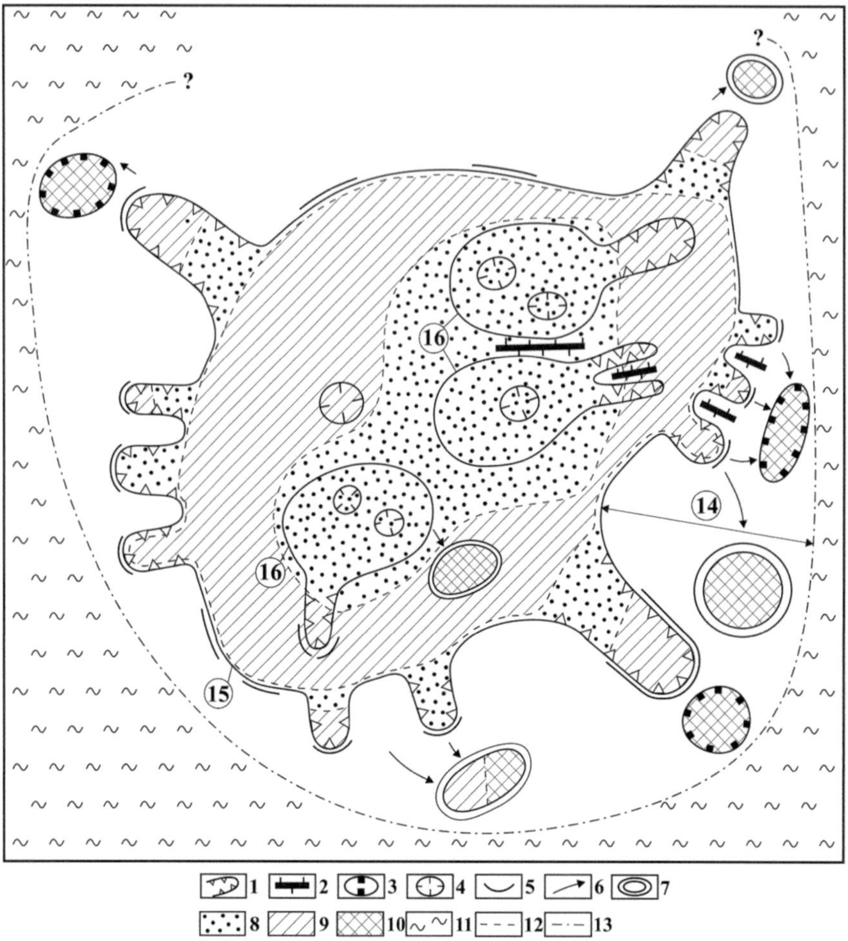

Fig. 6.19 Karst type pattern of the area of ice cap system (based on Adamson et al.'s data 2014). 1. Trough of outlet glacier, 2. combe-ridge, 3. polje, 4. giant depression, 5. terminal moraine, 6. material transport by meltwater, 7. fluvioglacial deposition on terrain exempt from karstic depression, 8. bare karst and soil-covered karst type assemblage, 9. soil-covered karst and concealed karst type assemblage, 10. allogenic karst type, 11. karst exempt from the effects of ice cap, 12. border of karst types, 13. border of fluvioglacial zone, 14. fluvioglacial zone, 15. the limit of the expansion of uniform ice sheet, 16. the limit of the expansion of smaller ice cap

former karstification. By now, concealed karst, allogenic karst or buried-like karstic patches (or karst-type assemblages that were formed from these) developed instead of bare karst or soil-covered karst.

The former ice caps of the Dinaric karst are surrounded by poljes as a wreath. This refers to the fact that the ice played a role in the development of poljes. Poljes are mainly situated at a low elevation such as in the Orjen Mountains. Therefore, the area of the polje belt is not a glacial erosional surface (ice penetrated into the

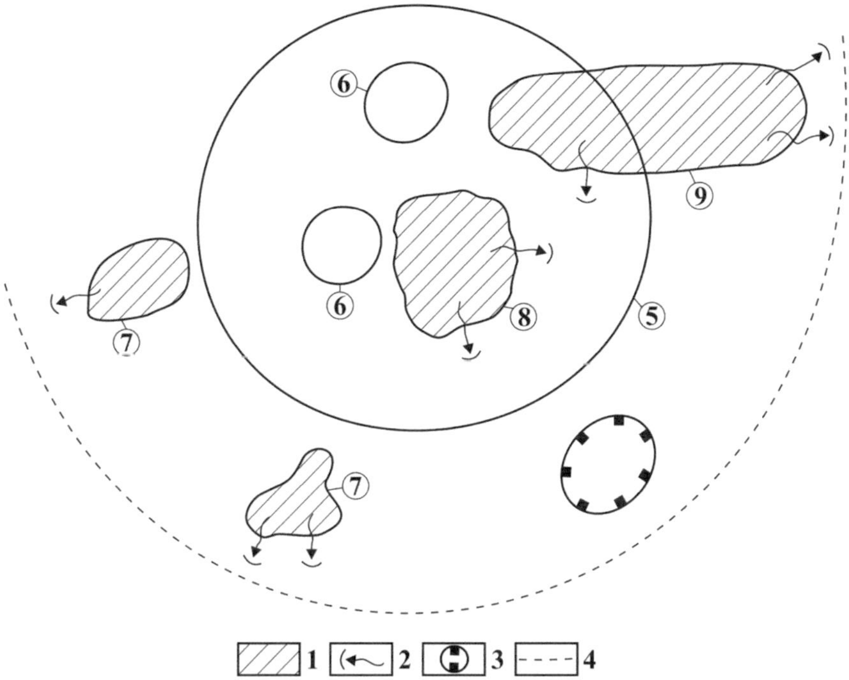

Fig. 6.20 Allogenic karst that was created by sporadic permafrost or by ground ice patches in the environment of ice cap. 1. allogenic karst (permafrost patch), 2. ponor, 3. polje, 4. border of fluvioglacial zone, 5. the limit of the maximum expansion of ice cap, 6. the limit of the expansion of smaller ice cap, 7. old allogenic karst, 8. younger allogenic karst on glacial erosional terrain, 9. composite allogenic karst with ponors of various age

poljes at most), but it is an accumulation one, where allogenic karst as well as soil-covered karst, concealed karst and allogenic karst types occur.

The majority of the meltwaters of former glaciers was received and conveyed into the karst by the fluvioglacial zone (mostly the terrains beyond ice cover with maximum expansion) which is proved by the accumulations with large expansion that developed here. Therefore, the glacial effect, as it has already been mentioned, has an indirect manifestation mainly in accumulation as well as in the complete upfilling of some landforms or maybe in their burial (dolines, uvalas), in their partial upfilling (poljes) as well as in the formation of subsurface karst features.

6.2.4.2 Karstic Pattern of Ice Sheet

Among karst features, karren are widespread, schachtdolines and small-sized solution dolines are absent and the solution dolines of the temperate belt are present (the younger varieties of these latter show similarities with small-sized solution dolines). A widespread karst type is soil-covered karst, bare karst and soil-covered karst-type assemblage and bare karst, soil-covered karst and concealed karst-type assemblage. Their distribution depends on altitude, on the thickness of the moraine and on the occurrence of non-karstic rocks. The pattern of karst types is not zonal, but they occur in wide, continuous expansion.

6.2.4.3 Karstic Pattern of Piedmont Glaciers

The expansion of glacial erosional surface increases in the area of piedmont glaciers. If there had been older, channel erosion, it was followed by areal erosion. In the area of piedmont glaciers, the intensity of glacial erosion was probably smaller than in the area of nourishing glaciers, therefore, in the area of piedmont glacier the dissected nature of the surface is negligible, and no trough develops. In some places, its surface may be dissected by elevations of roche moutonnée nature, the longer axis of which coincides with the direction of ice movement. Among elevations, a row of preglacial karstic depressions may occur. The karstic depressions are transformed by glacial erosion, there are moraines inside them. On such a small-inclined surface with a low altitude as compared to its glacial erosional environment, concealed karst type, bare karst and soil-covered karst-type assemblage as well as soil-covered karst and concealed karst-type assemblage occurs. The karst-type assemblages do not show zonality.

6.2.5 Karstic Pattern of Combe-Ridges

The karst is of linear, striped development. The karst type of combe-ridges is often not separated from the karst type of its environment. The area of more elevated and narrower combe-ridges is bare karst, while the area of combe-ridges with lower elevation and being wide is bare karst and soil-covered karst-type assemblage or soil-covered karst and concealed karst-type assemblage.

Karstic features depend on the width of combe-ridges. Thus, no karstification occurs on extremely narrow ridges, while on wider ridges, karren can occur, on much wider ridges there may be schachtdolines and small-sized solution dolines, in case of cover, suffosion dolines may also be present too (Fig. 6.3). On wide ridges, giant solution dolines could also survive (Žebre and Stepišnik 2015b). The intensity of karstification and the density and size of the features (for example that of karren) is smaller than in glacier valleys. A combe-ridge can show specific and rich karren formation under extreme circumstances. Thus, for example, on the island of Diego de Almagro, where a large amount of precipitation and high wind speed favours intensive karren formation on combe-ridges (Veress et al. 2006). On the island, karstification is mainly focused on combe-ridges. On relatively wide combe-ridges, giant grikes and shafts also occur next to large-sized karren features.

6.3 Karstic Pattern of Glaciokarst Types

Since the development of glaciokarst types (see Chap. 2) is different (which is primarily caused by various ice development environment), their karstic pattern will be different too. The pattern of types is summed up below.

Plateau karst was in a perched position as compared to its environment during karstification that took place at the beginning of glaciation and during glaciation. Because of this, if there was a cover, it was destroyed and the karst water level was under the surface. On plateaus, glaciation which later determined the pattern of karstification possibly took place in the following ways. (Ice primarily developed in karstic depressions.)

- Since glaciation was of small degree, the ice did not leave the karstic depressions. If no glacial erosion took place between the depressions, there are terrains exempt from glacial erosion there. If there are glacial erosional terrains, they were formed during older, more intensive glaciation. Thus, the karstic pattern of karstic depressions is younger than that of their environment.
- The ice left the karstic depression. In this case, a trough developed. One or several glaciers could leave the karstic depression. According to this, several troughs can branch out from a cirque. Probably there is a chance for the formation of more glacier snouts, and thus of troughs if the karstic depression is large and the glaciation is intensive. As it has already been mentioned, the pattern of troughs depends on the dip direction, dissectedness and elevation conditions of the bearing preglacial surface, while the density and size of karstic depressions depend on the intensity and duration of preglacial karstification. In case of several troughs, the troughs can branch off radially, can be parallel and coalescing. The chance of this latter increases if the density of preglacial karstic depressions is high. Schneeberg can be an example for this variety.
- In case of a lot of and large-sized depression and a karst with marine environment and adequate elevation, an ice cap develops. The shape and expansion of the ice cap, and thus the expansion of glacial erosion are also determined by the distance of the karst from the sea, by the dip direction of the surface of the karst and the area where preglacial karstic depressions are distributed. The karst types and karst-type assemblages develop zonally. This development is modified by outlet glaciers. To the effect of meltwaters, fluvioglacial zone developed around the former ice cap, the patches of allogenic karst could be formed on former permafrost patches. This variety includes, for example, the areas with ice caps of the Dinaric karst.
- If snow is accumulated on the leeward slope of an elevation, a ridge or of a plateau, a small snow pack and glacier snout develops. In case of an elevation or a ridge with small expansion, only one or few glaciers and troughs are formed. In case of a longer ridge, at the side of an older trough, at the side slope of a plateau (especially if there are rock shelters on the slopes), several, but smaller

troughs develop too. Because of high glacier density, the karst is dissected by glacier valleys. Probably, the trough pattern of the karst with ridges (Hochschwab) can develop in this way.

Because of the above-mentioned facts, on plateau glaciokarst, troughs are absent or they are only present in small quantity and size. Accordingly, karst represented by cirques may be present, the karst represented by troughs is absent or of a small expansion. On plateaus, the expansion of karst types and karst-type assemblages is wide and areal (expanded to all directions). If the plateau is of a larger expansion and its elevation is large, several karst types and karst-type assemblages develop, which have a zonal pattern. The zones of karst types and karst-type assemblages are interrupted by the karst types of troughs (these are organized into stripes).

There may be various types of ice development environment on semi-plateau glaciokarst, mountain glaciokarst and rock fortress glaciokarst. On one part of semi-plateau glaciokarst, a fluvial valley network developed, since karst water level was at the surface of the bedrock. On these parts of the karst, valley density is large, several tributary valleys are connected to the main valley, and the valley network is bunchy. Because of elevation only large valleys were able to keep up with the elevation in the inner areas of the karst, (or large valleys grew by glacial erosion), and thus to stay permanently near karst water level. (The position near karst water level and thus, fluvial erosional development at the margin of the karst could survive until these days.) The fluvial erosional development of the tributary valleys belonging to large valleys stopped then valleys were transformed into glacier valleys in the glacial. Subsequently (or preceding glacial erosion), karstification could take place even in the tributary valleys (valley floor dolines were formed), which did not modify the shape of the trough any more. Valley deepening continued by glacial erosion during the glacials (Fig. 6.21).

The most elevated part of this type was already a bare karst during glaciation, as a result of which karstic depressions were transformed into cirques. Main valleys may have hanging and non-hanging tributary glacier valleys. Non-hanging tributary glacier valleys are short, have a large gradient and do not become karstified (at most their cirques). Hanging glaciers are predominant on terrains between main valleys. In these valleys, postglacial karstification is dominant and all karst types and karst-type assemblages occur here. On combe-ridges between valleys karren formation and periglacial processes have an influence. The margin of the karst may also develop in an erosional way if karst water level is located at the valley bottom. Totes Gebirge is characterized by the features of this karst type.

In the area of mountain glaciokarst, the fluvial erosional deepening of the valleys inheriting from covered karst (buried-like karst) lasted longer. Although the valley net is dense here too, tributary valleys are less separated from the main valleys. These latter are also active from an erosional point of view. This type might also have had a bare karst part. The margin of the karst can also be developing in an erosional way if the karst water level is located at the valley bottoms. Its karstification is similar to that of the semi-plateau type. However, the tributary valleys connecting to the main valleys are of smaller gradient thus, the expansion of some karst types may be wider. Julian Alps may be an example for this karst type.

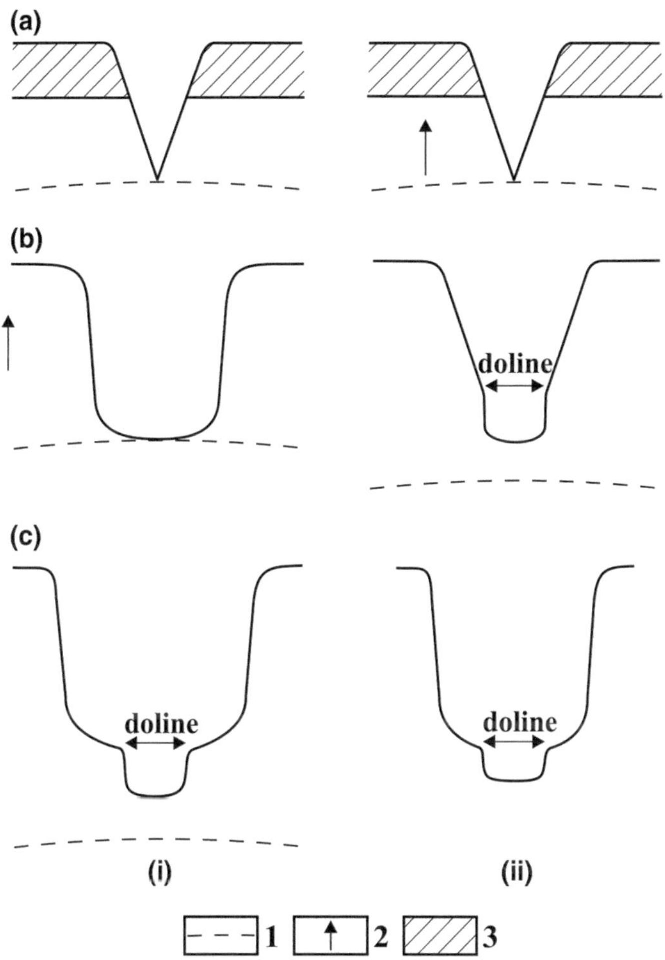

Fig. 6.21 Development of inheriting valleys. **i**(**a**) Development of fluvial erosional valley, **i**(**b**) trough development, **i**(**c**) doline development on the floor of trough, **ii**(**a**) development of fluvial erosional valley, **ii**(**b**) doline development, **ii**(**c**) development of glacial erosional trough, 1. karst water level, 2. uplift, 3. impermeable superficial deposit

On rock fortress glaciokarst, on covered karst, inheritance was replaced by karstification since the karst water level was located below the surface of the bedrock. (After inheritance valley bottoms got above the karst water level because the mountain became elevated.) The karstification of the valleys transformed the valleys since the process immediately began after the inheritance and it was intensive. Few valleys could develop in lack of streams, which favoured intensive glaciation in the valleys. The valleys are not hanging high above the surrounding terrain. This type may also have had a bare karst part. On combe-ridges between the valleys, karren formation and periglacial processes exert their effect. On valley bottoms, any karst type or karst type assemblage may occur. Its karstification takes place in valleys with varied morphology. The Durmitor can be an example for this karst type.

On complex glaciokarst, glacier valleys that were formed with inheritance and developed with fluvial erosional and then continued their development during the glacials (the floor of which is mainly located at the present karst water level) surround plateau-like areas. These can be plateaus (the plateau was bare karst), of semi-plateau and mountain nature depending on what the degree of the deepening of the valleys was which had developed on the covered karst of these areas (plateaus). In the area of plateaus, a complex valley network could also develop, while in the side of plateaus, a series of short glacier valleys with large gradient was formed. It is separated into isolated karstic terrains. Karstification may be represented by both the glacier valleys and the plain, plateau-like terrains between them. The development of the valleys between plateau-like areas is erosional at present too. On the valley bottoms of plateaus any karst type may occur. On the combe-ridges between the valleys, karren formation and periglacial processes have an effect. On plateaus and on wider terrain remnants between the valleys, bare karst type and soil-covered karst type, bare karst and soil-covered karst type assemblage or maybe bare karst, soil-covered karst and concealed karst type assemblage may occur depending on altitude. The Dolomites can be an example for this type.

In the area of ice sheet, there are few karst types with wide expansion. Their development can be interrupted by troughs, the karst type of which are not necessarily different from the karst types of their environment. The karstification of hardly dissected surfaces is homogeneous.

On island glaciokarst, the pattern of glacier valleys is mainly affected by the shape and size of the island. Because of the low elevation of the island, the karst water level was probably near the surface, whereby fluvial valley development could take place preceding glaciation. The valley network is dense and branching out radially. In case of an island with a position close to the pole, few karst types can develop in the glacier valleys. Karren formation is predominant on the relatively wide ridges between the valleys (Veress et al. 2006). The karstification of combe-ridges may be determinant, but that of troughs is less determinant since the karst water level may be situated at the bottom of troughs.

6.4 Conclusions

On glaciokarst, the pattern of karstification is affected by the elevation of the karst, its distance from the equator, the presence or lack and nature of former glacial erosion, the intensity of glaciation as well as the type and pattern of karst types and karst-type assemblages. However, the type and pattern of karst types and karst-type assemblages in itself depend on several factors too. These are the following: the elevation of the glaciokarst, the characteristic of glacial geomorphic activity, (erosion or accumulation happened and its pattern), the type and density of glaciers, preglacial karstification (its intensity, the density and pattern of the features) and the ice development environment.

Among various glacier types basically the size, density and pattern of valley glaciers determine the recent karstic pattern of glaciokarst. The pattern and density of valley glaciers depend on the former coveredness of the karst and on the position of the karst water level as compared to the bedrock during valley inheritance. In lack of superficial deposit (bare karst) the ice develops in karstic depressions. The developing glacier snout and thus, the size of the trough depends on the size of the depression and the intensity of glaciation. (However, if the ice did not leave the depression, no trough developed.) The number of troughs depends on the size of depressions, while their direction depend on the elevation and dip direction of the preglacial surface. On covered (on buried-like karst) terrain, if the karst water level was below the bottom of inheriting valleys, dolines developed in the valley. The density of valleys is small, their margin is arcuate and there is no separating cirque. Also, on covered terrain, if the karst water level was at the bottom of inheriting valleys, the glaciers developed in fluvial erosional valleys exempt from dolines. Cirques are connected to troughs, the margin of valleys is straight, the density of valleys and the number of tributary valleys is large.

In cirques, depending on their elevation, bare karst and concealed karst, bare karst and concealed karst-type assemblage, soil-covered karst and conc sealed karst-type assemblage as well as concealed karst and allogenic karst-type assemblage, soil-covered karst, concealed karst and allogenic karst type assemblage can develop in patchy expansion on their floor. On cirque walls, bare karren (wandkarren and schichtfugenkarren) are characteristic.

In troughs, the pattern of karst types and karst-type assemblages is striped, linear and elongated in the direction of the trough. The karst types and karst-type assemblages of the trough can develop homogeneously, patchy and zonally. Homogeneous development mainly characterizes short troughs with higher elevation. In this case, mainly bare karst type is formed, in the area of which bare karren are predominant. Patchy development occurs in the area of valley floor karstic depressions and of debris cones. In long troughs with a large bottom dip, karst types and karst type assemblages zonally develop. From the top to the bottom, the most frequent belts are the following: bare karst, bare karst and soil-covered karst-type assemblage, bare karst, soil-covered karst and concealed karst-type assemblage as well as soil-covered karst, concealed karst and allogenic karst-type assemblage.

Ice caps can be single-centred and multicentred, asymmetric, but smaller ice patches can also survive in preglacial depressions during deglaciation. The fluvioglacial zone is formed in a wide expansion in the environment of ice caps, where the stream load which is transported by meltwaters creates patches. Because of the cooling effect of the ice, permafrost patches may occur the areas of which operated as allogenic karst. In the area of ice caps, karst types and karst-type assemblages are widespread and create a ring-like pattern which are interrupted by the valleys or deposits of outlet glaciers. (The fluvioglacial zone is an exception for this where the karst types were formed on the patches of the superficial deposit.) Most common karst types and karst-type assemblages are bare karst, bare karst and concealed karst-type assemblage as well as soil-covered karst and conc sealed karst-type assemblage.

Glaciokarst types have three different patterns. In the area of plateau glaciokarst, cover was absent during glaciation, and the karst water level was below the surface of the bedrock. Glacier development could take place in a karstic depression. A trough only developed if the ice left the karstic depression. Glaciation, and thus glacier size was more significant if there was a large quantity of precipitation in the area of the karst and there were a lot of and large depressions on the karst. Glaciers could develop in large quantities on the leeward slopes of ridges and plateaus. The glaciokarst (karst with ridges) is densely dissected by valleys because of the large number of short glaciers.

There may be many kinds of ice development environment depending on the coveredness of the karst and the position of the karst water level. In the area of semi-plateau glaciokarst which was covered in large expansion, the former karst water level was located at the floors of inheriting valleys. A fluvial valley network develops on the karst the erosional deepening of which stopped during the rise of the karst (the karst water level got deeper thus, the water of streams percolated away). During glaciation, the development of valley continued by glacial erosion. On mountain glaciokarst, fluvial erosion lasted longer, and thus, partly, deeper valleys could develop in its area. The density of valleys is large in the areas of both the above-mentioned karst types. On rock fortress glaciokarst, the bottom of inheriting valleys had a further karstic development since the karst water level was below the valley bottoms. The density of valleys is small (tributary valleys only occur at those places where ice developed in former karstic depressions and then left them), valley margins are arcuate.

On complex glaciokarst, older, inherited but glaciated valleys surround plateaus and plateau-like areas which can be of semi-plateau, mountain or plateau character (no valleys developed) depending on the fact whether valleys developed on these terrains and on the extent of their deepening.

On island glaciokarst, the shape and size of the island determined the pattern of glacier valleys. Because of the small elevation of the island, the karst water level was close to the surface which favoured the development of valley network and high valley density. As a result of the closeness of the valley floors to the karst water level, the valley floors were less likely to become karstified. Karstification mainly takes place on combe-ridges.

References

Adamson KR, Woodward JC, Hughes PD (2014) Glaciers and rivers: Pleistocene uncoupling in a Mediterranean mountain karst. Quatern Sci Rev 94:28–43

Barrére P (1964) Le relief karstique dans l'ouest des Pyrénées centrales. Rev Belge Géogr Ed Soc Roy Belge Géogr Special Publ, Karst et Climats Froids 88(1–2):9–62

Corbel J (1957) Les karsts du Nord – Ouest de l' Europe et de quelques Regions de Comparaison: Etude sur le Rôle du Climat dans l' Erosion des Calcaires. Institut des Etudes Bhodaniennes de l' Université de Lyon, Mémoires et Documents, p 12

Davies D (2014) Antarctic Glaciers. www.antarcticglaciers.org/modern-glaciers/glacial-processes/

Djurovič P (2009) Reconstruction of the Pleistocene Glaciers of Mount Durmitor in Montenegro. Acta geographica Slovenica 49(2):263–289

Drew D (2004) Burren Glaciokarst, Ireland. In: Gunn J (ed) Encyclopedia of caves and karst science. Fitzroy Dearborn, New York, pp 169–171

Fels E (1929) Das Problem der Karbildung in den Ostalpen. Petermanns Mitt., Ergänzungsh, vol 202, pp 1–84

Ford DC (1996) Karst in a cold climate. In: Mc Cann SB, Ford DC (eds) Geomorphology sans Frontières. Wiley, Chichester, pp 153–179

Ford DC, Williams PW (2007) Karst hydrogeology and geomorphology. John Wiley and Sons, Ltd., Chichester, p 562

Goeppert N, Goldscheider N, Scholz H (2011) Karst geomorphology of carbanatic conglomerates in the Folded Molasse zone of the Northern Alps (Austria/Germany). Geomorphology 130:289–298

Gvozdetskiy NA (1965) Types of Karst in the U.S.S.R. Separatum, Prob. Speleol. Res. (Prague) 47–54

Hevesi A (1980) Adatok a Bükk-hegység negyedidőszaki ősföldrajzi képéhez (Data to the Quarternary Paleogeographical Features of the Bükk Mountains). Földtani Közlemények 110 (3–4):25–36 (in Hungarian)

Jakucs L (1977) Morphogenetics of Karst Regions. Adam Hilgar, Bristol, p 284

Miotke FD (1968) Karstmorphologische Studien in der glazialuberformten Hohenstufe der Picos de Europe' Nordspanien. Jahrbuch der geographischen Gesellschaft zu Hannover Sonderhef, vol 4, 161 p

Smart PI (1986) Origin and development of glacio-karst closed depressions in the Picos de Europa, Spain. Zeits f Geomorph 32(4):423–443

Van Husen D (2000) Geological processes during the Quaternary. Mitteilungen der Österreichischen Geologischen Gesellschaft 92:135–156

Veress M (2010) Karst environments. Karren formation in high mountains. Springer, New York, p 230

Veress M (2012) Glacial erosion and karst evolution (Karren formation on the surfaces formed by Glaciers). In: Veress B, Szlgethy J (eds) Horizons in earth science research, vol 8. Nova, New York, pp 1–94

Veress M (2013) A glaciális felszíntípusok karsztosodása (Karstification of glacial surface types). Földrajzi Közlemények 137(3):2–27 (in Hungarian)

Veress M (2016a) Covered Karst. Springer, 536 p. https://doi.org/10.1007/978-94-017-7518-2

Veress M (2016b) Postglacial evolution of paleodepressions in glaciokarst areas of the Alps and Dinarides. Zeitschrift für Geomorph 60(4):343–358

Veress M, Szunyogh G, Zentai Z, Toth G, Czopek I (2006) The effect of the wind on karren formation on the Island of Diego de Almagro (Chile). Zeitschrift für Geomorphologie 50(4):425–445

Žebre M, Stepišnik U (2015a) Glaciokarst landforms and processes of the southern Dinaric Alps. Earth Surf Proc Land. https://doi.org/10.1002/esp.3731

Žebre M, Stepišnik U (2015b) Glaciokarst geomorphology of the Northern Dinaric Alps: Snežnik (Slovenia) and Gorski Kotar (Croatia). J Maps. https://doi.org/10.1080/17445647.2015.1095133

Chapter 7 The Development of Glaciokarstic Surfaces

Márton Veress

Abstract This chapter deals with the geomorphic evolution of glaciokarst. The ways of surface denudation are presented on bare karst and soil-covered karst, on concealed karst and allogenic karst, then in light of them, the landscape evolution in the area of various glacial erosional surfaces and landforms will be outlined. The future geomorphic evolution of glacial erosional surfaces are also touched upon.

Keywords Turnover · Geomorphic evolution · Senile surface · Local Superficial dissolution · Peeling · Karren physical weathering

7.1 Introduction

In this chapter, the way of the denudation of glaciokarst is presented, an answer is also looked for whether the karst will have a senile state and if it will, how and under what conditions it will be achieved. It will be studied whether the denudation of glaciokarst is homogeneous or heterogeneous and if it is heterogeneous, how this affects the already existing features and the differences manifesting in landscape evolution on karst.

The landscape evolution of karstic terrains is partly similar to the evolution of surfaces built up of other rocks, but it also shows a specific character. The geomorphic evolution of glaciokarst is especially specific. Below, mainly those theories dealing with the evolution of karstic surfaces are presented which can be used at the landscape evolution of glaciokarst.

Early theories on karstic landscape evolution tried to apply the theory of Davisian cycle to karstic terrains. According to this, in the course of landscape evolution, during which the surface of karst becomes smooth in the end, various phases were distinguished. Later, opinions emphasized the specific, smooth becoming of carbonate terrains, which is only characteristic of them, to which no specific features belong (Patton 1964) and landscape development is of one-time and cannot be separated into phases (Jenko 1956). Thus, according to Jenko (1956), on the Dinaric

M. Veress et al., *Glaciokarsts*, Springer Geography,
https://doi.org/10.1007/978-3-319-97292-3_7

karst, there are no phases of landscape evolution and the landforms do not originate from each other. However, according to Thornbury (1965), there are specific features which can be associated with the phases of karstic denudation. Independent of the state of the karst surface, certain features occur on the karst if the conditions of their evolution are present (Sweeting 1973). However, denudation phases are also differentiated on karstic terrains by contemporary authors (Waltham and Fookes 2003). Probably, there is a greater chance for the karst to reach a smooth state in case of intensive karstification (tropical karst) and it is also likely that on karsts with various environment (mainly because of the climate), landscape evolution may be different, and if it is of low intensity, the stability of the surface increases.

There are various interpretations for surface denudation by karstification. Surface denudation can be described in connection with feature development, considering the way (nature) of surface denudation and the coveredness of the karst surface. Since the development of the Davisian cycle theory, three main stages of karstic landscape evolution have been distinguished: the juvenile stage, the mature stage and the senile stage. The karst gets into the senile state if its surface reaches (or approaches) the non-karstic bedrock (Cvijič 1918), the karst water level (Jakucs 1977; Balázs 1990), which results in the formation of a karstic plain on tropical karst (Salomon 2000). The final state of the senile surface, thus of the karstic denudation may be the intermountain plain (Waltham and Fookes 2003), the pediment (Jennings and Sweeting 1963) the surface covered with karren debris (Williams 1966), the local base level of erosion as the karst water level (Jakucs 1977), which is marked out by the surrounding, non-karstic terrain or the cave floor (Martini and Grimes 2012).

According to Sawicki (1909), during karstification, a difference has to be made between the development of superficial karst features and the landscape evolution of the karst, where the limestone is pure, the development of karst features is continuous and where the rock contains pollutants, enough weathering residue is formed during karstification for the karstic landscape evolution to be replaced by erosional landscape evolution. Then, since the limestone is exposed, a karstic denudation begins again. Later, this idea appears in a lot of other authors' work (Williams 1987; Cui et al. 2002; Waltham and Fookes 2003; Veress 2016a).

During geomorphic evolution that can be connected to feature development, the karstic features coalesce by their growth or the caves open up to the surface which results in the evolution of a newer surface. A surface evolution related to karren formation was described by Cvijič (1924) on Mediterranean karst and by Williams (1966) on glaciokarst. During both landscape evolutions, the initial, nearly plain surface is dissected by karren (however, in Cvijič's model, the growth of karren features can be followed by the development of larger, even non-karren features), then debris is formed by the denudation of the dividing walls and a newer plain surface develops at a lower level.

A geomorphic evolution that can be connected to the formation and development of solution dolines was described by Grund (1914). According to this, the karst surface reaches its plain character through surfaces with dolines, surfaces with uvalas and surfaces with karst hills. According to Cvijič's karst landscape evolution

(1918), after the covering, non-karstic rock denuded by fluvial erosion, the karst is denuded to the non-karstic rock constituting the bedrock with the growth of karst features. With the development of dolines, uvalas and poljes (their number and size increases), the karstic rock becomes more and more used up. The non-karstic rock which constitutes the bedrock is exposed (limestone survives at hum features), and thus superficial erosional denudation begins again. If there are also large-sized caves next to superficial karst features, the collapse of caves can also contribute to the development of senile surface (Bulla 1954).

According to the method of surface denudation, the karstic rock can become dissected vertically or it can be denuded without dissection at its surface only, which results in the decrease of surface elevation. During vertical dissection, the denudation of the karst takes place along fractures. Giant grikes develop. The development of labyrinth karst (Brook and Ford 1978) as well as the formation of inselberg karst (Paton 1964) are interpreted by a geomorphic evolution of this type. According to both ideas, increasingly wider, plain terrains develop on the karst by the decrease of the expansion of terrains between the giant grikes. However, the pinnacle-type karst also develops by the formation of giant grikes. The widening of giant grikes results in the development of stone forest karst (Song and Liang 2009). The coalescence of giant grikes and caves can also cause the dissection of the karst and the formation of pinnacle karst (Martini and Grimes 2012; Veress et al. 2008).

Ahnert and Williams (1997) studied the geomorphic evolution of the karst by digital model. When the runoffs converged on the surface of the model, polygonal karst developed and when they diverged, cone karst and tower karst developed.

Peng et al. (2007) described the landscape evolution of the karst dissected by pinnacles, with the presentation of the development of stone forest karst. The size of pinnacles can be of three kinds depending on the rate of subaerial erosion (A), soil-erosion (S) and subsoil erosion (C) as compared to each other. If the relation of the rate of the above-mentioned factors is $A < S < C$, both the visible height (height above the surface) and the actual height (as compared to the bedrock floor) of pinnacles increases and the soil thickens. If denudation factors are characterized by a relation of $A < C < S$, the visible height of pinnacles increases and the soil grows thinner and in the end, it disappears. The height of pinnacles decreases because of subaerial dissolution. If the relation of the rate of denudation factors is $S < A < C$, the visible height of pinnacles will be smaller and their actual height will be larger.

The surface of the karst can also become denuded superficially, without dissection. This occurs if debris develops during the dissolution of karstic rock, which results in the lowering of the bedrock, then as a result of the dissolution of debris parts, the surface becomes lower too (Veress and Péntek 1996). According to Balázs (1986), tropical karst is denuded superficially, but by dissection too.

On covered or on partly covered karst, the effect of the cover on karstification as well as the denudation not only of the karstic rock, but also of the surface of the cover are taken into consideration by theories of landscape evolution. In Waltham and Fookes's (2003) karst development model, five phases of surface denudation are distinguished. While the first two phases are characterized by a smaller distribution of covered karst, in the other three phases this is predominant. The fifth, senile phase (termed extreme karst by the authors) is the inselberg karst. Williams (1987) interprets the dissection of fengcong karst by inheriting valleys on tropical karst, while the development of cockpit karst by the karstification of the valleys at a lower level. Peng et al. (2007), as it has already been mentioned at the development of pinnacles, explains the development of stone forest with the dissolutional deepening of grikes being partly filled with cover. While the geomorphic evolution of pinnacles is determined by the rate of the three denudation (subsoil erosion, soil erosion and sub-aerial erosion) rates as compared to each other.

Starting from Büdel's (1957) etch-planation theory, according to Cui et al. (2002), the tropical karsts of China develop during planation. Double levelling surfaces are formed during the process. The lower surface is the surface of the bedrock, while the upper is the surface of the cover. The cover is the dissolutional residue of the limestone which is although denuded, is supplemented from below during the dissolution of the bedrock.

On covered karst, Veress (2016a) differentiated one-phase, two-phase and continuous landscape evolution. In case of one-phase geomorphic evolution, the development of the karst starts from a covered karstic state, while in the case of two-phase geomorphic evolution, karstic landscape evolution already takes place by the expansion of the cover (first phase), then it continues during its denudation (second phase). In case of a continuous landscape evolution, the cover develops during karstification, thus the karst functions at least partly permanently as covered karst. Veress (2016a) described the development of the covered karst of glaciokarst. Its characteristics are the following:

- Covered karst is characteristic in cirques, in troughs, but mainly in their depressions.
- The cover can be transported from glacial erosional features only into the karst. The material of the cover can be transported in the form of solution or debris.
- The transportation of the cover into the karst can take place through karren features, shafts and caves.
- The exhumation of karstic depressions depends on the proportion of inward and outward transport.
- During the denudation of the cover, the expansion of bare karst can mainly increase in the area of troughs at the expense of covered (concealed) karst.
- Since the cover is mainly transported away from karstic depressions, the dissectedness of troughs increases.

7.2 General Characteristics of Present Material Transport on Glaciokarst

On glaciokarst, geomorphic evolution can take place under positive turnover and negative turnover. In case of positive turnover, the present accumulation exceeds transportation away at a given site, while in case of negative turnover, exhumation takes place. It has to be mentioned that the physical weathering of the bedrock also results in the increase of turnover, however, this does not result in filling up, but at most, it results in the rise of the surface to a smaller degree because of the increase of pore volume.

In case of positive turnover, the surface becomes higher, debris cones and alluvial cones develop. Their material may originate from frost weathering, snow and rock avalanches (debris cone) and from water transport (alluvial cone). Their material can overlie a bare karstic surface or a moraine. The superficial deposit can be permeable or impermeable rock. During their denudation, the material of debris cones and alluvial cones is reworked, their areas expand and by this, they cover newer and newer terrain sections. In case of filling, concealed karst, buried-like karst or allogenic karst develops from bare karst. In case of negative turnover, the surface becomes denuded. This includes the erosion of the surface, suffosion and transport in solution too.

On glaciokarst—mostly in glacier valleys—material transport is limited on the surface because the transport of stream is impeded. This limited character is triggered and caused by karstic drainage, the depressions and terminal moraines which are situated on the surface of the karst.

The degree of material transport depends on the suitability for denudation and the sediment receiving capacity of the karst. Suitability for mechanical denudation depends on the inclination of the surface, on the quantity of water, the expansion of vegetation of the surface and on the grain size of the sediment on the surface. In case of greater inclination, more precipitation and superficial streams, smaller expansion of vegetation and smaller grain size, denudation is of larger degree. The sediment receiving capacity depends on the ability to transport into the karst (for example, on the maturity of the passages of ponors) and on the storing capacity of the karst.

The storing capacity depends on the density and size of the cavities of the karst, on their position as compared to the surface, on their type and on the degree they are open to the surface (Fig. 7.1). According to their position as compared to the surface, the cavities can be the cavities of the epikarst, of the vadose zone or of the phreatic zone. According to type, they can be karren features, solution passages, shafts, partly solution, partly erosional or completely erosional swallet caves, through caves and multi-floor caves. Primarily erosional caves, through caves and multi-floor caves can be of large size. The sediment receiving capacity is large in case of swallet caves with ponors or in case of caves that had such ponors earlier, partly because of their open nature to the surface and partly because their superficial streams ensured sediment transport. A sediment fill with significant water transport has been described from caves bearing water inflow formerly such as from the glaciokarst caves of Tasmania (Mt. Field Plateau) (Kiernan et al. 2001), from the Croatian (Velebit) Stirovica-ice cave (Bočič et al. 2012), from some caves of the Polish Tatra Mountains (Szczygiel et al. 2015) and from the Canadian Nakimu cave (Thomson 1976). (In the latter cave, there is a 120 m thick, well-sorted sediment accumulation.) The sediment receiving capacity is favoured by if the horizontal expansion of caves is large. The trough caves with stream can be mentioned from

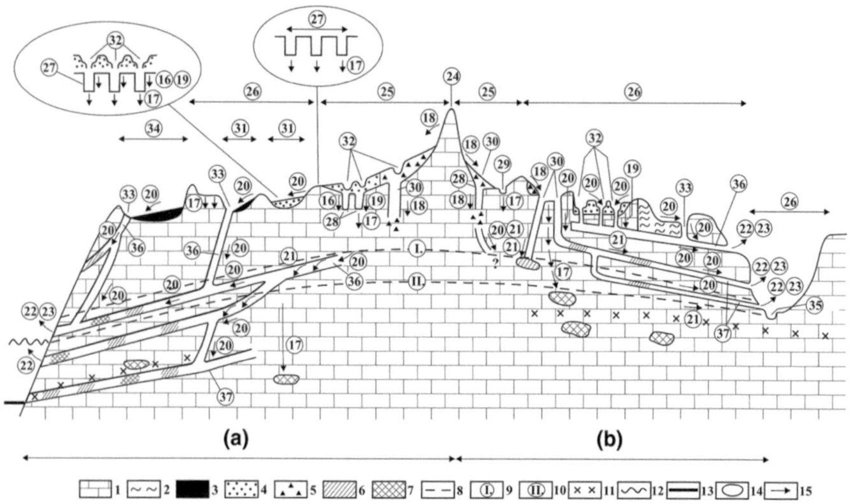

Fig. 7.1 Turnover and surface denudation on glaciokarst with marine origin (**a**) and land origin (**b**). 1. limestone, 2. non-karstic rock intercalated into limestone, 3. impermeable cover, 4. permeable cover, 5. material of mass movement, 6. fluvial accumulation, 7. material which got precipitated from solution, 8. karst water level, 9. high karst water level, 10. low karst water level, 11. former karst water level, 12. level of seawater, 13. former sea level, 14. dissolutional cavity, 15. material transport, 16. transportation of the dissolved material originating from the cover by percolating waters, 17. transportation of the dissolved material originating from the bedrock by percolating waters, 18. transportation of material of mass movement, 19. suffosional material transport, 20. sediment yield of former or present stream, 21. transport in solution by stream or by karst water, 22. material transport from the karst in solution, 23. material transport from the karst in the form of stream load, 24. horn, 25. cirque, 26. trough, 27. grike, 28. giant grike, 29. schachtdoline, 30. shaft, 31. giant solution doline, 32. subsidence doline, 33. ponor, 34. polje, 35. river, 36. recent allogenic cave, 37. paleokarstic allogenic cave

Totes Gebirge, which issuing from the inner part of the mountain reach the margin of the mountain (Bauer and Zötl 1972).

In the karst, a significant condition of storing capacity is the sediment transmitting capacity in the caves. This is large in case of caves with large floor dip and with intermittent or permanent streams. These conditions exist in case of swallet caves which developed in the vadose zone such as in the above-mentioned Stirovača ice cave (Bočič et al. 2012). An important condition of sediment transmitting capacity is that the karst should have an outlet. The sediment transmitting capacity is small in case of caves with vertical or horizontal position and caves situated below sea level. Sediment transport is impeded if the cave is located below karst water level since water flow is slower and the sediments become cemented (Bočič et al. 2012).

If the material is transported in solution, its transportation into the karst is independent of the caverned nature of the karst, but the outward transportation can be significant from such caves, for example, from the Stirovača ice cave (Bočič et al. 2012).

Turnover depends on the environment of former glacier and on glacial erosional features. While in case of glaciers with maritime environment, cavity development may be extremely deep penetrating as compared to the surface (Adamson et al. 2014) and it may be multilevel because of sea-level fluctuation, in case of glaciers with terrestrial environment, this is less characteristic. In case of the former case, storing capacity may be larger, but the sediment transmitting capacity is only limited. This can be significant within the caves, but since the entrances are below the water table, the sediment cannot leave the karst and it can only leave it in solution (Fig. 7.1). However, the sediment transmitting capacity is larger at the caves of glaciers with terrigenous environment (the inward transported sediment can leave the karst at springs).

As regards glacial erosional features, although there are a lot of similarities between the turnover of cirques and that of troughs, differences occur too (Figs. 7.2 and 7.3). Thus, at cirques, more sediment is deposited on the floor now, sediment accumulation expands on the whole floor, material transport is mostly completely limited, sediment transport is less characterized by redeposition, since there is only

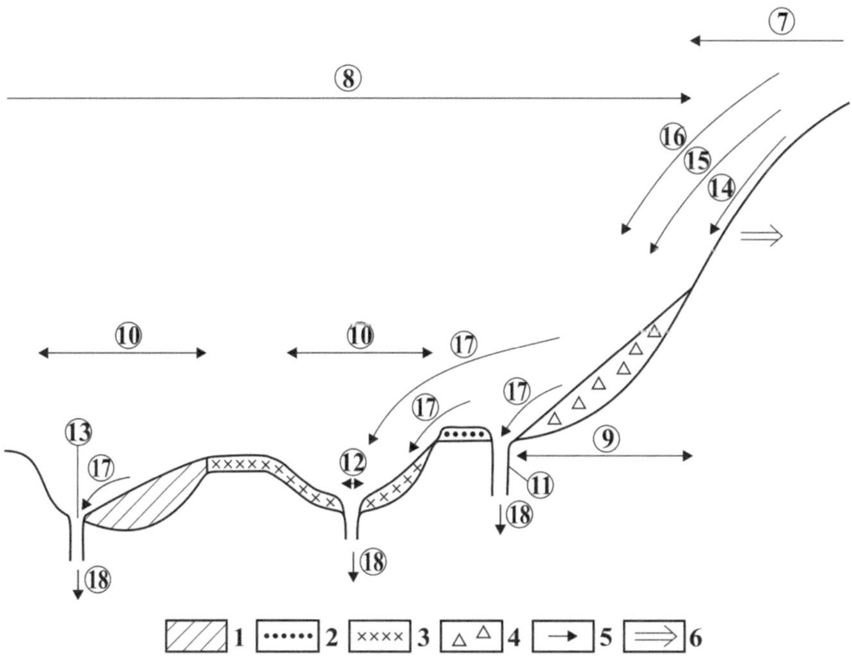

Fig. 7.2 Theoretical turnover of a cirque (karstic depressions are not aligned along the profile). 1. impermeable cover, 2. basal moraine, 3. reworked permeable moraine, 4. debris (material of frost weathering and avalanches), 5. material transport (water transport and in solution) into the karst, 6. retreat of the cirque slope, 7. combe-ridge, 8. cirque floor, 9. debris cone, 10. giant solution doline, 11. shaft, 12. subsidence doline, 13. ponor, 14. material transport from the rock wall by stream (suspended and in solution), 15. material transport by avalanches, 16. debris fall and collapse, 17. water transport from the cover (suspended, rolled, in solution), 18. material transport into the karst

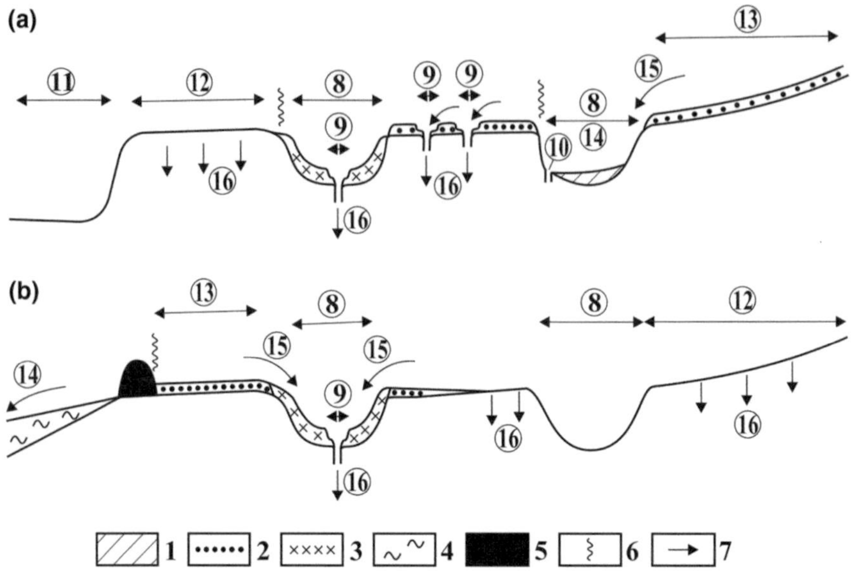

Fig. 7.3 Theoretical turnover of a trough, **a** hanging trough, **b** trough with terminal moraine (karstic depressions develop in the total width of the floor), 1. impermeable cover, 2. basal moraine (permeable), 3. reworked, permeable cover, 4. fluvioglacial sediment, 5. terminal moraine, 6. barrier of material transport, 7. material transport, 8. giant solution doline, 9. subsidence doline, 10. ponor, 11. main valley, 12. bare karst, 13. concealed karst, 14. allogenic karst, 15. water transport from the cover (suspended, rolled, in solution), 16. material transport into the karst

one sediment transportation obstacle because of the morphology and relatively small expansion of the cirque. In troughs, sediment accumulation expands to a smaller area, it is only characteristic of some valley sections. Material transport is not always limited (sediment can partly escape some valleys) and since there may be several sediment transportation obstacles in the trough and the valley is longer, sediment transport is divided into sections. Thus, on the valley floor, denudation and accumulation alternate and are repeated. In cirques, the ways of material transport are less varied as compared to troughs.

On glaciokarst, material transport and accumulation can be attained between various levels, which are the following (Figs. 7.1, 7.2 and 7.3):

- From combe-ridges, from horns to the floor of glacier valleys.
- From the slopes of glacier valleys to the floor of glacier valleys.
- From transverse steps into rock basins and to the floor of glacier valleys.
- From combe-ridges, from the slopes and floors of glacier valleys into depressions.
- From expanded erosional terrains (former ice caps, ice sheets) into karstic depressions.
- From the area of terminal moraines into depressions to the lower sections of troughs or into main valleys.

– From the area of cirques to the area of troughs.
– From the side slopes and from the environment of depressions to their floor.
– From the scarp fronts of stepped surfaces to the stepped surfaces situated below them.
– From the bedrock into the cover, during its physical weathering, on any glacial erosional surface.
– From glacial erosional features and surfaces into the karst. The transported material can get down into the epikarst, into the vadose zone or into the phreatic zone.
– From cavities and caves, through springs, into the valleys of the karst or outside the karst.

During glacier erosion, moraine and dissolved material develop while in areas exempt from glacial erosion or during the present denudation of the rock, dissolved material and debris are formed. Farther and farther from its development site, the moraine material is increasingly impermeable, which favours the superficial transport of the material. At a higher elevation (the upper part of the periglacial zone) and on steeper slopes, debris of frost weathering material develops in a larger proportion and the share of dissolved material is more subordinated. At a lower and lower elevation (lower part of the periglacial zone), the dissolved material and the debris that developed by karren physical weathering have an increasingly larger proportion. The cover material of the fluvial erosional zone of the temperate belt may be dissolved debris, debris that developed during karren formation or frost weathering debris or debris of erosional origin. Their proportion as compared to each other depends on the fact which subzone (dissolutional, erosional or frost weathering and mass movement) of a zone is considered Material transport is especially limited in the periglacial zone, while in the fluvial erosional zone of the temperate belt it can also take place farther, perhaps sediment is also transported from the area of glaciokarst. The transportation of the formed material can happen in the following ways (Fig. 7.4):

– In solution, the transporting water can be meteoric water and stream. On glaciokarst, transportation in solution can be the most significant and the material that is transported in other ways can also get into solution partly or completely subsequently. The distance of the transportation in solution is long and it is directed into the karst and outside the karst.
– As solid material by water transportation (floated, rolled). The material can originate from moraine, from debris of karren origin and from frost weathering material. The material transported by water is conveyed to a short distance, it is accumulated, and then it is redeposited again and again. Transportation takes place at the surface of the karst or into the karst. If it is transported into the karst, the frequency of the interruption of sediment transport (thus, the number of sites of accumulation) depends on the type and morphology of caves. (Thus, for example, at narrowing cave sections, accumulation occurs.)

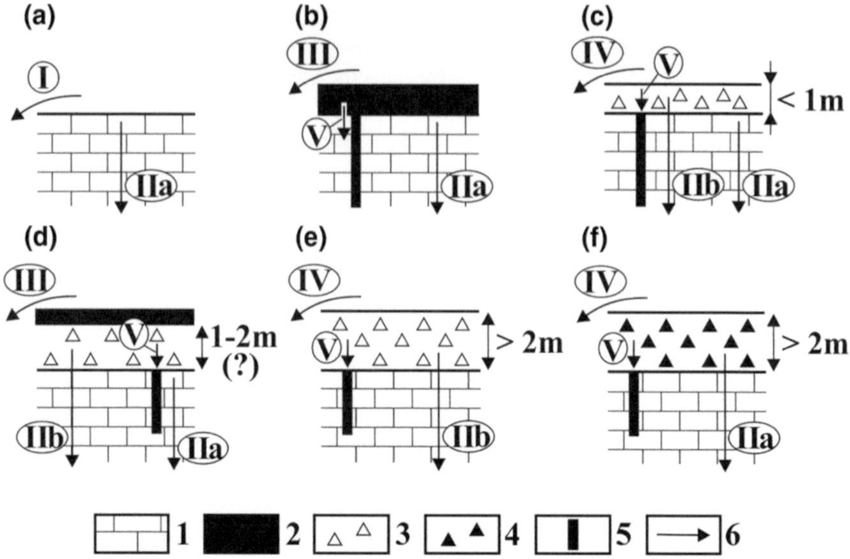

Fig. 7.4 Ways of material transport on glaciokarst. **a** bare surface, **b** limestone surface covered with soil, **c** limestone surface covered with calcareous debris with a thickness smaller than 1 m, **d** terrain covered with soil and calcareous debris with a thickness of 1–2 m, **e** limestone surface covered with calcareous debris thicker than 2 m, **f** limestone surface covered with non-calcareous debris, 1. limestone, 2. soil, 3. calcareous debris, 4. non-calcareous debris, 5. karstic passage (karstic passages in figures **d**, **e** and **f** are of subglacial origin), 6. material transport, I. material transport in solution on the surface, IIa. material transport in solution from the limestone, IIb. material transport in solution from the debris, III. transportation away of soil on the surface by meteoric water, IV. fine-grained material transport on the surface by meteoric water, V. suffosional material transport into the karstic (karren) passages

- By suffosion, when the finer grains of the cover are transported. The distance of transportation is short and it is directed into the karst.
- By mass movements (mountain failures, avalanches). The distance of transportation is short and it is confined to the surface.

The way of material transport into the karst can be diffuse and point-like. In case of diffuse transport, the width of transportation route can expand from the width of the fissures of the rock to the width of karren passages. The way of material transport mainly takes place in solution. The dissolved material can be transported by percolating water or by streams. The dissolved material can originate from the calcareous cover or from the bedrock. A point-like material transport can happen through a wider passage which can be a wider karren feature too. A condition of the process is that there should be a superficial stream. These streams connecting to the passages transmit their stream load into the karst. The sites of point-like inward water transport are mainly the ponors, but subsidence dolines, giant grikes and shafts can be sites of point-like inward transport too. When there is diffuse material transport, there is not necessarily a point-like material transport, while in the environment of point-like material transport there is always a diffuse one. On bare karst, diffuse material transport is present, while on concealed karst, both diffuse and point-like material transport are present. On allogenic karst, point-like material transport is predominant and diffuse material transport is subordinated.

7.3 Geomorphic Evolution Depending on Karst Type on Glaciokarst

7.3.1 Denudation on Bare Karst and on Soil-Covered Karst

In case of these karst types, material transport into the karst mainly takes place in the form of solution. The transportation of debris (or soil) is subordinated. The denudation of the surface is mostly areal or increasingly tends to be like this. On these karst types, karstification was interrupted by glacial erosion. After the retreat of ice, the process was renewed. Postglacial landscape evolution continued on the preglacial surface that was transformed by glacial erosion or it started on the surface that developed during glacial erosion. Superficial denudation takes place on bedrock karstic rock, but on these surfaces, it can be transformed into one having a double geomorphic evolution because of debris development (see Sect. 7.3.2). In this case, the lowering of the surface is directly caused by the denudation—first of all, the dissolution—of the cover with debris.

During dissolution occurring on the surface of the bedrock, karren are formed. Consequently, landscape evolution is strongly related to karren formation. The intensity of dissolution depends on the quantity of water, on the duration of dissolution and on the amount of CO_2. Furthermore, on bare surfaces, it depends on the type and thickness of the snow (Küfmann 2014). With the increase of elevation, the amount of precipitation (mainly snow) and of snow-water as well as the duration of dissolution increases, however, with the decrease of elevation, the amount of biogenic CO_2 increases. Depending on these conditions, the rate of superficial dissolution and the amount of dissolved material show significant differences in various environments. Denudation rates of dissolution origin show the following values in various environments.

In the temperate belt, on more elevated, bare glaciokarstic surfaces (at an elevation of 1800–2200 m) the rate of surface dissolution is 0.01–0.1 mm/year (Bögli 1971; Häuselmann 2008; Forti 1984; Kunaver 1979; Plan 2005), in case of precipitation of 1670–3500 mm. With the increase of the amount of precipitation, the rate of surface dissolution increases. Thus, in the area of Innerbergli in Switzerland, where the amount of precipitation is 1670 mm, the surface dissolution rate is only 0.014 mm/year (Häuselmann 2008), while in the Kanin Mountains, where the amount of precipitation is 3500 mm annually, the surface dissolution rate is 0.017–0.1 mm/year (Kunaver 1979). On the debris material of an experimental equipment, dissolution by melting snow is 14 mg L^{-1} (Küfmann 2014). On glaciokarst with extreme amount of precipitation, for example on the islands of the Patagonian archipelago (where the amount of precipitation is 8000 mm/year), on the island of Madre de Dios, it can be 0.14 mm/year, on the island of Diego de Almagro it can reach 0.09 mm/year (Maire et al. 2009). In areas with lower elevation, but with similar amount of precipitation (1500–2500 mm/year) such as on the British Islands, although a relatively higher dissolution rate also occurs thus, it is 0.04 mm/year (Sweeting 1966), extremely low rate values have been measured too. Thus, dissolution rates of 0.025–0.075 mm/year (Thomas 1970), and 0.003–0.004 mm/year (Trudgill 1975) have been reported. This refers to the fact that the surface dissolution rate depends on the duration of dissolution too, however, the dissolution capacity of meltwater is affected by the type and thickness of snow and the time when it develops (Küfmann 2014). As at a higher elevation, the snow survives longer, thus the dissolution effect exerted by melting snow is also of longer duration is, while on the British Islands, where the amount of snow is less and the snow survives for a shorter time, the duration of dissolution is shorter too. This is also proved by the dissolution rates of areas situated closer to the pole with a similar amount of precipitation, but also with a surface dissolution of long duration. Thus, at the Swartison glacier, at a precipitation of 2600 mm a rate of 0.03 mm/year (Lauritzen 1990), in Alaska, at a precipitation of 1750–2940 mm a rate of 0.04 mm/year (Allred 2004) was measured.

On soil-covered karst, the dissolution rates do not necessarily exceed the dissolution rate of surfaces with high elevation (where dissolution exists for a long time). Thus, below soil on Hochschwab at an elevation of 2000 m, at a precipitation of 2150 mm, the rate is 0.013–0.04 mm/year (Plan 2005), while in Vercors, at an elevation of 1060 m, at a precipitation of 1640 mm, the dissolution rate is 0.02 mm/year (Gams 1985). However, at sites with extreme CO_2 supply, for example, under peat, the dissolution rate can reach 5.0–8.2 mm/year (Sweeting 1966). The deepening of karren features depends on a large degree on the fact whether there is soil in the feature and of what nature it is. Thus, on the British Islands, if the rinnenkarren receives its water from peat, the dissolution rate is 6.3–11.5 mm/year (Sweeting 1966), while if the rinnenkarren receives its water from mineral soil, the dissolution rate is 0.01–0.015 mm/year (Trudgill 1985).

Karren geomorphic evolution can take place on non-stratified and on stratified karstic rock. On non-stratified rocks, plain surfaces and surfaces with roche moutonnées may have developed, while on stratified rocks, stepped surfaces and surfaces with steps and roche moutonnées were formed during glacial erosion. The denudation of the surface can happen by surface dissolution, by karren peeling and by karren physical weathering. In case of surface dissolution, sheet water covering the bare surface or the water percolating under the soil exerts a dissolution effect on the rock surface. In case of karren peeling, as a result of dissolution along bedding plane, bed sections are separated. In case of karren physical weathering, the growth and coalescence due to dissolution of karren features result in the physical weathering of the rock. Karren peeling and karren physical weathering exert their effect with frost weathering or they are efficient in this case. Karren physical weathering may include surface dissolution and karren peeling too.

7.3.1.1 Superficial Dissolution

Superficial dissolution can take place on surfaces with any inclination. A significant characteristic of the process is that no debris is formed. On bare terrains, the indicators of the process are karren tables (Bögli 1961), aeolian karren ridges (Jaillet et al. 2000) or relict ridges (Veress et al. 2006).

On glaciokarst made up of non-stratified rocks, megakarren can also contribute to superficial dissolution. Large-sized and coalescing kamenitzas result in the superficial dissolution of the surface (Veress et al. 2006). Superficial dissolution may take place both on the scarp fronts and bedding planes of stepped terrains. Its presence is proved by measurements on plain, bare surfaces (Cucchi et al. 1994, 1996; Häuselmann 2008; Hoblea et al. 2001), but by observations too. Since the schichtfugenkarren of scarp fronts are hanging over the bedding plane situated at the stem of the scarp front (Fig. 3.29). Its degree can change from site to site, which results in the mosaic-like denudation of the surface. The dissolved material is transported in karren features, mainly in giant grikes into the karst. As a result of this, the scarp fronts retreat and the elevation of bedding planes decreases. A consequence of scarp front retreat is that the parts of steps with bedding planes shorten and another consequence is that the giant grikes that developed at the stem of scarp fronts widen. (However, the widening of the latter contributes to the further shortening of terrains with bedding planes.)

During the dissolution of the bedding planes of the steps, the relative height of the scarp front that borders the bedding plane from above increases if the degree of dissolution increases in dip direction (Fig. 7.5a). Dissolution of increasing intensity in dip direction is proved by karren steps. These steps are slopes with a height of 1–2 dm, inclined opposite to the surface with bedding planes, which occur inside a surface with bedding planes at places where a giant grike with strike direction developed on the surface with bedding planes (Fig. 7.5b). If the surface with bedding planes is dissected by several giant grikes, a series of karren steps develops. Since superficial dissolution is of increasing degree, the uppermost layer thins out and wedges out in dip direction by which the surface that was formed by superficial dissolution cuts the beds (Fig. 7.5a.ii). Rinnenkarren becoming deeper in dip direction, which also develops on such surfaces with bedding planes go transversely through newer and newer beds. As a result, more and more bedding planes crop out in their sides. This contributes to the dividing up of ridges between the channels, and thus to karren physical weathering that will be described below.

Karren cells are also proof of superficial but local dissolution taking place on bedding planes since the area of karren cells creates a basin with small depth. In the area of karren cell, the dissolved material gets into the karst through pits and grikes. The large density of channels probably favours superficial dissolution, then the developing confinement further strengthens this. The thinning out of the bed and the large density of channels also favour physical weathering, and thus karren physical weathering. Superficial dissolution, since the bed constituting the surface thins out

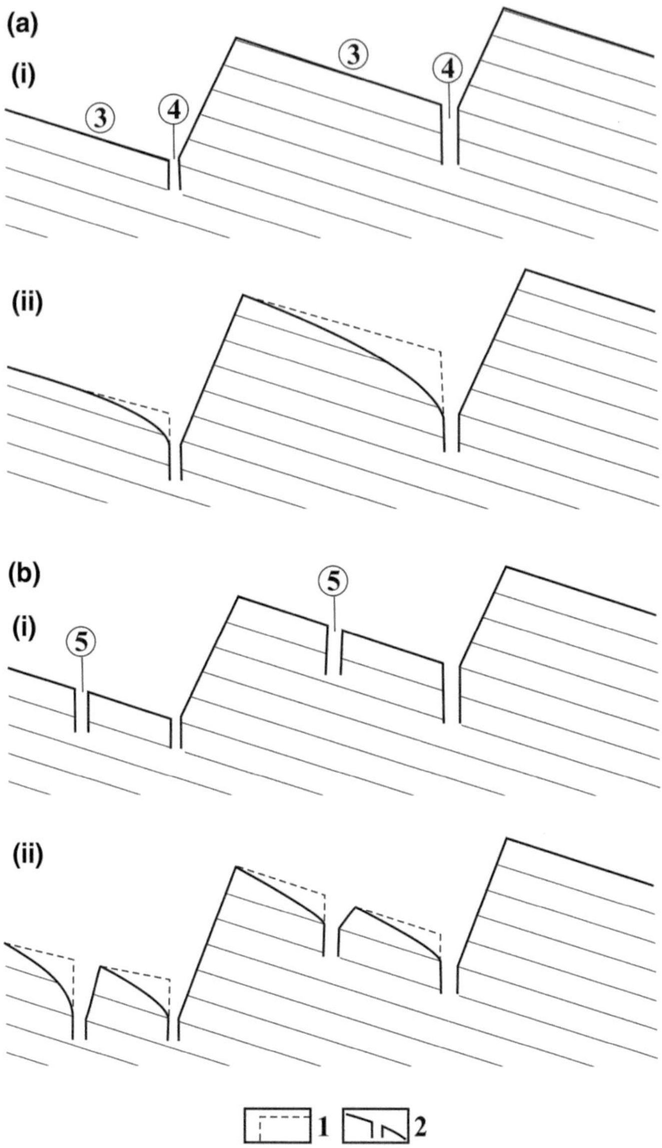

Fig. 7.5 Superficial dissolution on bedding planes, 1. destroyed bed unit, 2. giant grike, grike, 3. rinnenkarren and rinnenkarren systems occur on the bedding plane, 4. giant grike at scarp front, 5. giant grike inside the bedding plane, **a** the bedding plane is dissolved to an increasing degree towards the giant grike that developed at its margin, **b** the bedding plane is dissolved to an increasing degree towards the giant grike that developed inside it, **i**. original state, **ii**. developed state

to its effect, contributes to the efficiency of karren physical weathering. However, the peeling of the thinner bed takes place easier too.

7.3.1.2 Peeling

The peeling of the bed, and thus the instability of the karren surface are primarily contributed by grike-schichtfugenkarren assemblage (Fig. 7.6). The grikes of bedding planes break down and separate the bed into parts, while along the schichtfugenkarren developing on their walls if their lower part reaches the wall of another grike inclined in the direction of the bed inclination, the bed sections between grikes become separated from the ones below them (Fig. 7.7). The process is also contributed by the rinnenkarren (rinnenkarren and grike assemblage) if the floor of channels reaches a bedding plane or bedding planes. In this case, the ridges between the channels can become separated from the bedrock along bedding planes. Consequently, schichtfugenkarren contribute to the cutting up of beds in dip direction, while channels contribute to their cutting up in strike direction. It can be seen that karren peeling is significantly similar to subglacial peeling that has already been described in Chap. 3. A difference occurs in the fact that there is no glacial erosion (thus, newer steps do not develop) and meltwater is absent (thus, the epikarst is filled with water for a shorter time). Peeling on the same bedding plane can take place by the denudation of several beds simultaneously. Broken off bed sections continue their decrease mainly by frost weathering (but by dissolution too). If the grikes are inherited to the bedrock too, the process can be repeated: a newer bed becomes separated from the one below it (Fig. 7.7c). Peeling can happen in two ways: either a bed section of one (perhaps some) step(s) becomes peeled simultaneously or those of all steps. Its intensity and survival are increased and ensured by the following things:

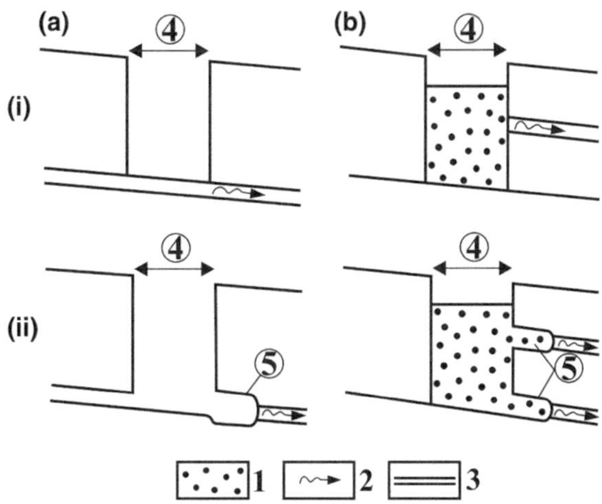

Fig. 7.6 Schichtfugenkarren development if the grike is without soil (**a**) and if it is filled with soil (**b**), 1. soil, 2. seepage, 3. bedding plane, 4. grike, 5. schichtfugenkarren, **i** original state, **ii**. developed state

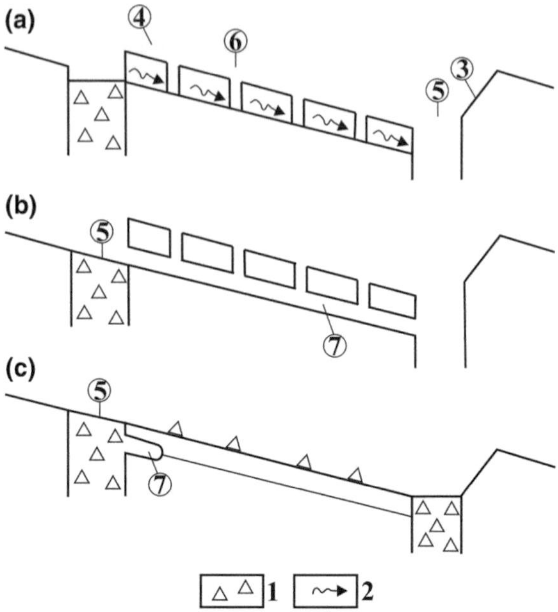

Fig. 7.7 Peeling of the bed unit of a bed in case of grike-schichtfugenkarren assemblage, 1. debris, 2. seepage, 3. scarp front, 4. surface with bedding plane, 5. giant grike, 6. grike, 7. schichtfugenkarren, **a** grike development, **b** schichtfugenkarren development, **c** the uppermost bed is separated into debris pieces, a newer schichtfugenkarren develops on the bedrock bed

- The grikes have to be situated in the strike direction of the bedding plane and their density has to be large. If their density is large, peeling takes place simultaneously, smaller pieces are formed from the beds.
- Soil in the fill of the grike, since the dissolution capacity of the water percolating into the rock increases. However, soil fill in case of thin beds also promotes the simultaneous development of several schichtfugenkarren, and thus also accelerates peeling (Fig. 7.6b): the soil does not only produce CO_2, but it also transmits water to the grike wall at various levels.
- Optimal bed thickness if bed thickness is large (1–2 m) the separated thick bed, the developed block is not able to get displaced, in addition, the bed is gone transversely by the grike only after a long time. In case of thin beds (some dm), the schichtfugenkarren, as already mentioned, can develop in several levels simultaneously if the grike is at least partly filled with soil. If the beds are extremely thin (some cm), the denudation of the beds can be so fast that so much debris is produced that is no longer able to be dissolved or transported away. Coveredness with debris restrains further karren formation and bed peeling. Karren terrain turns into concealed karst.
- Suitable inclination of the surface and the beds. The larger the inclination of the surface, the more easily the developed debris is transported into the sediment receiving features by mass movement. On a slope with an inclination of smaller than 10°–15° there is no debris movement or it is very little. Grikes do not develop on slopes with an inclination larger than 31° (perhaps 40°). Consequently, peeling can primarily happen on slopes with bedding planes with an inclination of 10° and 40°. However, the inclination of beds is also a

necessary condition of the development of schichtfugenkarren. If there is no bed inclination, the water percolates into the rock along fractures and not along bedding planes.
- Frost weathering promotes separation along schichtfugenkarren. Therefore, where its intensity is larger (for example, at higher elevation) it can substitute the smaller dissolution effect which is caused by the lack of soil.
- Debris receiving capacity is necessary for the transportation of the developed debris. It is large if there are large-sized features (karren, complex karren, giant grikes, shafts) close to debris development sites and their density is large too.

7.3.1.3 Karren Physical Weathering

The karren physical weathering of the karstic rock may involve karren peeling and frost weathering too (if the walls of karren features are denuded during this process), as well as all those karren processes during which debris is produced. From these, the most determinant is the coalescence due to the dissolution of karren features (Veress 2010). Karren features, and thus the development of karren debris can be different on bare karst and soil-covered karst. While on bare karst mainly rinnenkarren and grikes are characteristic, on soil-covered karst, grikes and kamenitzas are specific. During dissolution similar and different karren features can coalesce. Specific, coalescing similar karren features are grikes and pits. Pits can develop in a row or in groups. In the former case, grikes, while in the latter case, basins develop (Veress 2010). A lot of debris is produced in both cases. With the coalescence of grikes, giant grikes are formed which also results in the development of a lot of debris.

Physical weathering is also contributed by the growth of karren features. Thus, during the widening of channels, the dividing walls between them can be cut through, the so developed arches, and then the relics of dividing walls are destroyed too. Cutting through can also be contributed by meandering that develops on the channel floor: at meanders the dividing wall between channels is also cut through. As a result of the process the dividing walls between the channels are destroyed. Similarly, debris is produced when rinnenkarren develop between the cover and the bedrock or by the collapse of the ceiling of karren caves.

Karren debris is produced everywhere during karren formation where fracture density and bedding plane density is high and there is enough time available for the process. If the karren debris is produced on a slope with a larger inclination, it does not accumulate, but it gets into the giant grikes by mass movement that has already been mentioned at peeling. If it is not transported away, but it accumulates, saturation level is located at the contact of the limestone bedrock and the debris or in the debris because of the thickening of the debris. Bare karst is transformed into concealed karst (see chapter on concealed karst) and karren formation stops. However, the debris zone can grow thinner by the dissolution of debris pieces, by which the saturation level is translocated from the debris into the bedrock again.

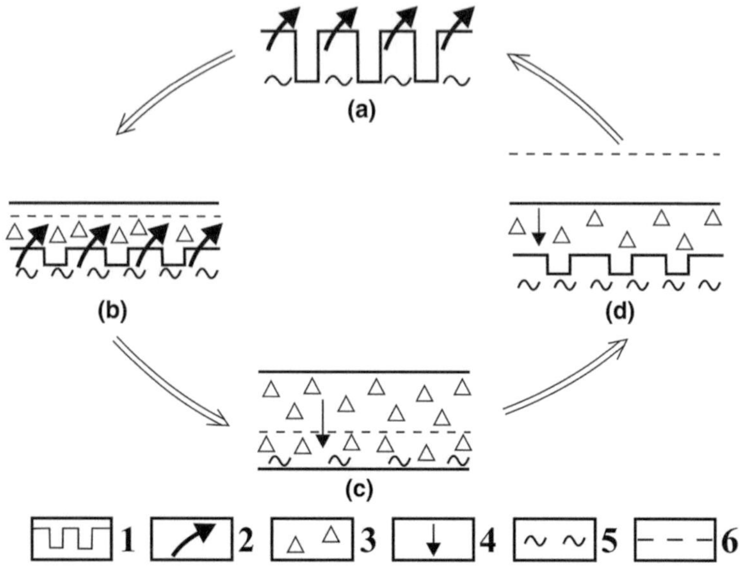

Fig. 7.8 Denudation cycle with karren formation, 1. grike, 2. formation of karren debris, 3. karren debris, 4. the material of the debris is transported away in solution, 5. saturation level, 6. original surface, **a** grike development, **b** formation of karren debris (the surface is a bit elevated), **c** the debris thickens out, the saturation level is located on the bedrock surface, **d** the debris becomes thinner since its material is transported away in solution, the saturation level is situated in the bedrock, a newer generation of grikes develops, then a newer karren denudation cycle begins with the using up of debris

Karren debris can be completely used up if the development of karren debris is restrained, bare karst (or soil-covered karst without debris) develops again (Fig. 7.8).

7.3.1.4 Karren Geomorphic Evolution

Karren Geomorphic Evolution on Unstratified Rock

In lack of stratification no (or only very little) karren debris is produced. Such karst can be found on the islands of the Patagonian archipelago, where the karstic landscape evolution of the terrains surrounded by troughs that developed during the last glaciation can be studied. The karren features are not destroyed, but they deepen. From grikes, giant grikes and from pits, shafts may develop. The process is favoured by the large amount of precipitation, which in itself results in the formation of megakarren (Maire et al. 2009). The surface of the karst is denuded by the growth of karren features. During this, it becomes dissected by deepening karren features (shafts, giant grikes) (Veress et al. 2003). The terrain, especially if it is in an elevated position, is separated into parts and may turn into one having a pinnacle nature. The surface of the unstratified rock can be dissected by roche moutonnées too. In this case, the denudation of roche moutonées primarily happens by superficial dissolution. Mainly their upwind side is denuded by dissolution (Veress et al. 2006). Roche moutonnées may affect the dissolutional denudation of the terrains between them. The water flowing down from their area exerts its dissolution effect on the terrain between them, which is primarily superficial dissolution. Such terrains do not become dissected, but they lower by superficial dissolution (Veress et al. 2003, 2006).

Karren Geomorphic Evolution on Stratified Rock

The initial surface can be a surface with steps or a surface with steps and roche moutonnées. Below the denudation (geomorphic evolution) of stepped surface is dealt with since this is more widespread on glaciokarst. In addition, the landscape evolution of the terrain with roche moutonnées does not significantly differ from that of the terrain with steps. However, roche moutonnées have some specific characteristic features which are the following (Veress 2012).

- Slopes of roche moutonnées with similar inclination are denuded in a similar way from their roof level to the stem of their slope. If the roof level is covered with vegetation, bare karst develop on the bare slope part, while subsoil karren formation takes place on the soil-covered part.
- On the upstream, gentle, smoothed slope of the roche moutonnées, there are more karrens (type B channels, rinnenkarren systems and grikes frequently occur), while the downstream slope is poor in karren features, it is a rough, glacial erosional surface (here, the occurring karren features are type A channels and kamenitzas, the density of which is small). Consequently, mainly the upstream slope is denuded. It becomes dissected with karren, and then it is affected by karren physical weathering. By this, the roche moutonée does not only lower, but the feature that developed during glacial erosion is increasingly modified: the feature having an asymmetric cross-section becomes more and more symmetric. Its lowering is primarily caused by subsoil dissolution at its roof level (if there is soil cover).

In addition to superficial dissolution, the denudation of stepped surface can take place during peeling, but by karren physical weathering too. While scarp fronts are mainly denuded by superficial dissolution (and by frost weathering of non-karstic origin), surfaces with bedding planes are destroyed by superficial dissolution, by peeling and by karren physical weathering too. As it has already been mentioned, peeling primarily takes place on surfaces with bedding planes having an inclination of 10°–40°.

The denudation of the steps of the step series constituting the stepped surface may happen in the following ways by peeling:

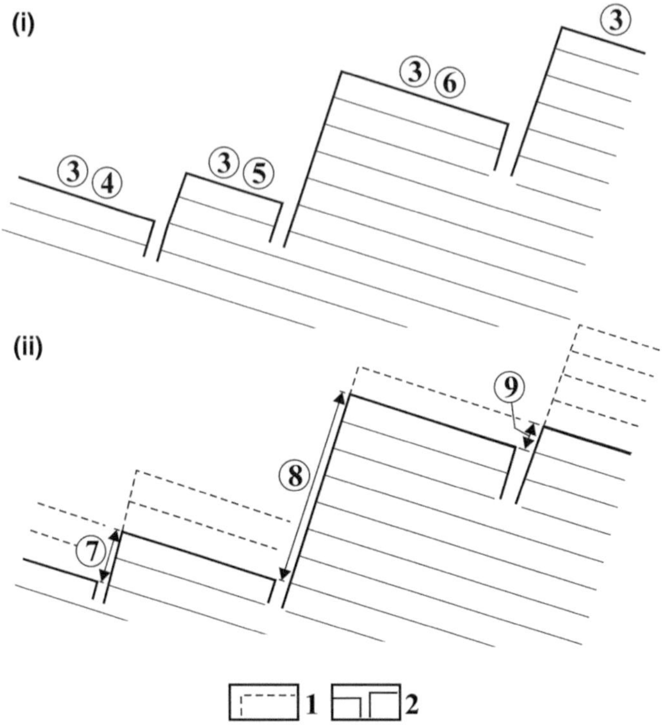

Fig. 7.9 The height changes of scarp fronts during the peeling of the beds of the bedding planes with various intensity, 1. destroyed bed unit, 2. giant grike, grike, 3. the denudation of the bedding plane is caused by grike-schichtfugenkarren assemblage, 4. peeling is similar to the one taking place on the adjacent step with upper position, 5. the degree of peeling is larger than the one taking place on the adjacent step with upper position, 6. the extent of peeling is smaller than the one taking place on the adjacent step with upper position, 7. the height of scarp fronts does not change during peeling, 8. the height of scarp fronts increases during peeling, 9. the height of scarp fronts decreases during peeling, (**i**) original state, (**ii**) present state

- The denudation rates of the adjacent surfaces with bedding planes constituting the step series may be different or the same. Where from two adjacent surfaces with bedding planes the denudation rate of the one with a lower position is smaller, the height difference of the scarp between them decreases (Fig. 7.9 right-hand part). The two surfaces with bedding planes having different elevations gradually develop into surfaces with the same level. Where from two adjacent surfaces with bedding planes the denudation rate of the one with a lower position is larger, the height difference of the scarp between them increases (Fig. 7.9 middle part), where the denudation rates are similar, the height of the steps does not change (Fig. 7.9 left-hand part).
- The denudation rates of the surfaces with bedding planes constituting the step series may increase or decrease in one direction. The height difference of scarp fronts continually decreases in the direction of the decrease (Fig. 7.10b).

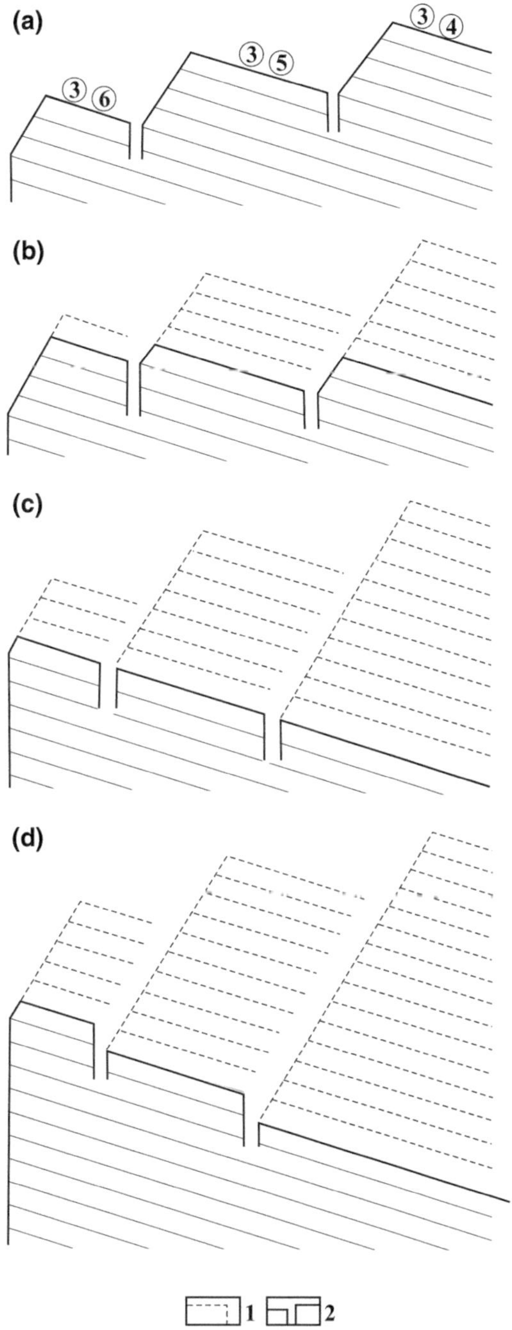

Fig. 7.10 The denudation of steps in case of bed peeling that has an increasing or decreasing intensity in one direction. **a** original state, **b** bed peeling decreases towards lower steps, by which the height of the scarp front of lower steps decreases, while that of higher ones decreases to a less extent, **c** by the different denudation of steps with various heights, the bedding plane surfaces of former steps constitute a common level, **d** the greater peeling rate of the beds of higher steps remains, higher steps will be the lower ones, 1. destroyed bed unit, 2. giant grike, 3. Grike—schichtfugenkarren on the bedding plane, 4. intensive bed peeling, 5. slow bed peeling, 6. bed peeling is extremely slow

Because of the faster denudation of the beds of steps with higher elevation, the bedding planes of those beds of the upper steps are exposed to the surface which constitutes the step surface at lower steps or they were closer to the former surface (Fig. 7.10c). The surface with bedding planes that was formed by peeling is becoming increasingly wider. The wider surface with bedding plane survives if peeling is of similar intensity in its whole area and when peeling reaches the lowest bed of the steps in the whole area of the step series. The so developed surface is hereinafter called base bedding plane surface (Fig. 7.11). If peeling keeps its intensity in the area of former higher steps, newer steps (karren steps) develop, where the dip direction of scarp fronts will be identical with the inclination of beds (Fig. 7.10d).

– Peeling can take place on a surface where the beds are inclined towards the lower steps (Figs. 7.11 and 7.12) and on a surface where they are inclined towards the higher (Fig. 7.13). Cases of the development of base surface with bedding planes may be the following:

  - When the beds are inclined towards lower steps and if peeling takes place on the beds of all steps (mainly if peeling is faster on increasingly higher steps), increasingly larger (wider) surfaces with bedding planes develop (Figs. 7.11 and 7.12i). An oblique, plain surface (base bedding plane) develops on the former stepped surface which is the bedding plane of the bed constituting the lowest step of the step series. If peeling occurs on a step or steps with lower position, although a wider bedding plane may be formed by the exposure of the lower beds, its width will be smaller than the expansion of the steps. However, the height of the scarp front of the step that is not (or only less) affected by peeling increases (Fig. 7.12ii). This restrains, hinders the evolution of base bedding plane surfaces.
  - When the beds are inclined towards the steps of higher position, peeling can take place in two ways too: It happens either at the lowest steps and it is the fastest here or at the uppermost. If it takes place at the lower steps, higher steps survive. Because of this, increasingly higher scarp front develops at the non-denuding step by the peeling of lower steps. The expansion of the bedding plane which is exposed by peeling is small, which is surrounded by a higher scarp front because of the denudation of lower steps. Above the latter, several steps can align (Fig. 7.13i). When peeling takes place at the uppermost step or it is the fastest here, following its denudation, the exposing bed of the step with lower position can be denuded at the higher step too. Consequently, the bedding planes of beds with increasingly lower position can be exposed. By this, the number of steps decreases, the surface with bedding planes of the lowest step will be increasingly wider. A base bedding plane surface develops which is only bordered by a steep surface (Fig. 7.13ii).

In case of karren physical weathering, dissection can not only take place along bedding planes. Surfaces formed by physical weathering will not necessarily be bedding planes (or only their larger and smaller parts). By this, surfaces enclosing

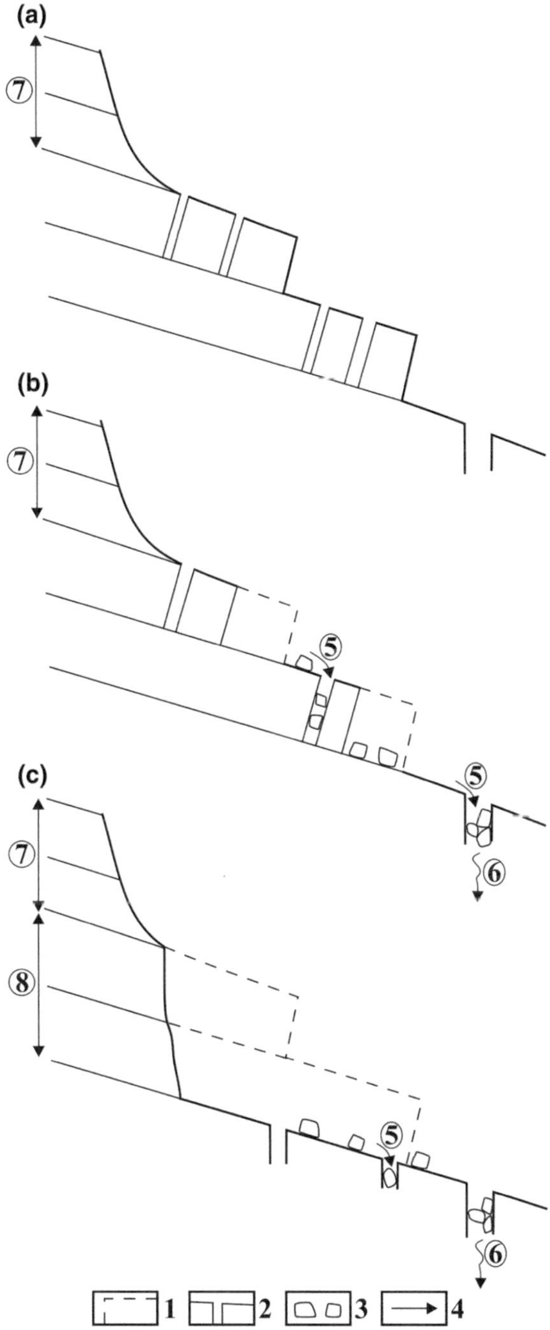

Fig. 7.11 Development of base bedding plane surface, when the beds are inclined towards lower steps, **a** grikes develop, **b** blocks are separated from the beds along grikes, **c** the base bedding plane surface develops, 1. destroyed bed unit, 2. grike, 3. debris, 4. material transport, 5. material transport in debris, 6. material transport in solution, 7. slope formed by glacial erosion, 8. slope developed by peeling

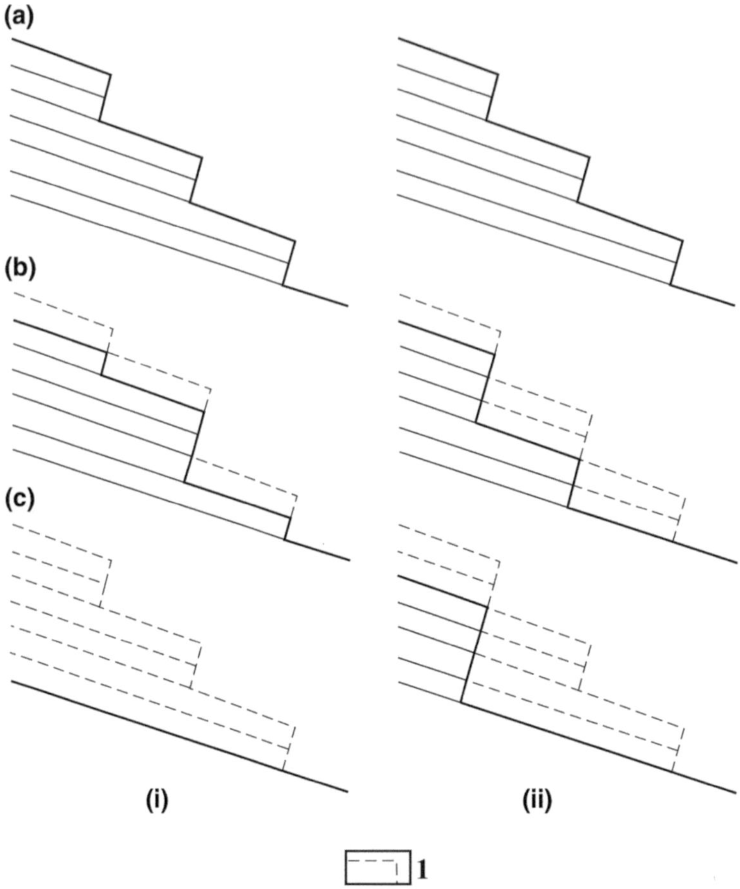

Fig. 7.12 Geomorphic evolution (peeling) if the beds are inclined towards the lower steps and the peeling of beds with upper position is faster (**i**) and when the peeling of beds with lower position is faster (**ii**), 1. destroyed bed unit, **a** original state, **b** the intensity of peeling is similar on the steps, the height of the scarp front does not change, **i**(**c**) the intensity of peeling increases towards higher steps, the height of the scarp front decreases, only one wide surface with bedding plane develops, **ii**(**c**) the intensity of peeling increases towards lower steps, the height of the surviving scarp front increases, the surface with bedding planes will be less wide

an angle with the bedding planes also develop. As a result of the process, the stepped surface is transformed: some former slopes are separated into several slope parts with different inclination. The scarp fronts of stepped surfaces are separated into several slope parts, they become gentler, while surfaces with bedding planes are dissected by slopes of various inclination. A plain, but dissected, oblique surface may be formed which is only partly bedding plane.

One part of rock surfaces are covered by karren debris. At the stem of steeper slopes, frost weathering debris appears too. During the formation of debris, the surface subsides, but the newer, younger surfaces (either the bedrock surface or the surface with debris) are different from those that developed during former glacial

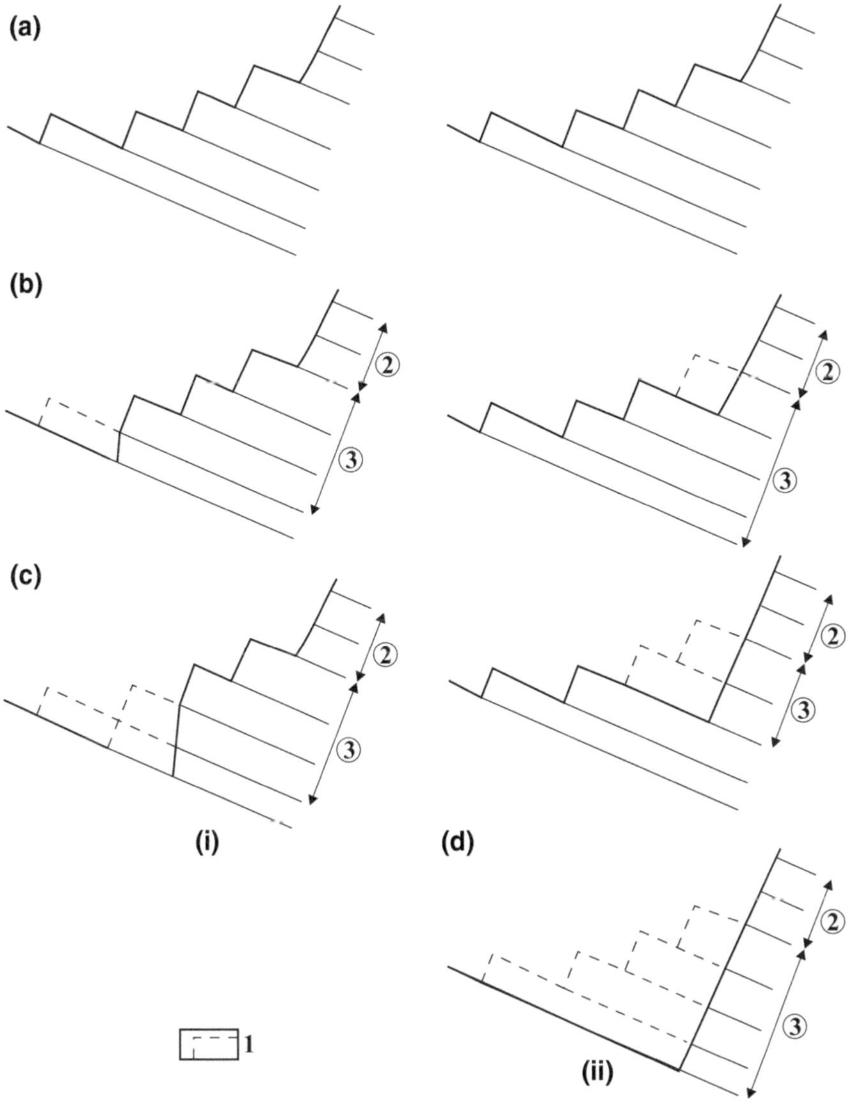

Fig. 7.13 Geomorphic evolution (peeling) if the beds are inclined towards the higher steps, but the peeling of beds with lower position (**i**) or with upper position is faster (**ii**). 1. destroyed bed unit, 2. slope formed by glacial erosion, 3. slope formed by the peeling of the bed, **a** stepped surface, **i**(**b**)–**i**(**c**) beds are destroyed at lower steps, during this a widening bedding plane and a high scarp front develops, **ii**(**b**)–**ii**(**c**) the beds of upper steps are peeled more intensively, by which wider and wider bedding planes are exposed, **ii**(**d**) only one bedding plane surface develops under the former stepped surface

erosion. The debris that developed as a result of karren physical weathering gives way to concealed karstification.

The difference from glacial erosional surfaces is not only caused by the development of surfaces with varied positions during karren denudation, but also by the fact that where the debris thickens out, the dissolution of the surface of the bedrock is interrupted or it is of small intensity, while where the debris is thin or absent, dissolution is more intensive. Consequently, surfaces with various elevation are formed adjacent to each other. Since peeling can be replaced by karren physical weathering, base bedding plane surfaces that developed during peeling can become dissected too.

Karren Geomorphic Evolution on a Surface that Is Becoming Bare

According to Kunaver (2009), subglacial karren can survive under the moraine. However, according to Williams (1966), there are no karren under the moraine, while there are karren on terrains exempt from moraines, and the size and frequency of karren increases farther from the margin of moraine cover. Taken the above-mentioned things into consideration, the denudation of the cover can be accompanied by the following karren geomorphic evolution.

- As a consequence of cover denudation, karren developing on bare surface or under a cover that is thinning out overlie subglacial karren. During the process, the frequency of karren increases, older karren may be fractured.
- No karren develop under the moraine. The denudation of the moraine can be partial or complete. In case of partial denudation, the moraine becomes thinner towards its margin. Under the increasingly thinner moraine karren formation takes place. Under the thinning moraine, karren of seepage origin develop (for example kamenitzas and grikes), which are overlain by karren of flow origin (trittkarren, rinnenkarren) of bare surfaces during the complete denudation of the cover. In case of complete denudation, the moraine is completely denuded at a given place, by which the margin of moraine cover retreats. The cover is shrinking, the bare terrain is expanding. In case of a complete denudation of the moraine, karren are increasingly older moving away from the present margin of moraine. There is a surface smoothed by dissolution at the margin of the present moraine cover, but increasingly farther from the moraine margin karren developing on bare surface occur in an increasingly larger size and in increasingly larger numbers. In the former case, the beginning of karren formation is not different (however the development environment is different), but in the latter case, it is.

7.3.2 *Denudation on Concealed Karst*

7.3.2.1 The Characteristics of Denudation on Concealed Karst

On concealed karst, material transport, and thus denudation takes place at three levels which are the following:

- The surface is denuded. The soil (if there is any) or the fine-grained, clayey sediment is mainly transported away on the surface by pluvial erosion or by fluvial erosion.
- The cover with limestone debris is denuded. After its dissolution, the cover becomes thinner as a consequence the surface becomes lower by subsidence.
- Undercover, the bedrock can be denuded by dissolution. The process can include the superficial dissolution of the bedrock surface and the development of karren, passages and pits if the cover is thin, but it has a calcareous content or it is thick, but and of non-karstic material. Superficial dissolution results in surface subsidence, while the formation of karren causes the development of subsidence dolines (Veress 2016a). The development of dolines also contributes to the erosional denudation of the surface.

Therefore, on concealed karst, material transport is of many kinds, and at various places, some ways of transport can be absent for shorter and longer time, in addition, the proportion of material transport ways may also be different at various places.

The material of the cover of concealed karst can originate from four sources which are the following:

- From primary or secondary (reworked) moraine.
- From fluvioglacial material.
- From the material of mass movements.
- From karren debris.

Concealed karst can lose its material in four ways. These are the following:

- Its material is transported in solution.
- By the reworking of the soil or of the cover, which takes place during water transport (pluvial erosion or fluvial erosion).
- Through subsidence dolines by suffosion and rock fall.
- By water transport that is directed towards subsidence dolines.

On concealed karst, the lowering of the surface is mainly contributed by the denudation of the cover. The material of the cover gets into the karst through subsidence dolines. Therefore, for the erosional lowering of the surface the presence of subsidence dolines is necessary, however, the dissolution of the bedrock is a precondition of the development of these dolines. According to Williams (1966), since the inward percolating water becomes saturated, under a moraine of calcareous material with a thickness of 45.72 cm, there is only a limited dissolution on the bedrock. According to Jones (1965), dissolution also occurs under moraine in case of CO_2 of biological origin. According to Sweeting (1973), no dissolution occurs on the bedrock in case of a cover with calcareous material thicker than 2 m. According to Trudgill (1972, 1975, 1985), under the cover, the effectiveness of dissolution depends on the pH value of the inward percolating water and the calcareous content of the cover. Consequently, under a cover with limestone debris, dissolution (the development of karren and subsidence dolines) depends on CO_2 production as well as on the thickness and composition of the cover. However, no dissolution is needed on the bedrock for the formation of subsidence dolines if features with sediment receiving ability had already developed preceding covering with debris.

The thickness of the epikarst can provide information on the depth of saturation level (thus, under which no dissolution occurs). This can even reach a thickness of 10 m on temperate karst (Al-fares et al. 2002). The thickness of the epikarst depends on several factors. Thus, on the amount of the inward percolating water, on the amount of CO_2, on the presence of the cover (Tooth and Fairchild 2003) and on the structure of the rock (Williams 1983, 2008). In Alpine areas, where fissures widened as a result of the elevation of the area, it can reach 30 m (Williams 2008). On glaciokarst, the thickness of epikarst may be some metres (Ford and Williams 2007). According to our opinion, saturation depth is well-represented by the depth of the karren features of bare karst. As the depth of pitkarren can even reach 2–3 m, the saturation level can be located at a depth of 3 m. This means that in case of a debris thickness of some metres, the saturation level can be situated at the surface of the bedrock. (Disregarding the fact that the surface of the debris can increase the length of percolation, and thus it can reduce the distance of saturation from the surface.) It is well-known that the amount of CO_2 is mainly of soil origin (Jakucs 1977; Trudgill 1985). Soil has a patchy development at a higher altitude. At a given place, in case of the same debris thickness, the saturation level is located under the soil patch in the bedrock or at its surface, while in the lack of soil, in case of the same thickness, it is situated in the debris. In case of increasingly lower altitudes, the inward percolating water can still keep its dissolution capacity under a debris of larger thickness, which can even be several metres because of the increasing amount of CO_2 of soil origin.

According to Veress (2009, 2016a), subsidence dolines can be syngenetic (the dissolution feature of the bedrock and the doline that developed in the cover over it are of the same age) and postgenetic (the features of the bedrock are older than the feature that developed in the cover). While syngenetic dolines are formed in a thinner cover (its value, as it has already been mentioned, can probably be several metres), postgenetic dolines can also develop in thicker cover. However, the increase of the thickness of the non-karstic cover also limits their development. However, according to literary data, the majority of subsidence dolines develops on a cover thinner than 10 m (Yuan 1987; Chen 1988) or even thinner than 30 m (Beggs and Ruth 1984; Currin and Barfus 1989).

7.3.2.2 Surface Denudation Depending on the Cover on Concealed Karst

The way of transportation and the amount of material that is transported away in various ways, and thus the denudation nature of the surface depends on the following factors:

- The material of the cover is calcareous rock, but its thickness has a degree that the saturation level is situated deeper than the lower surface of the cover and there are no subglacial karren on the bedrock. This state is called concealed karst with thin debris cover. In this case, karren are formed on the bedrock, while passages develop in the bedrock. Consequently, syngenetic subsidence dolines develop on the cover. The denudation of the cover is contributed by the erosional denudation of the surface, but by the transportation of the cover material in solution too. During erosional denudation, the material of the cover can be reworked on the surface, but it can also be transported into the developing subsidence dolines where it is either accumulated on their floor or it is transported into the karst through passages by suffosion and rock fall. Material transport can take place both from and out of the bedrock also because of dissolution occurring there. The denudation of the surface can happen by lowering (because of erosion), by subsidence (because of dissolution which can take place both in the cover and in the bedrock). If the thickness of the debris increases since toward transportation is of larger degree than transporting away (which can not only happen because of accumulation, but from below, from the bedrock by karren physical weathering too), dissolution ceases on the bedrock. This state is called concealed karst with thick debris cover. The thickening of the cover can also be caused by transportation away of small degree. Transportation away of small extent is contributed by that fact that in the epikarst, the size and density of cavities is small, large-sized caves are absent in the depth. This can contribute to the filling up of the passages of subsidence dolines and then to the upfilling of dolines (Fig. 7.14A). This results in the further thickening of the cover. If the debris is thinning again as a result of the dissolution of debris pieces, subsidence dolines can become reactivated or a newer doline generation develops on the filled surface. If the cover is further thinning (if the degree of transportation away exceeds the degree of towards transportation), the cover is used up. Covered karst is formed into bare karst.
- The cover is calcareous rock, its thickness has a degree that the saturation level is situated in the cover (or at the surface of the bedrock), but there are subglacial karren on the bedrock. Postgenetic subsidence dolines develop on the cover (Fig. 7.14B). The denudation of the cover is contributed by the erosional denudation of the surface, the transportation away of the material of the cover in solution. Erosional denudation is caused by the reworking of the material of the cover or its transportation into subsidence dolines. The denudation of the surface can happen by lowering (because of erosion), by subsidence (because of dissolution which only takes place from the cover) and by suffosion. The thinning

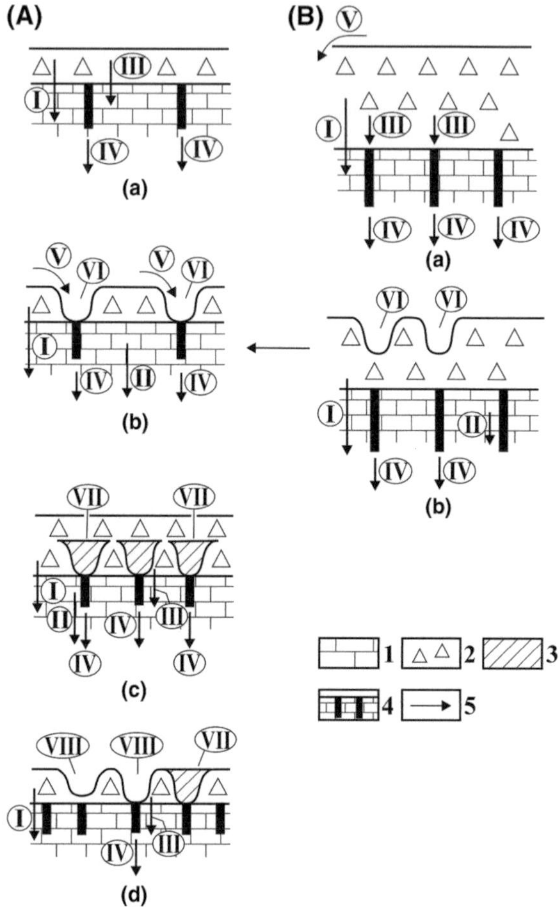

Fig. 7.14 The denudation of the surface during the development of subsidence dolines, 1. limestone, 2. moraine, 3. fill of subsidence doline, 4. karren, 5. material transport, I. transportation of dissolved material from the moraine, II. transportation of dissolved material from limestone, III. suffosional material transport, IV. forwarding of fill of suffosion origin in solution, V. sediment transport by meteoric water, VI. subsidence doline, VII. filled up subsidence doline, VIII. reactivating subsidence doline, **A** there are no subglacial karren in the bedrock, a. karren are formed b. subsidence dolines develop, c. the dolines are filled up and buried, d. filled up dolines are reformed, **B** there are subglacial karren under the moraine, a. the moraine becomes thinner by suffosion and by the dissolution of the debris, b. subsidence dolines develop above karren (pits)

of the cover may have a degree that syngenetic subsidence dolines can develop too. From this phase, the denudation process will be similar to the above presented (Fig. 7.14b–d).

- The material of the cover is calcareous rock, its thickness has a degree that the saturation level is located in the cover (or at the surface of the bedrock), no subglacial karren are present on the bedrock (Fig. 7.15a). This state is called concealed karst with thick debris cover. Concealed karst with thin cover and concealed karst with thick cover can be formed into each other depending on the proportion of material accumulation and transportation away. In case of thick cover, the denudation, and thus the thinning out of the cover are contributed by erosion (this is exclusively reworking of superficial material) and by the transportation of the material of the cover in solution (Fig. 7.15b). The surface becomes lower (by erosion) and subsides (by dissolution). With the thinning out of the cover (the saturation level is moved into the bedrock) dissolution takes place on the bedrock and karren are formed (Fig. 7.15c). In this case, even

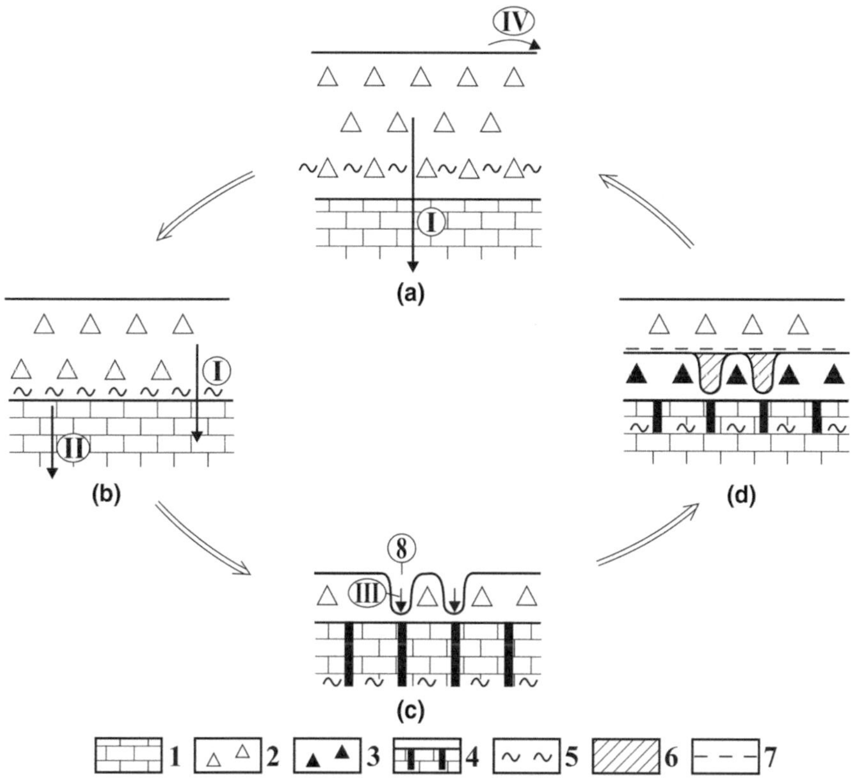

Fig. 7.15 Cyclical surface denudation on concealed karst, 1. limestone, 2. moraine, 3. karren debris, 4. karren, 5. saturation level, 6. fill of subsidence doline, 7. former surface of the bedrock, 8. subsidence doline, I. transportation of the dissolved material from the moraine, II. transportation of dissolved material from limestone, III. material transport by suffosion, IV. material transport on the surface by pluvial erosion, **a** the moraine becomes thinner by dissolution, **b** the saturation level is situated at the limestone surface, dissolution can take place on the limestone surface, **c** the cover is thin, the saturation level is below the limestone surface, karren features and subsidence dolines develop, **d** karren debris is formed, the cover thickens from below, which reaches the thickness of the cover at the beginning of the cycle

syngenetic subsidence dolines can develop too. From this phase on material transport away can take place by suffosion too. If the degree of transporting away is larger than that of towards transport, concealed karst can be transformed into bare karst. If the latter is larger, subsidence dolines become filled up and the debris thickens (Fig. 7.15d). In this case, the denudation cycle is repeated on concealed karst, but the surface and the surface of the bedrock will be lower as compared to the original.

– The material of the cover is non-karstic rock. Both syngenetic and postgenetic subsidence dolines can develop on the cover. The denudation of the cover is contributed by the erosional denudation of the surface and the dissolution of the material of the bedrock. During erosional denudation, the material of the cover is reworked on the surface or it is transported into the dolines. By these processes, the surface is mainly lowering and dissected, but it can also subside to a smaller degree. If the degree of towards transportation is larger than that of away transport, and non-karstic rock arrives at the cover, the terrain can develop into buried-like karst. If the degree of transportation away is larger than that of towards transport, the terrain can be transformed into bare karst. However, if limestone debris is formed from the bedrock, the karst can develop into concealed karst with limestone debris by the using up of the non-karstic cover.

- As a result of the appearance of ground ice, the cover with debris of the concealed karst can temporarily become impermeable (allogenic-like karst type). At increasingly higher altitudes, this state can survive for increasingly longer time. At this time dissolution is interrupted, but superficial erosion takes place. However, its intensity can be varying since superficial streams only exist during melting. But denudation is also influenced by the degree to which debris is frozen.

7.3.2.3 Geomorphic Evolution on Concealed Karst

On concealed karst, the denudation of the surface can take place by subsidence or by the denudation (removal) of the surface. Erosional surface denudation is characteristic of sites with subsidence dolines. The degree of denudation is determined by grain size, doline density, slope angle, and to which extent the karst has cavities. According to Veress (2016a), slope development (existence of a slope with suitable inclination) is necessary for denudation. This can originate from the original inclination of the surface or from slope formation triggered by developing dolines. Following the formation of dolines, pluvial erosion can start around them. The area subject to this effect becomes wider and the process is effective especially if the surface has original inclination. However, following the development of dolines, creeks and also gullies develop by retreating from their margins. These features dissect the surrounding plain surface. The side slopes of gullies are denuded by pluvial erosion (slope development directly caused by channel erosion). Meteoric water also flows into gullies from the surfaces surrounding the gullies. These surfaces are also destroyed by pluvial erosion. This is slope evolution indirectly caused by channel erosion (Fig. 7.16). After all, in the environment of subsidence dolines, the surface is dissected, but it is also destroyed superficially.

Surface subsidence of dissolution origin is of two kinds: If dissolution takes place in the cover, subsidence is direct, if it happens in the bedrock, it is indirect. In the latter case, the surface of the cover gets into a lower position without thinning. Subsidence is areal and its degree is controlled by the amount of CO_2.

On glaciokarst, terrains may occur where the surface is denuded by erosion only (for example, the cover is of non-karstic material). Such terrains may also occur where the surface only subsides since subsidence dolines have not developed yet, or they became filled up or they became denuded with the cover together. (Such areas

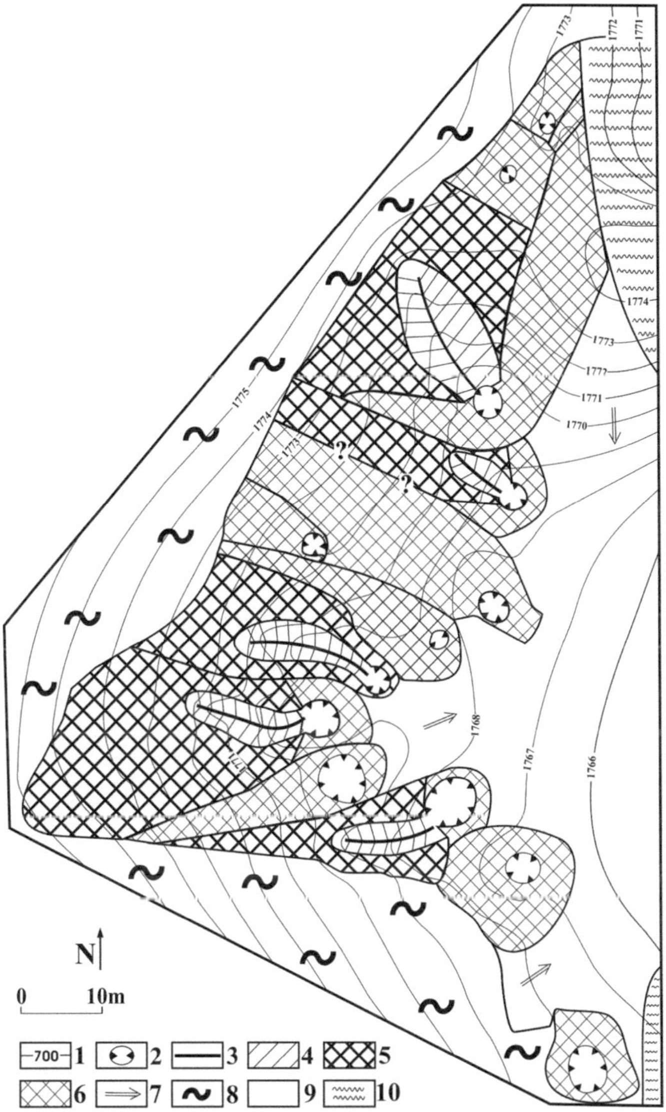

Fig. 7.16 Denudation map of the floor with concealed karst of a paleodepression (Hochschwab paleouvala—modified, Veress 2016a), 1. contour line, 2. suffosion doline, 3. linear erosion, 4. slope development directly caused by linear erosion, 5. slope development indirectly caused by linear erosion, 6. slope shaped by doline formation, 7. sediment transport on the surface, 8. upper accumulation level of the paleodepression, 9. concealed karst, 10. allogenic karst

are those where the calcareous cover is thick and there are no subglacial karren on the bedrock.) There may occur such sites where both subsidence and the denudation of the surface happen together, though their proportion may be extremely different. The proportion of the two kinds of denudation may change locally, but in time too (for example, the amount of CO_2 production or the amount of precipitation may be various seasonally). Such areas are those where the cover is thick, but there are subglacial karren on the bedrock.

It is probable that the degree of denudation (either it is of erosion or dissolution origin) decreases with the increase of elevation. It is caused by the decrease of CO_2 production, the increasingly larger grain size of the cover, the underdevelopment of the epikarst. With the decrease of altitude, the degree of denudation increases as a result of the increase of CO_2 production, the decrease of grain size, and the maturity of the epikarst. As a result of the change of cover thickness and that of CO_2 production as well as the local changes of the density of subsidence dolines, the degree of superficial denudation will not be zonal, but it will be mosaical.

7.3.3 Geomorphic Evolution on Allogenic Karst

Allogenic karst can develop in various depressions (giant doline, polje, rock basin). Consequently, allogenic karst is isolated, island-like on glaciokarst. The impermeable surface is constituted by the impermeable cover or the exposing non-karstic rock (Veress 2016b). The cover is thick and impermeable. Ponor(s) develops (develop) at rock boundary. A complex slope evolution occurs: slope denudation creates a slope inclined towards the direction of the depression, but the valleys that are connected to the ponor also control slope evolution (Veress 2016a).

A significant amount of sediment can be transported into the karst through ponors. The denudation of the surface is contributed by pluvial erosion, channel erosion and mass movements. However, streams can also transport dissolved material into the karst through the ponors.

On allogenic karst, landscape evolution is affected by ponor clogging or material balance (the proportion of inward transport and transportation away). The surface is dissected as a result of the downcutting of streams in case of opened ponor passages. If the ponor(s) become clogged, the denudation of the surface will be areal, and thus locally senile surfaces develop: the valleys become wider, the interfluves are used up, the relict ridges become lower. The unclogging of ponor may favour the erosional renewal of the surface. If the ponor does not unclog, the surface becomes stable. If a newer ponor is formed at a lower elevation, the denudation of the surface can be renewed.

If inward transport exceeds transportation away, the ponors become clogged, the surface rises by fill, and then they become covered. Gomance polje can be an example for this (Žebre et al. 2016). On a small-inclined karstic surface, the rock boundary and the ponor site are mainly relocated laterally, while on steeper slopes, they are mostly vertically relocated (Žebre et al. 2016; Veress 2016a). If transportation away exceeds inward transport, the surface is denuded. During denudation, the surface becomes lower and dissected. The rock boundary is located at a lower elevation, a newer ponor develops, such as in one of the poljes of the Northern Alps Molasse Zone (Goeppert et al. 2011).

As a result of surface denudation, the superficial deposit fill becomes thinner, is separated into patches and bare karst patches appear, in the end, the superficial deposit is completely transported into the karst. If the proportion of inward transport and transportation away changes as compared to each other, since the amount of inward transported material or that of away transported material changes, the surface may become lower, and then gets into a more elevated position (Veress 2016a, 2016b).

7.4 Geomorphic Evolution Depending on Glacial Erosional Features

7.4.1 Recent (Postglacial) Landscape Evolution

The side slopes of cirques and troughs retreat to a small extent, they become gentler and are dissected. Retreating is contributed by superficial dissolution and the development of wandkarren, but it can also be promoted by the crumbling of bed units above schichtfugenkarren. Dissection is contributed by avalanches and mountain falls. Combe-ridges may become lower and dissected by karren and other karstic and non-karstic features (rock shelters with circular margin).

If accumulation exceeds the transportation into the karst, filling up happens on the floor of cirques, debris cones may cover their complete area, and subsidence dolines may be buried. On the floor of cirques, mainly postgenetic subsidence dolines can develop. In this case, the extent of transportation away may exceed the degree of inwards transportation. Transportation away may be intensive since subsidence dolines are mainly formed above giant grikes and shafts (Fig. 2.10b). If the density of dolines is small, the floor of the cirque becomes dissected, if the density is large (especially if the cover is fine-grained) the floor becomes smoothed by denuding uniformly. During the denudation of the cover, closed depressions may develop in the debris, on the floor of the cirque, which are the pseudo Depressions of Superficial Deposit (DSD) (Veress 2016a). During debris formation, concealed karstic terrain parts are formed, while concealed karstic terrains may be transformed into bare karst for a longer–shorter time. Consequently, the pattern of the two karst types may be continually reorganized in the area of the cirque. Either on bare karst or concealed karst, especially on the floor of more elevated cirques, dissolution is small, the denudation of the surface is mainly caused by the transportation of the debris into the karst. Uneven glacial and preglacial floor may be exposed. The expansion of the floor is limited by the slopes of the cirque and the debris cones that are formed at the feet of the slopes.

On bare karst (but on soil-covered karst too), increasingly wider surfaces with bedding planes develop by the denudation of the steps in the trough. The surface of the bedrock becomes lower (by karren formation and by the dissolution of the surface), the valley floor widens. The denudation of the beds constituting the steps are limited by the side slopes of troughs. During peeling, increasingly longer and steeper valley slopes develop. However, valley floors are denuded into such surfaces that are less and less dissected by steps, the limit of the growth of which are determined by valley sides. The number of steps decreases, the expansion of surfaces with bedding planes increases. However, if karren debris is produced, the surface of the bedrock is formed into an irregular surface. The produced debris is transported into the karst by subsidence dolines.

On concealed karst, the denudation of debris of glacial origin takes place. On concealed karst, the bedrock and the surface with debris becomes lower. The lowering is caused by the subsidence of the surface (the dissolution of the debris, the dissolution of the bedrock) and by the erosional denudation of the surface. With the using up of the moraine, concealed karst is transformed into bare karst.

In the area of ice caps, peeling is not limited by the slopes that developed by glacial erosion. The surfaces with bedding planes of stepped surfaces will be much wider. As a result of the smaller height differences of stepped surfaces, the development of base bedding plane surfaces may take place by the denudation of fewer beds. On less dissected surface, the surface that was formed by karren debris formation is less different from the stepped surface that developed by former glacial erosion than in the troughs. The cover is formed by frost weathering and mass movement to a less extent, but it is rather supplied by karren physical weathering or during moraine reworking.

Depressions dissecting the glacial erosional surface are the sediment recipients of their environment. Consequently, they are the main sediment receiving and forwarding sites of glacial erosional surfaces. However, the depressions are also the sites of local erosional denudation of the glacial erosional surface, and thus these surfaces contribute to the different, heterogeneous superficial denudation of the glaciokarst. If accumulation exceeds transportation away in the features, they become filled up (Veress 2016a, b). This occurs if a lot of sediment arrives and there are no subsidence dolines or ponors on the floor. If there is no soluble debris, even transportation in solution is absent. However, according to Ford (1996), filled depressions also keep their hydrological function thus, they become exhumed sooner or later. If transportation into the depression is of smaller degree than transportation away, the depression is exhumed. Exhuming depressions are called real DSDs by Veress (2016a). The karst of DSDs can be concealed karst or allogenic karst. According to the degree of their exhumation, they may be exhuming to a small, medium, and large extent. The depression is exhumed to a large degree, if it is increasingly filled by impermeable cover (ponors and streams may develop). It is also exhumed to a large degree if more and more subsidence dolines develop on its floor to which a lot of or more and more streams are connected. However, exhumation of larger degree is contributed by if more and larger dissolution features (karren, giant grike, shaft) opening up to the surface are present in the bedrock.

In case of a material transport of large degree, the chance for the complete exhumation of the depression increases. The floor of the depression can be transformed into bare karst, and then into concealed karst during repeating inward transport or sediment formation. Since the cover is transported into the depressions from the surrounding terrains, the depressions can contribute to the partial or complete denudation of the cover of glacial erosional surfaces.

In case of infilling depressions, the dissected nature of the glacial erosional surface decreases during accumulation. Presumably, this is a temporary state since the cover of permeable material is later transported into the karst in solution, through subsidence dolines in the form of debris and in case of impermeable cover through the ponors. Ponors above the floors of depressions refer to the alternation of filling up and exhumation (Goeppert et al. 2011; Veress 2016a). Repeating exhumation results in the cyclical denudation of surfaces surrounding the DSD.

7.4.2 *Future Geomorphic Evolution*

The karstic development of glacial erosional surfaces is young, at most some thousand years old. On these surfaces, denudation is slow too. As a result of the young age and slow nature of denudation and since the position of these surfaces is relatively elevated as compared to the base level, there is a chance for the development of senile surfaces in their area only in the distant future. Their future development is described and separated into two stages. In the first stage, the glacial erosional features still survive, but they are partly modified, while in the second phase the denudation of the karst surface reaches the base level (senile state).

7.4.2.1 The First Phase of Landscape Evolution

In the first phase, the denudation of stepped surfaces and of surfaces bearing karstic depressions are outlined.

Geomorphic Evolution of Stepped Surfaces

The valley floors (and partly the slopes too) are formed into plain, the asymmetry of troughs increases, the valley side parts that developed by peeling will be much steeper. If the beds are inclined towards the lower steps (Fig. 7.12i) and the peeling intensity of upper beds is larger if the process lasts for a long time that is needed, the base bedding plane surface is formed. The floor of the trough is constituted by this surface, which is connected to the valley side that developed by the peeling of beds (Fig. 7.17ii.a). However, this valley side part is always of higher elevation than the plain valley floor with bedding planes and it is located at the upper termination of the latter. If the beds are inclined towards the more elevated beds and the peeling is larger at the top (Fig. 7.13ii) the base bedding plane surface also develop on the valley floor (Fig. 7.17ii.b). The valley side that developed by the

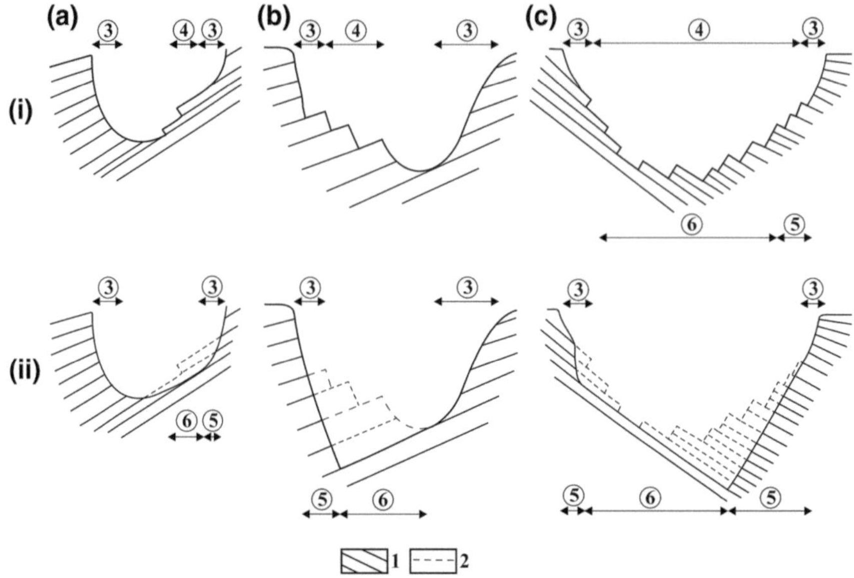

Fig. 7.17 The development of base bedding plane surface in troughs, 1. beds, 2. former surface, 3. steep valley side with rock wall, 4. stepped valley side and valley floor, 5. valley side that developed by the peeling of beds, 6. base bedding plane surface, **i** valley formed by glacier and its steps, **ii** glacier valley modified by peeling, a. development of base bedding plane surface, when the bedding planes of the steps are inclined towards the valley centre, b. development of base bedding plane surface, when the bedding planes of the steps dip towards the valley margin, c. development of base bedding plane surface, when the bedding planes of the steps dip towards the valley centre and valley margin too

peeling of the beds is only partly more elevated than the plain surface with bedding planes. This valley side is situated at the lower end of the surface with bedding planes that falls in dip direction.

Finally, if the beds are inclined to the lower steps on one slope of the valley, while they dip towards the higher scarps on the opposite side (Fig. 7.17i.c) a base bedding plane surface develops (Fig. 7.17ii.c) with a large expansion (since the lower parts of both side slopes of the valley took part in its development). The steeper valley side part that developed by the peeling of the beds is situated at the lower part falling in dip direction of the plain valley floor.

The above-described landscape evolution can only take place if the produced debris is not accumulated on the surface, but it is transported into the karst.

The Landscape Evolution of Surfaces Bearing Depressions

From glacial erosional surfaces, the cover is partly reworked into depressions. The cover is transported from depressions into the karst. If the transportation into the karst is of larger intensity, the surface subsides in the area of depressions and the cover can be completely denuded. Later, debris can be produced on the bare floor during karren formation (concealed karst with thin debris). If transportation into the depressions is larger, the depression can be filled up. Depending on the type of inward transported sediment, the depression floor can be transformed into concealed karst with thick debris or into buried-like karst (in case of non-karstic sediment).

If the surrounding terrain is transformed into bare karst and if its denudation is fast, the depressions can be truncated.

7.4.2.2 The Second Phase of Geomorphic Evolution of the Karst

The dissected nature of terrains with glacier valleys can increase or decrease. It increases if the material transport from the floor into the karst exceeds the amount of material originating from valley sides. It decreases if more material is transported from the valley sides to the valley floors than is transported from the valley sides into the karst. In this case, the valley floors become filled up and the combe-ridges become lower by denudation. The dissected nature of the surface can decrease. On the contrary, on free glacial erosional surfaces, the dissected nature decreases only to a small extent (this happens by the using up of scarp fronts), surfaces being parallel to themselves develop during the denudation of the surface. Concealed karst develops in increasingly larger areas. The surface of glaciokarst, on which concealed karst is increasingly widespread and its dissected nature is smaller and smaller, becomes lower.

On both glacial erosional surfaces, concealed karst with a cover of thin debris can be transformed into a concealed karst with a cover of thick debris. The two concealed karst states can be changed into each other several times, during which the surface of the bedrock becomes lower. The surface gets closer and closer to the karst water level. Concealed karst with a cover of thin debris is increasingly changed by that of with a cover of thick debris since the cavities of the karst become more and more filled. If the cavities are completely filled, a karst with a cover of thick debris becomes predominant. From this time on, material transport from the surface can only take place in the form of solution.

This phase of concealed karst with a cover of thick debris can be only interrupted by phases of concealed karst with a cover of thin debris for shorter and shorter times. At this time, subsidence dolines can develop because of the dissolution of the bedrock, the sediment forwarding capacity of which is small. However, they become destroyed within a short time since the features that developed on the bedrock are small, and thus they become filled with the material of the cover

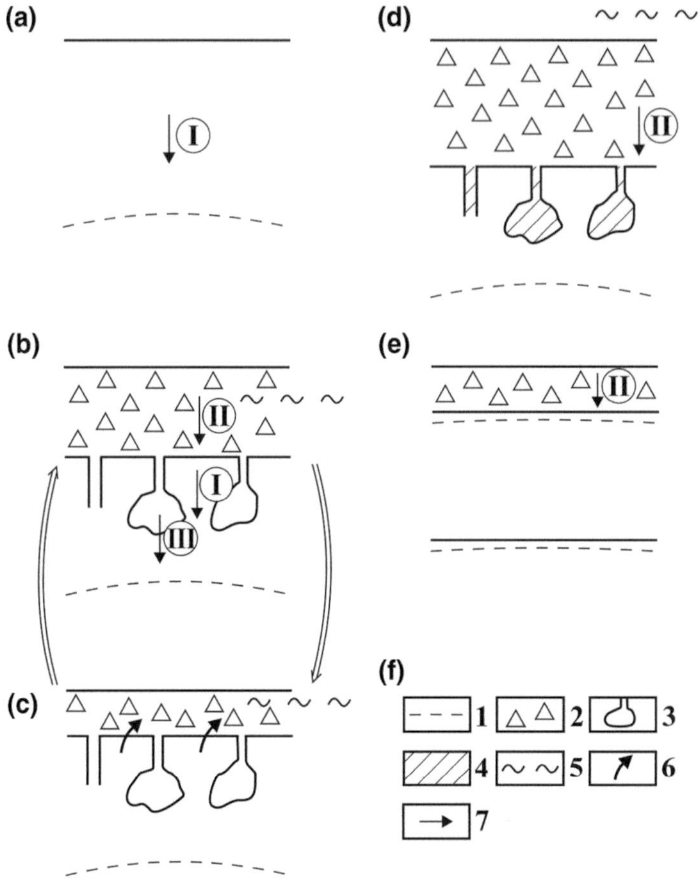

Fig. 7.18 Denudation of bare surface by karren formation until the karst water level, 1. karst water level, 2. debris of karren origin, 3. karren, cavity, 4. fill, 5. former surface of the bedrock, 6. formation of karren debris, 7. material transport, I. material transport from the bedrock, II. material transport from the cover, III. material transport from the fill of the passages of the bedrock, **a** bare surface, **b** concealed karst with thick debris, **c** concealed karst with thin debris, **d** the surface of the karst subsides, **e** the surface of the bedrock is at the base level, but there is debris on it, **f** with the dissolution of the debris, the karst becomes bare

quickly. When the surface of the karst reaches the karst water level, it is transformed into bare karst since the material of the surviving cover gets into solution, but it is not supplied from below anymore (Fig. 7.18). By this time, the karst has already lost its glacial erosional features. The karst can reach this state from the loss of the moraine (and if its elevation does not change during this time) in 50 million years with a denudation (dissolution) rate of 0.01 mm/year if the karst water level is in a 500 m-depth as compared to the surface. Under similar conditions, but at a rate of 0.1 mm/year this occurs in 5 million years.

7.5 Conclusions

On glaciokarst, denudation and geomorphic evolution are affected by the type of glacial erosion, the preglacial karstification, the glacial erosion (erosional features and moraine), altitude in the area of high mountains (a given place which is in a geomorphological belt). Landscape evolution is of short time on glaciokarst thus, glacial erosional features have hardly been transformed and survive for a long time. Both karstic and non-karstic processes take part in the denudation of the surface (in the area of free glacial erosion, mainly karstic processes, while in the area of linear glacial erosion, in addition to karstic processes, non-karstic processes can be significant too, material transport primarily happens into the karst (or more and more) and mainly in solution. In the beginning, especially mechanical material transport is more significant, where cover is present, then transportation in solution will be more and more predominant.

The degree of material transport depends on the suitability of the surface for becoming denuded and on the sediment receiving capacity of the karst. This latter is given by the material forwarding ability and storing capacity. Storing capacity depends on the size, density and sediment forwarding capacity of caves.

The denudation of the surface is mainly concentrated on glacier valleys. On glacial erosional surfaces, especially where altitude changes to a large degree, various karst types occur (though the karst types can be transformed into each other) and by this, the way and rate of surface denudation may be different on various parts of the karst. On glaciokarst, the denudation of the surface is mosaical. The karst type is transformed if debris is produced (bare karst is transformed into concealed karst), if the debris is used up (concealed karst is transformed into bare karst) or if the impermeable bed is denuded (allogenic karst is denuded into bare karst or concealed karst). Therefore, the way and rate of denudation can be different at the same place too.

At present, the denudation of the surface is accompanied by two results: on the one hand, the elevation of the surface decreases, on the other hand, the glacial erosional features and the preglacial karst features are denuded and transformed by the fact that postglacial features develop, grow, are denuded, and then regenerate.

On bare karst (but on soil-covered karst too), the denudation of the surface takes place by superficial dissolution, peeling and by karren physical weathering (in case of the latter two denudation ways, the process is connected to the development of karren features). During peeling, bed units are denuded, wider and wider bedding plane surfaces may develop if the beds of more elevated steps are denuded faster than the lower. Since the debris is not always transported away during their evolution, the stepped surfaces of the bare (soil-covered) karst are separated into more and more wider surfaces with bedding planes and into concealed kart terrains during the process. During karren, physical weathering, irregular surfaces are formed with the coalescence of various karren features, but stepped surfaces become more and more smoothed. Especially, in case of this latter process, there is a great chance of the transformation of bare karst or soil-covered karst into concealed karst.

On concealed karst, the landscape evolution can be one-phase (terrain covered by moraine), two-phase (terrain covered by reworking of moraine), and continuous (terrains of karren debris development). Following the using up of the former cover (moraine), continuous geomorphic evolution is increasingly predominant as a result of karren formation. On concealed karst, the cover can be denuded by erosion or by dissolution. The material of erosional denudation is transported into the karst through subsidence dolines. The denudation of the surface is dual: the cover with debris is denuded by dissolution and erosion, while the bedrock is lowering by dissolution. During the thinning out of the cover, concealed karst with thin cover develops. In this case, the debris is supplied from below by karren debris formation (but it can be supplied from above by reworking). Concealed karst with thin cover is transformed into concealed karst with thick cover with the thickening of the debris. The concealed karst of two types can be transformed into each other several times.

On glaciokarst, allogenic karst is island-like and lasts for a short time. With the denudation of the impermeable cover, it can be transformed into bare karst or concealed karst. Its surface is primarily denuded by erosion (fluvial and pluvial).

As a result of future geomorphic evolution, base bedding plane surfaces develop on the stepped surfaces of glacial erosional surfaces. These are formed at the lowest surface sections of glacial erosional features. These surfaces develop faster on free glacial erosional surfaces. The degree of debris cover especially increases on surfaces of linear glacial erosion. During landscape evolution, free glacial erosional surfaces create surfaces being parallel with themselves, and thus they become lower. Base bedding plane surfaces developing on linear glacial erosional terrains modify the shape of the troughs which is also contributed by the denudation of valley sides. In valleys where the transportation away of the material exceed accumulation (this is valid at the occurring sites of ponors and subsidence dolines thus, in valley sections with paleokarstic depressions), the dissected nature can increase, where it is smaller, it can decrease. On glaciokarst dissected by glacier valleys, as a consequence of the filling up of valley floors and the denudation of combe-ridges, the dissected nature of the surface decreases, but it becomes lower too. During lowering, in case of a tectonical rest of long duration, the surface can reach the karst water level.

References

Adamson KR, Woodward JC, Hughes PD (2014) Glaciers and rivers: Pleistocene uncoupling in a Mediterranean mountain karst. Quat Sci Rev 94:28–43

Ahnert F, Williams PW (1997) Karst landform development in a three-dimensional theoretical-model. Zeits für Geomorphologie, NF Suppl. 108:63–80

Al-fares W, Bakalowicz M, Guerin R, Dukhan M (2002) Analysis of the karst aquifer structure of the Lamalou area (Herdult, France) with ground penetraradar. J Appl Geophys 51:97–106

Allred K (2004) Some carbonate erosion rates of southeast Alaska. J Cave Karst Stud 66(3):89–97
Balázs D (1986) Kína karsztvidékei (karst regions of China). Karszt és Barlang II:123–132 (in Hungarian)
Balázs D (1990) A Dél-Kínai-karsztvidék főbb barlangtípusai (Main cave types of the Southern Chinese karst region). Karszt és Barlang I:53–60 (in Hungarian)
Bauer F, Zötl J (1972) Karst of Austria. In: Herak M, Stringfield VT (eds) Karst, important karst regions of the northern hemisphere. Elsevier, Amsterdam-London-New York, pp 225–265
Beggs TF, Ruth BE (1984) Factors affecting the collapse of cavities. In: Beck BF (ed) Sinkholes: their geology, engineering and environmental impact. Balkema, Rotterdam, pp 183–188
Bočič N, Faivre S, Kovacic M, Horvatincic N (2012) Cave development under the influence of Pleistocene glaciation in the Dinarides—An example from Štirovača Ice Cave (Velebit Mt., Croatia). Zeitschrift für Geomorphologie 56(4):409–433
Bögli A (1961) Karrentische, ein Beitrug zur Karstmorphologie. Zeitschrift für Geomorphologie 5:185–193
Bögli A (1971) Karstdenudation-das Ausmass des korrosiven Kalkabtrages. Regio Basiliensis 12(2):352–361
Brook GA, Ford DC (1978) The origin of labyrinth and tower karst and the climatic conditions necessary for their development. Nature 275:493–496
Büdel J (1957) Die doppelten Einebrungsflächen in dem feuchten Tropen. Zeitschrift für Geomorphologie 1:201–288
Bulla B (1954) Általános természeti földrajz II (Physical Geography II.) Tankönyvkiadó, Budapest 549 p. (in Hungarian)
Chen J (1988) Karst collapses in cities and mining areas, China. Env Geol Water Sci 12:29–35
Cucchi F, Forti F, Ulcigrai F (1994) Degradation by dissollution of carbonate rocks. Int J Speleol 16:125–138
Cucchi F, Forti F, Marinetti E (1996) Surface degradation int he carbonates of the karst of Trieste (Classical Karst, Italy). Fornos IJ, Gines Á (edits): Karren Landforms. Universitat de les Illes Balears, Palma de Mallorca, pp 41–51
Cui Z, Li D, Feng J, Liu G, Li H (2002) The covered karst, weathering crust and karst (double-level). Sci China 45:366–378
Currin WE, Barfus BL (1989) Sinkhole distribution and characteristics in Pasco County, Florida. In: Beck BF (ed) Engineering and environmental impacts of Sinkholes and karst. Balkema, Rotterdam, pp 97–106
Cvijič J (1918) Hydrographie suterraine et evolution morphologique du karst. Trav Inst Géogr Alpine 6(4):375–426
Cvijič J (1924) The evolution of lapiés a study in karst physiography. Geogr Rev XIV:26–49
Ford D (1996) Karst in a cold climate. In: McCan SB, Ford DC (eds) Geomorphology sans Frontières. John Wiley and Sons, Chichester, pp 153–179
Ford DC, Williams PW (2007) Karst hydrogeology and geomorphology. John Wiley and Sons, Ltd., Chichester, p 562
Forti F (1984) Messungen des karstabtrages in der Region Friul-Julisch-Venetien (Italien). Die Höhle 35(3/4):135–139
Gams I (1985) International comparative measurements of surface solution by means of standard limestone tablets. Zbornik Ivana Rakovca 26:361–386
Goeppert N, Goldscheider N, Scholz H (2011) Karst geomorphology of carbanatic conglomerates int he Folded Molasse zone of the Northern Alps (Austria/Germany). Geomorphology 130:289–298
Grund A (1914) Der geographische Zyklus im karst. Ges Erdkunde 52:621–640
Häuselmann P (2008) Surface corrosion of an Alpine karren field: recent measures at Innerbergli (Siebenhengste, Switzerland). Int J Speleol 37(2):107–111
Hoblea F, Jaillet S, Marie R (2001) Erosion et ruissellement sur karst nu en context subpolaire océahiguli lesiles calcaires de Patagonie. Karstologia 33(2):13–18

Jaillet S, Hobléa F et toute l'épiquipe Ultima Patagonia (2000) Une morphologie origanele liée au vent: Les – fusées – au crêtes édiennes de Lapiaz de l'Le Madre de Dios Actes de la 10° Denconte d'Octobre, Paris, pp 73–76

Jakucs L (1977) Morphogenetics of karst regions. Adam Hilgar, Bristol, p 284

Jenko, F (1956) Hidrologija in vodno gospodarstvo Krasa. URN:NBN:SI:DOC-T6N2EWFI from dLib.Si - Digitalna knjižnica Slovenije

Jennings JN, Sweeting MM (1963) The limestone ranges of the Fitzroy Basin, Western Australia. Bonner Geog Abh 32:1–60

Jones RJ (1965) Aspects of the biological weathering of limestone pavement. Proc Geol Assoc 76:421–433

Kiernan K, Lauritzen S-E, Duhig N (2001) Glaciation and cave sediment aggradation around the margins of the Mt. Field Plateau Tasmania. Aust J Earth Sci 48:251–263

Küfmann C (2014) Solution dynamics at the rock/snow interface during the ablation period in the subnival karst of the Wetterstein Mountains (Northern Calcareous Alps, Germany). Zeitschrift für Geomorphologie 58(1):37–57

Kunaver J (1979) Some experiences in measuring the surface karst denudation in high alpine environment. Actes du Symposium international sur l'érosion karstique, Aix en Provence 1979:75–85

Kunaver J (2009) The nature of limestone pavements in the central part of the southern Kanin plateau (Kaninski podi) Western Julian Alps. In: Ginés A, Knez M, Slabe T, Dreybrodt W (eds) Karst rock features – Karren Sculpturing Zalozba ZRC. Institut za raziskovanje krasa ZRC SAZU, Postojna. Carsologica, 9. Ljubljana, Eslovènia pp 299–312

Lauritzen SE (1990) Autogenic and allogenic denudation in carbonate karst by the multiple basin method: An example from Svartisen, North Norway. Earth Surf Proc Land 15:157–169

Maire R, Jaillet S, Hoblea F (2009) Karren in Patagonia, a natural laboratory for hydrogeolian dissolution. In: Ginés A, Knez M, Slabe T, Dreybrodt W (eds): Karst rock features. Karren Sculpturing Zalozba ZRC. Institut za raziskovanje krasa ZRC SAZU, Postojna, Ljubljana, Carsologica, 9. pp 329–348

Martini JES, Grimes KG (2012) Epikarstic maze cave development: Bullita cave system, Judbarra (Gregory Karst, Tropical Australia. Helectite 41:37–66

Paton JR (1964) The origin of the limestone hills of Malaya. J Trop Geogr 18:138–147

Peng J, Cai Y, Yang M, Liang H, Liang F, Song L (2007) Relating areial erosion, Soil erosion and sub-soil erosion to the evolution of Lunan Stone Forest, China. Earth Surf Proc Land 32:260–268

Plan L (2005) Factors controlling carbonate dissolution rates quantified in a field test in the Austrian alps. Geomorphology 68:201–212

Salomon (2000) Précis de karstologia. Presses Universitaires de Bordeaux, 251 p

Sawicki LS (1909) Ein Beitrag zum geographischen Zyklos im karst Geogr. Z (Vienna) 15(185–204):259–281

Song L, Liang F (2009) Two important evolution models of Lunanshilin karst. In: Ginés A, Knez M, Slabe T, Dreybrodt W (eds) Karst Rock Features. Karren Sculpturing Zalozba ZRC. Institut za raziskovanje krasa ZRC SAZU, Postojna. Carsologica, 9, pp 453–459

Sweeting MM (1966) The weathering of limestones. In: Dury GH (ed) Essays in geomorphology. Heinemann, London, pp 177–210

Sweeting MM (1973) Karst landforms. Columbia University Press, New York, p 362

Szczygiel J, Gaidzik K, Kicińska D (2015) Tectonic control of cave development: a case study of the Bystra valley int he Tatra Mts., Poland. Ann Soc Geol Pol 85:387–404. https://doi.org/10.14241/asgp.2015.015

Thomas TM (1970) The limestone pavements of the North Crop of the South Wales coalfield with special reference to solution rates and processes. Trans Inst Br Geogr 50:87–105

Thompson P (ed) (1976) Cave exploration in Canada. Edmonton, Canadian Caver, 183 p

Thornbury WD (1965) Regional geomorphology of the United States. Wiley, New York, p 609p

Tooth AF, Fairchild IJ (2003) Soil and karst aquifer hydrologic controls on the geochemical evolution of speleothem-forming drip waters, Crag Cave, southwest Ireland. J Hydrol 273:51–68

Trudgill ST (1972) The influence of drifts and soils on limestone weathering in N.W. Claire, Ireland. Proc Univ Bristol Speleol Soc 13:113–118

Trudgill ST (1975) Measurement of erosional weight-loss of rock tables. Br Geomorphol Res Group, Tech Bull 17:13–19

Trudgill ST (1985) Limestone geomorphology. Longman, New York, p 196p

Veress M (2009) Investigation of covered karst form development using geophysical measurements. Zeitschrift für Geomorph 53(4):469–486

Veress M (2010) Karst environments—Karren formation in high mountains. Springer, Dordrecht, Heidelberg, London, New York, p 230

Veress M (2012) Glacial Erosion and Karst Evolution (Karren Formation on the Surfaces Formed by Glaciers). In: Veress B, Szigethy J (eds) Horizons in earth science research 8. Nova, New York, pp 1–91

Veress M (2016a) Covered karst. Springer, 536 p. https://doi.org/10.1007/978-94-017-7518-2

Veress M (2016b) Postglacial evolution of paleodepressions in glaciokarst areas of the Alps and Dinarides. Zeitschrift für Geomorph 60(4):343–358

Veress M, Péntek K (1996) Theoretical model of surface karstic processes. Zeitschrift für Geomorphologie 40(4):461–476

Veress M, Zentai Z, Tóth G, Czöpek I (2003) Karsztos felszínfejlődési típusok Diego de Almagro szigetén (Chile) (Types of karstic geomorphic evolution on the island of Diego de Almagro (Chile). Karsztfejlődés VIII:213–229 (in Hungarian)

Veress M, Szunyogh G, Tóth G, Zentai Z, Czöpek I (2006) The effect of the wind on karren formation on the Island of Diego de Almagro (Chile). Zeitschrift für Geomorphologie 50:425–445

Veress M, Lóczy D, Zentai Z, Tóth G, Schläffer R (2008) The origin of the Bemaraha tsingy (Madagascar). Int J Speleol 37(2):131–142

Waltham AC, Fookes PG (2003) Engineering classification of karst ground conditions. Q J Eng GeolHydrogeol 36:101–118

Williams PW (1966) Limestone pavements: with special reference to Western Ireland. Trans Inst Br Greographers 40:155–172

Williams PW (1983) The role of the subcutaneous zone in karst hydrology. J Hydrol 61:45–67

Williams PW (1987) Geomorphic inheritance and the development of tower karst. Earth Surf Proc Land 12:453–465

Williams PW (2008) The role of the epikarst in karst and cave hydrogeology: a review. Int J Speleol 37(1):1–10

Yuan D (1987) Environmental and engineering problems of karst geology in China. In: Beck BF, Wilson WL (eds) Karst hydrogeology: engineering and environmental applications. Balkema, Rotterdam, pp 1–11

Žebre M, Stepišnik U, Colucci RR, Forte E, Monegato G (2016) Evolution of a karst polje influenced by glaciation: the Gomance piedmont polje (northern Dinaric Alps). Geomorphology 257:143–154

Chapter 8
Case Studies on Glaciokarst

Gábor Tóth and Márton Veress

Abstract This chapter involves five case studies, in which the landscape and the relation of karstification and glacial erosion are described. Three case studies deal with the sample sites of the Alps (Northern Calcareous Alps, Julian Alps and Bernese Alps). One area is situated in the Durmitor (Dinarides), while the fifth area is a special site. It is a subarctic karst in the southern part of Patagonia developing under extreme climatic circumstances.

Keywords Tsanfleuron · Totes Gebirge · Julian Alps · Durmitor
Diego de Almagro · Patagonia · Karren · Karst depression

8.1 Karrenfield in the Foreground of Lapiés de Tsanfleuron (Switzerland)

8.1.1 General Description of the Lapiaz de Tsanfleuron

The Lapiaz de Tsanfleuron is a calcareous plateau expanded on 9 km^2 tilted on eastwards (Fig. 8.1). The rock of the plateau is essentially the outcrop of slope side of the Diablerets nappe. The upper part of the karrenfield is covered by a rapidly receding glacier—Glacier of Tsanfleuron—the lower part is constituted by the Tsarein and Genievre karrenfield. High lithological diversity of the rock and the glacier erosion result in a large variety of karst features, especially the different forms of the karren features (Maire 1977; Reynard 1997). Compared with other karrenfield, it can be stated that the glacier activity bears a main role on the karren formation. For this reason, the karrenfield is separable into two great parts: the upper part developed under intense relation with the glacier and a part behind the moraine of the Little Ice Age (Maire 1976). In the foreground of the glacier, the karstic terrain is dominated by features of glacial erosion. The altitudinal amplitude of the karst spread between 1400 and 2800 m including the Tsarein and Genievre karrenfield, where Maire (1976) propose a bioclimatic classification.

M. Veress et al., *Glaciokarsts*, Springer Geography,
https://doi.org/10.1007/978-3-319-97292-3_8

Fig. 8.1 General view of Lapiaz de Tsanfleuron (photo by Tóth)

8.1.2 Methods

The documentation of karren forms raises several problems that necessitate both the traditional field surveys and applying geoinformatical methods (also surveys). All the data have been gathered by field and follow-up surveys applying the methods mentioned below. Among these methods photogrammetry, which gave the majority of data, was applied in mapping karren surfaces.

- Making a morphological section and gathering morphometrical data,
- Geomorphological mapping with GPS (description, observation),
- Distance surveys on digital relief models,
- Applying photogrammetry (distortion, orientation) and
- Calculation of spatial specific solution for forms and cells.

To survey the data, a digital morphological map of every karren cell studied was compiled. The spatial specific solution was calculated according to the data surveyed on the map. The value given in percentage tells the proportion of the dissolved forms—or any selected form—on the entire karren cell and on a partial area of it too. The data obtained this way serve as a basis for the classification of the cells.

$$\text{The spatial specific solution is}: \frac{\sum_{i=1}^{n} t_i}{T}$$

where t_i means the area of the ith form ($1 \leq i \leq n$), while T means the area of the karren cell.

8.1.3 Relationship Between the Distance of the Glacier and the Development of Karren Forms

In the foreground of Lapiés de Tsanfleuron, we investigated the specific solution of the karren cells. The distance between the examined terrains and the glacier proved to be 1.5–3.2 km (Fig. 8.2). The value of spatial specific solution significantly increased parallel with the distance from the glacier; consequently, the development of cells probably started after the regression of the ice (Tóth 2007).

Investigating the effect of the glacier, the especially low dissolution value (2%) of the first area has overriding importance. This value is one order magnitude smaller than that of specific dissolution in other areas and twentieth part in comparison to the highest value.

This disparity is hardly interpretable on the base of 200 m distance which indicates only 20–25 year difference in development calculating the mean glacier regression time. The terminal moraine of Little Ice Age (Reynard 2008) is situated on this 200 m distance.

The other cells are beyond the terminal moraine, where one can take into consideration a 10,000–11,000 years ice-free dissolution. The first area is the only one occurring from the glacier side of the moraine barrier, where barely before 40–50 year

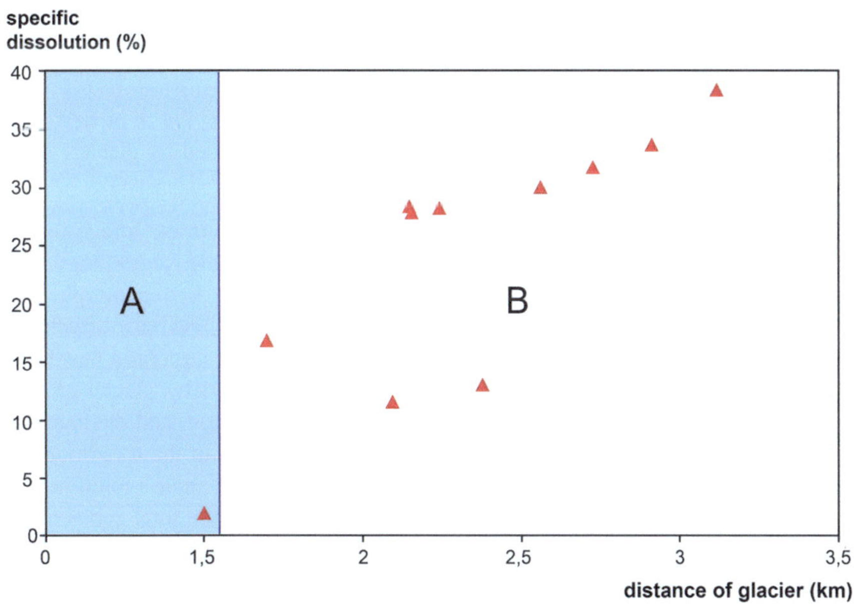

Fig. 8.2 The specific dissolution plotted against the distance of glacier. The blue zone is the place of terminal moraine of Little Ice Age. The "A" zone between terminal moraine and glacier where one can take into account a 0–150 year solution period. The "B" zone beyond the terminal moraine where karren formation lasts since 10,000–11,000 years

ice cover could be found. Thus, despite the small distance, the difference in the duration of the dissolution explains the significant change of the karren forms and the dissolution.

8.1.4 Karren Morphology of Lapiés de Tsanfleuron

During the investigations of karren cells, we have outlined the karren morphological map of the karrenfield of Tsanfleuron. We distinguished the karren terrains according to the general morphological map of Reynard (1997) by which there are four morphological zones. These are closely related to the retreat and shape of glacier. The zones can be distinguished on the base of karren morphology, state of development and genetics (Tóth 2008).

- Zone 1: An area between the glacier and the terminal moraine divided by its morphology into two parts.
  Its southern side is composed by karst on structural bench (*Schichttreppenkarst*), while at the northern side is a younger small valley evolved during glacial erosion; accordingly, the karren landscape is also different.
  - Terrain 1/A is characterized by undeveloped, rudimentary forms, first of all types of grooves. Between the uncovered bedding planes, the bedding karren (*Schichtfugenkarren*) take forms.
  - On the terrain 1/B, line up gigantic grooves which are 15–20 m long, evolving on the base of dissolution, glacial and fluvial erosion, respectively (Fig. 8.3). These polygenetic grooves (channels) are about 1 m deep and 1.5–2 m broad. Their sides and surroundings are polished by glacier. Their karstic development could be started at most before 10–15 years.
- Zone 2: This zone is composed of karst on structural bench in the foreground of the terminal moraine. The terrain is characterized by variable morphology with karren cells of distinctive exposition and slope angle. Its karren morphology contains mature zonal cells with the full collection of all alpine basic structures.
- Zone 3: This morphology is characterized also by karst on structural bench but the karren forms are filled by soil and partially destroyed by glacial erosion (Fig. 8.4). The denuded steps are cut down by glacial erosion, and the here and there emerging Nye channels are mixed with karren forms in the foreground of the glacier. Most karren forms are in their ruined phase of their evolution.
- Zone 4: The karren formation takes place in almost all of the total zones under the soil. The sporadically rising bare rock surfaces show that karren forms are testimonies of the activity of bare karren evolution.

Fig. 8.3 Giant groove (rinnenkarren) in the near of the end of Tsanfleuron glacier (photo by Tóth)

8.2 Examination of Karren Surfaces in the Foreland of the Glacier Below Triglav

8.2.1 Terrains in Front of the Glacier

We can assume that the terrains in front of the glacier (from where the ice retreated) have become deglaciated for the past decades, since 1850, the end of the Little Ice Age. Here, the initial phase of karstification can be reached, which may be of particular interest in determining the speed and intensity of the karstification. The investigated surrounding sample sites lost their ice cover about 50 years ago (Gams 2002) . The glacier forelands are areas where the glacial erosion ended and another process begins, the karstification. By measuring the parameters of karst microforms,

Fig. 8.4 The steps are devastated by the glacier (photo by Tóth)

we can deduce the development of glaciokarsts and the interaction of the two processes. The appearance of karren forms is indicative of the initial stage of karstification. The rillenkarren appear the fastest. Their centimetre versions are formed after 1–2 years in ideal conditions. The kamenitzas can be identified after 7–10 years. Also, a short time (about 10–15 years) is required for the formation of the rinnenkarren.

8.2.2 Terrains with Grikes

The grikes are longitudinal, dissolution forms that are bordered by vertical walls. Almost all of the authors dealing with grikes agree that the longitudinal elongation of the feature is the cause of the rock structure (Bögli 1976; Jennings 1985; Ford and Williams 1989). The faults on rocks favour seepage, which promotes the solution widening of fissures and then of grikes. Intense deepening is contributed by the presence of snow and soil in the fissures (Haserodt 1965; Trudgill 1985; Howard 1963).

It is important to note that the grikes are one of the earliest forms of karstic forms. They have a rapid development on the surface becoming deglaciated, especially if this is favoured by rock structure. Thus, in the foreland of glaciers, they occur in large numbers.

When the karren cells were examined, the grikes occupied three positions within the cells. In one case, they appeared anywhere on the cell, in a disordered way, irrespective of slope direction and the slope angle. The second option is that they appear in the lower zone of the cell in strike direction while in the third case, they appear as lattice-like karren. At this time, they cover most of the whole cell and form a characteristic form of it (Fig. 8.5).

8.2.3 Geographical Position of Sample Sites

The stratigraphic and geological characteristics of the Southern Limestone Alps differ at many points from the usual structure of the Alps. Their nappes dip southwards. The significant fault system of the mountain range also affects morphology. Figure 8.6 shows that the largest glacier valleys follow the direction of faults (e.g. Vrata valley). The base of the Julian Alps is formed by Triassic Main Dolomite which is 1000 m thick. The Triassic Main Dolomite is overlain conformably Dachstein Limestone with a thickness of 700 m. Its development period is late Norian–Rhaetian and the average thickness of the layers is 2.5 m (Hinnov 2003).

Fig. 8.5 A special case of the grike: grike surface on the Lapiés de Tsanfleuron glaciokarst (photo by Tóth)

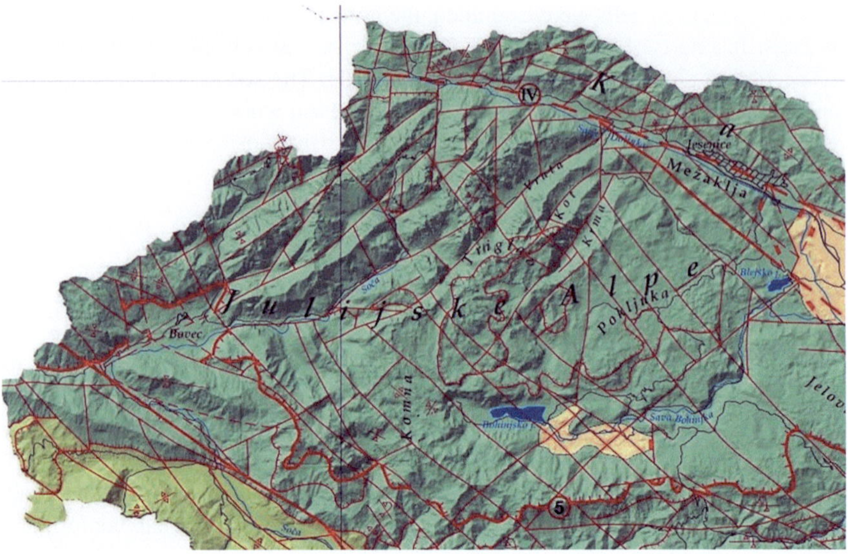

Fig. 8.6 Structural–tectonic map of the Julian Alps (Slovenian Geological Service) (photo by Tóth)

The research area is located in the glacier valley north of the Triglav peak in the Julian Alps at a height of 2200–2300 m (Fig. 8.7). The glacier valley's bottom is divided by steps and rock basins, and the height of the steps can reach 10–15 m. There is considerable debris coverage in the area, which can originate from frost weathering and moraines. The valley, and thus, the research area, has recently been covered with ice, and its ice-free period can be up to 50 years old. Among other things, this is due to the fact that the dissolution features of the area are underdeveloped. On the slopes with bedding plane slopes of uncovered steps, young karren surfaces are formed with primitive form. The karren are grikes, rinnenkarren, rillenkarren and pits, among which the grikes formed along the faults have evolved to the greatest extent (Veress and Zentai 2004). In addition to the grikes reaching the width of 15–20 cm, there is a large number of microfissures. There are Schichtfugenkarren on scarp fronts.

8.2.4 *Methods and Parameters*

Studies were carried out in three surfaces (Figs. 8.8 and 8.9). All three areas are surfaces rising above their environment, so they could not get water from elsewhere, just from rainfall to their surface. When choosing the sites, we took care of that the slope angle will be permanent within the mapped area. The slope angle of the two medium sloping bedding planes fell between 15° and 20°.

Fig. 8.7 The studied area in the Julian Alps with the Vrata valley in the background (photo by Tóth)

Fig. 8.8 The remote image and rectified orthophoto of the sample area I with the drawn grikes (photo by Tóth)

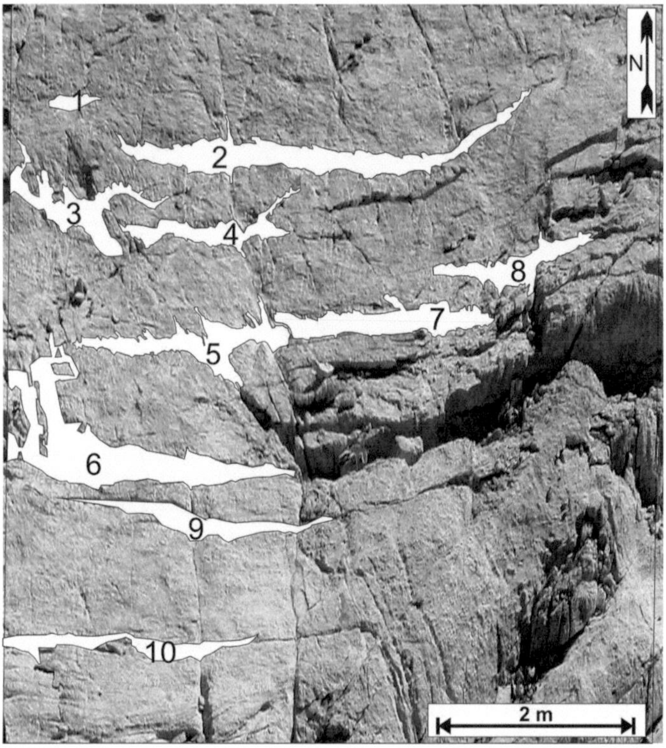

Fig. 8.9 The grikes of sample area II are photographed from the cuesta above it (photo by Tóth)

Using the features of the terrain, orthophotos were made of each karren surface. We could make measurements after rectification and orientation. The area of the forms drawn in orthophotos was determined and supplemented by field–depth measurements. Along the cross section, the features were graded. We measured the depth and direction of the features and the fault number and the fault directions in each sample area. We were looking for a relationship between the depth and the width of the grikes. For this, we have used the results of morphometric measurements in the field. We measured the width and the belonging depth along the grikes every 10 cm. According to the measurements, the width of the grike is a linear function of the depth.

$$y = 2.8818x + 0.0041,$$

where y means the depth, while x means the width.

The degree of scattering remains within the acceptable value of morphometry: $R^2 = 0.8814$, so the correlation is appropriate to assign depth to the width measured at any point of the fissure. As shown in Fig. 8.10, the straight line moves to the origin with a minimum deviation, i.e. 0 width belongs to 0 depth.

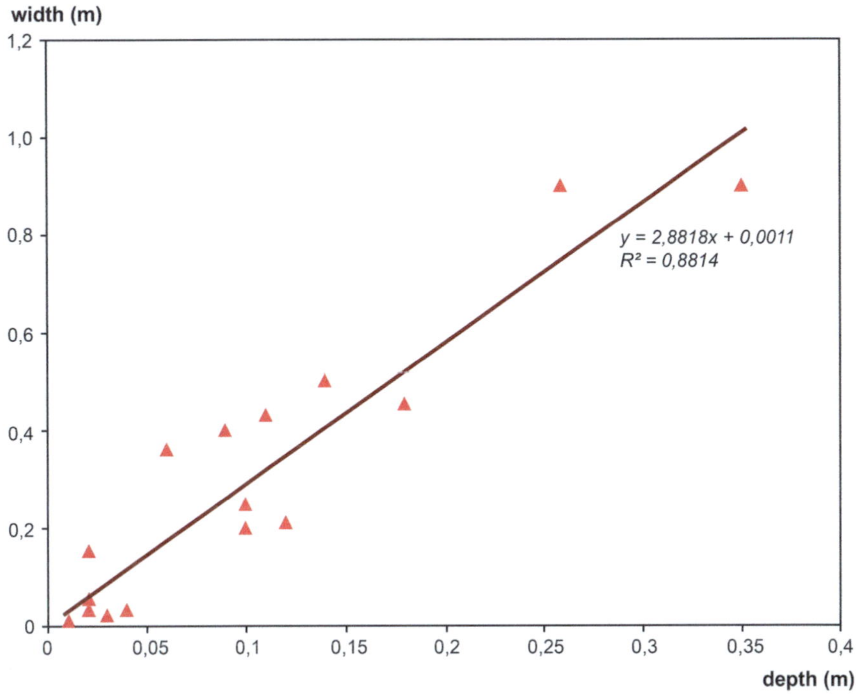

Fig. 8.10 The regression relationship between depth and width of the grikes (photo by Tóth)

In the next step, we use the information provided by the orthophoto as follows:

- We can determine the area and length of the features with the help of ArcView program. As a ratio of the two data, we get the average width.
- Based on the function, average depth is assigned to the average width.
- Knowing the average width, depth and length of the feature, the volume of forms can be determined.

By this method, each fissure is transformed into a rectangular cuboid, and the volume of these gives a good approximation of the amount of dissolved material. The required data can be measured on the orthophoto, and the depth can be calculated using the function.

8.2.5 *The Development of Grikes*

There is a relationship between width and depth; accordingly, larger depth belongs to a wider segment of the grikes. The middle segments of the measured grikes are broader while their edges are narrower; moreover, the depth is also decreased towards the edges of the grikes. Consequently, dissolution of preformed grikes developed in this way begins at only one point and then progresses in two directions along the grike.

8.2.6 Morphogenetic Grouping of Grikes

The Alpine grikes can be grouped into two large groups based on their genetics and their morphology. We can distinguish grikes among fault and grikes, which coalesce from the pit. From the genetic point of view, the lattice-like karren do not form a separate group, forming a special type of grikes among fissures.

In each of the three processed sample areas, we found that the size of the grikes exceeded the size of the other karren forms by an order of magnitude, and the fracture number (piece/cm) was twice that of Totes Gebirge or Dachstein (Veress et al. 2001). This observation demonstrates the hypothesis that the increase of fracture number favours the development of grikes.

The edges of the grikes that developed along the fracture are straight lines following the direction of preforming fracture. The grikes of the area show a narrowing tendency towards their end points and at both ends they end up punctually (Fig. 8.8). Their deepest and widest point is in their middle. From this, it can be concluded that their development started from the middle, that is, from one point. This seems to be supported by the fact that proportionality can be detected between depth and width. The features are reduced from the centre, but not only from the width but also from the depths. This is only possible if the deepening is parallel to their widening. In case if at least in two different directions fracture systems are formed, then the lattice-like karren is formed by dissolutional widening of the fracture. The morphological concordance of the lattice-like karren and the grikes shows that the formation of lattice-like karren is also from one point. These points are the intersection of different fracture directions. When the lattice-like karren is formed, it becomes easily the dominant form of the karren surface or cell, since by meshing the surface, it facilitates the rapid seepage of the dissolvent (Fig. 8.5).

The other large group of grikes coalesces from the pits, which can be well distinguished from the grikes among fractures both in their location and morphology. They are often located in the lower zone of the karren cells and they lead the precipitate fell to the surface of the cell into the karst. Their sidewalls are also vertical, but their margins are not straight. The edge of the form keeps its origin all the time, and its edge has an arcuate form (Fig. 8.11). These curved nooks can also be traced along the vertical sidewalls.

A third transient type of grikes can be considered asymmetric grikes, which are located at the intersection of a bedding plane of a step and at the front of the step above it. The asymmetry of such forms is caused by the rinnenkarren coming from the slope to dissect one side of the fissure with more intensive dissolution (Fig. 8.12). The surface projection of these shapes, therefore, consists of curved nooks on the sloping side, while on the other side there is a straight line.

Fig. 8.11 Grikes, which coalesce from the pits (photo by Tóth)

Fig. 8.12 An asymmetric grike formed on the bottom edge of the karren cell. The grikes at the junction of the steps lead the water of the cell to the bottom (photo by Tóth)

8.3 The Karstification of a Depression of Totes Gebirge

8.3.1 The General Description of the Depression

The giant doline that is described in the case study is located between the Wildgössl and Einserkogel peaks in the Totes Gebirge (Fig. 8.13), north from Lake Vd. Lanngang. From the south, it can be reached along the tourist track number 213 which runs along its eastern margin. The tourist track number 201 branching out from the previous tourist track runs along its southern margin. The marginal parts of the depression are at an elevation of 1915 and 2066 m. The cliff constituting its southern margin is built up of Jurassic limestone, while the total part of the depression was formed in Dachstein limestone. Its diameter is about 2 km in NS direction, but it is somewhat more in WE direction. Three troughs are connected to the depression: at its southern margin, the trough below Scheiblingkogel (I), at its eastern margin, the trough under Hochkogel (II) and at its western margin, the trough under Wildgössl (III). The origin of the troughs can be explained in three ways which are the following:

- The glaciers of the three troughs left the depression that had been transformed into a cirque.
- From the three troughs, the glacier of the trough marked II did not tap the depression, but it fed it. This may be indicated by the fact that the floor of the trough marked II slightly dips towards the depression.

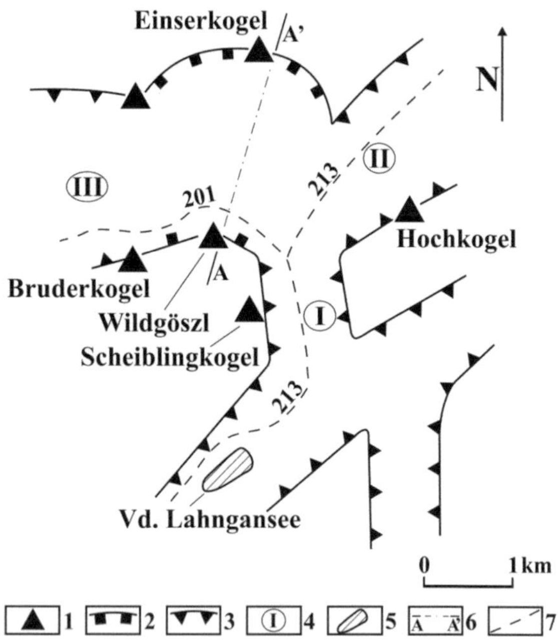

Fig. 8.13 The giant doline that was transformed into cirque and its environment. 1. horn, 2. cirque, 3. trough, 4. identification code of the trough, 5. lake, 6. site of profile and 7. tourist track

– The combination of the two above-mentioned cases. First, the glacier of the trough marked II flowed through the depression in any of the glacials, feeding the glacier of the large-sized trough marked III. This is referred to by the stepped character of the northern slope of the depression (the stepped character is not widespread in cirques because of the lack of the significant glacial erosion of the slopes). Following this, a glacier flowed out into the trough marked II from the area of the depression in a later glacial.

The floor of the depression is separated into several part-depressions with different sizes in EW direction. (Consequently, it can also be regarded as an uvala.) The part-depression of its eastern part is the largest, which is marked "A". Further part-depressions are arranged in a western direction from this.

8.3.2 Morphology

Both the giant doline and the part-doline marked "A" within this have an asymmetric cross section in NS direction: the upper part of the southern slope is a steep (nearly vertical) cliff, to which a debris slope is connected that developed from debris cones. The northern slope is predominantly constituted by a stepped surface, the inclination of which is smaller than that of the cliff.

In the area of the part-depression marked "A", the debris slope goes into an alluvial cone which occupies the floor of the giant depression (Figs. 4.46 and 8.14). The surface of the alluvial cone is dissected by several gullies. The gullies were formed by the streams arriving from the cliff and from the debris cone. Although the streams are intermittent, the water can flow in them for several weeks. The water can originate from the intermittent springs of the cliff, from the permanent springs of the debris cone and from snowmelt. The gullies have smaller and smaller depth towards the margin of the alluvial cone. They do not reach the rock boundary, but they terminate in the superficial deposit (blind gully). The material of the alluvial cone originates from the material of the debris slope. Since the intermittent streams are incised into the alluvial cone, they transport stream load to its lower part (the alluvial cone becomes higher) and to its margin (the alluvial cone becomes wider).

The surface of the alluvial cone dips mainly in northern direction, but some of its parts can also be inclined to a small extent to the west as a result of its fan-like development. Its margin (the rock boundary) can be of EW and NS direction because of the different positions of the side slopes of the depression. Three types of the rock boundary can be distinguished according to the inclination of the surface with superficial deposit, and the relation between the direction of rock boundary and the dip of the bedrock layer.

– Type 1: The direction of the rock boundary is nearly perpendicular to the inclination of the bedrock layers (its direction is almost of NS), and the surface of superficial deposit slightly dips towards the rock boundary (Fig. 8.15a).

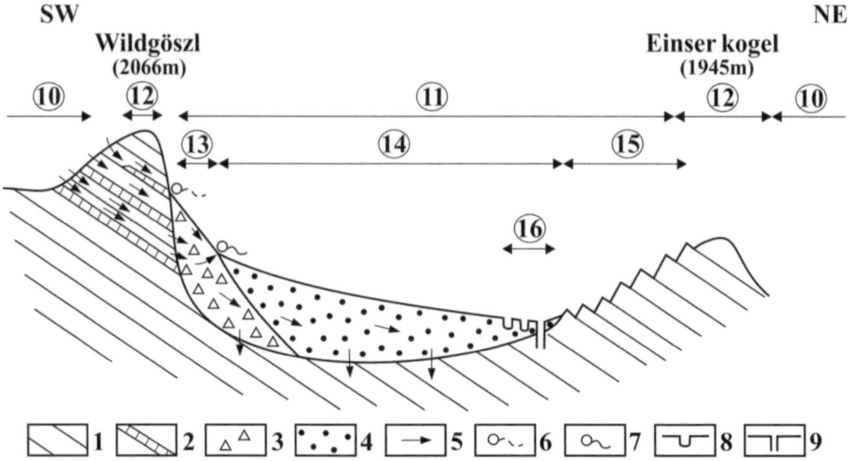

Fig. 8.14 The main parts of the giant solution doline. 1. limestone, 2. sandstone, 3. debris of frost weathering origin, 4. sediment that was formed by water transport, 5. water percolation, 6. intermittent spring, 7. permanent spring, 8. suffosion doline, 9. covered karst ponor, 10. cirque, 11. giant solution doline, 12. horn, 13. debris cone, 14. alluvial cone, 15. stepped surface and 16. covered karst zone

- Type 2: The rock boundary is nearly perpendicular to the dip of the bedrock layers, the dip direction of the surface of superficial deposit coincides with the direction of the rock boundary (Fig. 8.15b, left part).
- Type 3: The rock boundary is almost parallel with the dip direction of the bedrock layers, the surface of superficial deposit dips towards the direction of the rock boundary. At this type, the rock boundary has a wavy pattern since the superficial deposit penetrates into the lower parts of the surfaces with bedding planes of the scarp fronts like fingers (Fig. 8.15b, middle part and right part).

Along the rock boundary, on the uncovered part, bare karst features (shafts, giant grikes) occur, while on the covered part, there are covered karst features (suffosion dolines and covered karst ponors, Veress 2016). The latter are the pits and pit groups situated at the end of the floor of gullies.

In case of the development of the rock boundary Type 1, the covering process of bare karst can be recognized. Here, such karst features can be found, one part of which is on bare karst, but the other part of them is on covered karst. This is possible because the feature of bedrock also exists under cover, but this section became covered after its formation.

In case of Type 2, covered karst features do not have a bare karst part. They are elongated parallel with the rock boundary since the suffosion of waters that flow down and percolate in the dip direction of the surface expands in this direction too. (The development of elongated dolines is favoured by the fact that the cover can be transported into grikes with strike direction situated close to the margin of the cover since over it, the thickness of the cover is uniformly small.)

(a)

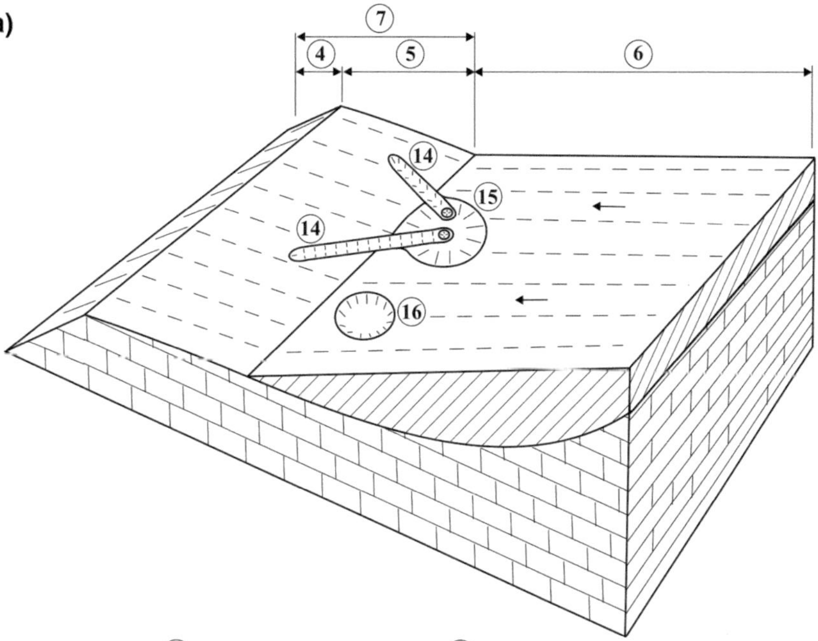
7
4
5
6
14
15
14
16

(b)

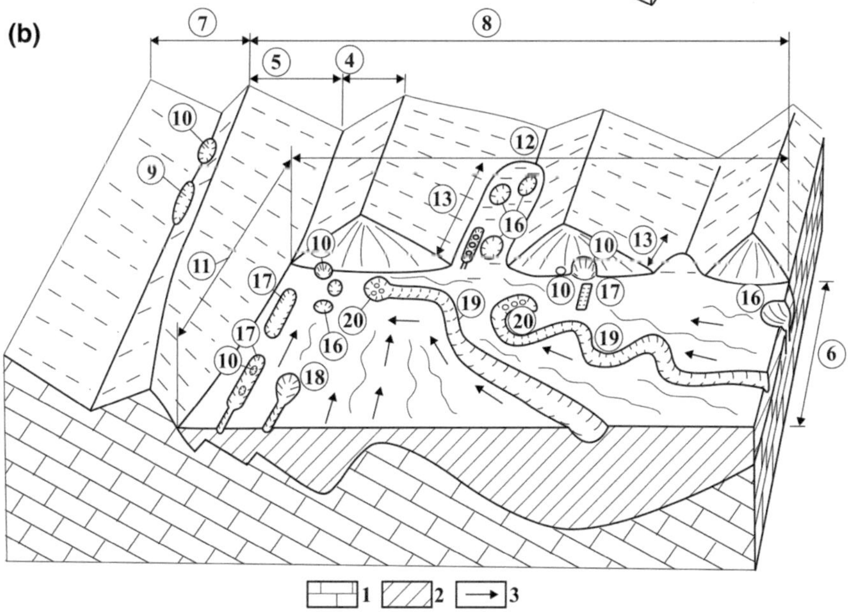
7
8
5
4
10
12
9
13
16
11
10
10
13
17
10
17
16
19
20
20
17
16
19
6
10
18
1
2
3

◂**Fig. 8.15** Karstification on rock boundary in a DSD of a giant doline if the cover overlies a stepped surface (utilizing the morphological properties of the depression shown in Fig. 4.46) **a** strike of rock boundary perpendicular to bed dip, the surface with superficial deposit slightly slopes towards the rock boundary, **b** strike of rock boundary coincides with bed dip, the surface dips towards the rock boundary (middle and right parts of the figure), perpendicular to that, but the dip of the surface is parallel with the rock boundary (left part of the figure) 1. limestone, 2. cover, 3. dip direction of the covered karst terrain, 4. scarp front, 5. bedding plane, 6. gently sloping surface with superficial deposit, 7. uncovered stepped surface, 8. partly covered stepped surface, 9. giant grike, 10. shaft, 11. the dip of limestone beds is perpendicular to the strike of the rock boundary, 12. the dip of limestone beds coincides with the strike of the rock boundary, 13. covered karst terrain intruding into the lower part of the stepped surface, 14. giant grike with partly uncovered, partly covered environment, 15. suffosion doline developed in the environs of giant grikes, 16. suffosion doline, 17. elongated suffosion doline, 18. subsidence doline with gully, 19. blind gully and 20. covered karst ponors and pits

In case of the Type 3 of the rock boundary, karstification is intensive. Since the surface of superficial deposit primarily dips towards the rock boundary of this type, the gullies of the alluvial cone are also oriented towards this place. The abundance of water will be the largest at this type of the rock boundary which is proved by the covered karst ponors that developed near the rock boundary. However, most of the sediment is also transported here during the reworking of the cover. This is proved by the covered karst zone that developed in a relatively wide stripe and by the fact that the ponors are not directly situated on the rock boundary. As a result of continuous sediment accumulation on the bare karst part, the rock boundary "leaves" the termination of the blind gullies (at the same time, favourable conditions are created on the thin cover for the development and redevelopment of suffosion dolines along the shifting rock boundary). This is also referred to by the wavy pattern of the rock boundary.

8.3.3 *The Development of the Depression*

The giant doline was transformed by glacial erosion. From the cliff that borders the depression from the south, debris was formed, creating a debris cone. The waters of the cliff created springs. The waters of the springs reworked the debris, creating an alluvial cone. The alluvial cone expanded, the transported material became finer by physical weathering, the debris of the sandstone became disintegrated, and thus the impermeable character increased in its area. Consequently, the expansion of the alluvial cone was much more intensive since its material was reworked. With its expansion, the gullies lengthened, the rock boundary shifted at the expense of bare karst (which is the northern slope of the depression) and the depression filled (fills) up partly. The covered karstification started and is still in progress at the rock boundary. The cover was (is) partly transported into the karst.

8.4 The Karstification of a Depression (Mlječni Do) in the Durmitor

8.4.1 General Description

The Mlječni do is a depression in the Durmitor. The mountains are built up of Triassic and Cretaceous limestones and Cretaceous flysch, but sandstones, sandstone–limestone and andesite also occur here (Zivaljevic et al. 1989).

The Mlječni do is an old (of preglacial age) giant doline transformed by glacial erosion, with the doline of Urdeni do it is situated on the floor of the glacier valley constituting the continuation of Surutka between the mountains of Šareni pasovi and the Bobotov Kuk (Fig. 8.16). Glacial erosion has taken place twice in the area of Mlječni do (Djurovič 2009). The giant doline is elongated in NS direction, its longitudinal axis is 700 m, while the shorter is about 500 m (Fig. 8.17). The elevation of its floor is between 1940 and 1950 m and its marginal parts are about 2000 m.

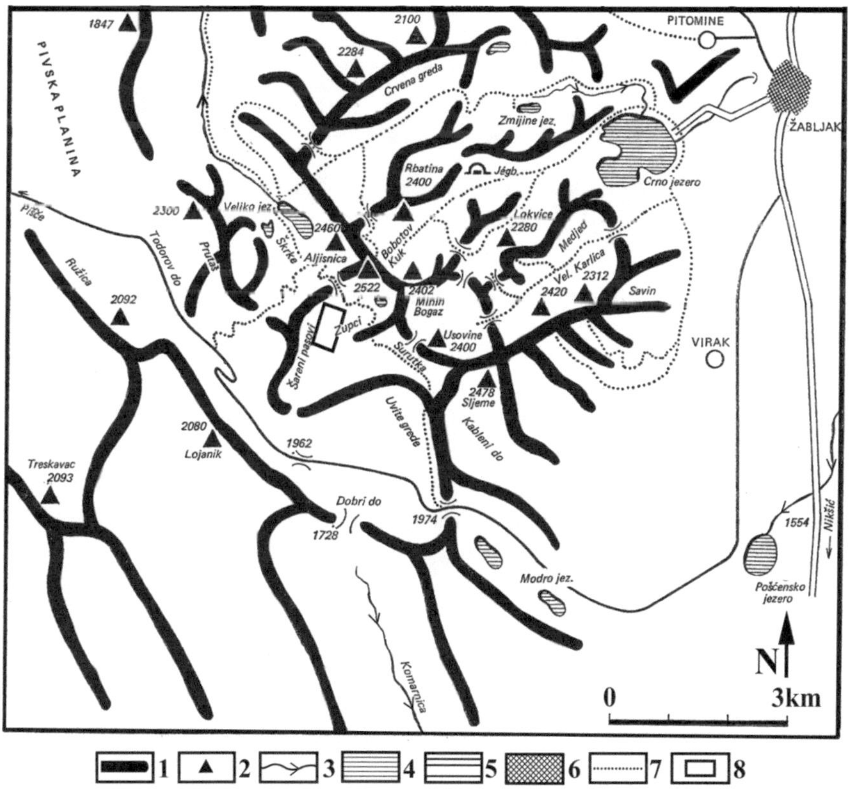

Fig. 8.16 The site of Mlječni do in the Durmitor 1. combe ridge, 2. horn, 3. stream, 4. lake, 5. road, 6. settlement, 7. tourist track, 8. research area

Fig. 8.17 Mlječni do. 1. Šaroni pasovi, 2. Mlječni do and 3. thick (3a) and thin (3b) beds of the doline floor destroyed to a wavy character (photo by Veress)

8.4.2 Geological Structure

It is situated in the root zone of the recumbent fold, which is exposed on the slope of Šareni pasovi (this is the side slope of the bearing trough) (Fig. 8.18). The strike of the beds of the root zone is 90°–270°, their inclination is large and their dip angle is between 39° and 87° according to our measurements (Veress 2016). In the area of the root zone, and thus in the area of Mlječni do, thick and thin beds alternate. Thick beds, the thickness of which exceeds 5 m at some sites, among which several beds occur built up of thin beds with a thickness of some cm and dm (Figs. 8.17, 8.18 and 8.19). Thin beds contain silica which can develop in nodules or in layers.

8.4.3 Morphology

In the area of the giant doline, features of karstic and non-karstic origin can be distinguished (Veress 2016, 2017, Fig. 8.19). Hogbacks, rock basins and till are of non-karstic, glacial erosional origin, while debris cones are of frost weathering, mass movement origin.

Fig. 8.18 Šaroni pasovi from Surutka. 1. Samar col, 2. Mlječni do (behind it the recumbent fold of Šaroni pasovi can be seen) (photo by Veress)

Hogbacks developed at the head of beds of thick beds in a way that thin, siliceous beds were destroyed to a larger degree during glacial erosion (Fig. 8.20). The hogbacks marked I, II, III and IV rise 1–2 m above their environs. Their height increases towards rock basins (mainly towards that which is situated in the East). The height of the hogbacks marked V, VI and VII is smaller, and they are only higher with 1–2 dm than their environs. The side slopes of the latter are less steep than those of the former.

The rock basins (three can be distinguished) are elongated in NS direction. Thus, both the surface and the heads of bed are of wavy character (Fig. 8.21).

The till covers the floor of the giant doline in a carpet-like, patchy development. The patchy occurrence is only present on the head of beds of thin beds and it is absent on the heads of bed of thick beds (on hogbacks). The debris cones originate from the slope of Šaroni pasovi and they are also developing at present (Fig. 8.20).

The karst features of Mlječni do are schachtdoline (1 schachtdoline, Fig. 8.22), small-sized solution dolines (19) and subsidence dolines (9), which are of suffosion origin (Veress 2017) and the caves that developed along some bedding planes in the side of Šaroni pasovi (Figs. 8.19 and 8.20).

Both solution dolines, but mainly subsidence dolines, are of small size and some of them are elongated. The longer diameter of subsidence dolines does not exceed 3 m, while their depth is not larger than 1 m except one (with a depth of 2.0 m) and

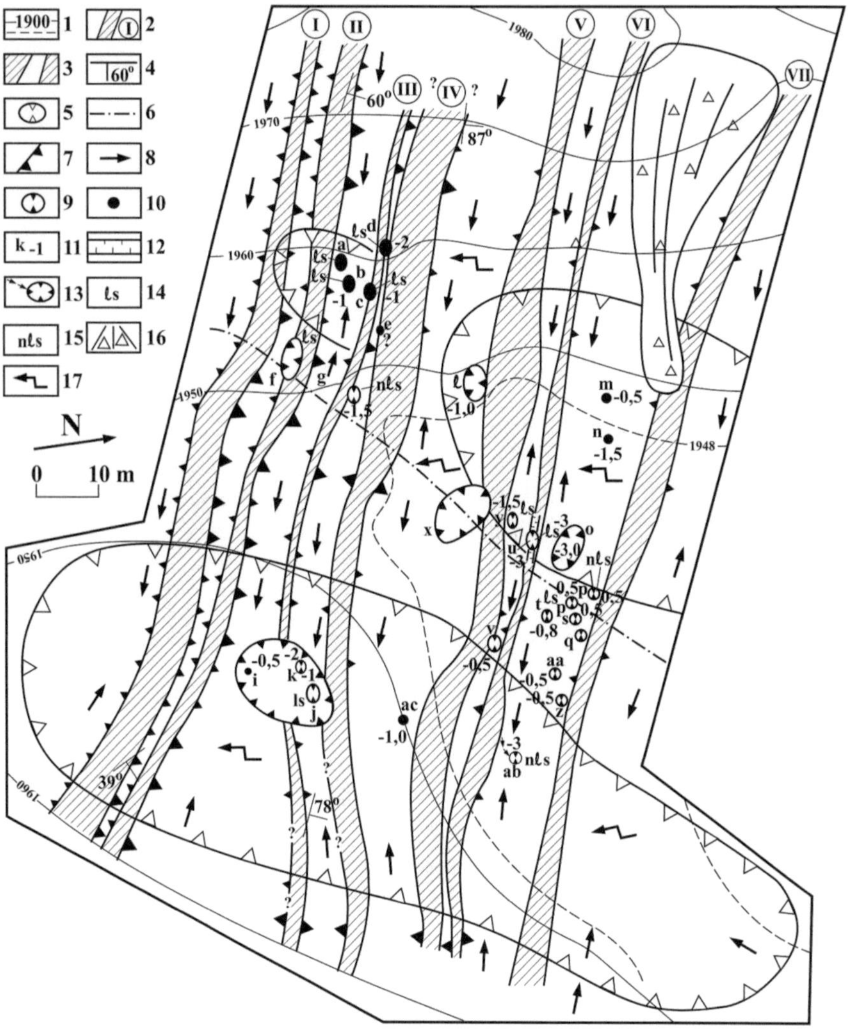

Fig. 8.19 Structural and morphological map of a part of Mlječni do (modified Veress 2017). 1. contour line, 2. thick bed, 3. series of thin siliceous beds, 4. strike and dip of bed, 5. rock basin, 6. top of roche moutonnée, 7. hogback (rock blade), 8. dip of surface, 9. solution doline, 10. suffosion doline, 11. code and depth of doline, 12. grike, 13. solution doline functioning as ponor, 14. doline elongated in strike direction, 15. doline not aligned in strike direction, 16. debris fan and 17. direction of ice motion (photo by Veress)

50% of them is elongated. The longer axis of these dolines can be three times larger than the shorter. Solution dolines have a larger size. The longer axis of five dolines may be about 10 m (Fig. 8.23). The longer diameter of the other dolines (14 dolines) is even smaller than 3 m, while their shorter diameter is below 2 m. Their depth is smaller than 3 m. Among solution dolines, there are a larger number of

Fig. 8.20 A hogback from a short distance, farther the caves of Šaroni pasovi that developed along bedding plane (photo by Veress)

Fig. 8.21 Eastern rock basin (photo by Veress)

Fig. 8.22 Schachtdoline (photo by Veress)

Fig. 8.23 Solution doline (photo by Veress)

elongated dolines, and only four dolines have a circular ground plan. Some solution dolines contain an open passage. It may happen that a depression is connected to the doline; in spite of its solution origin, it can function as a water feeder too.

Dolines (mainly subsidence dolines) mostly occur in rock basins. According to the beds, their position can be the following:

- Their longer axis (or that situated close to it) coincides with the strike direction of the beds.
- The dolines are situated on the heads of bed of thin beds (Figs. 8.19, 8.22 and 8.23).
- However, they often occur at the contact of thick and thin beds too.
- However, in case of only one solution doline, it can be experienced that it developed both in the area of thick and thin beds.

In case of the three types of dolines (thus, including schachtdoline too), the development is controlled by bedding (Fig. 8.22). The solution resulting in doline development mainly takes place along bedding planes (Fig. 8.22).

Thus, doline development can occur in the area of thin beds because the high density of bedding planes favours seepage along the bedding plane, and thus solution along the bedding plane. (Consequently, thin beds experience physical weathering to the effect of solution, and thus a depression and solution debris is generated.) The contact between thick beds and thin beds also promotes solution since the water flowing down from hogbacks seeps away at the contact of thick and thin beds (Fig. 8.22). An exception is only one solution doline (as above mentioned), the development of which was affected by the bearing rock basin since the direction of elongation of the two features coincides, and the solution along bedding plane takes place here too.

Karren formation is subordinate in the area of Mlječni do. Bare karren only occur on the slightly inclined surfaces of hogbacks. On these surfaces, there are pitkarren, grikes and rinnenkarren connecting to the latter (Fig. 8.24). The reason for the occurrence of small degree of karren is the coveredness of the surface and its covering also happening currently and the intensive denudation of beds with a small thickness by physical weathering. Another reason is that the surfaces with hogbacks, being capable of karren formation, are of small expansion.

8.4.4 Geomorphic Evolution in Mlječni Do

The doline has reached its current state during the following geomorphic stages (Fig. 8.25):

- The Mlječni do develops by preglacial solution.
- Rock basins and hogbacks are formed by glacial erosion. (It is probable that hogbacks had already developed during the first glaciation and they only continued their development during the second glaciation.)

Fig. 8.24 Hogback with karren surface (photo by Veress)

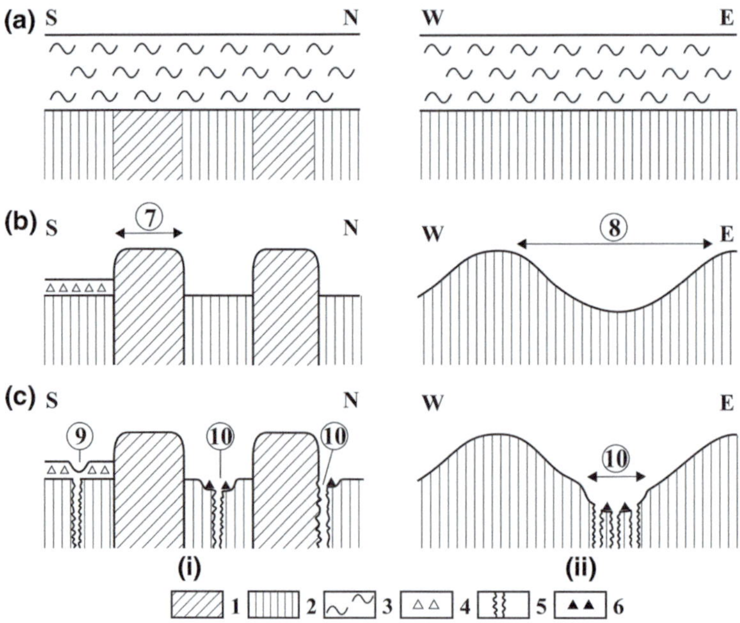

Fig. 8.25 Geomorphic evolution in the area of Mlječni do. i Profile along NS direction, ii profile along EW direction, 1. thick bed, 2. beds of thin beds, 3. glacier, 4. moraine, 5. grike that developed along bedding plane, 6. debris that developed during solution, 7. hogback, 8. rock basin, 9. subsidence doline and 10. small-sized solution doline (photo by Veress)

- Following the retreat of ice, cover patches were formed from ground moraine between the hogbacks. During the second glaciation, the floor of the doline subsided further, the hogbacks were renewed or regenerated. The karst features being older than the second glaciation of the doline were probably destroyed.
- After the second glaciation, karren formation took place on hogbacks and solution dolines (bare surface) and subsidence dolines (surface covered by moraine) developed on the heads of bed of thin beds between the hogbacks by solution along bedding plane.

8.5 Karren Formation on a Combe-Ridge (Diego de Almagro, Chile)

8.5.1 General Description of the Island

Diego de Almagro is an island of the Patagonian archipelago with an area of 375.6 km^2 (Fig. 8.26). The rocks building up the island underwent metamorphosis during the tectogenesis at the end of the Carboniferous period (Forsythe and Mpodozis 1983). Metamorphosed lamprophyre intrusions also wedged in into the marble that was formed during metamorphosis. The unstratified marble occurs in tracks with a width of 2–3 km with NS direction on the island (Forsythe and Mpodozis 1983).

The formation of the present surface of the island can be related to two generations of glaciers (Maire 1999). During the LGM, the ice sheet around the South Pole reached the island and scoured the surface (from this surface only the former nunataks are exposed). The glaciers that developed during the retreat of ice sheet dissected the surface with glacier valleys. The valleys became partly filled with seawater and created fjords.

The below-described terrain section is situated on a combe-ridge between two fjords. The marble stripe of the ridge, where the marble terminates, goes into the non-karstic metamorphic rock emerging from below the marble by steep denudational step since the latter denudes faster. The width of the ridge built up of marble is 100–200 m and it has an expansion of 1–2 km in NS direction. From the side, it is bordered by steep (nearly vertical) slopes. The surface of the outer edge of the ridge slightly dips towards the fjords, while the inner surface part is almost plain, but dips in various directions (Fig. 8.27). The ridge is a smoothed glacial erosional surface, which is dissected by roche moutonnées. At some places, the non-karstic metamorphic intrusions rise weathered out over the well-soluble, thus, lower marble surface. Considering the elevation of weathered out non-karstic intrusions and the time of deglaciation, the denudation rate of the marble is 0.09 mm/year (Maire et al. 2009). The disintegrating pieces of the non-karstic metamorphic rock modify the karstification taking place on the surface of the marble. Although the lower parts of the island are overlain by vegetation, the area of the combe ridge is a bare karst without vegetation and soil.

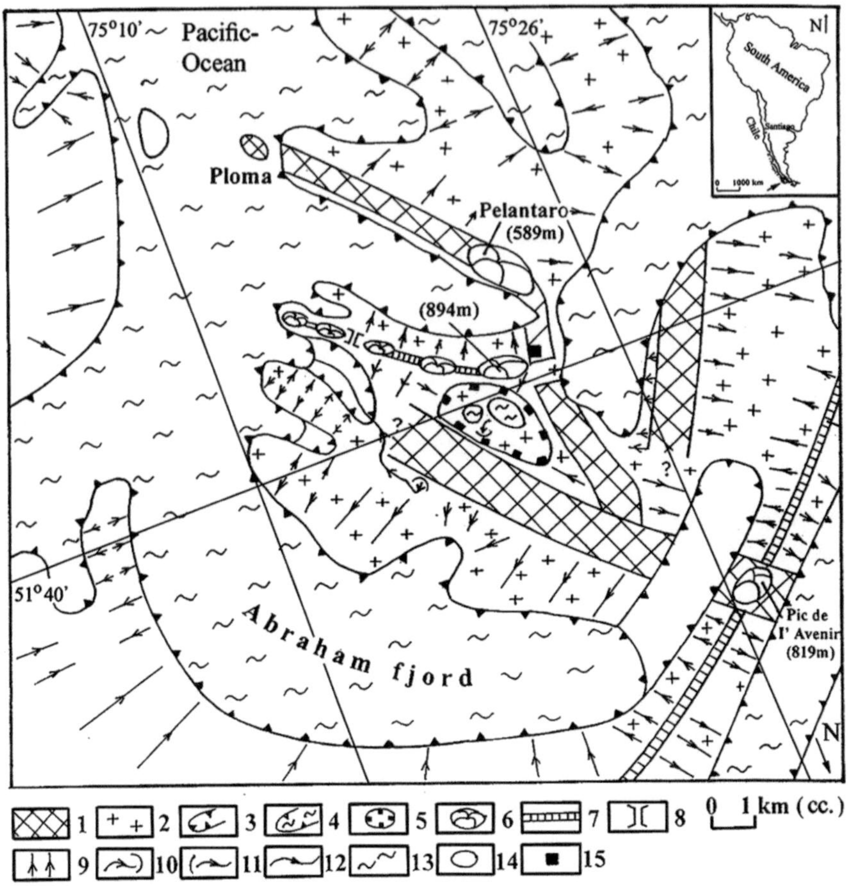

Fig. 8.26 The research area (Veress et al. 2006—modified). 1. marble, 2. non-carbonate metamorphic rock, 3. glacier valley, 4. fjord, 5. rock basin, 6. horn, 7. combe-ridge, 8. saddle, 9. slope with dip direction, 10. ponor, 11. karst spring, 12. river, 13. sea, lake, 14. island and 15. research area (photo by Veress)

The island is characterized by extreme climatic conditions. The amount of precipitation exceeds 8000 mm, the average wind speed is 60–80 km/h, but gusts with a speed of 150–200 km/h occur too (Zamora and Santana 1979).

8.5.2 Karst Features

The combe ridge is dominated by karren and relict landforms. According to their size, the here occurring karen can be megakarren and mezokarren. The characteristic megakarren (Figs. 4.4 and 8.27) are giant grikes and pits (or shafts),

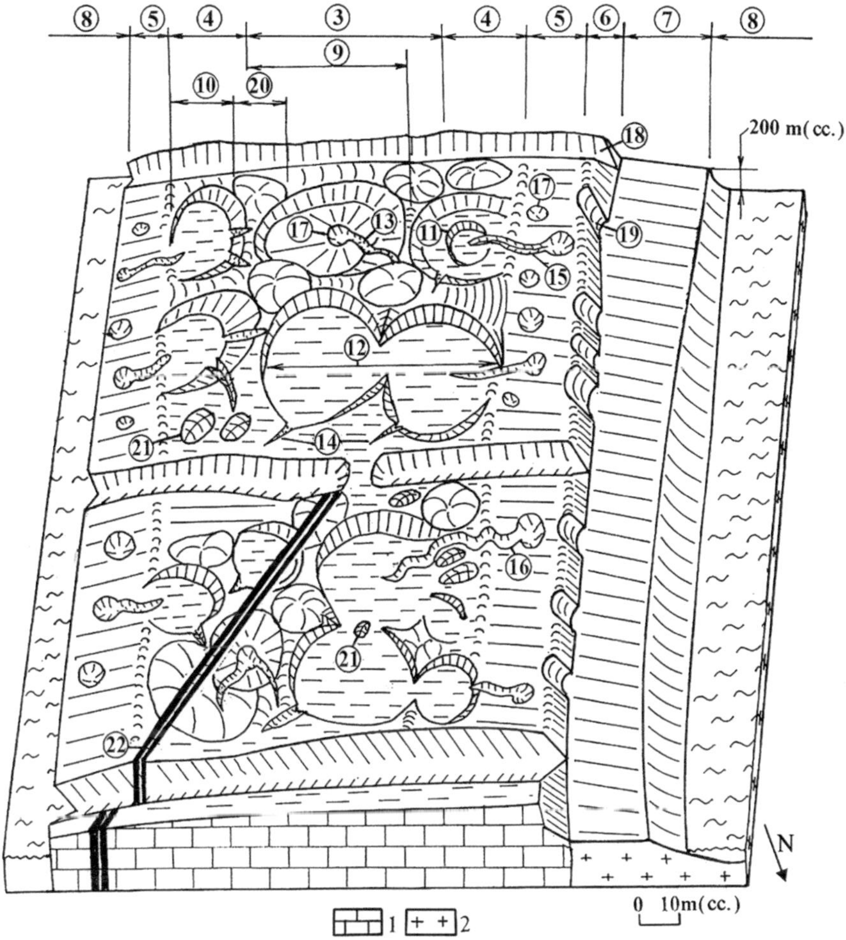

Fig. 8.27 Theoretical figure of the karren forms group of the marble zone (Diego de Almagro island) (Veress et al. 2006—modified). 1. marble, 2. non-carbonate metamorphic rock, 3. middle part of the inside zone, 4. margin part of the inside zone, 5. margin zone, 6. karren formation on the front of the step, 7. slope development on non-carbonate metamorphic rock, 8. fjord, 9. inner dissolution basin, 10. marginal dissolution basin, 11. bottom basin, 12. uvala dissolution basin, 13. rinnenkarren which feeds a pipe, 14. rinnenkarren which feeds a dissolution basin, 15. decantation runnel, 16. meandering decantation runnel, 17. pit, 18. grike, 19. Wandkarren, 20. roche moutonnée, 21. karren 'inselberg' and 22. non-carbonate metamorphic intercalation which has weathered out (photo by Veress)

meanderkarren and kamenitzas (dissolution basins). These karren are regarded as the largest karren features in the Earth (Maire 1999; Maire et al 2009). There is no sharp line between pits and shafts. They mainly, but not exclusively occur on the marginal part of the combe ridge. These drain the surface waters that flow in linear karren features.

Among megakarren, kamenitzas are the most widespread (Figs. 8.27, 8.28 and 8.29). The characteristics of kamenitzas are the following (Veress et al. 2006):

- They can be situated among roche moutonnées, more rarely on the top of roche moutonnées.
- Their diameter is 10–30 m and their depth is 1–2 m. Greater kamenitzas (uvala kamenitza) are developed by the coalescence of kamenitzas.
- Their western slope is gentle and long, while their eastern slope is short and steep. Roche moutonnées may mainly occur at their western margin. At their western slope, some small-sized rinnenkarren (type A channel), while at their eastern margin (and on the surface that dips in this direction) larger sized rinnenkarren (type B channel) occur. These latter are stepped channels.
- Their floor is plain where a smaller inner basin with a depth of 1–2 dm may be present too.
- Their water is drained into the karst by pits or it is tapped by rinnenkarren and meanderkarren.

The giant grikes have EW direction and they dissect the combe ridge into well-separating parts. On their walls, smaller sized wandkarren and relics of pits occur. These latter were formed by the coalescence of pits.

Meanderkarren are complex features, they are meanders and the lower part of which is remnant meander. They are karren features tapping kamenitzas.

Rinnenkarren and ripple karren can be mentioned among mezokarren. Varieties of rinnenkarren are type A channels and type B channels. Type A channels occur on the western side slopes of kamenitzas. Their width and depth are smaller than those of type B channels (the latter may have a width and depth of several decimetres), but their length does not differ from the length of type B channels. The floor of type B channels is mostly dissected by scarp front with a height of some decimetres (stepped rinnenkarren) and they occur on the eastern side slopes of kamenitzas and on the slopes which border them from the east. Type A and type B channels of kamenitzas are not connected to pits, but they terminate on the floor of kamenitzas (type B channel) or above the floor (type A channel).

Ripple karren (Fig. 8.30) are stepped features. The height of steps is 1–2 cm, and their length is 1–2 dm. This karren type is a trittkarren variety. They are widespread, mainly on the eastern slopes of kamenitzas. Where they occur, their density is very high (they show a continuous development).

Relict landforms are roche moutonnées, tail-dune karren buttes (Fig. 2.12) and whaleback-dune karren buttes (Fig. 8.31). Roche moutonnées are glacial erosional features with a height of some metres, the western side of which is gentler and it is more dissected by karren features, while their eastern slope is steeper and less dissected by karren. Tail-dune karren butte is a feature with a height of some dm and even with a length of 1–2 m occurring behind and in an eastern direction from the metamorphic rock block on the marble. Its height is smaller and smaller farther

Fig. 8.28 Uvala dissolutional basin which is on the margin. 1. part of a basin, 2. giant grike, 3. roche moutonné, 4. whaleback-dune karren butte and 5. direction of wind (photo by Veress)

from the rock block. Whaleback-dune karren buttes are arranged in rows with NS direction, the direction of their longer axis is WE. Between these features there occur surfaces (corridors) that are dissolved to plains with small expansion.

8.5.3 *Karren Formation*

On the combe ridge, karren formation and the peculiarity of karren features is manifested in the following things:

- Large size of some karren features.
- Asymmetric cross section of kamenitzas.
- Different karren formations of the slopes bordering the kamenitzas. There are other karren features in smaller density on western slopes than on eastern slopes. Thus, the slopes of kamenitzas show heterogeneous karren formation.
- The presence of relict landforms.

The formation of large-sized karren features is made possible by the large amount of precipitation, which is well proved by the already mentioned high corrosion rate.

The above-mentioned other karren formation characteristics are contributed by glacial erosion, western wind with large speed and non-karstic rock debris. Lee develops at the eastern margin of roche moutonnées. While farther from them, in

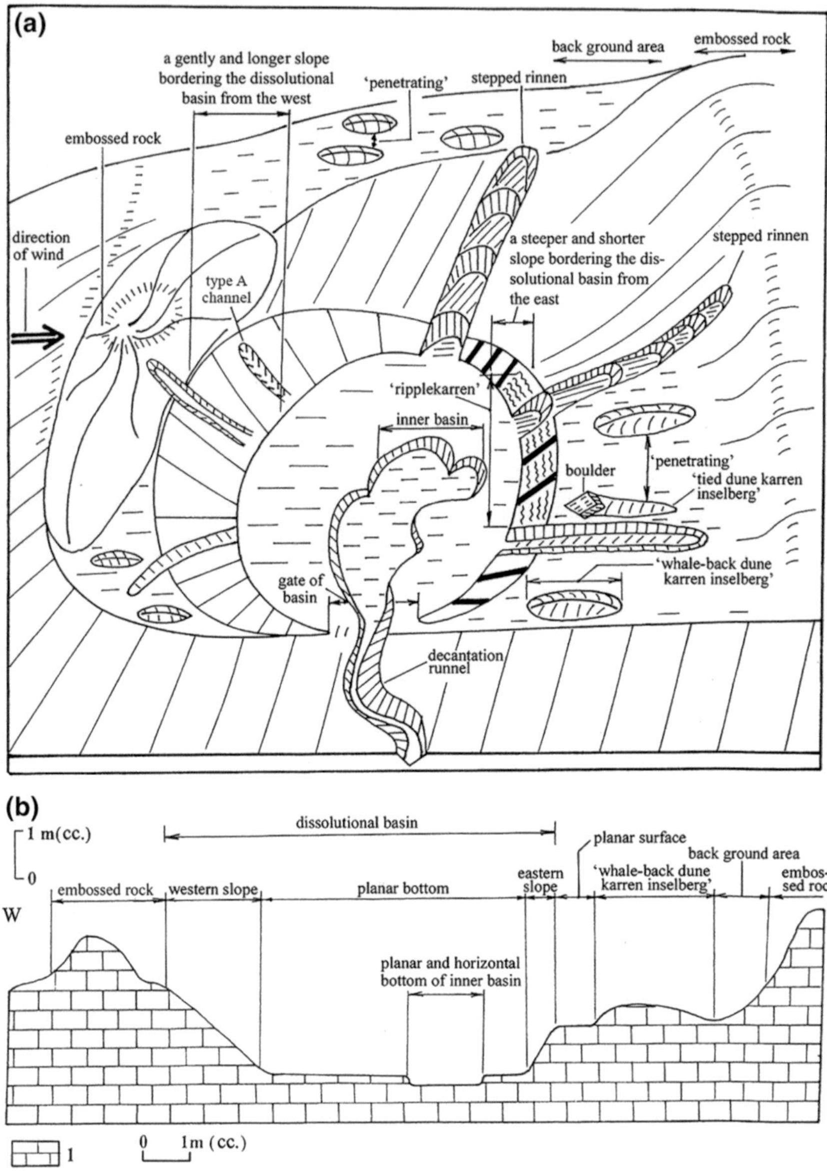

Fig. 8.29 Theoretical figure of a dissolutional basin in space (**a**) and cross-sectional (**b**) (Veress et al. 2006—modified). 1. marble (photo by Veress)

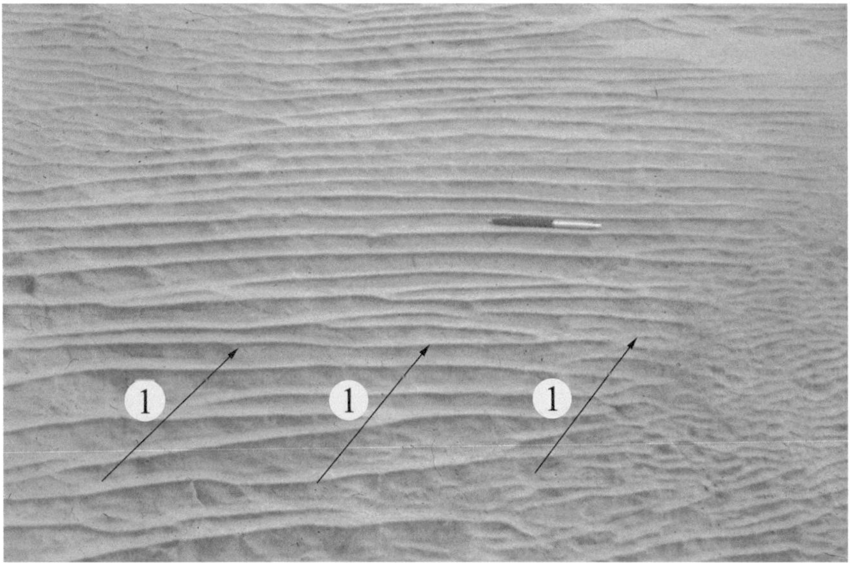

Fig. 8.30 "Ripple karren". 1. direction of wind (photo by Veress)

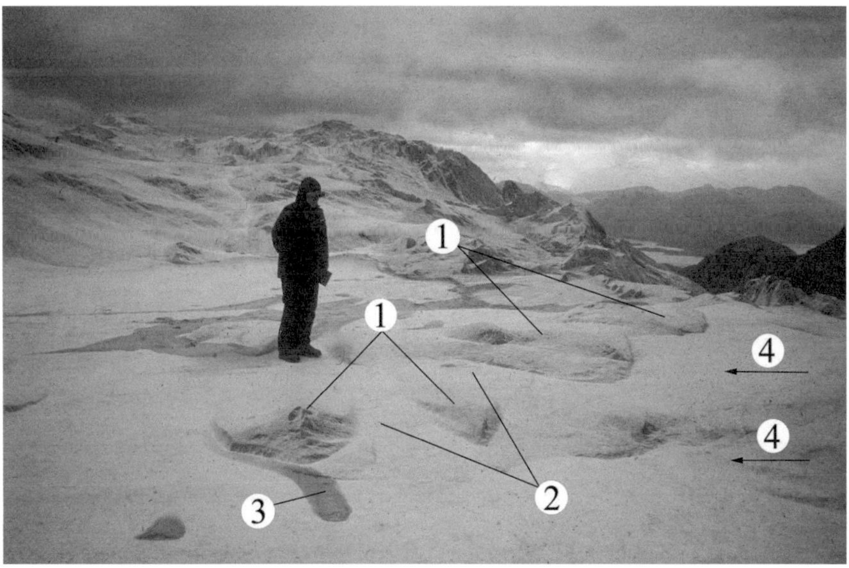

Fig. 8.31 Whaleback-dune karren buttes. 1. whaleback-dune karren butte, 2. corridor, 3. rivulet and 4. direction of wind (photo by Veress)

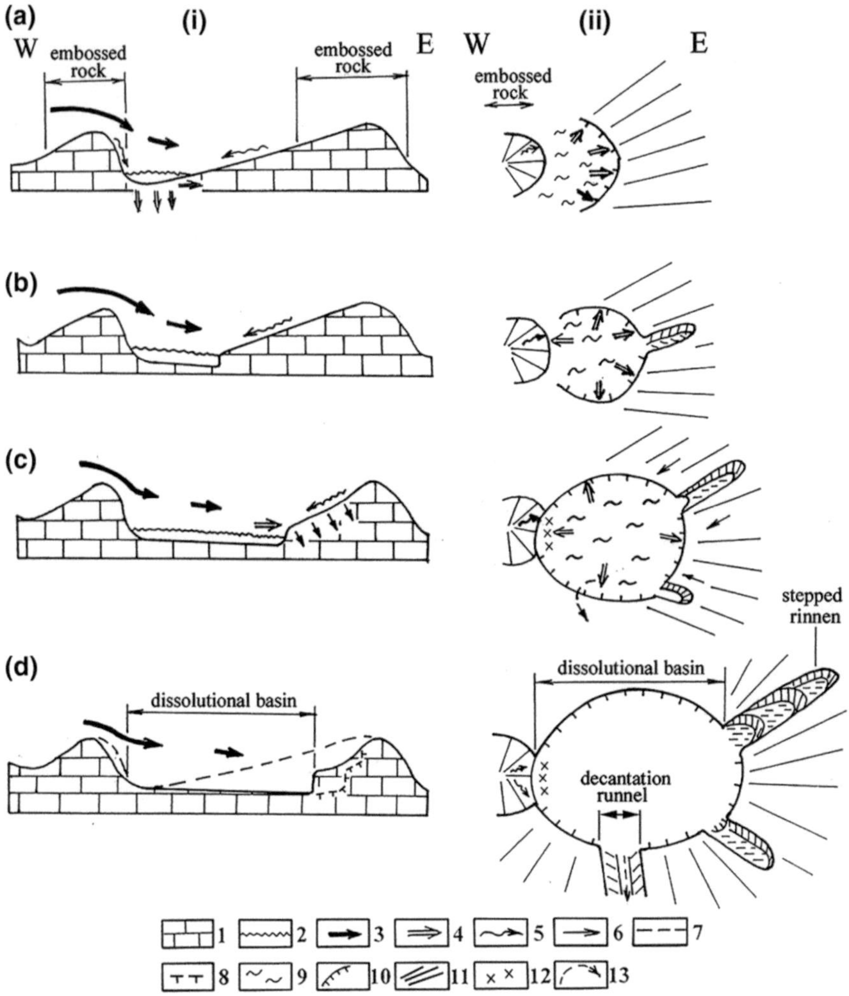

Fig. 8.32 The development of dissolutional basins (Veress et al. 2006—modified). **a** The water of the lake which developed between some embossed rocks will be moved from east to west by the wind, **b** a step develops because of the dissolution of the surface, the step retreats parallel with itself, **c** the background area which borders the step from east deepens because of dissolution and rinnenkarren develop on this surface, **d** the decantation runnel of the basin develops, cross section (i): 1. marble, 2. lake, 3. wind, 4. dissolution in the lake, 5. sheet water, 6. dissolution under sheet water, 7. original surface, 8. the bottom of the stepped runnel, from above (ii), 9. lake, 10. the margin of the dissolutional basin, 11. background area, 12. the original margin of the embossed rock and 13. the decantation water of the basin

eastern direction, where there is the western slope of another roche moutonnée (embossed rock) or other slightly rising surface sections, there will be upwind slopes, and thus they will be surfaces exposed to wind. On upwind slopes, since raindrops and snowflakes will have downwind component, rainwater will have a greater amount on a unit surface than on leeward slopes. As a result of this, according to Szunyogh (2004a, b), if the dip direction of the wind and that of the slope are identical, there is no dissolution, while if they are opposite to each other, dissolution takes place. Wind also causes pressure increase on the upwind slope, by which atmospheric CO_2 can enter the water film that covers the rock, which also results in the increase of dissolution.

In addition to the large amount of precipitation, the development of large-sized kamenitzas is also favoured by western wind. The largest kamenitzas develop behind roche moutonnées since the wind moves the water that is collected behind the roche moutonnées in an eastern direction. Consequently, the kamenitza increases in eastern direction by dissolution (Fig. 8.32). The water moved by the wind is pressed to the slope and dissolves it at its lower part making it steep and at some places overhanging too. The steep eastern slope of the kamenitza is formed. As a result of high wind speed, but also because the wind keeps the water on the slope (consequently, the duration of dissolution increases), the water film of this slope performs a significant dissolution, by which mezokarren (ripple karren, type B channels) develop in large density here and on the bordering surface.

In the development of tail-dune karren butte, the wind, the non-karstic rock block and the surface smoothed by glacial erosion play a role. In the lee of the rock debris, dissolution can only take place by falling precipitation. However, outside the lee, the sheet water being moved by the wind performs dissolution on the plain surface. Here, it is worth mentioning that the dissolution rates presented by Maire et al. (2009) are not directly caused by falling precipitation, but by the sheet water originating from the falling precipitation that is moved by the wind.

In the development of whaleback-dune karren buttes, the wind and the surface being smoothed by glacial erosion have a role. At this time, to the effect of the wind, the sheet water is separated into rivulets and it dissolves the surface in stripes moving on the smoothed surface which dips upwind. Between the rivulets, elongated elevations develop downwind. At their downwind end, the elevations are separated from their environment where the rivulets coalesce. Coalescence can happen where the inclination of the surface is identical with the wind direction, and thus lee can develop. In the lee, the rate of water flow decreases; therefore, the widening rivulets can coalesce with each other (Fig. 8.33).

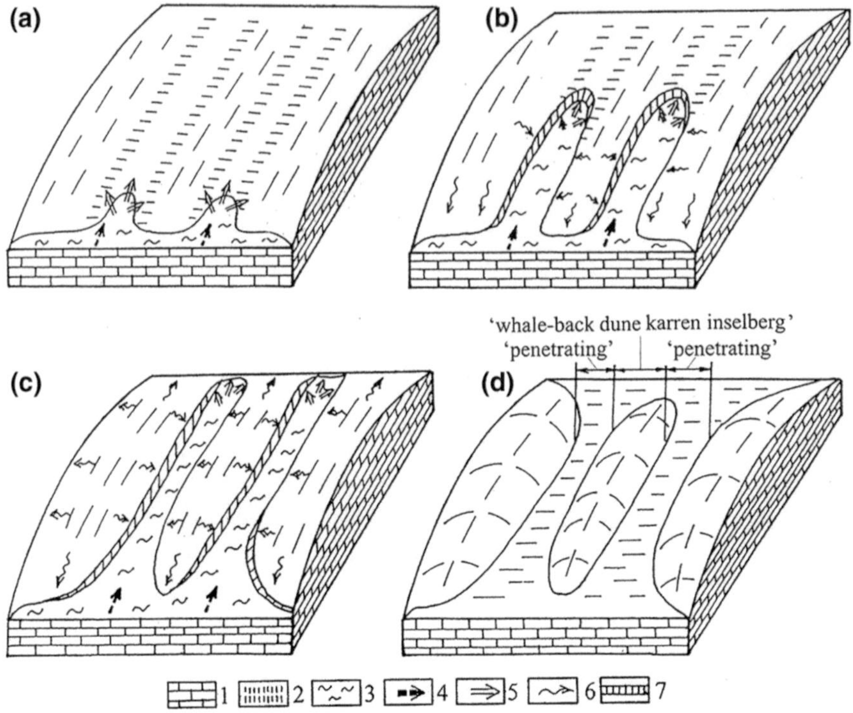

Fig. 8.33 The development of the "whaleback-dune karren inselberg" (Veress et al. 2006—modified). **a** The wind moves the rain water in the WE depressions to the east on the marble surface; therefore, the rock dissolutes to the eastern direction along the depressions, **b** the water streams with the updip of the slope; therefore, the developing channel is narrow, **c** the dissolutional surface cuts through the highest surface part, the dissolution continues on the dip of the slope, for this reason the water, which moves to the east spreads, and the channel can widen, **d** because of the widening of the dissolutional channel the surface parts which could not dissolute between the channels will budd, 1. marble, 2. depressions' margin, which could develop non-dissolutionally (for example, glacial strius), 3. lake, 4. direction wind, 5. dissolution, 6. dissolution at sheet water and 7. the wall of the dissolutional channel

References

Bögli A (1976) Die Wichtigsten Karrenformen der Kalkalpen. Karst Processes and Relevants Landforms. ISU Comission on Karst Denudation, Ljubljana, pp 141–149

Djurovič P (2009) Reconstruction of the Pleistocene Glaciers of Mount Durmitor in Montenegro. Acta Geogr Slov 49(2):263–289

Ford DC, Williams PW (1989) Karst geomorphology and hydrology. Unwin Hyman, London

Forsythe R, Mpodozis S (1983) Geologia del basamento pre-jurasico superior en el archipelago Madre de Dios, Magallanes, Chile-Servicio National de Geologia y Mineria, Chile, Bol., pp 39–63

Gams I (2002) Changes of the Triglav Glacier in 1955-94 period in the light of climatic indicators. http://ai.ijs.si/mezi/personal/triglav/

Haserodt K (1965) Untersuchungen zur Hohen - und Altersgliederung der Karstformen in den nördlichen Kalkalpen - Münchner Geogr. H. 27

Hinnov LA (2003) Lofer Cyclothems in the Dachstein Limestone of the Julian Alps. Geol Soc Am, abstracts with programs 35(6):426

Howard AD (1963) The development of karst features. Bull Nat Speleol Soc. 25:45–65

Jennings JN (1985) Karst geomorphology. Basil Blackwell, New York, p. 293

Maire R (1976) Recherches géomorphologiques sur les karsts haut-alpins des massifs de Platé, du Haut-Giffre, des Diablerets et de l'Oberland occidental. Thèse, Université de Nice

Maire R (1977) Les Karsts haut-alpins du Platé, du Haut-Giffre et de la Suisse occidentale. Rev Géogr Alpine 65:403–425

Maire R, l'Equipe Ultima Esperanza (1999) Les "glaciers de marbre" de Patagonie, Chili. Un karst subpolaire océanique de la zone australe. Karstologia 33:25–40

Maire R, Jaillet S, Hoblea F (2009) Karren in Patagonia, a natural laboratory for hydrogeolian dissolution. In: Ginés A, Knez M, Slabe T, Dreybrodt W (eds) Karst Rock Features. Karren Sculpturing Zalozba ZRC. Institut za raziskovanje krasa ZRC SAZU, Postojna, Ljubljana, Carsologica, 9, pp. 329–348

Reynard E (1997) Carte géomorphologique des Lapiés de Tsanfleuron (Hautes Alpes Calcaires, Valais). Bull Soc Neuch Géogr 41:23–38

Reynard E (2008) Le lapiaz de Tsanfleuron: un géomorphosite glacio-karstique à protéger et à valoriser. In: Reynard E, Hobléa F, Delannoy J-J (éds) Karsts de montagne: géomorphologie, patrimoine et ressources, Chambéry/Lausanne, Cahiers EDYTEM, No. 7, p 168

Szunyogh, G. (2004a) Talajnélküli mészkőfelszínek leoldódási idejének elméleti vizsgálata (Theoretical study of the corrosion time of limestone surfaces without soil) – Karsztfejlődés IX. BDF Természetföldrajzi Tanszék, Szombathely, pp 35–51 (in Hungarian)

Szunyogh G (2004b) Theoretical investigation of the duration of karstic denudation on bare, sloping limestone surface. Acta Carsol 34(1):9–23

Tóth G (2007) A mérsékelt övi mészkő magashegységek fedetlen karros celláinak osztályozása és fejlődése (The classification and development of the uncovered karren cells of the limestone high mountains in the temperate belt), Department of Physical Geography, Berzsenyi Dániel College, Szombathely, p 118 (in Hungarian)

Tóth G (2008) Actes colloque AFK. In: Reynard E, Hobléa F, Delannoy J-J (éds) Karsts de montagne: géomorphologie, patrimoine et ressources, Chambéry/Lausanne, Cahiers Savoisiens de Géographie et Travaux et Recherches de l'Institut

Trudgill ST (1985) Limestone geomorphology. Longman, New York

Veress M, Tóth G, Zentai Z, Kovács GY (2001) Study of a new method for characterising karren surfaces based on alpine researches. Revue de Géographie Alpine 89:49–62

Veress M, Zentai Z (2004) Karros lejtőfejlődés a Triglav északi előterében (Karren slope development in the northern foreground of Triglav). Karsztfejlődés IX, pp 177–196 (in Hungarian)

Veress M (2016) Covered Karst. Springer, Dordrecht, Heidelberg, London, 536 p. https://doi.org/10.1007/978-94-017-7518-2

Veress M (2017) Solution doline development on glaciokarst in alpine and Dinaric areas. Earth Sci Rev 173:31–48

Veress M, Szunyogh G, Tóth G, Zentai Z, Czöpek I (2006) The effect of the wind on karren formation on the Island of Diego de Almagro (Chile). Zeitschrift für Geomorphologie 50:425–445

Zamora E, Santana A (1979) Caracteristical climaticas de la costa occidental de la Patagonia entre las latitudes 46°40′ y 56°30′s. Anales Inst Patagonia 10:109–154

Zivaljevič M, Vujučič P, Stijovič V (1989) Tumačza list Žabljak K. 34-37. Osnovna geološka karta 1:100000, Beograd

Chapter 9
Notable Glaciokarsts of the World

Tamás Telbisz, Gábor Tóth, Dmitry A. Ruban and Jaroslav M. Gutak

Abstract In this chapter, notable glaciokarsts of the world are presented. Geographical location, geologic and tectonic settings, climatic conditions, glaciation phases as well as surface and underground karst landforms are presented about each selected region. Obviously, the areal extent, the degree of exploration and the amount of publicly available information are different in each case. Historically, the first glaciokarst studies were based on the Alps, the Pyrenees, the Dinaric Alps and the British Isles, and they have remained in the focus since then. Hence, these regions are presented here in more detail, but even these presentations can be considered only short overviews. Some other glaciokarst terrains, such as Scandinavia or the Rocky Mountains, have also been thoroughly studied but later in history; nevertheless, there are abundant internationally available publications about them. Certain parts of the Balkan Peninsula, the Apennines or even Anatolia received high attention more recently and novel methods have been used to investigate their glaciokarst terrains. The Carpathians and the Appalachians, which are also discussed in this chapter, are extensively studied mountains in general, but glaciokarsts occupy a relatively small proportion in them. On the other hand, there are still regions, which are difficult to access, where glaciokarsts are poorly explored, and/or the available literature is limited (or the publications are only in Russian, for instance). Some of them, namely, the Altai Mountains, the Greater Caucasus, the Tian Shan, the Pamir and the Patagonian archipelago, are also briefly presented here. Finally, it is noted that our selection does not contain all glaciokarsts of the world because it is beyond the scope of this chapter.

Keywords Alpine glaciokarst · Arctic glaciokarst · Glacial phases
LGM · Microglacier · Cosmogenic nuclide chronology · U-series dating
Exposure age · Isostatic rebound

M. Veress et al., *Glaciokarsts*, Springer Geography,
https://doi.org/10.1007/978-3-319-97292-3_9

9.1 Introduction

Glaciokarst terrains are widespread on Earth. In the followings, we try to present the variety of glaciokarst landscapes. Historically, the first studied glaciokarsts were the Alps, the Dinarides, the British Isles and the Pyrenees. All of them contain extended glaciokarst terrains with abundant landforms and caves. Lately, the Rocky Mountains (especially the Canadian parts) as well as Scandinavia were also thoroughly explored and some novel methods, such as U-series speleothem dating, were just demonstrated using these regions as examples. More recently, the glaciokarsts of the southern parts of the Balkan Peninsula, the Apennines and Anatolia were also analysed by modern techniques such as cosmogenic nuclide methodology, so these regions can be presented also based on a wealth of new data. The Appalachians and the Carpathians are also presented in this chapter. They are well-explored mountains in general, but the area of glaciokarsts is relatively small in them. In addition, some less easily accessible regions, such as the Altai Mountains, the Greater Caucasus, the Tian Shan, the Pamir and the Patagonian archipelago, are also presented here by researchers having field experiences. Nevertheless, there are still many other glaciokarsts on Earth that are not discussed in this review due to the lack of available information, or field experience or other reasons. Just to mention some of them, there are significant glaciokarsts in the Himalayas, in Siberia, in Papua New Guinea, in New Zealand, in Tasmania and many places in Canada (out of the Rocky Mountains and the Appalachians, which are presented here).

In the regional overviews of this chapter, the following scheme is generally applied with modifications in some cases. First, the factors influencing glaciokarst development are discussed, namely, the geologic and tectonic settings, the relief and the climatic conditions. Second, the style and phases of glaciations are briefly described, this information has quite different level depending on the given region and finally, surface and underground glaciokarst features of are presented, occasionally listing the most remarkable study areas within the region.

9.2 Alps

The *Alps* are the most thoroughly studied mountains of Europe and probably of the whole world in all aspects. This is mainly due to the fact that their research history is the longest. During 300 years of research in "modern" terms, the literature has been enriched with a number of phenomena, which are among the basic recognitions of geomorphology (or geology). In geological terms, we can mention the significance of the cover structure and the tectonic windows, as well as the description of a number of Mesozoic rock formations. The science of geomorphology was also enriched by several significant discoveries. From the point of view of our study, it is worth mentioning the description of mountain glacial forms and high mountain karst features. We discuss the glaciokarstic forms according to the geological zones of the mountain range.

9.2.1 Relief Settings

Europe's centrally located mountain range is about 220,000 km^2 in area, its length is 1200 km and its average width is 200 km. There are 82 peaks over 4000 m asl (almost exclusively in the Western Alps), with Mt. Blanc (4809 m) as the maximum, and the Alps are considered the highest mountain in the continent. (Although the highest peak of Europe, Mt. Elbrus, is located in the Caucasus, the whole mountains are not considered European.) There is a significant difference in the peak and mean elevation between the Eastern and the Western Alps (Fig. 9.1).

9.2.2 Geologic Settings

Due to the geological complexity of the Alps, the geological conditions are presented in this section only in a nutshell. Instead of a complete evolution history, the distribution of the carbonate rocks on the surface is emphasized.

We divide the mountain range based on its geological features into two major morphogenetic units, the Western and the Eastern Alps. The Eastern Alps are located east of the Upper Rhine valley. From their three structural units, the north

Fig. 9.1 Map of notable glaciokarst locations in the Alps with the extent of glaciations

and south are built up of carbonate rocks. The southern is considered an independent geological unit by some experts, called the Southern Alps. The central part of the Eastern Alps is part of the Penninic Unit, built up of highly metamorphosed rocks. There are no rock formations suitable for karstification. In both the northern and southern carbonate units, Mesozoic limestones are predominant (e.g. Dachstein Limestone, Wetterstein Limestone, Hierlatz Limestone). As a typical constituent of the Southern Alps, it is important to mention dolomite, which occurs in huge quantities. Approximately, two-thirds of the Eastern Alps area is built up of karstic rocks, which offer significant potential for glaciokarstic evolution, even if these are not among the highest mountain ranges within the Alps.

The geological structure of the Western Alps is more complicated. It is worth considering internal and external sedimentary traits in terms of karstification. The inner limestone mountains are located largely in Switzerland, while the outer mountain ranges are found in France.

The formation of most of the carbonate rocks in the Alps can be dated to the Mesozoic. Without being exhaustive, the major carbonate rocks in the Alps are the following:

The *Hierlatz Limestone* (Jurassic, Upper Sinemurian–Pliensbachian): limestone of skeletons of crinoids and brachiopods. The *Kössen Formation* (Upper Norian–Rhaetian): dark grey marl, clay marl and calcareous marl rich in organic matter. It was formed in closed basins under anoxic conditions. The most important is the *Dachstein Limestone* (Upper Karnian–Rhaetian): lagoonal reef facies limestone with microbial fabric and Megalodus, Lofer cycles, often of grey colour. It was formed in the peritidal and subtidal section of extensive carbonate platform. The *Main Dolomite* (Upper Triassic, Karnian–Norian): thick sediment sequence formed on carbonate platform, light grey to grey colour, mostly thick-banded, early diagenetic dolomite with Lofer cycles of alternating peritidal and lagoonal facies. The *Hallstatt Limestone* (Upper Triassic, Karnian–Norian): pelagic basin facies, predominantly of pink or red colour, generally well-stratified, locally thick-banded, fine-grained limestone. The *Wetterstein Formation* (Ladinian): light grey, massive reef limestone with calcareous sponges, corals and hydrozoa. Its lagoonal facies is light grey or dark grey, thick-banded, with Lofer cycles of intertidal and subtidal members. The *Steinalm Limestone* (Anysian): shallow marine limestone with Dasycladaceans algae; white, greyish white or light grey, thick-banded. Cyclicity of intertidal and subtidal laminae is emphasized by Dasycladaceans algae. And finally, the *Gutenstein Formation* (Anysian): carbonates of high organic content, formed in shallow marine anoxic environment, dark grey or black, fine-bedded, bituminous (with white calcite veins) limestone alternating with (dark) grey stratified, bituminous dolomite and thin (1–2 cm) marl intercalations.

The Western Alps also consists of several rock formations that are prone to karstification. The *Urgonian Limestone* is worth mentioning, which is the building rock of the Vercors, Chartreuse and Bauges Mountains. This formation of Early Cretaceous is massive and thick-banded with a thickness of 300 m. It is excellent for karstification. The similar *Tithonian Limestone* was deposited in a pelagic environment at the very end of the Jurassic. It occurs in several mountain ranges of the Western Alps and in the French Prealps, too.

9.2.3 Climatic Settings

The Alps as Europe's central mountains are located at the meeting point of climatic provinces, from which they rise as an island. Climatic impacts reaching the foothills are either blocked or allowed to penetrate the mountain interior. In the Western Alps, the Atlantic and Mediterranean influences are predominant, but in the case of the Eastern Alps there is noticeable continental influence as well. For instance, Grenoble in the French Alps (219 m asl) has 11.2 °C mean annual temperature and 18.5 °C mean annual range of temperature, whereas Vienna (186 m asl) in the Eastern Alps has 9.9 °C mean annual temperature and 20.5 °C mean annual temperature range, showing a bit more continentality. Larger differences are in precipitation, while Grenoble receives 856 mm a year, Vienna gets only 623 mm (https://en.climate-data.org). Naturally, as one goes higher, the temperature decreases, and precipitation increases. The average lapse rate is 0.65 °C, which is significantly affected by geographical location, exposure and vegetation cover. This value is significantly lower in winter and rises to around 1 °C in summer. The Alps play a very important role in the European water cycle, freshwater supply, runoff and river flow recharge (Isotta et al. 2013).

9.2.4 Glaciation in the Alps

The classical terminology of alpine glaciations (Günz, Mindel, Riss and Würm) was introduced by Penck and Brückner (1909). Since the second half of the twentieth century, these general terms have been replaced in many cases by local chronological systems. Nowadays, oxygen isotope stages (OIS) can be used more precisely in Quaternary studies due to modern dating techniques; however, the classical terms are still often used in reviews.

The earliest glaciation took place at the end of the Pliocene that corresponds to the 96–100 OIS. The last glacial maximum (LGM) occurred in the Late Pleistocene, represented by OIS 2 (Martinetto and Ravazzi 1997; Uggeri et al. 1994, 1995; Bini 1997). Alpine glaciation events fall between these two endpoints with many climatic fluctuations.

The maximum alpine glaciation was reached during the *Würm* period in many places of the Eastern Alps and mostly in the *Riss* Glacial in the western Alps (Fig. 9.1). The ice cover of the Alps generally spread from northwest to southeast. The main reason is the distribution of the precipitation, which caused the glaciation of the Atlantic side of the mountains already in the Early Würm. Afterwards, the glaciation shifted towards the Eastern Alps and reached its maximum there. Glaciers up to 700–800 m thickness appeared in low-lying valleys. The large amount of till attest glaciers which moved down to the Piedmont. For instance, glaciers reaching the southern Piedmont of the Alps at Lake Garda or Tagliamento river had terminals at 100–200 m asl.

As for the glaciations, there were significant differences among certain parts of the Alps. This is especially true for the Würm period. At the beginning of Würm, the western and northern regions were the most humid. However, there was a change at the end of the Würm, when the zone of the highest precipitation was more in the Mediterranean Region, and Southern Alps were the wettest (Ivy-Ochs et al. 2008). Nevertheless, the Eastern Alps and the central areas of the mountain range received all the timeless precipitation than the peripheral areas. Due to these climatic conditions, the glaciation of peripheral areas reached significant dimensions, even at relatively low altitudes.

The change of alpine glaciers in the Holocene is still an important research topic. Numerous studies concerned Holocene environmental indicators identifying several advance and regression cycles in the Alps. These are mostly short-term changes indicating the unstable nature of glaciation. The Holocene ice cover was significantly reduced, which made glaciers sensitive climate indicators. Variations in glacier size are often due to local climatic factors.

According to morphological position, Alpine glaciers can be divided into two groups. The first group includes large glaciers filling fluvial or tectonic valleys. Due to their lower elevations, they are not in direct contact with glaciokarst plateaus; however, they indirectly influence karstification. The valley floors are between 500 and 1000 m asl in the Austrian Alps. This is the level where springs issue from the large karst massifs.

The other group of glaciers was formed in the higher (2000 m) zones, commonly on high plateaus and they are decisive factors in glaciokarst formation. Due to their relatively smaller volumes, they responded sensitively to Pleistocene climate changes. Due to their permanent changes, they had a spatially and temporally variegated effect on the karstic surface.

There are three remarkable steps based on the magnitude of the glacial effect. Naturally, the Late Pleistocene ice had the largest extent, and its retreat is generally dated to 11–12 ka. However, there are temporal variations, for instance, in the westernmost and lower lying areas of the Alps (e.g. in Vercors) glacier retreat happened 15 ka ago, while on the northern plateaus of the Eastern Alps (e.g. in Totes Gebirge) it occurred about 10 ka ago. From a morphologic point of view, this is an important milestone. The second step is the Little Ice Age, which lasted from 1650 to 1850. Its effects are more pronounced in the Western Alps, where this short cooling event resulted in several kilometres of glacier advance. This period can be relatively well reconstructed from moraine ramparts. Finally, a significant area became ice-free in the last half-century, and hence carbonate rocks became exposed to "fresh" karstification. At these sites, subglacial and fresh subaerial dissolution forms are mixed today. The joint result of the above steps is a very complex and interactive glaciokarst development.

9.2.5 Glaciokarst Areas in the Western Alps

The structural evolution of the Western Alps started in the Eocene–Oligocene. The main period of orogeny was the Miocene, but lasts even now. Thus, the Western Alps reached their maximum elevation in the Quaternary.

Structurally, two ranges of the Western Alps, the exterior and the interior sedimentation zones, are constructed of carbonate rocks. Although the average elevation of these two ranges is about 1000 m lower than that of the two large granite massifs of the central zone of the Western Alps, their altitudes are high enough for the formation of glaciokarsts. Glacial features dominate the Alps, especially the western part. We can see glacial valleys and moraine ramparts from 500 to 600 m of altitude. Due to the extent of ice cover, glaciation affected all karst areas above 600 m. Glacial erosion has transformed the surface to such an extent that the identification of pre-Quaternary landforms is hardly possible.

The largest extent of ice cover was reached in the Riss, and the glacier terminus was west of Lyon. Subsequently, a slight retreat (10–40 km) is observed in the Würm (Coutterand and Buoncristiani 2006). Due to the complete ice cover, the western part of the Alps caused a strong glacial destruction. Hence, the Pliocene or Early Pleistocene morphological environment is very difficult to reconstruct. It is certain that at the end of the Pliocene there was a mild, wet (slightly subtropical) climate in the area. The conditions favoured intensive karstification. Not only surface landforms were formed during this time, but extensive cave systems also took shape at relatively high altitudes. During the glacial periods, a connected valley glacier network covered typically the Western Alps. In addition, in the Vercors and Chartreuse (French Prealps) a minor, westward extending glaciated area was also found. The snow limit was at around 1400 m in the Vercors and around 1200 m in the Chartreuse (Chardon 1984).

The most common glaciokarst forms in the Western Alps are glaciokarst cirques. Over-deepened glacial depressions serve as relatively large catchment areas, so intense karstification takes place in them. Such features were found in the Rochers de Chalves Chartreuse, but glaciokarst depressions were described by Maire (1990) as well from the surrounding area of Tsanfleuron and Diablerets. The glaciokarst depressions evolve in densely fractured limestone, and there is debris in their bottom, or debris slopes in their sides. In winter, snow is accumulated in them, and snowmelt or rainwater may occasionally form small lakes. In these places, recent karstification deepens the interior of the form. Shafts and vertical caves are opened at the bottom and side of glaciokarst depressions. Their openings are usually sealed by moraine or debris. Large depressions occur in the irregular valleys of the higher zones, where the preglacial valleys were underdeveloped, differently from the lower regions. The valley network of the Western Alps shows a strong structural preformation. These are the places where glacial erosion was the most efficient.

On the other hand, glaciers exerted less intensive erosion on the plateaus. Areal denudation may be significant, but new glacial forms are less remarkable. In these places, the erosion of ice was similar to the polar icecaps. The surface is similar to the fjells of the Scandinavian Mountains. Glacier striae demonstrate the work of plucking erosion. Typically, they are found on Désert de Platé, Lapiés de Tsanfleuron or Dévoluy in the Vercors Mountains. On the plateaus, the remains of the former ground moraine and the debris from the fresh attrition are mixed. The gelifraction is supplemented by recent karstification from the meltwater. As a result, karren fields cover a large part of the plateaus. Well-developed karren fields were formed: Lapiés de Tsanfleuron (Tóth and Reynard 2011, Fig. 9.2), Desert de Platé (Maire 1990) or Märenberg plateau (Bögli 1960). In some places, where the ice exerted a major transformation, escarpment karsts were produced. In general, the ice emphasizes the structural and bedding properties of limestone. In the case of Märenberg and Tsanfleuron, we can find escarpment karsts, and Talour (1976) described similar features on Charmant Som in the Chartreuse Mountain.

Glacial valleys are characterized by regular glacial forms. Parabola cross-sectional valleys, roche moutonnée and terminal moraines are abundant. Large amounts of glacial and periglacial accumulation often complicate the development or occasionally the exploration of karsts.

Fig. 9.2 Karrenfield in the foreland of Tsanfleuron glacier (photo by Tóth)

9.2.6 *Eastern Alps*

9.2.6.1 The Northern Limestone Alps

The range of the Northern Limestone Alps is the most extensive glaciokarst area of the Eastern Alps. These mountains represent the entire range of alpine glaciokarsts. The diversity is due to the fact that it is not a continuous chain, but a series of mountains separated from each other by deep valleys. Thus, each of them had a distinct evolutionary path having both common and individual traits. One of the common features is that they have a plateau-like surface. Average ridge elevations are between 2000 and 3000 m asl, the highest peak is in the Dachstein (2995 m). Another common feature is the intensive karstification relative to the temperate zone (Seefeldner 1961), accompanied by significant sediment accumulation in the foothills. Since the Pliocene–Pleistocene boundary, several glaciations affected the area. The glaciers began to destroy karst macroforms at an early stage (Weingartner 1983). At the beginning of the Pleistocene, there was a slow uplift in the area. As a consequence, the mountains of the Northern Limestone Alps were gradually transformed to a high mountain karst. Due to climatic deteriorations, this is process was particularly fast. The zone above 2000 m began to be converted to plateaus. The mountains were separated by the glacier network filling the deep river valleys (e.g. Salzach, Enns).

Tennengebirge

The *Tennengebirge* is a member of the Salzburg Limestone Alps, and its highest peak is Raucheck (2430 m). It is a typical Eastern Alpine plateau mountain. The plateau itself extends over 3 km^2 area and rises above 2000 m asl. It shows an entirely glaciokarstic form assemblage. The Tennengebirge is built up of Dachstein Limestone of more than 1000 m thickness. The well-known cave systems of the mountain range developed since the Miocene (e.g. Eisriesenwelt, Fig. 9.3), and some old caves are now in a ruined state (ruinenhöhlen).

Surface karst is composed of poljes, dolines formed at the beginning of the Pliocene karstification. Since its inception, it has had a significant fluvial effect. With the Pliocene uplift, the rivers flowed further away and the karst became an autochthonous system. The uplift is proven by the vertical development of underground networks.

In the higher regions, true glaciokarst evolution took place. Karstified roches moutonnées, dolines and shafts replaced the Tertiary landforms. Shafts are formed under snow fill. The karren surfaces of the glacial period are mixed with the previously formed karren, which have been thoroughly modified by recent mechanical weathering. Bauer and Zötl (1972) identified the plateaus as *cuesta karst*, which is divided into ice-transformed hollows and elevations aligned along faults. The pre-Pleistocene forms were largely destroyed. Either the ice has

Fig. 9.3 Eisriesenwelt Ice Cave in Tennengebirge (photo by Egri)

devastated them, or they are buried under till. Postglacial karren formation was controlled by precipitation (snow/rain ratio). Audra (1994) places the level of rundkarren formed by covered and semi-covered karren formation between 1700 and 2100 m. Above 2100 m, bare karren are the most typical, but only in a very narrow zone. From 2100 m upwards a thin debris layer covers the surface, and karren forms are missing due to intensive frost shattering.

Dachstein

The *Dachstein Mountains* are a typical Alpine glaciokarst, covering an area of 500 km^2. Its surface is covered by the Upper Austroalpine Nappe and the main building rock is the Upper Triassic Dachstein Limestone. Among the three mountain ranges (Tennengebirge, Totesgebirge and Dachstein), it has the highest plateau. The average altitude of the plateau is over 2000 m. The southern, higher part of the plateau is still covered with ice. Like other Alpine glaciers, it is in a significant regression state. In the forefront, there are recent karren, trittkarren and the rinnenkarren being the most recognizable. The rinnenkarren are especially primitive, and their widths are less than a few inches (Tóth 2004, 2009). There are glaciokarst dolines on the north side of the plateau. In the Riss glaciation, the plateau was covered with ice (Van Husen 2000), which heavily eroded the plateau.

Therefore, the preglacial forms can be only hypothetically reconstructed. According to Veress (2016a), giant dolines were formed during the Riss-Würm interglacial. Dolines effectively helped snow and ice accumulation.

Totes Gebirge

This 1100 km^2 area mountain is located in a central position within the Northern Limestone Alps. Like the Dachstein and the Tennengebirge, Totes Gebirge also towers above its surroundings. The highest peak of Totes Gebirge is Grosser Priel (2515 m). The Totes Gebirge is also mainly built up of Dachstein Limestone, which is a part of the Upper Austroalpine Nappe. Werfen Limestone has a smaller extension, and at the western side of the mountain, Jurassic Limestone is also found. The plateau of the mountain ranges between 1600 and 1800 m, and it was formed also by Pleistocene glaciers. The glacial valleys inside the mountain range are connected to the peripheral valleys by steep steps. The preglacial forms consist of shafts, ponors and giant dolines. The geomorphologic structure of the plateau is made up of topographic steps of 10–15 m height (Fig. 9.4). On the relatively lower plateau surfaces, dwarf ponds and soil patches are found. The topographic step dips are variable that plays an important role in the development of the karren form assemblage. The Dachstein Limestone has more than 1000 m thickness, which is ideal for karstification. While on the neighbouring Tennengebirge built up of Wetterstein Limestone, only primitive karren are found due to the intense frost shattering, there are abundant microkarstic features in the Totes Gebirge. There are several paleodolines on the plateau, which were deepened by glacial erosion as well, but their modern karstification is only indicated by smaller forms and karren in their interior. Thus, their entire surface is not actively karstified (Tóth 2003; Veress 2016b, 2017). Common forms are vertical shafts, the origin of which was investigated by Gruber et al. (1998).

Fig. 9.4 Tilted beds (Schichtreppenkarst) in Totes Gebirge, the most specific glaciokarst forms in the Northern Limestone Alps (photo by Tóth)

9.2.7 The Southern Alps

The Southern Alps are usually not considered an independent unit of the Alps but they have some distinct geologic, tectonic and climatic settings, so it is worth discussing them separately. The highest point of the mountain range is Ortler (3905 m asl) found in South Tyrol (Italy).

The glaciation of the Southern Alps has been studied for a long time. The formation of valleys and lakes has been explained by the events of glaciation (Garda, Como, Orta, Iseo lakes, etc.). Thirteen Pleistocene glaciations have been identified in the area (Bini et al. 1998). In general, there is very little difference between the spatial extent of the large glaciations. It is likely that the lowest altitude glaciokarsts in the Alps are found here.

The southern foreland of the Alps is shaped by a valley system from north to south. In the valleys, there is an advanced endokarst network, which opens at the valley sides. The relationship between the karst and the glaciation was summarized in a model by Baker (1968). First, glacier width and thickness of is growing, while its terminus is lowering. In the second stage, the ice completely fills the valley and suppresses the karstic systems. In the third stage, glacier recession begins and the karst becomes ice-free again. The fourth stage is characterized by postglacial evolution. In the transitional stages (stage 1 and 3), the movement of ice may be oscillating, additionally a periglacial environment may be evolving (Washburn 1979). Naturally, glaciations cause the shift of vegetation zones as well. More importantly, Alpine forests are retreated during glacial advance periods. This significantly reduces the general aggressivity of karst waters.

Large systems of endokarst have been considered by some to be of glaciation origin (Delannoy 1986), while others associate them with normal erosion (Chardon 1989). In general, it can be accepted that new caves were only exceptionally created during the glaciations. Changes in the flow regime may enlarge passages during glacials, but dissolution is less effective (Bini 1998). Significant cave sections could become completely inactive during glaciations.

9.2.7.1 Julian Alps

The Julian Alps is 4400 km^2, the larger part of which lies in the northwestern corner of Slovenia. One-third of it falls to Italy. The stratigraphic and geological characteristics of the Southern Limestone Alps differ in many points from the usual structure of the Alps. In the Southern Alps, nappes have a southern vergence, making a transition to the Dinarides. The core area of the Julian Alps is formed by the 1000-m-thick Triassic Main Dolomite, which is overlain concordantly by 700 m of Dachstein Limestone. Its formation period is late Nori–Rhaetian and the average bed thickness is 2.5 m (Hinnov 2003). A number of well-developed, U-shaped glacial valleys are found in the Julian Alps (Fig. 9.5). There are two well-known areas in the mountains in terms of the glaciokarst. The Seven Lakes

Fig. 9.5 Characteristic U-shaped valley (Krma) in the Julian Alps (photo by Telbisz)

Valley, located in the Eastern Julian Alps, crowned by Triglav peak (Fig. 9.6), is rich in glaciokarst forms. In the western part of the mountain, the Canin plateau partly stretches over Italian territory, too. In the glacier valley north of Triglav peak, there is an area at 2200–2300 m, full of actively forming karren (Veress 2010). The floor of the glacial valley is dissected by escarpments and rock basins, the height of escarpments is up to 10–15 m. There is considerable debris coverage in the area partly due to frost shattering and partly due to remaining moraine materials. This section of the glacier valley has been covered with ice even recently, and it became ice-free only some 50 years ago. Hence, karren features of this area are underdeveloped. On the bedding plane slopes of uncovered escarpments, embryonic features can be observed. Karren are represented by enlarged cracks, rinnenkarren, rillenkarren and vents. There are also a large number of micro-gaps beside crevices reaching 15–20 cm width. The scarps are dissected by cracks formed by dissolution (Schichtfuggenkarren).

The *Canin Massif* (Kanin in Slovenian) is located in the western part of the Julian Alps, with its highest point being 2587 m asl (Fig. 9.7). Its two main building stones are the Triassic Main Dolomite and the overlying Dachstein Limestone. Both of them have more than 1000 m of thickness. Quaternary sediments (glacial and fluvioglacial) occur in the valleys (Carulli 2006). Thick carbonate rocks and active structural movements both promote the formation of surface and endokarst systems. The glacial periods of the Pleistocene strongly reshaped the mountain, and its glaciokarst morphology was mapped by Kunaver (1983). Telbisz et al. (2011) presented some phases of glaciations in the Italian side

Fig. 9.6 North face of Triglav Peak (Julian Alps; photo by Telbisz)

Fig. 9.7 The Italian side of the Canin Massif: plateau surface and the main ridge (photo by Telbisz)

Fig. 9.8 Large glaciokarst depressions and karren (Canin, Italy; photo by Telbisz)

Fig. 9.9 Rugged topography of Kanin plateau (Slovenia) with snow dolines and shafts. See the house in the top-centre for scale (photo by Telbisz)

of the massif. Based on the erratic blocks, Audra (2000) calculated that postglacial decay was in the order of 200 mm. Similar values can be found in other glaciokarsts of the Alps. The plateaus of the mountain are fine glaciokarst complexes (Fig. 9.8). The dominant forms are escarpments, karren and shafts (Fig. 9.9). Caves are also extremely well-developed, notably the world's deepest vertical pitch (603 m) in Vrtiglavica Cave is also found here.

9.3 Altai Mountains

The Altai Mountains (Fig. 9.10) are parts of an extensive mountain range in the southern part of Western Siberia. It is the highest part of the Altai–Sayan mountain area and is located between 48°N and 53°N latitudes and 82°E and 90°E longitudes. In the north, the Altai Mountains are neighbouring with the West Siberian Lowland. In the west, they extend to the lake Zaysan basin, and in the south and southeast it is linked to the Mongolian Altai. In the east, through a range of medium–high mountains it borders with the mountainous regions of Tuva. It has conventional boundaries with the Western Sayan and the Mountain Shoria in the northeast. Administratively, the area is a part of the Russian Federation (Republic of Altai and the Altai Territory). The western part of the area of the Altai Mountains belongs to the Republic of Kazakhstan (the East Kazakhstan Region). The highest point of the Altai Mountains is Belukha (4509 m) in the Katun Range on the border of Russia and Kazakhstan. Modern glaciation of the Altai Mountains is traced in the south-eastern part within the South Altai (122 glaciers), the Katun (391), the North-Chuya (168, Fig. 9.11), the South-Chuya (219), the Karaalahinsky (32), the Terektinsky (7), the Kurai (15), the Saylyugemsky (18), the Chikhacheva (49), the Ivanovsky (36) and the Sumultinsky (5) ranges. At the beginning of the 2000s, the total area of the glaciers slightly exceeds 890 km^2. The leading role in the study of modern glaciation of the Altai Mountains belongs to scientists of the Tomsk State University (Sapozhnikov, Tronov, Revyakin and others). Most of their works were published in Russian, among them Tronov's publication is to be emphasized (Tronov 1948, 1966).

9.3.1 Geologic Settings

Geologically, the Altai Mountains were mainly formed by Paleozoic Orogenies (the Caledonian and the Hercynian) but they were also affected by the later orogenies during the Mesozoic and Cenozoic Era (Gutak 2015a). The oldest rocks, in the modern view, belong to the end of the Neoproterozoic (the Vendian System of Russia). They are the deposits of oceanic carbonate platforms and consist of marbled limestones interbedded with silicilytes and dolomites. In southeast of the Altai Mountains, they are allocated in the rank of the Baratalskaya formation

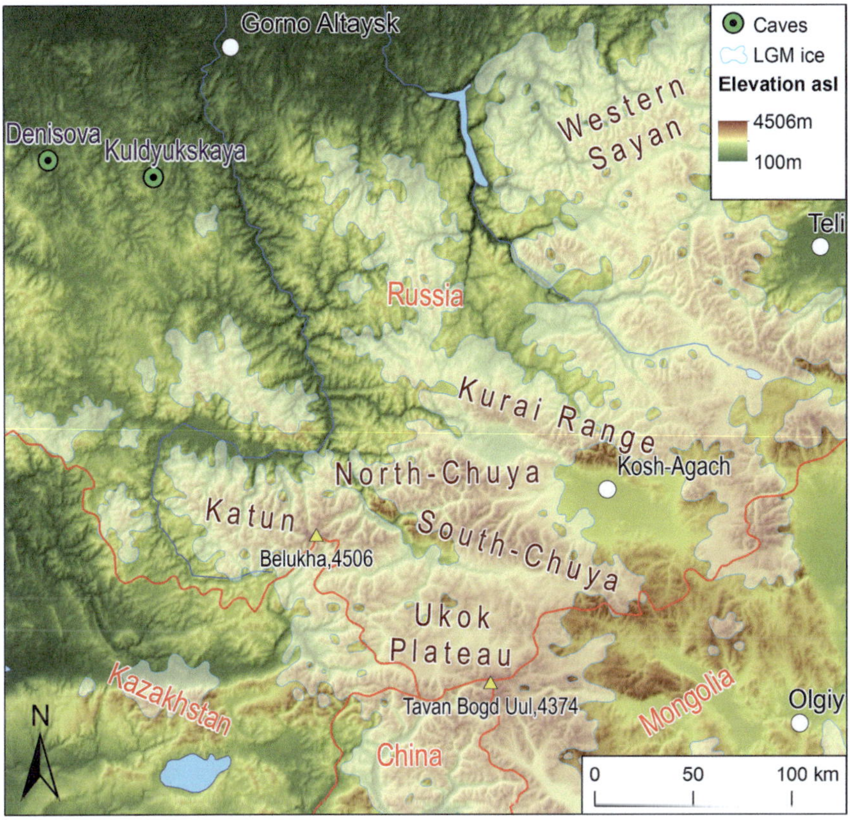

Fig. 9.10 Map of the Altai Mountains. Glaciation data from Ehlers et al. (2011)

(Turkin and Fedak 2008). The foundation of the carbonate platform is composed of basaltoids. The Early Paleozoic deposits are represented by a variegated set of rocks of island arc associations including deposits of a deep-sea trench (the Mountain Altai formation), volcanic islands and retro-arc basins. In the Ordovician–Silurian–Early Devonian time in the retro-arc part of the basin on the shelf of the ancient Siberian continent, terrigenous carbonate formations containing reef massifs were formed (Gutak 2015b). The Devonian stage is marked by andesitic–dacitic volcanism typical of island arcs and by the formation of coastal marine and terrigenous sediments (Gutak 2002). In the late Paleozoic, marine sedimentation conditions were replaced by continental (terrigenous coal-bearing molasses were formed). The Mesozoic activation of the region became apparent in the formation of carboniferous mountain trenches (Gutak et al. 2001) and in the presence of subvolcanic and granitoid intrusions. At that time, the Altai Mountains uplifted along an arcuate line and the surface underwent peneplain formation. The Cenozoic period of the region was marked by the end of the uplift and by the formation of large mountain trenches (the Uimonskaya, the Chuya and the Kurai). During the Pleistocene, the territory

Fig. 9.11 The North-Chuya Mountains (photo by Gutak)

was subjected to largely extended glaciation. The large Altai Ice Sheet was formed (Fig. 9.10), which occupied a large part of the modern Altai Mountains (Butvilovskiy 1993; Rudoy 2005; Gutak et al. 2008). Its formation played an important role in the karst development of the Altai Mountains. Carbonate rocks of the Neoproterozoic Era (the Baratalskaya formation and its analogues) and reef formations of the Early and the Middle Paleozoic are exposed to karstification. In general, most of the well-known karst caves in the Altai Mountains (including the world-famous Denisova cave) are located around the periphery of the large Pleistocene ice sheet. Glacial streams of the latter, apparently, played a huge role in their formation.

9.3.2 Climatic Settings

The climate of the southeastern part of the Altai Mountains is severely continental. In Kosh-Agach village (in the Chuyskaya trench 2000 m asl), the average temperature is −32 °C in January and 13.8 °C in July. The mean annual temperature is −6.7 °C. The minimum temperature reaches −62 °C and summer maximum is 31 °C. Humidity highly depends on the exposure of the mountain slopes. The slopes exposed to the north, as a rule, are more humid than slopes exposed to the south. In mountain basins, rainfall depths are minimal. In Kosh-Agach village, it is less than 110 mm a year, while in the area of the Aktur glacier (in the North-Chuya ridge) the annual depth of rainfall is over 700 mm. The amount of precipitation is related to vegetation cover. The landscape of the mountain basins is mostly desert or semi-desert, whereas on the humid mountain slopes, dense tree vegetation grows. The average elevation of the snow line is at 2200 m. Above it, alpine meadows and alpine tundra are predominant.

9.3.3 Glaciation of the Altai Mountains

As for the modern glaciers of the Altai Mountains, there are seven major glaciation centres: the Belukha massif, the Bish Iordu mountain ridges in the North Chuya ridge, the Taldurinsko-Akkelsky in the South Chuya ridge, the South Altai, the West Katun and the East Katun in the Katun range. The seventh largest centre of the Altai Mountain glaciation is located at the conjunction of Chinese, Russian and Mongolian territory with the Potanin glacier, which is the longest in Mongolia. The largest Russian glaciers of the Altai Mountains, the Sophiysky and the Grand Taldurinsky are located in the Taldurinsko-Akkelsky sector, in the northern side of the ridge. Morphologically, the modern glaciers of the Altai Mountains are characterized by considerable diversity. Many of them are cirque glaciers. Many of the glaciers are hanging glaciers and niche glaciers. The largest glaciers are of the kettle hole type (the Taldurinsky glacier and the Yadrintsev glacier in the South Chuya ridge, the Alahinsky in the Southern Altai ridge and the Ioldoayry in the Katun range). The Alpine glaciers are also variegated. Among them, the largest are the Large Maasheysky (Fig. 9.12) and the Tronovy Brothers' in the North Chuya ridge. There is a single flat-summit glacier formed on a relatively flat watershed of the North-Chuya ridge (the Dome of Three Lakes glacier). The modern centres of glaciation of the Altai Mountains are very popular among tourists, mountaineers and rock climbers. Mountain Altai glaciation began at the beginning of the Quaternary period and got its peak in the Pleistocene period. At present, glaciers of the Altai Mountains are in process of degradation, which is directly linked to global warming.

9.3.4 Glaciokarst Forms

There is no significant relationship between modern glaciation and karst processes because current areas of glaciers do not coincide with the area of carbonate rocks. However, in the period of maximum glaciation, the ice sheet of the Altai Mountains directly interacted with rocks subjected to karst formation and its glacier streams played a major role in karst formation. Most of the known karst caves in the Altai Mountains are located around the periphery of the Altai Ice Sheet. Most of them

Fig. 9.12 Valley of the Maasheysky glacier (photo by Gutak)

have the status of protected territories and are included in the Red Book of the Republic of Altai and the Altai Territory (Krasnaya kniga Altayskogo kraya 2009). The total number of karst caves in the Altai Mountains is 1200, and many cavers, both amateurs and professionals, consistently state that caves in the Altai Mountains are more beautiful and deeper than all the other caves in Siberia.

There are numerous vertical or steeply inclined caves, i.e. natural shafts in the karst landscapes of the Altai. Caves are mainly formed in the Paleozoic reef limestones, which build up steep massifs with vertical or near vertical margins. The average depth of the shafts is around 100 m. The Ecologic Cave (the Tektash) reaches the maximum depth of 345 m. It is the deepest shaft in Paleozoic structures of all Russia for over 20 years. In 1996, this karst cave was approved a natural monument of the Republic of Altai. Another glaciokarst phenomenon in the Altai Mountains is the underground glaciation in a number of caves. The Kuldyukskaya Cave near Cherga village in a tributary valley of the Kuldyuk river in the Northern Altai can be mentioned as an example. Underground ice is found in this cave, though it is at a much lower elevation than the regional snowline. The cave ice is located at 750–800 m asl, while surface glaciers nowhere stretch below 1970 m asl. The Kuldyukskaya Cave ice mass covers an area of 510 m^2 and at present, it is the largest known underground ice mass in all Russia. Century-old layered masses of ice reach 17–20 m thickness. Ice mass volume is estimated to be 7200 m^3. Currently, the cave ice is fed by the infiltrating water of the aeration zone, but at low temperatures it is blocked because of the ice freezes. The cave lies in a thick bed of Middle Cambrian marbled limestone. The cave length is 150 m.

Surface karst forms in the Altai Mountains are sparse. The reason for this is the widespread distribution of Quaternary deposits. Surface karst processes are active only in areas where Pleistocene glaciers removed the covering sediments. Naturally, these forms cannot be compared to the classic manifestations of glaciokarsts in the Dinarides. The Barlaksky karst massif can be taken as an example (in Shebalinsky district, Altai Republic near Barlak village). This massif is considered a geological natural monument of Russia (Karpunin et al. 1998). It has a cliff of 100 m height and an area of 1 km^2 of Neoproterozoic marbled limestone, in which karren are developed along the fractures.

In addition to the above-mentioned glaciokarst landforms of the Altai Mountains, buried karsts can be also mentioned. These landforms of preglacial origin can be traced in the upper part of the river Eskongo (tributary of the Kadrin river) in the watershed of the river Karakudyur, a tributary of the Bashkaus river, in the northwestern side of the Aygulaksky ridge. They are subsidence dolines formed on the surface of the marbled deposits of the Neoproterozoic Baratalskaya (the Eskonginskaya) formation (Gutak et al. 2004). The karst depressions are covered by the moraine of the Pleistocene glacier. Buried karst was detected during the exploration for mercury in the early nineteenth century. The Eskongo plot is a slightly hilly highland plateau with an absolute elevation of 1900–2200 m. The topography shows traces of Paleogene erosion and the subsequent Pleistocene glaciations. There are moraine hills, erratics and also glacial striations on the rare rocks outcrops. Karst is represented by funnels of 3–5 m depth. They are filled with the erosional remains of weathering crusts of the Karachumskaya formation. This formation is composed of yellowish-brown loam with some debris of the Baratalskaya formation rocks (e.g. limestone, siliciclasts and microquartzite). Volume of clastic components in the Karachumskaya formation occasionally reaches 60% (but the average is 2–7%). During exploration of the funnels, pieces of cinnabar ores, often in industrial concentrations, were recorded. They were transported into the funnels from the nearby mercury deposits. Cinnabar is represented by grains and fragments of 1–5 mm size. Mercury content in the karst funnels reaches 1.5%. Buried karst is covered by the main moraine of the Altai Ice Sheet. The moraines comprise dirty grey loam with inclusions of boulders and debris representing the geological composition of the territory (marbled limestone, metamorphic rocks, granites and volcanics). Thickness of the moraine cover according to the deep pits data is 8–12 m in average. In this case, the glacier serves a conservator of the Paleogene relief, covering ancient karst relief by moraine. Data from the deep pits and the exploration drillings prove that depths of karst caverns at the plot reach 50–60 m. Maximum depth of karst cavities is suggested to be 100–200 m.

As a summary, karst forms in the Altai Mountains can be divided into two groups: open karst of Pleistocene–Holocene origin and buried karst of pre-Quaternary age. For the first type, the Pleistocene Altai glacier acted as a solvent supplier and was directly involved in karst formation, and for the second it acted as a conservator by covering ancient karst formations with moraine. The modern glaciation of the Altai Mountains has no influence on the development of glacial karst because of the long distance between the modern glacier fields and the carbonate rocks.

9.4 Anatolia

Anatolia, also known as Asia Minor, is basically the Asian part of Turkey (Fig. 9.13). It has variegated topographic and geologic settings. In Anatolia, there are largely extended karst terrains, especially in the southern regions, and since the elevations are also high with lots of mountains above 2500 m, glaciation took place in several places during the Pleistocene in spite of the relatively low geographic latitude of the area. Some smaller glaciers exist even today (Kurter 1991). Thus, preconditions for the formation of glaciokarst landforms are satisfied in several mountains of Anatolia. Local studies principally focused on the relationship of glacial morphology and climate, and there are only few studies addressing directly the issue of glaciokarst (Bayari et al. 2003; Klimchouk et al. 2006; Çılğın et al. 2014).

9.4.1 Geologic and Tectonic Settings

Asia Minor was formed by the collision of the Eurasian, the Anatolian and the Arabic plates. These tectonic units are separated by active strike-slip faults in North and East Anatolia. As a result of the collision, most part of Anatolia has uplifted to high elevations during the Alpine Orogeny (Yilmaz et al. 1998; Dilek 2006).

Fig. 9.13 Map of glaciokarst locations in Anatolia

As for the karstifiable rocks, thick sequences of limestones were deposited in the Thetys Ocean (both in the Palaeo- and Neo-Thetys), from the end of Paleozoic to the Early Tertiary (Bozkurt and Mittwede 2001). The Cenozoic uplift became even more significant in the Miocene period. The amount and rate of uplift were variegated (Ekmekci 2003). In general, rates were higher towards the East. There are some karst terrains in Anatolia that have gone through continuous karstification since the Miocene. Around 65% of the caves of Taurus mountain range was formed in Mesozoic carbonates (Bayari and Özbek 1995).

9.4.2 Relief Settings

As a result of tectonic uplift, the relief of Anatolia, in general, gradually increases from west to east. In the north, the Pontic mountain range (or Pontic Alps) is found, where magmatic and metamorphic rocks prevail and limestone is found on the surface only in smaller patches. On the other hand, in the South, the Taurus mountain range is dominated by limestones and consequently, by karst terrains. In Central Anatolia, there are extended volcanic mountains, but some carbonate terrains are also present. The highest peaks of all segments are found in the far east of Anatolia, the Kaçkar Dağı (3937 m) is the most elevated mountain in the Pontic Alps, the Mount Ararat (5137 m), called Ağrı Dağı in Turkish, is the highest volcanic peak, and the Cilo (Buzul) Daği (4135 m) is the highest mountain in the Hakkari Range, which is the eastern extension of the Taurus range.

9.4.3 Climatic Settings

The coastal areas along the Mediterranean Sea are characterized by Mediterranean climate, whereas the Black Sea coasts have a marine climate. Coastal mountains receive high amount of precipitation (more than 2000 mm/a) in both cases (Kurter 1991). Naturally, the precipitation in the upper zones of the coastal mountains is significantly higher than in the lower areas lying directly at the coast due to the orographic effect. In turn, the Central Anatolian territories have a (semi-)arid climate. The degree of continentality increases towards east. The above trends were roughly valid during the Pleistocene glacial periods as well, and thus, the snow line elevation increased towards east and from the coasts to inland areas. However, the relief factor has also an eastward increasing trend with a higher gradient than that of the snow line elevation trend, and consequently, the largest glaciers existed (and still exist at a very reduced size) in the eastern parts.

9.4.4 *Glaciation Characteristics and Phases*

The first scientific studies dealing with the present-day glaciers were already published in the nineteenth century (Sarıkaya et al. 2011), whereas traces of the Pleistocene glaciations in Anatolia were recognized and investigated since the 1940s (Erinç 1944 in Sarıkaya et al. 2011). Numerical datings of glacial landforms have been carried out recently, and now there are a number of OSL and cosmogenic isotope exposure age data usually measured on moraine materials. In karst terrains, a ^{36}Cl cosmogenic isotopes were used, while in other study areas, ^{10}Be and ^{26}Al isotope methodology was applied. These studies evidenced that the most extended and best recognizable glacial landforms are connected to the LGM (cca. 20 ka BP) in Anatolia (Sarıkaya et al. 2008, 2014; Zreda et al. 2011; Çiner et al. 2015); however, pre-LGM moraines have also been identified in some places, namely, in Akdağ Mountains, where 35 ka BP old moraines, and in Kaçkar Mountains, where 57 ka BP old moraines have been found. Besides, post-LGM glaciations were also demonstrated. Notably, glacial phases in the end of the Late Pleniglacial (15 ka BP), in the Younger Dryas (12 ka BP) and in the Middle Holocene (5 ka BP) were detected. At present, the snow line is at around 3000 m asl in the Pontic Alps, and at cca. 3600 m asl in the southeast (Cilo Daği), where 20 smaller (i.e. less than 4 km long) glaciers still exist (Kurter 1991). During the LGM, the temperature could be 9–11 °C lower, while the precipitation in the coastal mountains could be similar or even higher than today, and thus, the ELA was 1000–1400 m lower in these places than now (Zreda et al. 2011; Sarıkaya et al. 2014).

9.4.5 *Study Areas*

In Anatolia, the northernmost mountains with an extended glaciokarst are the *Munzur Mountains* (3462 m). Although they are closer to the Black Sea than to the Mediterranean Sea, they belong to the northeastern range of the Taurus Mountains, and they are found south of the North Anatolian fault. Thick-bedded Upper Triassic and Cretaceous neritic limestones build up the Munzur Mountains in more than 1000 m thickness. As karst processes have been active here since the Oligocene, the karst morphology is well-developed. The uplift rate increased during the Pliocene that resulted in a more intensive vertical karstification since that time. Dolines, uvalas and fluviokarst terrains landmark the surface. Fluviokarst is present where the impermeable ophiolite and limestones are next to each other. At present, there is one microglacier somewhat below the theoretical snow line (3600 m). During the LGM, the ELA was at 2750 m asl that made the formation of small plateau glaciers possible. Outlet valley glaciers extended down to as low as 1400 m asl. These glaciers are documented by moraines and cirques. Preglacial karstic depressions favoured the formation of cirques. The majority of cirques have northern aspect. In the postglacial period, glacial sediments deposited in uvalas have been dissected by streams. The local erosion base of these streams is determined by inflow caves. As a summary, the Munzur Mountains are characterized by a complex, polygenetic glaciokarst morphology (Çılğın et al. 2014; Bayrakdar et al. 2015).

All other glaciokarst terrains are situated relatively close to the southern Mediterranean coast. The westernmost of them are the *Akdağ Mountains* (3016 m). The karst of Akdağ is formed mainly on Jurassic and Cretaceous carbonates. Beside dolines and uvalas, there are some larger, polje-like depressions as well, their strike and edges are apparently predetermined by tectonic faults. The karstic plateau with depressions obviously provided favourable conditions for snow accumulation and glaciation. Several phases of glaciations have been documented here, namely, at 35 ka, 21.7 ka and 15 ka BP. There was a contiguous ice cap at the plateau and large, compound cirques as well as valley glaciers were also formed, principally towards northern or northeastern directions. The lowermost moraines are found at 1950 m asl, and the corresponding ELA was at 2520 m asl. Terminal, lateral and bed moraines are all present in the Akdağ. On some moraine boulders, solution grooves are also observable (Sarıkaya et al. 2014).

The *Geyikdağ Mountains* (2877 m) are important because they bore the largest extent (20 km^2) Piedmont (Malaspina) type glacier in Anatolia. The area of the former Piedmont glacier is now filled with well-preserved hummocky moraines. In the upper plateau (2350–2650 m asl), the plateau ice cap had an even larger extent (40 km^2). In the valleys starting from the high plateau, there were alpine-type glaciers. Polished rocks, lateral moraines and occasionally terminal moraines mark the existence of former glaciers. Large cirques are also found in the upper zone. Again, the typical aspect of glacial forms is north, similarly to the previous study areas. ^{36}Cl cosmogenic isotope exposure ages suggest three glacial phases here, at 18 ka, 14 ka and at 5.2 ka BP. These ages are considered as minimum ages. Due to the closeness of the coastline, the Geyikdağ receives lots of precipitation, thus the ELA was as low as 2000 m asl during the LGM. Glaciofluvial landforms are basically absent because of the well-developed karst of the Cretaceous and Eocene limestones. On the other hand, some periodic lakes do exist, whose bottoms are filled with fine-grained sediments (Çiner et al. 2015).

Probably, the most spectacular glaciokarst terrain within the Taurus mountain range is found in the *Aladağ mountains* (3756 m), which are built up mainly of Mesozoic limestones (Fig. 9.14). The Aladağ has been uplifting since the Late

Fig. 9.14 Aladağ mountains seen from the west (photo by Nagy)

Oligocene. The most pronounced plateau, the Yedigöller (Fig. 9.15), hosted a 40 km^2 extension ice field during the LGM. The deeply incised Hacer valley was filled with an alpine-type glacier at a length of 14 km. The highest ridges are sharp with jagged peaks, and cirques are also present. The lowermost moraines are found at 1100 m asl. Fluvioglacial sediments, erratics and polished rock surfaces are widespread in the valleys. Based on cosmogenic exposure ages of moraines, the researchers could estimate the rate of ice withdrawal. It was 4.25 m/a between 10.2 ka and 9 ka BP, but as high as 17.1 m/a between 9 ka and 8 ka BP. These rates are extremely high and are similar to or even higher than modern alpine glacier withdrawal rates. It means that the Early Holocene warming took place at a quick pace here (Zreda et al. 2011). Meanwhile, the ELA increased from 2080 m asl in LGM to the present-day 3510 m asl. In the highest zone of the mountains, there exists still a 1 km^2 area microglacier, mostly covered by rock debris and dotted by kettle holes (Bayari et al. 2003). In the eastern part of Yedigöller plateau, the moraines cover a cockpit type former tropical karst (Bayari et al. 2003). In the higher levels, all preglacial landforms have been strongly eroded by glacial exaration, while negative landforms have been filled up by debris (Fig. 9.16). Due to the intense uplift, shaft caves are very typical in the Aladağ Mountains. Many of them have been explored by glacial erosion or by postglacial mass wasting processes. Shafts, whose side became open, are called by Klimchouk et al. (2006) *unwalled shafts* and they can be mostly observed in the vertical mountain slopes. They are ephemeral phenomena in geological terms, but there are a number of them

Fig. 9.15 Yedigöller Plateau with several cirques, small lakes, moraines, arête and a characteristic nunatak (photo by Nagy)

Fig. 9.16 Kayacuk valley in Aladağ Mts with a 500-m-long rock glacier at 3200–3300 m asl (photo by Nagy)

in Aladağ. Further on, Klimchouk et al. (2004, 2006) described some other forms of shafts according to their relationship with the surface. Namely, they mention *ponor shafts*, which, at present, function as sinks, *epikarst shafts*, which were formed at the bottom of the epikarst, but where the epikarst was erased by ice, *decapitated shafts*, where the ice wiped off not only the epikarst but also some parts of the bedrock, and finally, *collapse shafts*, which are open due to the collapse of the upper part of the shaft. In several of the deep, open shafts, there is a snow and ice plug of more than 100 m thickness.

9.5 Apennines

The Apennine Mountain Range rises along the axis of the Italian Peninsula in a northwest–southeast orientation (Fig. 9.17). As it is more or less orthogonal to the prevailing wind directions, it receives large amounts of precipitation both at present and in the glacial periods. However, as the maximum elevation of the mountains is slightly less than 3000 m asl, and as the mountains are located at relatively low latitude within Europe, the glaciation of the Apennines was restricted, and only

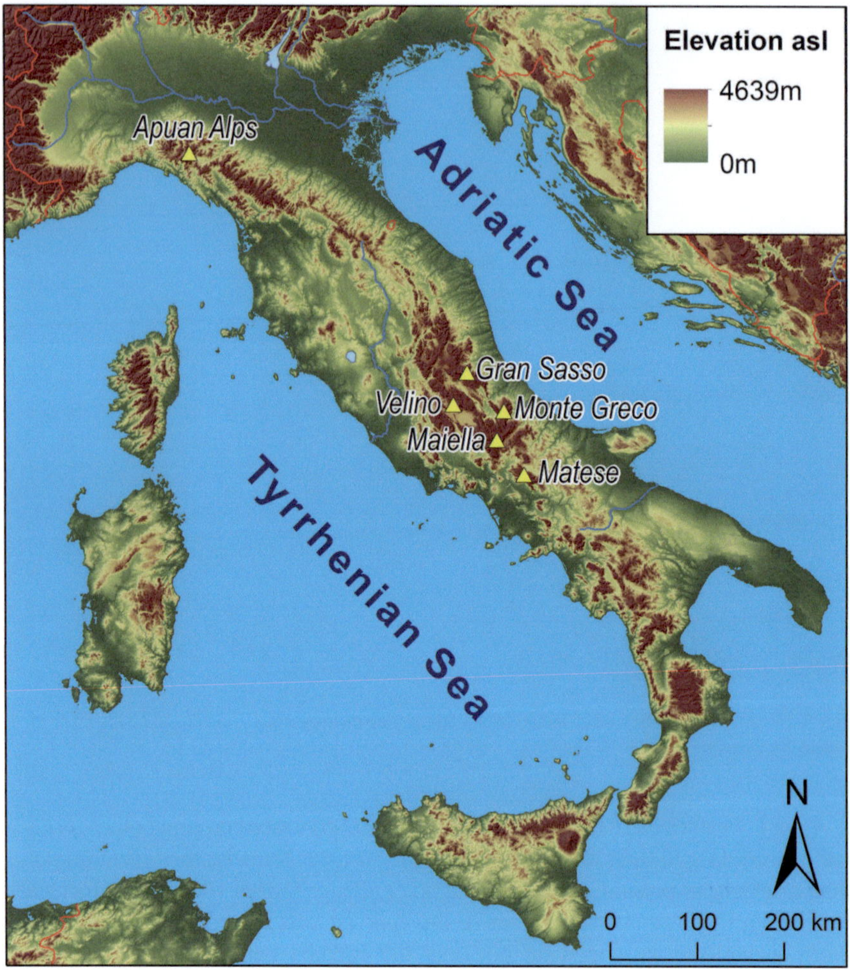

Fig. 9.17 Map of glaciokarst locations in the Apennines

smaller valley glaciers existed in the massifs higher than 2000 m asl. Extensive plateaus are less characteristic of the Apennines than the Dinaric Alps, and thus plateau icefields were much smaller in size. The first studies on the glaciation of the Apennines were carried out back in the nineteenth century (e.g. Stoppani 1872 in Baroni et al. 2015). The most remarkable glacial traces are dated to the end of the Pleistocene, but as it was recognized in the 1980s, there were significant glaciations even during the Middle Pleistocene, when the extension of glaciers was occasionally larger than in the LGM (Giraudi and Giaccio 2015). At present, there is only one microglacier in the Apennines, the Calderone glacier, which is found in the Gran Sasso Massif. It is one of the southernmost glaciers in Europe (Giraudi

2005; Grunewald and Scheithauer 2010). As carbonate rocks are widespread in the Apennines, glaciokarst morphology is relatively frequent on formerly glaciated terrains.

9.5.1 Geologic and Tectonic Settings

The Apennines are even younger than the Alps since the orogeny forming the Apennine mountain range began in the Late Miocene. There are two main tectonic regimes in the Apennines, one at the eastern side, where the Adriatic Plate subducts, implying a compressional stress field. As a result of compression, sedimentary rocks of mainly Thetyan origin were folded in the eastern segment, and thrust-and-fold structures were created, with high proportion of metamorphic rocks in some places. Here, we focus on marble among metamorphic rocks, which is widely distributed in the Apuan Alps, most notably, the famous Carrara Marble is also located here. Due to the ongoing compression, tectonic uplift has been very important in the whole mountain range during the Pliocene and the Quaternary alike. On the other hand, the western, Tyrrhenian side is characterized by extension with fault-block structures. Grabens and horsts equally occur in this zone, and tectonically uplifted blocks are found at different levels. There are several northwest–southeast oriented, elongated polje-like depressions, which are locally called *tectonic-karst basins* (Fig. 9.18). They are usually bordered by fault displacements. Recent neotectonic deformations are remarkable, and thus vertical surface movements are not negligible since the time of glaciations. This fact must be taken into account when ELAs are estimated (Giraudi and Giaccio 2015).

The Northern Apennines are basically composed of flysch rocks, although there exist some mountains built up of limestones or marble. The most important of them are the Apuan Alps. As for the Central and Southern Apennines, they are predominantly of carbonate rocks. A further special feature of the Apennines is that tephra sediments originating from the young, Tertiary and Quaternary volcanoes towering at the western side of the peninsula can be found in many places and these tephras are suitable for dating.

9.5.2 Relief Settings

Based on topographic features, the Apennines can be divided into three parts. First, the Northern Apennines (whose highest peak is Monte Cimone, 2165 m asl) should be mentioned, which are closer to the Tyrrhenian Sea. They only exceed 2000 m asl in few locations. The Central Apennines (with Gran Sasso, 2912 m asl) represent the highest and widest part of the whole mountain range, and, in contrast to the northern and southern sections, they are closer to the Adriatic Sea. Finally, the Southern Apennines (with Pollino, 2233 m asl) are lower than the central part and they are again near the Tyrrhenian Sea.

Fig. 9.18 M. Velino and Campo Felice—a karst–tectonic basin. 1. LGM moraines (MIS 2), 2. older moraines (MIS 6), 3. oldest moraines (Middle Pleistocene), 4. outwash fans, 5. cirques. Elevation data from SRTM, satellite image from Google Earth, glacial data from Giraudi et al. (2011)

9.5.3 Climatic Settings

The climate of the Apennines is partly Mediterranean, partly mountainous. Precipitation maximum is in winter, which is very important from the viewpoint of snow accumulation. Mean annual temperature is 4–5 °C in the highest zones, and around 14 °C near the sea level (Costantini et al. 2013). Temperatures could be 7–8 °C cooler during the LGM (Giraudi 2015). Due to the proximity of the sea, and the northwest–southeast strike of the mountains, which is almost at right angles to the prevailing winds, the annual sum of precipitation may exceed 1800 mm in the peak zone of the mountains. One of the wettest places in Europe is found here, in the Apuan Alps, where the main ridge gets ca. 3000 mm precipitation a year (Piccini et al. 2008). At present, the theoretical snow line is at 3200 m asl and this elevation is higher than for any other peak in the Apennines. However, the snow line was at much lower elevations during the cold and wet periods. LGM snow lines were at 1200–1500 m asl in the north, at 1550–1900 m asl in the centre and at 1600–1800 m asl in the south. In general, snow line altitude increased from north to south and from west to east (Giraudi 2015).

9.5.4 Glaciation Characteristics and Phases

In the course of glacial research, it became obvious step by step that glaciers were formed in the Apennines in each glacial phase since the Middle Pleistocene, namely, during MIS 14, 12, 10, 8, 6, 3–4 and 2 (Giraudi 2012, 2015; Giraudi and Giaccio 2015). The most extended glaciation took place generally in the LGM, but in certain locations, the Middle Pleistocene moraines demonstrate an even larger extent glaciation (Giraudi and Giaccio 2015). The substances used for dating were partly moraines, partly polje sediments. U-series dating of the carbonate cement of moraines provided the first results, which numerically proved the existence of pre-Würmian glaciations in the Apennines (Kotarba et al. 2001). As for the polje sediments, radiocarbon and tephra Ar/Ar methodology was used to infer glaciation phases (Giraudi 2012). However, to date, numerical age data are only known from a limited number of sites, for instance, from the Gran Sasso and Campo Felice areas. Based on the available data, it seems likely that local LGM, which took place in 27–28 ka BP, somewhat preceded the LGM of the Alps (Giraudi 2012). A feasible explanation for this fact is that although the climate remained cold after the local LGM, it turned out to be more arid at the same time. That implies that the minor and sensible Apennine glaciers began to retreat.

Based on the deposits in the tectonic-karst basins, it was claimed that warm periods were characterized by little energy, alluvial and colluvial sediments, whereas in the cold periods, lake sediments as well as glacial and glaciofluvial sediments with high carbonate content were deposited in these basins. The reason for this phenomenon is that during the warm periods, karstic infiltration was highly effective in the mountainous zones and there was no direct, i.e. superficial material transport into the lower basins. In turn, during the cold periods, frost shattering and debris production intensified, and glaciers as well as glacier-fed streams transported large amounts of subrounded or subangular limestone clasts, cobbles and boulders into the basins (Giraudi 2012). These regularities were observed on other glacio-karst terrains as well, namely, by Adamson et al. (2014), who called this process *uncoupling*.

9.5.5 Study Areas

In the North Apennines, the most pronounced glaciokarst terrains are found in the *Apuan Alps*. Although they are parts of the Apennines and are lower than 2000 m asl (the highest peak is 1947 m asl), they are called "Alps" on the basis of jagged ridges, pointed peaks, cirques and the alpine morphology in general. The hypogean karst of the Apuan Alps is of utmost importance, the Corchia Cave (with length of 60 km, depth of 1187 m) can be mentioned as the most significant landmark. As for subsurface morphology, there are extensive horizontal cave passages up to 1400 m asl, formed as phreatic segments. The higher zones are dominated rather by large, vertical shafts and vadose passages. Speleothems have been evolving here for more than 1 Ma, which provides a very good opportunity for paleoclimate reconstructions. It is observed that during glacial–interglacial transitions there was a lag in the growth of speleothems. This lag is related to the time necessary for the formation of soils after glacial termini (Piccini et al. 2008). Surface karst landforms are also remarkable, and some of them were formed under preglacial conditions, when the terrain was much lower and less dissected. The prevailing bedrock of karst features is marble, but limestones and dolostones are also present. There are a variety of dolines, doline fields, limestone pavements and karren. Cirques do exist as well, but their lower edges are usually less pronounced. U-shaped glacier troughs are short (less than 1 km long), and in certain places, Roche moutonnées can be observed on the surface. Within the whole Apennine mountain range, terminal moraines reach their lowermost elevation just here, at around 900 m asl (Baroni et al. 2015).

In the Central Apennines, there are several mountains with glaciokarst morphology. Naturally, the most outstanding of them is the *Gran Sasso Massif* (2912 m asl; Giraudi and Giaccio 2015; Fig. 9.19). The relief of Gran Sasso is characterized by northwest–southeast oriented tectonic ridges and elongated, tectonic-karst basins in between. One of the most remarkable basins is the Campo Imperatore lying southwest of the main ridge (Fig. 9.20). Glaciers departing from the higher zones reached these basins and deposited loads of moraines here in forms of terminal moraines and hummocky moraines, as well as glaciofluvial sediments. It is the area where pre-Würmian moraines were first dated (within the Apennines) by Kotarba et al. (2001). In addition, post-LGM phases, such as the Younger Dryas and Holocene, stages can also be investigated here.

The *Velino Mountains* (2486 m asl) look like the mirrored version of Gran Sasso. The tectonic-karst basin of Campo Felice is found at the northeastern side of the highest zone of Velino (Fig. 9.18). The basin has a thick sediment fill. At the top, moraines cover extensive areas. Deeper strata have been thoroughly studied by 9 drillings down to a depth of 125 m (Giraudi et al. 2011; Giraudi 2012). Limnic, glacial and glacifluvial sediments of cold periods alternate with alluvial and colluvial deposits of warmer periods which demonstrate the details of Pleistocene and Holocene climate changes. Small ponors are found along polje edges.

The *Maiella Massif* (2795 m asl) is found southeast of Gran Sasso. It has a wide plateau, which is a rare phenomenon in the Apennines. As a consequence, the mean

Fig. 9.19 Doline dotted area in Gran Sasso (photo by Mari)

Fig. 9.20 Campo Imperatore near Gran Sasso (photo by Mari)

elevation of Maiella is even higher than that of the Gran Sasso. Massive Cretaceous and Paleogene limestones build up the mountains. Due to the special relief settings of Maiella, there was a 30 km^2 plateau glacier during the last glacial period—a unique situation in the Apennines (Blasi et al. 2005).

Another formerly glaciated member of the Central Apennines is the *Monte Greco Massif* (2285 m). With the Aremogna tectonic-karst basin, they have a similar relief and tectonic structure to the Velino-Campo Felice area. Pre-LGM moraines were also found here, suggesting that pre-LGM glaciers were much longer in this case than LGM glaciers. Moreover, there are also some neotectonic fault lines in Monte Greco Massif, which were active even during the Holocene. Thus, it is likely that the Middle Pleistocene terrain determining the shape and extension of glaciers could be remarkably different from the present-day topography (Giraudi and Giaccio 2015).

Finally, the *Matese Mountains* (2050 m asl) are located already in the Southern Apennines. Comparably to the previous mountains, the upper zones of the mountains are built up of a northeast verging overthrust, which consists mostly of carbonate rocks. On the other hand, the western segment of the mountains is characterized by elongated tectonic-karst basins created by the extensional stress field. Several erosional paleo-levels can be recognized in the mountains, which are now at different elevations due to block tectonics. Drainage is predominantly karstic, but a poorly developed surface stream network is also present. Cirques are found around the main peak, and short glacial valleys start from them. LGM moraines end at 1200–1400 m asl in the Campitello basin, but older, Middle Pleistocene moraines are found at as low as 800 m asl. Here again, ponors are found along polje edges, whereas dolines are common features of the levelled erosion surfaces. Compound glaciokarst depressions are present in the highest zone, where small karren features decorate the bare rock surfaces polished formerly by glaciers (Aucelli et al. 2013).

9.6 Appalachians

The Appalachian Mountains are a 2400-km-long mountain system, which stretches parallel with the Atlantic coast in North America. The strike of the Appalachians is roughly northeast–southwest, thus the mountain range has a remarkable north–south extension from 69°N (in Canada) to 32°N (in Alabama State, USA). As a consequence, climatic and geographic conditions are very much different in the northern and southern parts, and it was true during Pleistocene cold periods as well. Given the elevation range of the Appalachians, they can be thought of as medium mountains. The highest peak, Mt. Mitchell (2037 m asl), is found in the South. Glaciation reached only the northern territories, and the southern terminal of the continental ice sheet during the maximum extent was at around 41°N (Ridge 2004; Fig. 9.21). Karst terrains are present at many locations in the Appalachian (see Weary 2008), but glaciokarst terrains are restricted to the northern karstlands due to

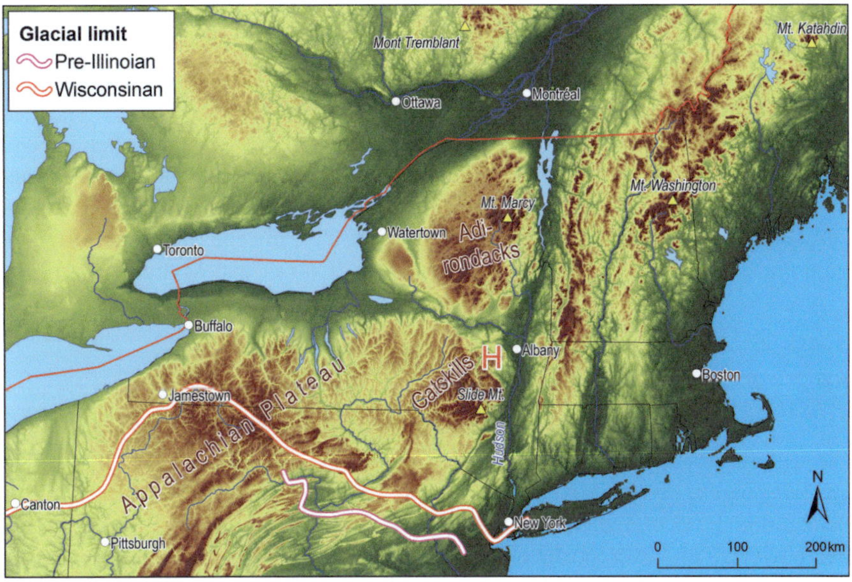

Fig. 9.21 The Northern Appalachians with Pleistocene ice. H: Helderberg Plateau. Glacial limits are after Garrity and Soller (2009)

the above-mentioned glacial limit. Appalachian glaciokarsts are mostly related to the continental ice sheet, and thus their morphology is different from alpine-type glaciokarsts, which are typical in the Rocky Mountains in North America (Werner 1979). Appalachian glaciokarsts are found predominantly in the so-called "Northwest" of the USA, which means New York State and New England. These karstlands are relatively poor in surface karst landforms; instead, their morphology is dominated by glacial features. However, caves are numerous and of diverse types. The longest cave of the area is McFail's Cave (New York State) with a length of 11.5 km (www.caverbob.com). The main topic of local karst research is the relationship of speleogenesis and glaciation, and especially, whether caves are of preglacial or of postglacial origin (Cooper and Mylroie 2015).

9.6.1 Geologic and Tectonic Settings

Appalachian glaciokarsts were prevailingly formed on Cambrian–Devonian limestones and on Proterozoic Marble. Some glaciokarsts belong to the Appalachian Plateaus Province, notably to Catskills Mountains. They are characterized by non-deformed, low tilt (1°–2°) limestone strata (Weary 2008). In turn, the zone along the Hudson River valley underwent more intensive tectonic deformation, and a fold-and-thrust belt was created during Acadian and Allegheny Orogeny. The orientation of structural features in this zone is basically north–south. The youngest mountains of the area are the Adirondacks, which are still actively doming, but strangely, their rocks are the oldest, including the 1.1-billion-year-old Grenville Marble, which is suitable for karstification (Cooper and Mylroie 2015). The relatively old bedrocks, especially in the western and northern parts, are often covered by some 10-m-thick glacial sediments, tills (Weary 2008). In the Mesozoic and Cainozoic era, erosion and peneplanation characterized the Appalachians. The formation of the present karst terrains began in the Pliocene, but karst processes were occasionally interrupted by glacial periods (Cooper and Mylroie 2015).

9.6.2 *Relief Settings*

The highest peaks of the Northern Appalachians are at 1300–1900 m asl, namely, Mt. Washington (White Mts) is 1916 m, Mt. Marcy (Adirondacks) is 1629 m, Mt. Katahdin is 1605 m and Mt. Slide (Catskills Mts) is 1270 m high. Karst terrains are, however, at lower positions. The most thoroughly investigated Heldelberg Plateau (Schoharie County, New York State) is found next to the Hudson Valley. The valley is just above sea level, while the plateau reaches some 650 m asl. There are some other karst terrains in the Northern Appalachian that reach even 1000 m asl (Cooper and Mylroie 2015). A low-elevation (100–150 m asl) glaciokarst terrain is found near Ontario Lake, in the New York Lowlands. The topography of low-elevation terrains was dominantly shaped by glaciers, and, as a result, tectonic structures are emphasized, too.

9.6.3 *Climatic Settings*

At present, the mean annual temperature ranges from 5 °C in Quebec to 16 °C in Atlanta, when moving from North to South in the Appalachians (https://en.climate-data.org). The mean annual temperature of Albany, a city next to Helderberg, is 8. 8 °C. Annual precipitation values range between 800 and 2000 mm (http://prism.oregonstate.edu). Naturally, temperatures decrease, whereas precipitation values increase as one moves upwards. For instance, the mean annual temperature at the top of Mt. Washington is −2.6 °C, while the annual sum of precipitation is 2460 mm, and this latter value is also due to the proximity of the ocean.

9.6.4 *Glaciation Characteristics and Phases*

In the Appalachians, Pleistocene glaciation began 2.4 Ma ago, but up to 0.85 Ma ago, the impact of glaciation was limited both in space and time (Braun 1989; Ridge

2004). Nonetheless, after this date, there were 10 large ice advances in North America, 8 of which reached the Northern Appalachians as well (Braun 1989). From 0.85 Ma to the present, in 15% of total duration, ice sheet extent was close to its maximum, in the Late Wisconsinan, and in 25% of total duration, in the Early Wisconsinan, the continental ice sheet was somewhat smaller in extent, but still large enough (Braun 1989; Fig. 9.21). The LGM lasted from 28 to 23.7 ka BP, and it was followed by glacier retreat. In New York State, complete deglaciation took place by 12–13 ka BP (Ridge 2004; Cooper 2014). During the LGM, the ice limit reached the New York City—Jamestown (NY)—Canton (Ohio) line, but in some places, pre-Wisconsinan glaciation ended even further south by ca. 30–50 km (Braun 1989; Garrity and Soller 2009). The continental ice sheet reaching the area was basically created by the growth of the Laurentide Ice Sheet, but in the northwestern part, in Canada, there existed the so-called Appalachian Glacier Complex, which extended to the shelf area as well. During the LGM, these ice sheets coalesced (Dyke et al. 2002; Dyke 2004; Stea 2004).

Markers of the former glaciations are different types of glacial deposits, namely, tills, kames, drumlins as well as glacially deepened valleys, polished rocks and glacial striae. The latter features demonstrate that the movement of the ice sheet was predominantly from the north to the south (Cooper and Mylroie 2015), but locally, northeast to southwest motions due to topographic obstacles are also recognizable (Lauritzen and Mylroie 2000; Weremeichik and Mylroie 2014). Glacial striae are generally quickly destroyed by dissolution on karstifiable rocks; however, they can be preserved, where they are covered by some sedimentary layers (Cooper 2014). Glacial lakes existed near the rim of the ice, and they had an impact on karstification as well (Weremeichik and Mylroie 2014). In the Northern Appalachians, the amount of erosion was highly remarkable during the Pleistocene,; notably, the thickness of the eroded material was around 120–200 m, as testified by marine sediments deposited in the Atlantic Ocean. The reason for this high erosion rate in a relatively low area is that the surface was covered by thick and warm-based glaciers with higher erosion capacity than cold-based glaciers (Braun 1989). Nevertheless, the erosion of high resistance, hard bedrocks was more restricted.

9.6.5 Study Areas

The ***Helderberg Plateau*** is the finest glaciokarst terrain in the Appalachians. It is found in New York State, southwest of Albany (Fig. 9.21). It hosts the largest caves, notably the above-mentioned McFail's Cave with a length of 11.5 km, the Skull Cave (7.3 km) and the Barrack Zourie Cave (5.2 km). Larger caves with wider passages are supposed to be of preglacial origin. In order to test this hypothesis, U/Th age datings were performed by Lauritzen and Mylroie (2000), and, in several locations, they found speleothems older than the last glacial period. Caboose Cave, Schoharie Caverns and Barrack Zourie Cave are among these locations, and speleothem ages were older than 350 ka BP, which was the measurement limit of U/Th datings at that time, so it must be a minimum age of the studied speleothems. According to the above measurements, the main speleothem growth periods were at 60 ka BP, between 105 and 75 ka BP, and between 235 and 135 ka BP. It is noted by the authors, however, that sampling of the speleothems was not representative. Further on, it is also stated that most of the preglacial caves were at least partly remodelled during the glacial periods. It has several reasons. First, there were significant changes in base levels during glacial phases; second, new flow paths were created; and third, postglacial isostatic rebound generated a number of fractures in the bedrock that contributed to subsequent cave evolution.

On the other hand, smaller size, mostly maze-type caves are also present in the area. They are likely to be of postglacial origin because they are controlled by the postglacial drainage network, and because they lack glacial sediments. Moreover, calculation of the time necessary for their formation based on the actual size of passage cross sections, the estimated hydrologic regime and dissolution rates proved that there was sufficient time for the emergence of such forms since the end of the last glacial period. Caves of this type are, for example, Hannacroix Maze, Merritts Cave, Skips Sewer (on the Helderberg Plateau), Big Loop Cave (in the Adirondacks) and Glen Park Labyrinth (at Watertown, New York State). Nonetheless, there are maze segments in the preglacial caves, too, which could be formed in the late glacial or postglacial phases as a result of floodwater conditions due to the damming effect (Cooper and Mylroie 2015).

Surface glacial sediments may influence speleogenesis in several ways. They may fill up former sinks, new sinks can be generated along the edge of drumlins, and where glacial sediments thickly cover the valley, the karst water table can be increased (Cooper and Mylroie 2015). Surface karst landforms are only incidentally referred to in the literature, but sinkholes, grikes and "*karst windows*" (i.e. stream sinks) are present on the Helderberg Plateau (Cooper 2014).

The Schoharie glacial lake is a thoroughly studied ice-marginal lake (Weremeichik and Mylroie 2014; Fig. 9.22). It existed in the Late Wisconsinan from 23 to 12 ka BP. Its extent oscillated at least four times, and it had different water levels from 213 to 366 m asl. These phases are reflected in the sediment depositions of eight caves on the Helderberg Plateau. In addition, these phases are recognized only in these caves. This means that it was in fact a local effect. The related sediments show a lithofacies of three units, the bottom is a white-grey, finely layered mud and clay containing significant amount of calcite. It was certainly deposited during the stagnant lake phase. The middle unit contains poorly sorted, matrix-based gravels that were formed by large discharge flows during the terminal period of the lake. Finally, the upper unit is a brown mud and clay with thin layers, supposed to be the result of anthropogenic impact, preserving the traces of soil erosion following the settlement of European colonizers coming to the area some centuries ago.

Besides, some smaller area karst terrains are also found east of the Hudson river valley, especially on marble bedrock, but to the present, these areas are only poorly explored (or presented) by the karst community.

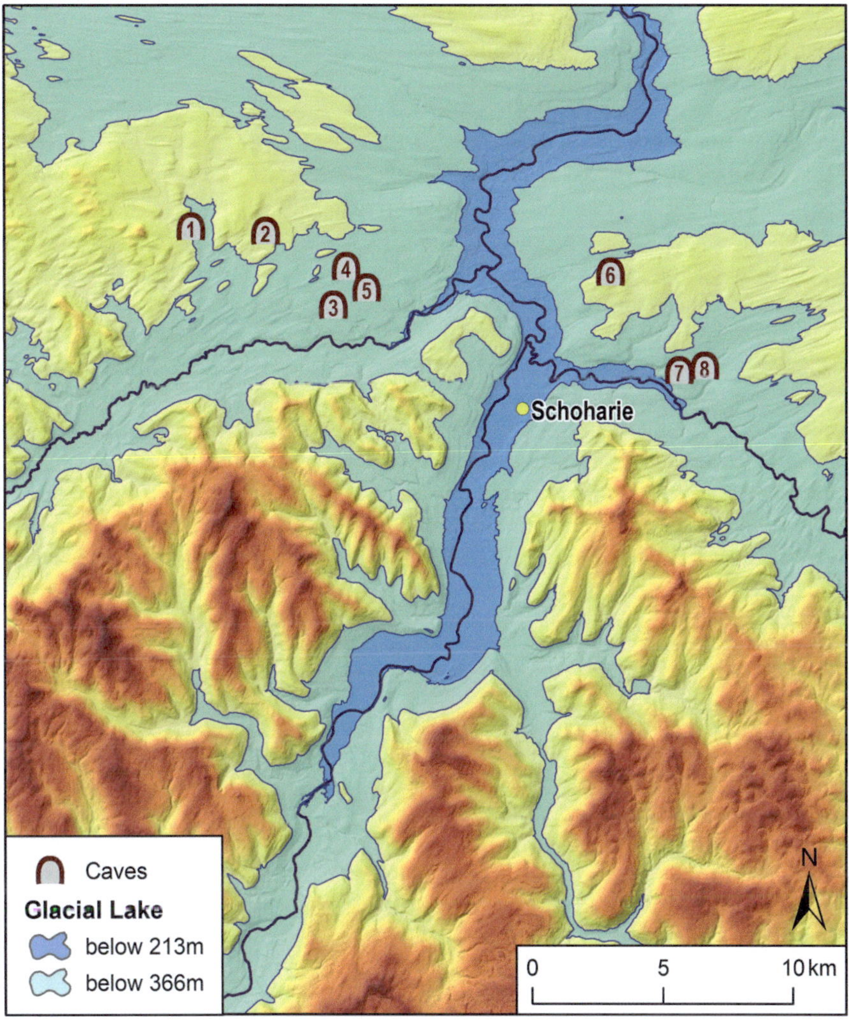

Fig. 9.22 Caves of the Helderberg Plateau and Schoharie glacial lake. Water level limits are at 213 and 366 m a.sl. The present topography may slightly differ from the LGM relief. Caves: 1. Barrack Zourie, 2. McFail's, 3. Howe, 4, 5. Secret-Benson, 6. Gage, 7. Schoharie, 8. Caboose. Data are from Weremeichik and Mylroie (2014)

9.7 Balkan Peninsula

The Balkan Peninsula is one of the large peninsulas of Southern Europe; its area is 667,000 km^2. Almost the whole peninsula has a rugged topography, and the highest peak is Musala (2925 m asl) found in the Rila mountains (Bulgaria), but many other mountains exceed 2000 m elevation. The Balkan peninsula is situated at a lower latitude than the Alps, between 36.4°N and 45.8°N. Both mean and the maximum elevations are lower than those of the Alps. The above factors imply that glaciation was much more limited in the Balkan than in the Alps, and there was no contiguous icefield covering large areas; instead, smaller ice caps and valley glaciers existed here during the glacial periods. Traces of Balkan glaciations were recognized by local scientists as early as at the turn of the nineteenth and twentieth centuries (Cvijić 1899, 1900; Grund 1910). Local researchers also found that at many locations, glaciation was surprisingly more extensive than what could have been expected taken into consideration the southern position and the relatively lower elevation of the mountains in the Balkan.

The geology of the region is dominated by limestones and dolostones, especially in the western mountain ranges of the Dinarides and the Hellenides. Since carbonate rocks are so widespread, most of the glaciated areas are also glaciokarst terrains. As mentioned above, the early descriptions of glaciokarst terrains date from the first part of the twentieth century (e.g. Cvijić 1899; Grund 1910), but later on, glaciokarst studies were sporadic during the second half of the century. The twenty-first century, however, has brought a boom in Balkan glaciokarst research (see Hughes and Woodward 2017 and references therein) due to new methodology (GIS, digital elevation models, new geochronological methods) and better accessibility of the area as a result of political changes.

9.7.1 Geology and Tectonics

Most of the glaciokarst terrains in the Balkans are built up of marine carbonates deposited from the Permian to the Eocene periods. These carbonates are mostly high purity limestones and dolostones. The full thickness of the carbonate formations even reaches 8000 m at some locations (Vlahović et al. 2005). The collision of the Adriatic and Eurasian plates caused the uplift of the Dinarides and the Hellenides since the Late Eocene (Pamić et al. 1998) and it formed the fold-and-thrust structure of the mountains (Dilek 2006; Bennett et al. 2008; Faccenna et al. 2014). The inner zones, namely, the Vardar Belt, are rather characterized by ophiolites and older, non-karstic rocks. In the outer zones, along the Adriatic Sea, carbonate rocks are locally intermixed with flysch belts (Pamić et al. 1998). The mountain ranges of the Eastern Balkan Peninsula were formed as a continuation of the Carpathians, also during the Cenozoic era.

The uplift involved denudation and the limestone debris deposited in the valleys and basins was often cemented secondarily. These secondary carbonates are widespread throughout the Balkans, but they are found usually at lower elevations (except where they are of glacial origin). Therefore, they are not important from the viewpoint of glaciokarst. The highest terrains are dominated by limestones, but an important exception is the Rila Mountain, which is built up of silicate rocks.

Tectonic uplift has been a major control of glaciation because, among other factors, the total uplift determines whether a given mountain reaches the threshold elevation of glaciation or not. At present, the ongoing tectonic uplift is most significant in the southern and western regions. There is an active subduction along the Hellenic Arc, where the uplift rate may reach or even exceed 1–1.5 mm/a. Uplift rates of Crete Island, Peloponnese Peninsula and Olympus Mountain are discussed by Bathrellos et al. (2015) and references therein, Smith et al. (2006), Faccenna et al. (2014) and by Pope et al. (2015). Comparably, high uplift rates also occur in the central part of Balkan Peninsula in the Rila and Pirin mountains. Beside thrust tectonics, right-lateral strike-slip motions became also significant since the end of Pliocene, but they are not accompanied with high uplift rates (Faccenna et al. 2014). However, strike-slip processes favour the formation of pull-apart basins, which are very typical in the Dinarides and, as a result of karstification, these basins are often transformed into poljes (Ford and Williams 2007). The uplift may cause differences in the extent of glaciation between different stages. If the 100 ka time scale is considered, and 1 mm/a uplift rate is taken into account, then the sum of uplift is in the order of 100 m. It is not fully negligible, particularly, in the case of mountains which were near the glacial threshold elevation during earlier glaciations. In general, it is argued that Early Pleistocene glacial periods were somewhat more limited from the viewpoint of glaciation if only the relief factor is considered. Of course, there are other factors to consider, too.

9.7.2 *Karstic Relief Settings*

Preglacial relief influences the possibility of snow and ice accumulation. Given the mostly karstic nature of the Dinarides, one can find lots of elevated plateaus and corrosional plains at different altitudes. Plateaus are usually drained by subsurface karst systems, and thus their fluvial dissection is minimal, although fossil dry valleys exist at many locations (Mihevc et al. 2010; Bočić et al. 2015), and there are active gorges as well, where incision keeps pace with uplift rate. Closed depressions on the plateaus, namely, dolines and uvalas, serve as "snow traps" contributing to the accumulation of snow and ice. The steep walls of plateau margins are often of tectonic origin and they may also help snow and ice accumulation on the leeward side of mountains. Poljes are found at lower elevations, so they do not have a direct impact on glaciation. However, they often served as outlet zones of glaciers and fluvioglacial sediments are frequently deposited in poljes (Adamson et al. 2015).

9.7.3 *Climatic Settings, Sea Level Changes*

The climate is Mediterranean along the coastline, continental in the inner parts, whereas the higher terrains are characterized by different subtypes of mountain climate.

The mean annual temperature is 10–15 °C near the sea level, but as we get higher, there the temperature decreases, and thus, Žabljak, for instance, the highest town in the Balkans at the foot of Durmitor Mountains has 5.7 °C as mean annual temperature. Sv. Jure peak in Biokovo (1762 m asl) has 3.9 °C. The above data demonstrate that the temperature had to be significantly lower during the glacial periods in order to meet the conditions of glaciation. The drop of temperature in the glacial periods has been estimated by Hughes et al. (2007) for the Pindus Mountains. They calculated that, during glacial periods, mean summer temperature was at least 11 °C colder than at present.

The other critical factor from the viewpoint of glaciation is the amount of precipitation. The humidity of air masses coming from westerly directions significantly increases above the Adriatic Sea and, when it reaches the coastal mountains, precipitation is formed. This process is especially emphasized by the fact that the orientation of coastal mountains is by and large perpendicular to the prevailing wind. As a consequence, the high mountains of the Adriatic coast receive large amounts of precipitation (>2000 mm). In addition, given the present Mediterranean climate, most of this precipitation falls during wintertime as snow. The wettest place in Europe is found just here, Crkvice village in the Orjen Mountains (Montenegro) holds the record of annual precipitation with 8036 mm (measured in 1937), but even the mean value is close to 5000 mm (Ducić et al. 2012). On the other hand, the amount of precipitation quickly decreases towards the inner parts of the peninsula which exerts a negative effect on both glaciation and karstification. It is noted that Olympus Mountain, which is found at the eastern part of the peninsula, also gets high precipitation due to the proximity of the Aegean Sea (Styllas et al. 2015).

Naturally, such nominally high precipitation was not possible during the glacial periods because of the cold climate. Moreover, sea level was significantly (100–120 m) lower than today, and a large proportion of the Adriatic Sea and the Aegean Sea was land that also influenced the precipitation regime. However, as the sea near Montenegro is deep enough, the area was water-covered even during glacial periods and airflow directions could have been similar to the present situation (Kuhlemann et al. 2009). As a result, Montenegrin mountains received *relatively* high precipitation even during the glacial periods.

Sea level changes along the Balkan coastlines were influenced not only by global sea level changes but by local tectonics as well as briefly discussed above (Lambeck 1995). In addition to the actual elevation of mountains, sea level also influences the vertical extent of karst systems' evolution.

9.7.4 Characteristic Landforms and Sediments

First, we mention the landforms directly related to the ice, moraines. Lateral and terminal moraines are found in several mountains of the Balkan Peninsula. In certain locations, recessional moraines are also recognizable. Basal moraines are less characteristic, because much of the material of basal moraines was denuded after glacials partly dissolved, partly washed into sink points. Nevertheless, other types of moraines are surprisingly well preserved at many sites, where, due to karst processes, postglacial fluvial erosion was subordinate and unable to dissect the surface. In some places, hummocky moraines are also preserved. The material of moraines is usually poorly sorted diamiction, with cobble to debris size limestone clasts in fine-grained matrix. Clasts are angular or subangular and sometimes striated.

Glacially polished surfaces, roche moutonnées are common glaciokarst landforms in the Balkan Peninsula, although not as widespread as in the Alps. Given the present climatic conditions, soil covers the surface at most places below 2000 m, where natural conditions prevail. Naturally, the soil layer is usually thin and more importantly, the soil has been eroded to a large extent due to anthropogenic impact. Polished surfaces are usually ornamented by different types of karren landforms (runnels, meanderkarren, kamenitzas, karr pavements, etc.), although on the highest terrains, where periglacial conditions prevail today, frost weathering may hinder the formation of small-scale dissolution features (Fig. 9.23).

Drumlins were described by some authors in the Dinarides, notably in the Velebit Mountains (Croatia) by Velić et al. (2011, Fig. 9.24) and in the Prenj Mountains by Lepirica (2008). However, as drumlins located in full alpine setting are extremely rare (van der Meer and van Tatenhove 1992), it is questionable whether the mentioned landforms can be considered true drumlins.

Fig. 9.23 Limestone pavement in Mt. Tymphi (photo by Telbisz)

Fig. 9.24 Debris hillock at the bottom of a large uvala with two small depressions (photo by Telbisz). These are marked on the map by Velić et al. (2011) as drumlin and kettle holes

Relatively small diameter, but occasionally very deep, shaftlike depressions, called kotlići in Slovenian and Schneedolinen in German, are present at many locations (Fig. 9.25). Some researchers state that these shafts form at the sink points of supraglacial meltwater streams (Stepišnik et al. 2016), and thus, they may be thought of as the karstic version of glacier mills (moulins). Another explanation for their formation may be that these vertical depressions are the shafts of former dolines of the preglacial surface without the upper bowl-like part of the depression, which could have been destroyed by glacial erosion. Kunaver (1983) attributed the formation of kotlići to the meltwater infiltration of snow patches. This means that these forms are the results of processes active in the present climate.

Traces of glacier exaration are also obvious on terrains, where the smaller interdoline ridges are rounded (Fig. 9.26). Glacier striae are usually less clear, because small-scale dissolution features (karren) often overprint these glacial traces. At certain locations (e.g. in the Sinjajevina Mountains/Montenegro), larger grooves (in the scale of 100 m length) are recognizable on satellite images. These forms could be incised into the surface by a plateau glacier moving towards the edge of the plateau (Telbisz 2010b, Fig. 9.27).

U-shaped glacier troughs also occur in several mountains in the Balkan Peninsula. However, valley glaciers were much shorter than in the Alps. The longest valley glacier (which is convincingly demonstrated) is found in the Prokletije Mountains (Albania-Montenegro) with 12.5 km length (Milivojević et al. 2008, Fig. 9.28). In some places, valley glaciers reached surprisingly low

Fig. 9.25 Snow-plugged shaft in the Durmitor Mts (photo by Telbisz)

Fig. 9.26 Rounded interdoline hills in the Sinjajevina Mts (photo by Telbisz)

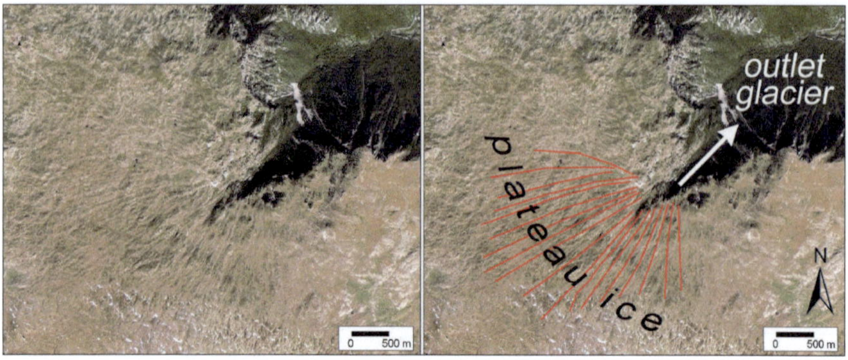

Fig. 9.27 Grooves created by the Sinjajevina plateau glacier. The ice slowly moved towards the outlet valley glacier in the northeastern part of the image. Left: the original GoogleEarth image, right: red lines show the directions of grooves

Fig. 9.28 The longest glacial valley (Ropojana) in the Prokletije Mts with a dried up periodic lake (photo by Telbisz)

elevations, for instance, they ended at as low as 500 m asl in the Velebit Mountains (Croatia, Bočić et al. 2012), at 600 m asl in Montenegro (Hughes et al. 2011) and there exist some extreme views (Marjanac and Marjanac 2004, 2016), which state that some glaciers could even reach the sea level (see later the discussion of this controversial issue).

There is an important difference between karst and non-karst terrains. Although glacial valley profiles are stepped in both cases, lakes are typical only in non-karst setting. On well-developed karst terrains, in turn, mountain lakes are sporadic features due to the high intensity of infiltration. Nevertheless, lakes do exist on some of the Balkan high karst terrains (Fig. 9.29). In some places (e.g. Ropojana Valley/Prokletije), lakes are intermittent, usually filled with water during the wet season. Cirques are much more common phenomena on glaciokarst terrains than lakes. Their formation and deepening are partly due to glacial erosion, and partly due to karst processes. As it was already mentioned, where depressions of karstic origin already existed on the preglacial surface, the accumulation of snow and ice was promoted. During post- and interglacial periods, cirques were further deepened by karst processes and their edges were not dissected by streams. In spite of the relatively large number of cirques, lakes only occur in them where the rocks contain some impermeable material, either within the limestone strata (e.g. flintstone) or as separate layers (e.g. flysch). Due to the prevailing westerly or southwesterly wind direction and the effect of insolation, cirques are typically open towards north, northeast and east. However, it is to be noted that this orientation is far from being exclusive (e.g. Milivojević et al. 2008; Hughes et al. 2007).

Steep, almost vertical walls at the edge of plateaus and at valley sides are not specific to glaciokarst, but the high physical stability of limestone, especially in the case of thick-bedded limestone, emphasizes the formation of these walls, and so they are common features throughout the Balkan glaciokarst terrains. Where

Fig. 9.29 One of the rare examples of permanent mountain lakes in the Balkan glaciokarsts, the Trnovačko jezero in the Maglić Mts (photo by Telbisz)

limestone strata have a large dip, the steep or occasionally vertical rock surfaces are typically dissected by runnels (wandkarren).

Canyons (or gorges) are in general typical landmarks on karst terrains. They may be formed basically by cave collapse or by continuous incision in different environments. A further type of formation is in connection with glaciation. The erosion of subglacial water flows can also create gorges (e.g. Nevidio canyon/Durmitor, Montenegro—Djurović and Petrović 2007). It is also notable that glacier-fed streams also have high incision capacity and may form gorges in certain cases. In other places, however, these streams are quickly swallowed after leaving the glacier, or even the water sinks below the glacier (Gremaud and Goldscheider 2010).

Glacier-fed streams usually transport high amounts of subangular and poorly sorted sediment load, which is deposited in the vicinity of the glacier front. Deposition is typical at sites where the slope abruptly decreases, for instance, where the stream reaches the polje floor. Smaller depressions can be completely filled in by these sediments, and thus, the stream may cross them and increase its flow path (Žebre and Stepišnik 2015b). The interpretation of fluvioglacial sediments is ambiguous in certain cases. It is not always clear whether a sediment was deposited by the glacier itself or it was deposited by fluvial processes. This is the reason why some low-elevation sediments cause problems in the delimitation of glaciation. Certain researchers assign low-elevation sediment accumulations to glacial processes and describe extreme long valley glaciers, whereas others claim that these sediments are of fluvioglacial origin and so they do not mark glacier terminus. Now, some examples are presented to the aforementioned critical locations: Lepirica (2008) described glacier terminus at 250 m asl in the Prenj Mountains (Bosnia and Herzegovina), Marjanac and Marjanac (2004, 2016) mentioned erratic blocks on terrains near the sea level (at Velebit and Risnjak mountains in Croatia) and even from areas which are today on islands (Krk, Rab and Pag islands also in Croatia). However, these views are under criticism and not accepted as glacial evidence by most of the researchers (Hughes et al. 2006; Hughes and Woodward 2017).

The sediments either of glacial or of fluvioglacial origin have been cemented locally during postglacial and interglacial periods, but not everywhere. Cementation, where present, is important because, first, it highly increases the resistance of the rocks, and, second, the cement material can be efficiently used for dating the sediment (see later).

Glaciokarst terrains in the Balkans, as most other glaciokarsts of the world, are characterized by medium-sized closed depressions, i.e. depressions larger than a typical solution doline, but smaller than a polje. These depressions are usually compound forms with several sub-depressions and uneven bottom. The depth may exceed 100 m. Local names for these landforms are konta or do. Their formation is due to favoured snow and ice accumulation in the originally smaller preglacial karstic depressions. Glacier erosion often implies the coalescence of smaller depressions. Kranjc (2006) mentions that there are often unconsolidated glacial sediments at the bottom of kontas that may form a local aquifer and feed springs.

Glaciation highly influenced the evolution of caves as well. These interactions were thoroughly studied in the Velebit Mountains (Croatia) by Bočić et al. 2012.

Due to low temperatures and frost shattering in glacial periods, the amount of debris was increased. Meltwater streams transported more debris into the caves that could lead to a periodic or permanent clogging of certain cave passages. Upstream of the blocked passages, lakes, could form and even some passages could be completely filled with water, i.e. phreatic conditions prevailed in these parts of the cave. Near cave entrances, frost shattering and collapse processes were more active. At the onset of warm periods, the amount of incoming water significantly increased in most places, though not everywhere, since the recharge points could be different than before. The increased postglacial discharge started to clear the passages and to transport away the debris deposited during the glacial period. Further on, the horizontal passages could become deeper in interglacial periods. However, where the water recharge was more diffuse, the cementation of cave sediments was favoured and flowstones were forming more actively.

9.7.5 Glaciation Characteristics and Phases

Due to the climatic settings, the glaciers of the Balkan Peninsula were warm-based glaciers. Consequently, they intensively eroded the surface and subglacial melt-waters could infiltrate into the karstified rocks even during glacial periods.

Plateaus are typical karst landforms which favour the formation of plateau glaciers (e.g. in the Lovčen, Sinjajevina and Snežnik Mountains/Montenegro, Slovenia). These icefields had outlet glaciers towards the marginal valleys. As the outlet glaciers were fed by a plateau glacier, they extended relatively far into the valleys and reached low elevations. In the high, uplifted mountains without plateau surfaces, alpine-type valley glaciers could form following the preglacial valley network (e.g. Durmitor Mountains/Montenegro). Where valley glaciers reached a plain area, which was still at high elevation (e.g. a lower plateau or a polje surface), they created large, fanlike glacier tongues which are called Piedmont or Malaspina-type glaciers (Djurović 2009).

Even early researchers of the area recognized that several glacial phases took place in the highest mountains of the Balkan Peninsula. In most of the thoroughly studied sites, 3–5 phases have been identified since. First, there are the traces of the most extended glaciation (MEG), which is assigned to the lowermost moraines and other glacial landforms. Second, proofs of a much smaller, but still quite extended glaciation, can be observed. In some places, this second phase of glaciation could be divided into two separate sub-phases. Third, there was a highly reduced glaciation, which was limited basically to cirque valley glaciers. Finally, there is the most recent phase, which means some still surviving microglaciers. These forms are today at the verge of final disappearance due to the ongoing global warming. Microglaciers are found in Durmitor (Montenegro), Prokletije (Albania) and Pirin (Bulgaria) mountains (Gachev et al. 2016; Grunewald and Scheithauer 2010).

Based on glacier-related landforms, especially the elevation and location of moraines, the equilibrium line altitude (ELA) of former glaciers can be determined using different methods, namely, the *maximum elevation of lateral moraines*, the *accumulation–area ratio*, and the *area altitude balance ratio* (Žebre and Stepišnik 2014). Results of these calculations demonstrate that west–east gradients are more significant in the regional changes of ELA values than north–south gradients. This circumstance underlines that precipitation had a crucial role in the extent of glaciation in the mountains of the Balkan Peninsula. Naturally, the ELA acquired higher and higher positions in the subsequent glacial phases mentioned above.

For long, the age of glacial phases could be determined only by relative methods and by analogies or using correlative sediments and soil remnants from interglacial periods (Smith et al. 2006). However, in the last two decades, numerical dating became also available. The Little Ice Age glacial changes were studied by lichenometry (Hughes et al. 2007), whereas the minimum age of older glaciations could be identified based on the glacial (fluvioglacial) sediments cemented by carbonates. The age of the carbonate cement can be measured by U-Th radiometric methods (Hughes et al. 2006). ^{10}Be exposure age studies were also performed in some of the Balkan Mountains (Kuhlemann et al. 2009, 2013). These cutting-edge studies helped to clarify the chronology of glaciations in the Balkan Peninsula. The MEG took place mostly in MIS 12, and the ages of the medium phases are MIS 6 and MIS 5d. Finally, the last major glaciation occurred roughly simultaneously with the LGM in other European areas. In some places, the LGM was the most extensive glaciation (e.g. in the Central Balkans) and it overprinted the forms of earlier glacialiations (Kuhlemann et al. 2009, 2013). Nevertheless, to date, these numerical datings have only been completed in some selected mountains of the Balkan Peninsula and further research is necessary to gain a more precise image.

As for the real extent of the MEG, there are still some debates, conservative or excessive estimations. Most plausibly, the extent of the ice caps was the largest in the middle Montenegrian mountains, where the elevation of the terrain is high, karst plateaus are large and the amount of precipitation is also abundant. According to the well-established observations of Hughes et al. (2011), the maximum area of contiguous ice caps could be as large as 1500 km^2 in this area. Marjanac and Marjanac (2016) stated that in the Middle Pleistocene, there existed an enormous icefield of 5400 km^2 area with a focus in the Velebit and Risnjak mountains, extending to some of the coastal islands. Nonetheless, this theory is not accepted by most researchers (see Hughes et al. 2006).

It is important to mention from the viewpoint of glaciokarst that the interaction of glacial and karst processes was highly variable and climate-dependent during the Quaternary (Adamson et al. 2014). During glacial periods, the intensity of karstification was strongly reduced on ice-covered terrains, although some infiltration could occur below the warm-based ice bodies. Due to the far-extending glaciers and the meltwater-fed streams, coarse sediments were transported from high elevations to lower terrains, i.e. there was an active transport connection between the higher and lower regions. With glacier retreat, the intensity of karstification increases again, fluvial sediment transport and erosion diminish, or even cease to work. As a consequence, the high and low parts of karst mountains become gradually disconnected during interglacials. This process is called *decoupling* by Adamson et al. (2014).

9.7.6 *Study Areas*

9.7.6.1 Dinaric Karst

The Dinaric Alps have northwest–southeast strike and they run parallel with the eastern coast of the Adriatic Sea and the total length of the mountain range is ca. 650 km (Figs. 9.30 and 9.31). The elevation gradually increases from northwest to southeast, and the highest peak, Mt. Jezerce (2694 m) is found in the Prokletije Mountains (Albania, Fig. 9.32). As for the glaciokarst landforms, and in general, karst landforms, the Dinaric Alps are among the best studied landscapes. Several mountains within the Dinarides have been thoroughly investigated from these points of view. Selected mountains are presented in the followings from northwest to southeast. Here, we note that there are numerous other glaciokarst terrains in the Dinaric Alps that are still partly unexplored or less thoroughly examined.

The glaciokarst landforms of *Snežnik* (1796 m, highest elevation is given for the highest peak in brackets) and *Gorski Kotar* (1528 m) have been studied by Žebre and Stepišnik (2015a), who presented a detailed geomorphological map and a thorough study, too (Žebre et al. 2016). These mountains are found at the

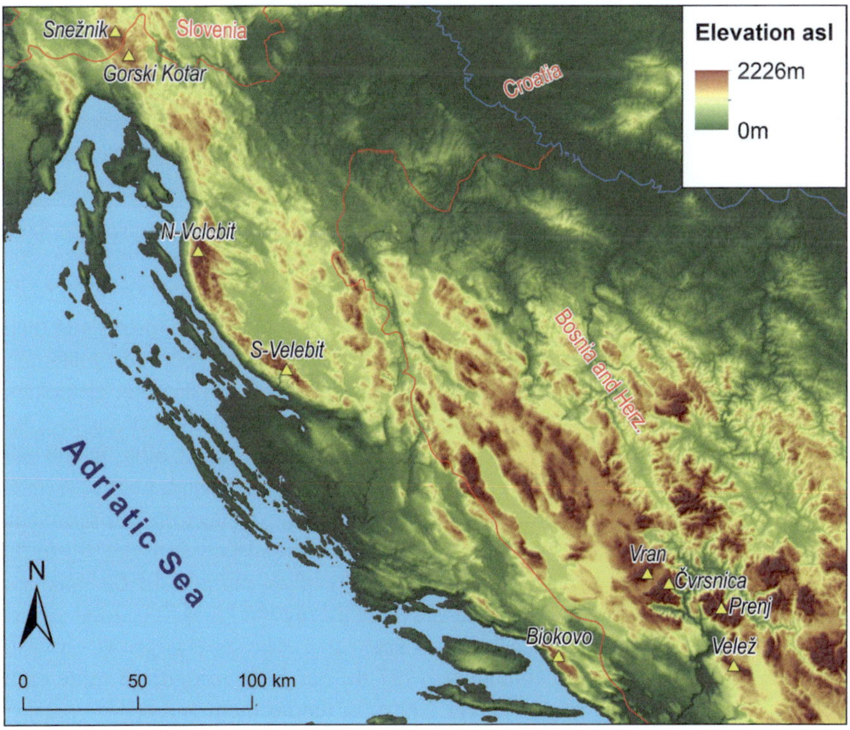

Fig. 9.30 Map of glaciokarst locations in the North Dinaric Alps

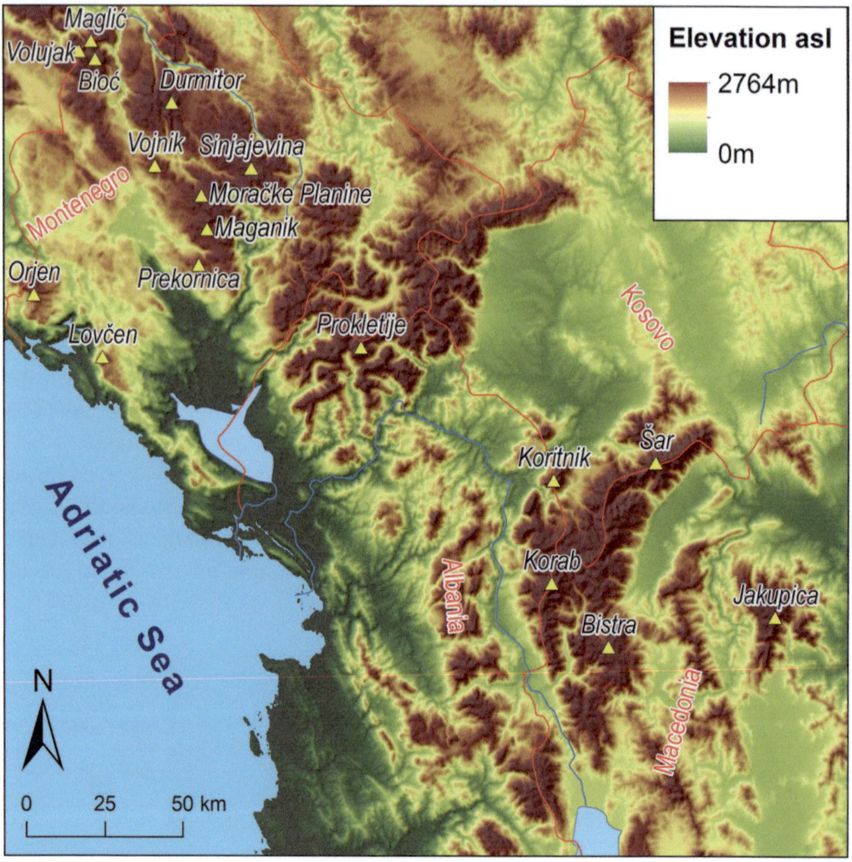

Fig. 9.31 Map of glaciokarst locations in the Southern Dinaric Alps

northeastern Dinarides. At present, their area is covered mostly by forests and there is no surface drainage network due to the high intensity of karstification. Quaternary glaciations remarkably shaped the surface of these mountains. The main glaciokarst landforms are the large-diameter, closed depressions (with diameters up to 2 km and depths as large as 110 m), glacial valleys, rounded conical hills, lateral and terminal moraines (up to 70 m height) occasionally with hummocky surface. Glacial sediments are found at as low as 580 m elevation. Glaciofluvial sediments cover parts of the neighbouring Piedmont polje as alluvial fans and they fill in the small depressions of the polje bottom resulting a plain surface. In Snežnik and Gorski Kotar, glaciation affected the largest area during the LGM. The total area of glaciation in these mountains reached 140 km^2.

Glaciation of the *Velebit* Mountains (1757 m) has been recognized quite early (Hranilović 1901 in Bognár and Faivre 2006). In the northern and central parts of the mountains, the snowline could be at 1300 m, whereas it was at 1200 m in the

Fig. 9.32 Jezerce Peak in the Prokletije Mts (photo by Telbisz)

southern segment (Bognár and Faivre 2006; Bočić et al. 2012). A detailed chronology with numerical age data is still missing from this area, so it is only supposed that glaciation could cover the largest area in the Middle Pleistocene. The deep karst depressions of the plateau certainly played an important role at the onset of glaciation. Several types of glaciers emerged in the Velebit Mts: plateau glaciers, valley glaciers and cirques (Fig. 9.33). Former plateau glaciers have been replaced by deep (occasionally more than 100 m deep) dolines and uvalas. Abrupt breaks in the valley floor, typical of valley glaciers, are also present in the area. In some places, tectonically predetermined, preglacial doline rows controlled the formation of valley glaciers. The longest glacier (Šatorinski) was 4.5 km long and reached down to 500 m asl (Bočić et al. 2012). In several large and relatively low-lying depressions (uvalas and dolines), lakes formed during the glacial periods, but the sediments are not always preserved as the material was mostly limestone, which could be dissolved and transported away through the karst system during interglacials. Lateral, terminal and ground moraines, as well as glaciofluvial deposits, were observed in many places in the Velebit Mts. Quaternary changes in the evolution of Štirovača Ice Cave have been presented by Bočić et al. (2012).

The *Biokovo* Mountains (1762 m) have an elevation range similar to that of the Velebit Mts. The plateau is characterized by deep, densely packed and connected dolines. This type of landscape is called polygonal karst (Telbisz et al. 2009, Fig. 9.34). As for the older (Middle Pleistocene) glaciation of the Biokovo, there is no precise data, but it is assumed that most part of the plateau was covered by ice (Žebre et al. 2013). Locally, there are cemented sediments, which may be attributed

Fig. 9.33 A typical glacial valley (Lomska Duliba) in the Northern Velebit (photo by Telbisz)

Fig. 9.34 Polygonal karst of the Biokovo Mts (photo by Telbisz)

to this older glacial period. The younger (Late Pleistocene) glaciation, in turn, was limited to a small, ca. 1 km^2, area northeast of the main peak, where two cirques have survived as glacial landforms. The ELA calculated from these landforms was 1515 m asl, that is, 200–300 m higher than in the nearby coastal mountains (e.g. Velebit or Orjen Mts, Žebre et al. 2013). One explanation for the higher ELA can be that precipitation was lower in the Biokovo, but this issue needs further clarification.

The high karst of Bosnia–Herzegovina is more elevated than the coastal mountains, but it is farther from the sea and obviously receives less precipitation. Traces of glaciation and glaciokarst landforms were studied here also at the beginning of scientific karstology (Cvijić 1899; Grund 1902 in Stepišnik et al. 2016). Investigations became quite dangerous in 1992–1995, when Bosnian War broke out because of the military mines laid there. The glaciokarst landforms of Dugo Polje (Blidinje) and the neighbouring mountains (*Vran*, 2074 m, *Čvrsnica* 2226 m) have been studied by Stepišnik et al. (2016), who also drew a geomorphological map. The *Prenj* Mountains (2115 m) have been investigated by Lepirica (2008) from a geomorphological point of view, also taking into account glaciokarst features. The *Velež* Mountains (1969 m) are found more to the south, and there are also glaciokarst terrains presented by Žebre and Stepišnik (2015b). All of the above mountains have similar characteristics: alpine landscape with sharp ridges and jagged peaks above 2000 m asl. Cirques and U-shaped glacier troughs with stepped profiles also occur there. The coalescence of uvalas into a glacier valley can also be observed locally. Well-preserved, fluvially undissected moraines are present, too. The Dugo polje is a typical Piedmont polje. The valley glacier ending in this polje created a fanlike lobe and a three-km-wide terminal moraine. Glacial and periglacial sediments, as well as the dolomite rocks, are partly impermeable, and thus small streams could form on them, and these streams are swallowed, where they reach the karstified limestone bedrock. Blidinje Lake, found today in Dugo Polje, had a much larger extension as reconstructed from the surrounding abrasion terraces. Lepirica (2008) describes roche moutonnée with karren features, truncated shafts, erratics and drumlins as glacial or glaciokarst landforms. As for the presence of drumlins, see Sect. 9.7.4. The lowermost elevation of erratics and moraines is 250 m asl (Lepirica 2008). However, taken into consideration reliable data from other parts of the Dinarides (e.g. Hughes et al. 2006, 2011), these results are highly questionable, and the given landforms are most probably not of glacial origin. For the Bosnia–Herzegovinan karsts, the geomorphological map of Stepišnik et al. (2016) marks the lower terminus of glaciokarst sediments at 1200 m asl, and Žebre and Stepišnik (2015a, b) observed terminal moraines at ca. 1000 m asl in the Velež Mts. These data are acceptable and are in good agreement with other regional studies. Further on, Lepirica (2008) presents gorges created by subglacial erosion and glaciofluvial sediments, i.e. strongly cemented conglomerates in major valleys.

In southeastern Bosnia and Herzegovina, the adjacent mountains of *Volujak* (2336 m), *Bioć* and *Maglić* (2386 m) were also glaciated. Milivojević (2007) identified three glaciation stages here. During the most extended glaciation, the ELA was at 1700 m asl, and relatively large valley glaciers could form, locally even covering the mountain passes, i.e. glacial bifurcation occurred. The glacier with the lowermost terminus was the Mratinje glacier extending down to 1100 m asl. On the other hand, the longest glacier with 7 km length was the Trnovački glacier (Fig. 9.35). In the second and third glacial stages, only cirque glaciers could exist as the ELA was 100–200 m higher. Curiously, the largest cirques are found in southern exposure in this area. It is due to the fact that these landforms were predetermined by preglacial depressions and tectonic fractures. The westerly or

Fig. 9.35 The longest glacial valley of Maglić Mts. Arrow points to the location of Trnovačko jezero (photo by Telbisz)

southwesterly winds accumulated snow in the northeastern parts of the main ridges, but the cirques there are somewhat smaller than the largest, southerly exposed cirques. In one of the largest, compound, cirques, there is a relatively spacious lake, the Trnovačko Jezero, which is a rare phenomenon in the high karst area.

The high karst terrains of Montenegro have been most thoroughly studied for glaciokarst since the pioneering work of Cvijić (1899). The coastal mountains, i.e. the *Orjen*, have been studied recently by Stepišnik et al. (2009), the *Lovčen* (1749 m) by Stepišnik et al. (2011), Stepišnik and Žebre (2011) and Žebre and Stepišnik (2015a, b). The inner parts, namely, the *Durmitor* (2523 m), have been investigated in recent times by Djurović (2009), the *Sinjajevina* (2277 m, Fig. 9.36) by Telbisz (2010a, b), the *Moračke Planine* (2226 m), *Maganik* (2139 m), *Prekornica* (1927 m), and *Vojnik* (1998 m) together with the previously mentioned mountains have been included in a detailed synthesis by Hughes et al. (2010, 2011). During the maximum glaciation, all of these mountains could be characterized by extensive and thick (200–300 m) plateau glaciers, which had outlet glaciers descending towards the lower valleys and poljes. The thickness of certain valley glaciers amounted to 700–800 m in the Durmitor. Valley glaciers reached down to as low as 600–700 m asl around Šavnik, but in other places, the lowermost terminus was rather at around 1000 m asl. The largest extension of the Orjen ice cap was 165 km^2, whereas the Lovčen glaciers covered 40 km^2. The ELA was at 1260 m in both mountains. The inner mountains had a total 1500 km^2 area of icefield at the maximum glaciation, and the ELA increased gradually towards the inner parts up to 1600 m asl. Detailed geochronological studies were carried out by Hughes et al. (2011). U-series datings were performed on the carbonate cement of glacial tills. The ages so obtained can be thought of as minimum age for the formation of moraines. These results clarified that in these mountains, maximum glaciation occurred in MIS 12, and the subsequent glacial stages in MIS 6 and MIS 5d-2. During the MIS 6 glaciation, the ice-covered area was about one half of the maximum, but during the MIS 5d-2, it was only one-30th of the maximum! Thus, this latter glaciation was limited to cirques. In the Little Ice Age, there were some microglaciers, of which the Debeli Namet (in Durmitor) still exists, though in a stage of retreat (Hughes 2007, 2008).

The highest member of the Dinaric Alps is the *Prokletije* Mountains (2694 m), which tower along the Albanian–Montenegrian border (Fig. 9.37). These mountains have alpin- type glacial landforms, which clearly showed, even for the early researchers, that the Prokletije was glaciated during the Pleistocene (Cvijić 1903 in Milivojević et al. 2008). More recently, the glaciokarst morphology of these mountains has been studied by Milivojević et al. (2008) and by Petrović (2014). The western part had a large, contiguous ice cap (of 180 km^2) with nunatak peaks, while the eastern section was characterized by valley glaciers, including the Ropojana glacier, the longest in the region. There are several estimations for its length, but the most reliable study claims that it reached 12.5 km at its maximum (Milivojević et al. 2008). On higher terrains, large and compound cirques are the typical glaciokarst landforms. In the western section, it is clearly observable how

Fig. 9.36 Plateau of the Sinjajevina Mts with glacially polished surface in the foreground and cirques in the background ridge (photo by Telbisz)

Fig. 9.37 Jagged peaks in the Prokletije Mts (photo by Telbisz)

the preglacial uvalas controlled glacial valley formation. Roches moutonnées are widespread in the area. Moraines are also present and as the eastern valleys are built up of non-karstic rocks (phyllite and schists), it is comparable, how the limestone-based moraines are much better preserved than their non-karst-based counterparts. The ELA during maximum glaciation was at 1750 m asl, and the lowermost terminal moraines are found at 920 m asl. It is in good agreement with other studies regarding that the Prokletije Mts are somewhat farther from the sea, so the climate is slightly more continental with less precipitation than coastal mountains. In some cirques of the eastern parts, tiny rock glaciers and a microglacier still survive. At present, some smaller permanent and periodic mountain lakes are found in the Prokletije, but morphological traces from the Early Holocene suggest that once larger lakes also occurred here.

The *Korab* (2764 m), *Šar Planina* (2748 m) and *Koritnik* (2393 m) mountains rise at around the triple frontier of FYR Macedonia, Kosovo and Albania. Sensu stricto, they are not parts of the Dinarides, but they are practically the spatial continuations of the Dinarides with similar morphological and geological settings. Given the high elevation of these mountains, they were significantly affected by glaciation (Menković et al. 2004; Kuhlemann et al. 2009). During the LGM, the ELA was at 1900 m and 2200 m asl in the northern and southern slopes, respectively. Using ^{10}Be exposure ages, the LGM was found to be the time of maximum glaciation here (Kuhlemann et al. 2009). The highest terrains are characterized by sharp ridges, cirques and valley glaciers. Glaciers extended down to 1300–1500 m asl as testified by moraines. From the ELA values, it is remarkable that the glaciation had a higher limit here than in the mountains closer to the sea. This is due to lower precipitation, which in turn is explained by more inland position. Limestones are less dominant in the geological composition of these mountains, or they are even subordinate. Therefore, glaciokarst morphology is not mentioned in the publications. However, there are some high karst mountains in FYR Macedonia, for instance, *Jakupica* (2540 m) with its marble karst (Bosák et al. 2015), or *Bistra* (2163 m), which would certainly deserve to be studied for glaciokarst.

9.7.6.2 Hellenides

The Hellenides mountain chain is the continuation of the Dinarides on the territory of Greece and partly in Albania. Due to lower latitude, the glaciation threshold is usually higher in the Hellenides. However, there are lots of karst mountains with elevations above 2000 m, and thus, glaciokarst terrains are also numerous here. Papers focussing specifically on glaciokarst landforms are few, except the well-explored *Tymphi* Mountains (Hughes et al. 2006, 2007), because the main issue of glaciation studies in the Hellenides was rather the relationship between climate and tectonics (Fig. 9.38).

The highest peak in Greece is Mount *Olympus* (2918 m), well-known from Greek mythology, where 24 cirques are found (Smith et al. 2006). During the most extended glaciation (MEG), even valley glaciers could form reaching the Piedmont area. From indirect evidence, Smith et al. (2006) proposed that the most extended glaciation occurred in MIS 8. However, Hughes et al. (2006) presume MIS 12 for the MEG, based on the analogy of the Tymphi Mountains. Smith et al. (2006) also state that Olympus glaciers had 20–25 km length and reached the present-day sea level with an ELA of 500–600 m asl. Nevertheless, these data seem to be highly exaggerated, even if the relatively high tectonic uplift rate (1.6 mm/a) is taken into consideration. A second, Middle Pleistocene, as well as a third, Würmian, glaciation was also observed by Smith et al. (2006). Most recent research (Styllas et al. 2015) plausibly points out that on several occasions during the Holocene, for instance, in the Little Ice Age, there existed a small microglacier in the most remarkable, north-facing cirque of Mt. Olympus, called Megala Kazania. The potential effect of limestone on glacial landforms is mentioned only in the context of gorge-like, deep, incisions and the formation of large, vertical rock walls.

The most thoroughly studied and well-developed glaciokarst of the Hellenides is found in the Tymphi Mountains (2497 m), which belong to the Pindus Range. The Tymphi is built up of high resistance, pure, crystalline limestones and, to a lesser extent, of flysch rocks. Hughes et al. (2006) emphasized that Tymphi is "*one of the finest examples of a Mediterranean glaciokarst landscape*". Its glaciokarst morphology includes limestone pavements formed on glacially abraded terrains, Schichttreppenkarst (Fig. 9.39), which means cuesta-like pavement, shafts, cemented and unconsolidated moraines (Fig. 9.40), large depressions (Waltham 1978). Doline density is the highest along former glacier margins, due to the pattern of sinking points of glacial meltwater. In the higher cirques, periglacial rock

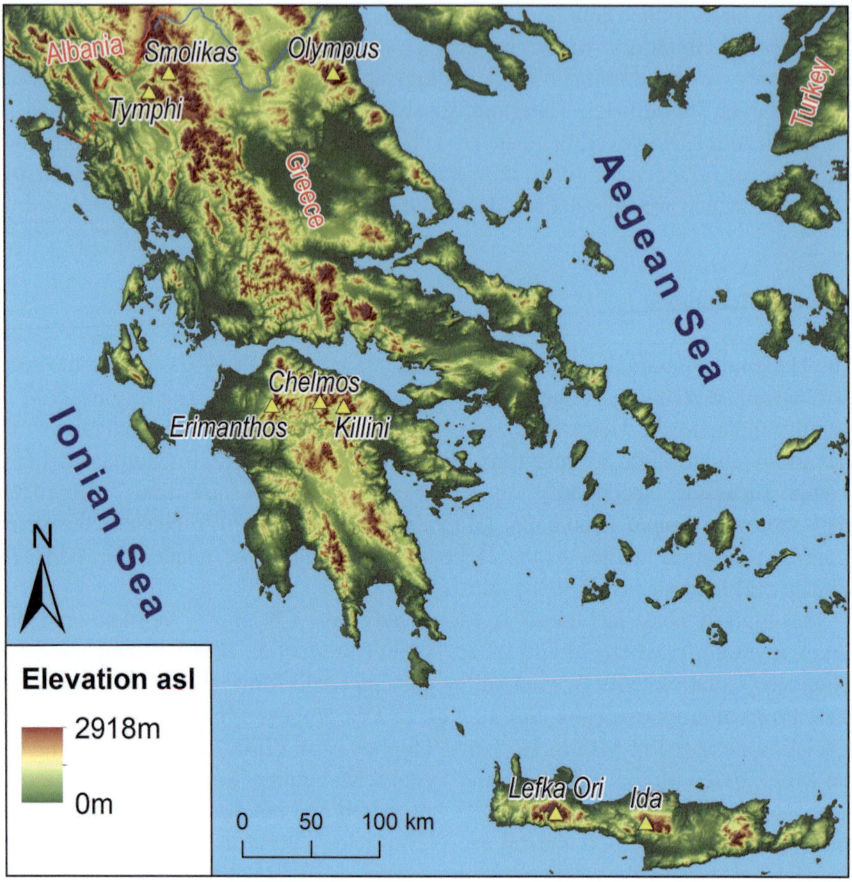

Fig. 9.38 Map of glaciokarst locations in Greece

glaciers are also found, which are considered today as fossil landforms (Hughes et al. 2003). Three glacial stages were identified here as well, and it was the first location, where numerical U-series dating of the carbonate cement of glacial sediments was performed. Moreover, fossilized soils were also dated (Hughes et al. 2006, 2007). As a result, the ages of three glacial stages were determined: MIS 12, MIS 6 and MIS 5d, the first having the maximum extension. The ELA shifted higher and higher during the subsequent stages from 1741 m through 1862 m to 2174 m. In MIS 12, the time of the most extended glaciation, summer temperatures were at least 11 °C colder than now (Hughes et al. 2006, 2007). The first stage was characterized by a plateau glacier and outlet glaciers, the second stage by cirque and valley glaciers, and the third stage by only cirque glaciers. Comparing Tymphi to its neighbourhood, the mostly ophiolite-based Smolikas Mountains (2632 m), Hughes et al. (2006, 2007) managed to convincingly demonstrate that the karst terrains of Tymphi had a much more extended glaciation (59 km^2 vs. 12 km^2)—although the

Fig. 9.39 Schichttreppenkarst in Mt. Tymphi (photo by Telbisz)

Fig. 9.40 Moraines between two ephemeral lakes in Mt. Tymphi (photo by Telbisz)

elevation and the position of the two mountains are practically similar. In addition, they also presented that the orientation of cirques in Tymphi is less dependent on exposure than in the Smolikas. In the latter case, most cirques have a northern or northeastern exposure, but cirques on Tymphi have a more variegated aspect distribution. The reason for the above facts is the more complex topography of karst that favours snow accumulation. Nonetheless, it is noted that these facts about cirque orientation do not have a general validity (see later).

Hydrologic connections between the glaciated high karst terrains and the surrounding lower areas have been studied by Woodward et al. (2008) using Pindus Mts. as an example. They found that surface runoff was higher during the Middle Pleistocene glaciation than during the interglacials, because the infiltration on limestone terrains was hindered by the ice and by increased rock debris accumulation. Meanwhile, during interglacials, karstic infiltration became more significant at the expense of surface runoff, and consequently, the hydrologic connections between the rivers of the lower areas and the high karst terrains became more indirect. This phenomenon is called "uncoupling" by Adamson et al. (2014).

Bathrellos et al. (2015) analysed 227 cirques found in different parts of Greece, among others in the aforementioned Olympus and Pindus Mts, as well as in the high mountains of the Peloponnese Peninsula and Crete Island. They emphasized, too, that karst plateaus provide an excellent environment for the formation of icefields. According to their analysis, 87% of cirques are connected to limestone terrains.These terrains can be practically conceived as glaciokarsts. A great majority of cirques have northern or northeastern exposure, especially in areas lying at lower latitudes (Stera Hellas, Peloponnese, Crete).

Detailed glaciological investigations took place in the ***Chelmos*** Mountains (2355 m, Peloponnese Peninsula) by Pope et al. (2015). Although the mountains are partly built up of limestone, the authors do not explicitly mention the notion of glaciokarst. Moraines, glacially polished rock surfaces and glaciofluvial sediments testify three glacial stages here. The first of these stages was the most extended glaciation, yet undated. The ELA of this stage was as high as 1967 m asl, and a plateau icefield with outlet glaciers existed in this period. On the other hand, the late Pleistocene glacial sediments have been dated using ^{36}Cl cosmogenic exposure age technique, a more precise dating than that of U-series. Based on these sediments, they could determine two phases of glacial advance (40–30 ka and 13–10 ka BP) with ELAs of 2046 and 2173 m, respectively. An interesting fact is that during these phases, the ELA of the Chelmos was similar to the ELA of Northern Greek mountains, whereas in the earlier stages, the ELA was significantly higher in the south than in the north. It suggests that climate was not the same in different glacial periods. Namely, in the late Pleistocene, the Peloponnese could receive relatively more precipitation than Northern Greece. The mountains of *Taygetos* (2404 m), *Killini* (2374 m) and *Erimanthos* (2124 m) of the Peloponnese have not yet been studied, but they are all potential glaciokarst terrains.

As for the glaciation on Crete Island, there has been no detailed geomorphological analysis yet. Based on Bathrellos et al. (2015) and Hughes and Woodward (2017), however, it is likely that Mt. Ida (2456 m) and Lefka Ori (2453 m) were glaciated at least during the Middle Pleistocene. Cirques, moraines and glaciofluvial sediments are also mentioned in some guidebooks. The deepest cave of Greece, Gourgouthakas (depth: 1208 m), as well as some deep gorges (e.g. Samaria Gorge) are also found in Lefka Ori.

9.7.6.3 Central Balkans

In the Central part of Balkan Peninsula, two glaciated high mountains, the partly karstic *Pirin* (2915 m) and the neighbouring *Rila* (2925 m) built up of silicate rocks (Fig. 9.41), are located. The karst of Pirin developed in a 2000-m-thick marble sequence (Lóczy et al. 2012). Although glaciations during earlier stages of Pleistocene have been observed by some researchers, the most obvious, well-developed glacial forms, namely, moraines, cirques and U-shaped glacial troughs were all dated to the LGM (Lóczy et al. 2012). Based on these landforms, the age of maximum glacier advance was established by ^{10}Be cosmogenic exposure method in the silicate rock of the Rila Mts by Kuhlemann et al. (2013). They found glacier advance stages at 24–23 ka BP and at 18–16 ka BP with a short retreat in between. This period of retreat occurred in the coldest, but at the same time the driest, times. The ELA was at 2200 m asl during maximum glaciation and glaciers reached down to 1150 m asl in the Rila Mts. The Pirin glaciers were shorter

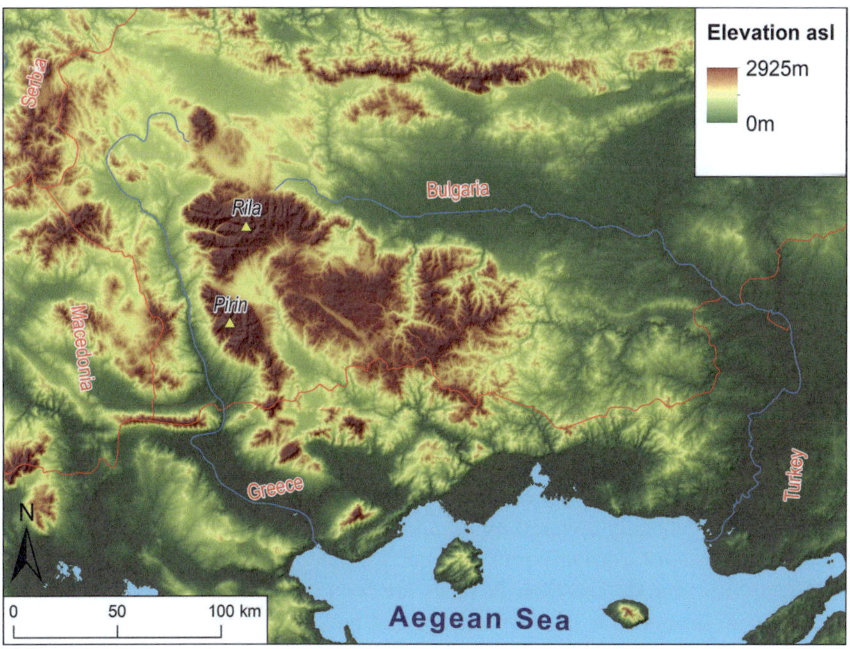

Fig. 9.41 Location map of Rila and Pirin mountains in Bulgaria

because subglacial meltwaters of the valley glaciers were “tapped” by karstic infiltration (Lóczy et al. 2012). Nonetheless, cirques of the Pirin marble karst are deeper than in the Rila Mts. The largest vertical rock wall (420 m relative height) of Bulgaria is found just here at the side of Vihren peak. Such deep cirques may promote the survival of microglaciers. Although microglaciers ceased to exist during the warmer periods of Holocene, they were renewed in colder phases. There is a still existing *microglacier*, called Snezhnika, which is the southernmost of its kind in Europe (Grunewald and Scheithauer 2010). Given the present-day climatic setting, periglacial conditions prevail, meaning intense frost shattering. Thus, karren features are less developed. Mountain lakes occur only in the non-karstic parts of the Pirin. The karstified terrains contain deep shafts, but these are not yet fully explored speleologically.

9.8 British Isles

Pleistocene glaciations covered almost the whole area of the British Isles, except the territory south of the Bristol–London line (Fig. 9.42). Although karstifiable rocks are not prevalent, there are many places, where they are at the surface, so the conditions for glaciokarst are satisfied at many locations in the British Isles. The arctic glaciokarst terrains of the British Isles have different characteristics than alpine glaciokarsts because of two reasons. First, the relief is relatively low, and second, thick continental glaciers covered the area in the Pleistocene cold periods distinct from alpine-type valley glaciers or dinaric-type plateau glaciers. The most typical features of the surface morphology are limestone pavements and pedestals, but cave evolution was also remarkably influenced by the presence of the ice sheet. The finest glaciokarst terrains are Yorkshire Dales and Burren. Today, anthropogenic impact is also an important factor in the evolution of British glaciokarsts, namely, forest clearing and mining can be mentioned among other processes. At present, in order to protect glaciokarst terrains against human impact, many of these terrains enjoy some kind of nature protection.

9.8.1 Geologic and Tectonic Settings

In Ireland, the proportion of limestones is roughly 40%, which is higher than in Britain. The central parts of Ireland are predominantly built up of limestones, but the bedrock is generally covered by till and peat bogs (Gunn 2004). Limestones in Ireland are mostly of Carboniferous age, but a smaller proportion is chalk formed in the Cretaceous period. These chalk rocks are found in Ulster.

In Britain, limestone is not so widespread. Chalks of Cretaceous age are present in Southern England, whereas Carboniferous limestones are typical in Wales and Northern England, and they bear the finest British glaciokarst terrains. In Scotland,

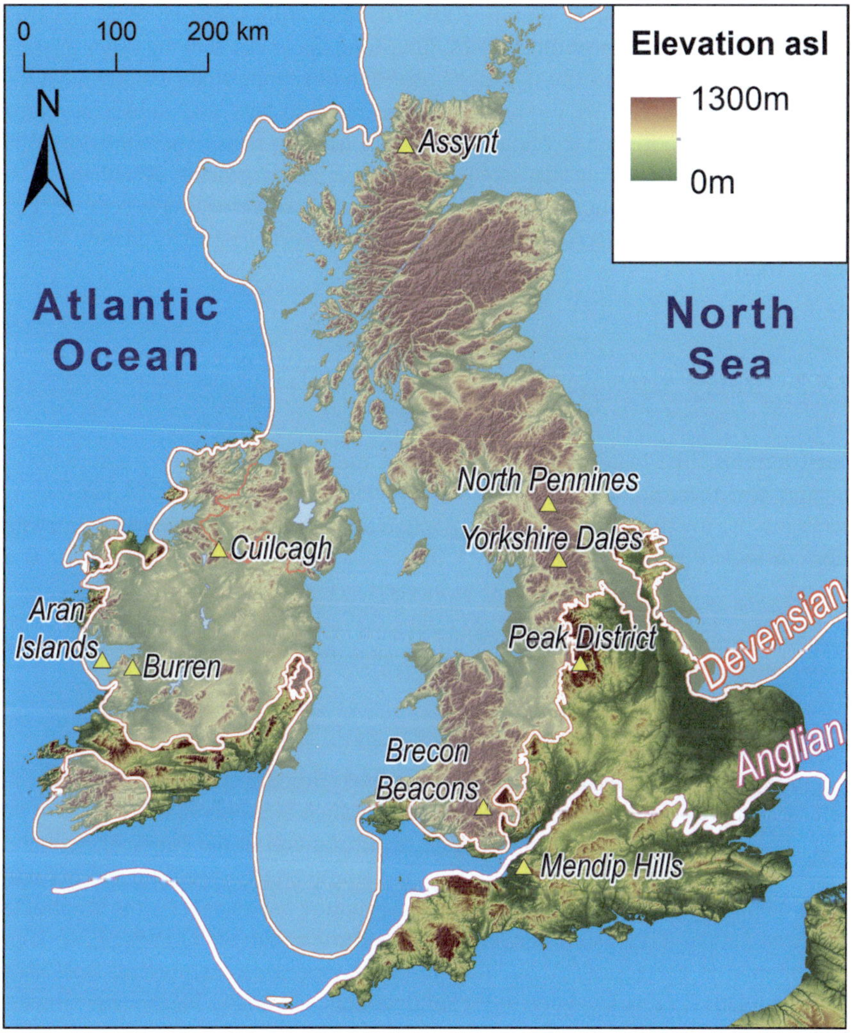

Fig. 9.42 Map of glaciokarst locations in the British Isles, with Devensian and Anglian glacial limits (glaciation data from Ehlers et al. 2011)

karstifiable rocks crop to the surface only in smaller patches. The most prominent area of them is in the North, near Assynt, where Cambrian dolomites are at the surface.

Chalks have high porosity and weak structure, and thus they are less appropriate for cave evolution, and even surface karst landforms are scarce on them, except dry valleys, which are characteristic to chalk landscapes. On the other hand, Carboniferous limestones are rich in both surface and subsurface karst features. In some cases, namely, at Mendip Hills and in Peak District, the thickness of the limestone exceeds 1000 m (Gunn 2004: Mendip Hills 1000 m; Peak District: 1800 m). However, in most other cases, limestone thickness is in the order of some 100 m (Waltham et al. 1997). It is also noted that limestone strata usually have low dip angles, which has its consequences for both surface and subsurface morphologies. An important exception is the area of Mendip Hills, where the dip of the limestone beds is high due to intensive tectonic displacements (Waltham et al. 1997). Finally, it is mentioned that some karst terrains of Britain have been affected by Late Glacial loess deposition as well (e.g. the Northwest of England; Wilson et al. 2013).

9.8.2 Relief Settings

Karst terrains, like many other landscapes in the British Isles, have low relief in general, and there are almost plain areas as well. In Ireland, 75% of karst terrains belong to the lowland category. In England and Wales, karst terrains are somewhat more dissected and have higher relief, but elevations are still well below 1000 m asl, and relative relief is below 330 m in each case (Waltham et al. 1997).

9.8.3 Climatic Settings

At present, the British Isles, including the karst terrains, are characterized by marine climate. It means high amount of precipitation, which is evenly distributed over the year. The wettest karst terrains are found in Burren and in the North of Scotland, where 1500 mm of precipitation falls a year. Central and Eastern England get only about one half of it. The annual course of temperature has relatively low amplitude, the mean annual temperature range is 9 °C in Burren, but 14 °C in the East. High precipitation is a favourable condition for karst processes. At present, frost shattering is relatively restricted due to the mild winters. Naturally, Pleistocene climates provided quite different conditions for karstification. Areas near glaciated terrains were characterized by periglacial conditions that highly influenced karst processes as well. The role of vegetation is also an issue in karst processes and it is related to climate. At present, most of the British karst terrains are either bare or covered only by a thin soil layer, and forest cover is uncommon. This poor vegetation and soil cover is, however, partly the result of anthropogenic impact, and several researches are in agreement that significant forest clearings took place already in the Bronze Age, for instance, in Burren (Drew 1983; Moles and Moles 2002). Nevertheless, even a thin soil layer may have a significant effect on the dissolution processes causing a more intensive enlargement of grikes, for instance. In the cold phases, soil was absolutely missing. In the interglacial periods and in the Holocene, in turn, thicker soils and more widespread forest vegetation occurred occasionally.

9.8.4 Glaciation Characteristics and Phases

Glaciation of the British Isles has been studied since the nineteenth century. During the Quaternary, there were at least 30 glaciation phases (Farrant et al. 2014). The last glacial period including several phases is called *Devensian*. It matches the Würmian period in alpine terminology, and it took from MIS 5d to the middle of MIS 1 (Merritt et al. 2003). During this period, the southern part of Ireland as well as Central and Southern England remained ice-free. The most extended glaciation took place earlier, in the *Anglian* period (the same as Mindel in the Alps, or MIS 12). That time, only the southernmost belt of England, the area south of the Bristol–London line remained ice-free and the whole Ireland was covered by ice. The thickness of the ice exceeded 1 km in many places, and due to high pressure, a significant part of the ice sheet was warm-based (Murphy et al. 2015). In case of warm-based glaciers, the fluvial erosion and sediment transport capacity of subglacial water flows are remarkable, but the equilibrium concentrations related to calcite dissolution are very low, and thus subglacial waters become quickly saturated. Glacier striae present in limestone terrains demonstrate that ice movement directions at a given location could be variegated among different glacial stages (Moles and Moles 2002; Wilson et al. 2013). Karst terrains without ice cover were also influenced by the nearby glaciers. Due to periglacial conditions, draining passages were often blocked by ice or by sediments, and hence surface runoff became active and fluvial valleys were formed in these karst areas, namely, in Peak District, Mendip Hills and the chalk karst terrains (Waltham et al. 1997; Gunn 2004). Some authors (e.g. Murphy et al. 2015) emphasize the speleogenetic role of ice dammed lakes, which were formed during the glacier melt period. First, the ice-dammed lakes, which were created during the glacier withdrawal phase, increased the water level in the karst, and second, they poured unsaturated water into the karst system, that made the widening of passages more intense. In contrast, subglacial waters were less effective as they became quickly saturated. Nevertheless, it is noted that these lakes were geologically short-lived and had highly variable water levels, so their precise relevance in speleogenesis is yet to be clarified.

9.8.5 Study Areas

In Britain, the finest example of glaciokarsts is *Yorkshire Dales* (Fig. 9.43). Beside dolines, stream sinks, gorges and rocky plateau edges, one may find here a large number of limestone pavements (precisely, 219 sites are known here; Fig. 9.44), which are deemed to be the most typical glaciokarst landforms (Viles 2003). Previously, it was thought that limestone pavements were completely formed since the end of the last glaciation, but recently, some researchers believe that some pre-Devensian forms could survive glacial exaration (Goldie 2006). There are

Fig. 9.43 Yorkshire Dales landscape with limestone pavement in the foreground (photo by Mari)

several possible reasons, which may explain the conservation of these preglacial karren features. First, where glaciers were cold-based, the erosion was less intensive. Second, some elevated rocky hills could remain ice-free even in the coldest phases. Third, where ice flow was diffuse, its eroding power was significantly decreased. Erratic blocks and limestone pedestals are also characteristic features in Yorkshire Dales, for instance, near Norber, and these forms may help in quantifying denudation rates. The substance of erratics can be either limestone, or granite, and these latter rocks originate from neighbouring terrains. These erratics precisely date the onset of glacier retreat, because they are deposited when the glacier melts. Age determinations showed that glacier withdrawal took place in Yorkshire at 17–18 ka BP (Wilson et al. 2013). The Giggleswick Scar is a rocky edge partly predetermined by tectonic displacements. Lots of relict phreatic caves are found there, whose formations are supposed to be a consequence of the aforementioned ice-dammed lakes (Murphy et al. 2015). Yorkshire Dales are famous not only because of the limestone pavements, but caves are also abundant here, the total length of explored caves is cca. 300 km (Viles 2003) and the longest cave of the UK is also found here. The Three Counties System stretches around the triple boundary of Cumbria, Lancashire and Yorkshire, and its full length is 86 km. Based on radiometric dating of speleothems, Waltham and Lowe (2013) proved that the oldest parts of the cave were formed before the last glacial period. Nonetheless, there are other passages, where water flows are still active. Murphy et al. (2008) carefully investigated the sediments of Gaping Gill stream sink cave, and they concluded that cave collapses principally occur in interglacial periods, grain sizes typically decrease in glacial

Fig. 9.44 Clint and grike features in Ingleborough (photo by Mari)

periods due to frost shattering, and finally, extreme flood discharges are the most characteristic at the end of glacial phases (during deglaciation) and they transport and distribute the fine-grained sediments in the cave.

In the *North Pennines*, limestone pavements are also widespread. Besides, suffosion sinkholes are found here that were formed on glacial till. These latter forms are locally called *shakeholes* (Waltham et al. 1997). Among other small-scale dissolution features, rundkarren, i.e. rounded runnels, are common landforms in the North Pennines. They were formed under soil cover.

Peak District is usually categorized as a fluviokarst (Gunn 2004); however, this area has been also affected by former glaciations except the last one, the Devensian. Remarkably old, i.e. more than 730 ka years old, glaciofluvial sediments have been described from here (Gunn 2004). Bare limestone terrains are rare, because the area is generally covered by soils, which were formed on loess deposits. Dry valleys are typical features in Peak District. These landforms are epigenetic features, that is, the valleys were originally formed on clastic cover layers and then printed over the karst by inheritance. Especially, Pleistocene periglacial conditions favoured the evolution of fluvial valleys. Some of the valleys have steep, gorge-like rock walls. There are extended cave systems, in which sediment deposition and occasional clogging of the passages took place in the glacial periods. Fine-grained sediments could penetrate the interior parts, whereas the passages close to the entry host glacial till in some places (Waltham et al. 1997).

The second longest cave of the UK is found in *South Wales*, and it is called Ogof Draenen and has a length of 70 km. This area is a typical inter-stratal karst, with maze-like passages. In the Late Devensian, this area, also called as Brecon Beacons, was close to the southern terminal of the continental ice sheet. Till and morainic deposits evidence that valley glaciers and small cirques were also formed in South Wales (Farrant and Simms 2011; Farrant et al. 2014). The above authors outline that a kind of *hydrological see-saw effect* worked in the area, that is, the direction of hydrologic gradient changed several times during the Pleistocene as a consequence of variable ice thickness and incision depth in the neighbouring valleys. In case of thicker glaciers, the upper cave passages became active, and flow direction in them was different from directions during ice-free stages. The situation was even more complicated by the fact that passages were occasionally clogged by sediments transported here during glacial, or especially during deglaciation periods. Another influencing factor was the type of the glacier, whether it was warm-based or cold-based. As for the surface landforms of South Wales, dolines are the most common features here (Waltham et al. 1997).

Mendip Hills are found south of the previous area, but a significant difference is that they were not glaciated, and therefore it is not a real glaciokarst terrain, but a fluviokarst. Nonetheless, periglacial conditions were important in the formation of dry valleys and gorges, the most remarkable of them being the Cheddar Gorge (Viles 2003). Cave evolution was also influenced by decreasing sea levels during glacial periods, as well as by clastic sediments transported into the caves in cold periods that caused occasionally the clogging of cave passages (Gunn 2004).

In Scotland, the only significant glaciokarst terrain is found around *Assynt*. This area was the last territory in Britain that became ice-free at around 11.5 ka BP, and the thick and long-lived ice sheet virtually bulldozed all preglacial karst features in Assynt, thus the present karst landforms were almost totally formed during the Holocene. Dolines and limestone pavements are only rare phenomena; instead, dry valleys characterize the karst surface (Viles 2003). However, some cave passages are of preglacial age, i.e. they are much older than the present surface morphology.

In Ireland, the finest glaciokarst terrain is found in *Burren* (Williams 1966). At present, this area is mostly bare land or covered only by a thin soil layer. Likely, the unsoiled nature of Burren is a consequence of soil erosion following Bronze Age (4–2.6 ka BP) forest clearings. However, non-anthropogenic reasons, namely, climate deterioration, could also play a role in the vanishing of soils (Drew 1983; Moles and Moles 2002). Although the Burren was affected by the two last ice advances (from 79 to 56 ka BP and from 35 to 13 ka BP), the highest parts could remain ice-free during the LGM (Viles 2003). The best-known features of Burren are the limestone pavements (Fig. 9.45), though dolines are also widespread in the area (there are more than 1500 of them), and stream sinks occur as well (Gunn 2004). Further on, there are some tens of largely extended, closed depressions (similar to uvalas or small poljes), which are periodically filled by lakes. The local name for this kind of depression is *turlough*. Smaller caves are also found in Burren, but due to their low elevation, they are usually filled with water, and thus they are poor in speleothems. Most of the simple caves with active streams are of Holocene age, but U-series datings proved that certain caves of the Burren did exist already in the last interglacial period (Gunn 2004). The spectacular morphology of *Aran Islands* is essentially similar to that of Burren, but glaciokarst processes are intermixed here with abrasion effects. The *Cuilcagh Karst* of Ireland was also glaciated, but neither limestone pavements nor turloughs are present there. A knoll-like topography characterizes the surface, which is dotted by small-scale

Fig. 9.45 Limestone pavement in Burren (photo by Mari)

subsidence dolines. A high proportion of the area is covered by glacial till (Viles 2003). Caves are also found here, the most significant of them being the Marble Arch Cave, which is still actively shaped by a stream.

9.9 Carpathians

The Carpathians are among the longest mountain ranges in Europe. The arcuate mountain range with a total length of 1500 km extends over several countries of Central and Eastern Europe between 44.5°N and 50°N latitudes. The elevation of the mountain range is the highest in the north (the main peak being Gerlachovský štít, 2655 m asl) and in the south, whereas all other parts are lower than 2500 m asl. As the Carpathians are in the middle of the continent, the amount of precipitation is moderate. The wettest zones are the Tatra Mountains and the Northeastern Carpathians with 1600–1700 mm annual precipitation (http://www.carpatclim-eu.org). As a consequence of the above settings, glaciation was much more limited in the Carpathians than in the Alps. Only the higher zones of the Tatra Mountains, the Northeastern Carpathians and the Southern Carpathians were glaciated during the glacial periods of the Pleistocene (e.g. Urdea 2004; Lindner et al. 2010; Ehlers et al. 2011). Further on, these mountains are predominantly built up of granitoid, metamorphic and to a lesser extent, volcanic rocks, and thus glaciokarst conditions are only exceptionally satisfied in the Carpathians. Hence, there is just one minor area in the carbonate zone of the High Tatras, where glaciokarst morphology is present, and in a more limited meaning, the Low Tatras also host some glaciokarst related landforms (Gadek and Litwin 1999; Litwin and Andreychouk 2008).

9.9.1 Geologic and Tectonic Settings

The core of both the High and the Low Tatras are built up principally of Paleozoic granite, metamorphic schists and gneiss (Fig. 9.46). These crystalline rocks are covered by Mesozoic sedimentary nappes. Since the Miocene, the mountains have been uplifting, and as a result, the sedimentary rocks were largely eroded and they remained at the surface only in smaller patches (Jurewicz 2005; Žák et al. 2009; Křížek and Mida 2013; Makos et al. 2014). In the High Tatras, Mesozoic limestones and dolomites are found in the western and northern parts. In case of the Low Tatras, limestone is present only in narrow, highly tectonized stripes at the northern and southern sides of the main ridge (Žák et al. 2009).

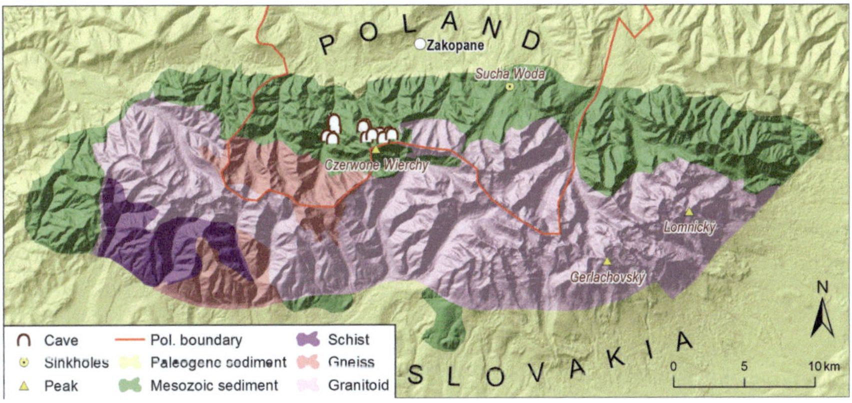

Fig. 9.46 Simplified geology of the High Tatras with the location of glaciokarst sites (Czerwone Wierchy, Sucha Woda). Geology is after Zasadni and Kłapyta (2014)

9.9.2 Relief Settings

The High Tatras are densely dissected high mountains, where glaciokarst morphology is observable in the western and northern parts. In the western parts, the ridge elevation slightly exceeds 2100 m asl, whereas in the northern foreland, where the valley profiles are gentler than near the main ridge, the elevation is between 900 and 1200 m asl. The higher cave openings are above 2000 m asl, the highest of them is at the Slovakian side of the main ridge, at the Czerwone Wierchy (Červené vrchy) mountain (2122 m asl). Karst terrains in the Low Tatras extend up to 1750 m asl, south of Ďumbier peak (Žák et al. 2009). The northern valleys also host karst features, but at lower elevations.

9.9.3 Climatic Settings

At present, the mean annual temperature of the High Tatras is 5–6 °C in the northern foreland (Zakopane) and in the southern Poprad basin, but it decreases to −4 °C at the Lomnický štít. The average yearly precipitation changes between 1100 and 1800 mm, whereas the southern slopes of the mountains receive 40% less precipitation in general (Makos et al. 2014). During the LGM, the temperature could be 11–12 °C colder, the precipitation 40–50% less than now (Makos et al. 2014).

In the Low Tatras, at the level of the highest caves, at around 1700 m asl, the mean annual temperature is 0.8 °C, and the mean annual precipitation is 1300 mm according to the measurements (Žák et al. 2009).

9.9.4 Glaciation Characteristics and Phases

In the High Tatras, there were relatively large valley glaciers during the Pleistocene (Fig. 9.47), at least 8 times. However, moraines of older glaciations are not preserved, and only glaciofluvial sediments exist from older periods (Lindner et al. 2003). Glaciations took place in the following periods according to the alpine and marine isotope chronology: Mindel (MIS 12), Riss (MIS 6-10) and Würmian (MIS 5d-2, Křížek and Mida 2013). The Riss glaciation had a slightly larger extent here than the Würmian (Makos et al. 2013). Most glaciers were relatively thin (50–100 m), but the largest forms reached a thickness of 300–400 m (Zasadni and Kłapyta 2014). Glaciers in the Tatras were warm-based (Gadek and Litwin 1999; Litwin and Andreychouk 2008).

ELA values during the LGM were between 1460 and 1700 m asl with significant differences among valleys (Křížek and Mida 2013; Makos et al. 2014). In the Early Holocene, at 9.2 ka BP, the ELA increased to 1950 m asl, and thus small cirque glaciers existed at that time. Today, the theoretical ELA surpasses the elevation of the mountains (Křížek and Mida 2013). Cosmogenic ^{36}Cl exposure ages testify that the local LGM had its maximum at 21.5 ka BP. It was followed by glacier retreat, but at 17 ka and 12 ka BP, short advance phases interrupted the long-term process of withdrawal (Makos et al. 2013).

Meanwhile, in the Low Tatras, glaciation affected only the terrain around the highest peak, and small cirque glaciers existed at the northern side of the main peak (Ďumbier vrch, Hercman et al. 1997; Žák et al. 2009).

Fig. 9.47 Glaciation of the High Tatras during the LGM (extension and contour lines of the glaciers are after Zasadni and Kłapyta (2014)

9.9.5 Study Areas

Czerwone Wierchy (Červené vrchy) is the largest karstic mountain in the High Tatras. It contains several caves, the longest and the deepest (Jaskinia Wielka Śnieżna/Great Snowy Cave with 23.6 km length and 824 m depth, www.sktj.pl) among others (Szczygieł 2015). There are two paleophreatic levels in the caves connected by vadose passages. The older level (Czarna Cave) is more than 1.2 Ma old, but it shows already an adaptation to the present-day drainage settings. The lowermost level (Zimna Cave) is at least 120 ka BP old (Szczygieł 2015). Several of the vadose passages are located along the former glacier margins or at rock boundaries. Decapitated shafts are rare (Szczygieł 2015). Cave passage directions are oriented by tectonic fissures (single-plane faults and multiple fault cores) and by bedding planes as well. Ongoing tectonic uplift at a rate of 1 mm/a also plays a role in the opening or closure of tectonic features (Szczygieł 2015). There are shafts, which are parallel with the valley walls and are separated from the open surface by a narrow rock mass only. In the formation of these shafts, dissolution processes were the most active agents, and postglacial gravitational stress release had only a subordinate role (Szczygieł 2015). Based on datings carried out on speleothems from Wielka Śnieżna Cave, four cold periods were demonstrated at 70–65 ka BP, 55 ka BP, 40–35 ka BP and at 23–10 ka BP, as the growth of speleothems was interrupted during these phases (Gadek and Litwin 1999).

In the High Tatras, there exist only few cirques on limestone rocks. They are situated between 1450 and 1880 m asl (Gadek and Litwin 1999; Křížek and Mida 2013). At the bottom and sides of the cirques, glacially polished roche moutonnées are found.

The Mała Łąka valley descends from the Czerwone Wierchy mountain. Although the valley is built up mainly of Mesozoic limestones and dolomites, surface karst landforms are uncommon (Gadek and Litwin 1999). On the contrary, glacial features are obvious: cirques, moraines and polished rocks are typical in this valley (Litwin and Andreychouk 2008). Nevertheless, some dolines and open shafts are also present, as well as some karren features (grikes, runnels, subsoil rundkarren, kamenitzas and heelprint karren) can be also observed, but limestone pavements characteristic of many other glaciokarst terrains are absent here (Gadek and Litwin 1999; Litwin and Andreychouk 2008).

In the Polish part of the High Tatras, in the Sucha Woda valley, there are a number of dolines, partly on karstified Middle Triassic limestone, and partly on Quaternary morainic sediments as covered karst landforms (Birkenmajer 2008). The material of the moraines originates from higher areas; therefore, mostly crystalline debris builds up the moraines (Makos et al. 2014). At the confluence of Sucha Woda and Pańszczyca valleys, glaciofluvial sediments and different type moraines (terminal moraine, lateral moraine, recessional moraine and dead ice moraine) are found. The lower section of the two valleys was formed on Mesozoic carbonates; hence, it is much wider and shallower than the neighbouring valleys. It is feasible that the difference in the valley form is due to the well-developed karst system, which was capable of draining the subglacial meltwaters (Makos et al. 2014). Similar explanation is given by Gądek and Litwin (1999) for the formation of Mała Łąka valley.

One of the terminal moraines was sampled and ^{36}Cl exposure age was determined. The result, 11.6 ka BP can be thought of as the minimum age of the sampled moraine unit. Moreover, there are even younger moraines in this area (Dzierżek et al. 1999 in Birkenmajer 2008). There are about 20–25 closed karst depressions, with 15–30 m diameter and 5–12 m depth (Birkenmajer 2008). Moreover, a 200-m-wide and 400-m-long, arcuate depression is also observable. It was filled by a lake during a part of the Holocene. The above author calls it a polje (Birkenmajer 2008), but obviously, it is not a polje in the classical meaning (see, for example, Mihevc et al. 2010).

In the Bystra valley (lying also in the Polish part of the High Tatras), 62 caves are known. The longest of them, Kasprowa Niżna with a length of 3 km, is proved to be older than the Eemian interglacial period (Szczygieł et al. 2015). In some caves, there are clear traces of sedimentary filling due to glacial meltwaters (Szczygieł et al. 2015). A flowstone covering this type of sediment was dated, and its age was 165 ka BP. This age as well as the position of the cave implies that valley deepening was not too intensive here during this partly glacial period (Szczygieł et al. 2015). Caves of this area are particularly strongly constrained by tectonic fissures, which are related to the uplift.

Either in the Low Tatras, or in the High Tatras, it is typical that several horizontal levels are evolved in the caves. In certain cases, the number of levels is even more than 10 (Hercman et al. 1997; Szczygieł et al. 2015). One of these caves is the longest cave of Slovakia, the 36-km-long Demänova Cave System (http://www.sss.sk, 2013), whose position is lower than the former glacier limit (Hercman et al. 1997). In this cave, four periods of intense speleothem growth were recognized by U/Th measurements, notably from 685 to 410 ka, from 170 to 140 ka, from 104 to 70 ka and from 5.6 ka BP to now. Level IV of this cave was dated as 780–685 ka old, and upper levels (up to Level IX) are necessarily older than this (Hercman et al. 1997).

From a methodological point of view, it is particularly interesting that not only glacial, but periglacial impact in caves can be also recognized by examining cryogenic cave calcites (CCC, Žák et al. 2012). These formations consist of irregular and small size (1–40 mm) crystal grains adhered to each other, which precipitate from freezing water and they indicate permafrost conditions within the cave. In addition, the age of these formations can be determined. Using them as climate markers, both the existence and the age of permafrost conditions could be evidenced in several caves of the Low Tatras (Žák et al. 2009), namely, in the Jaskyňa Studeného Vetra (Cold Wind Cave) and in the Jaskyňa Mŕtvych Netopierov (Dead Bats Cave), as well as in the High Tatras. In the Mesačný Tieň Cave, whose entry is at 1800 m asl, the permafrost extended down to as low as 285 m depth during the LGM (Žák et al. 2012; Orvošová et al. 2014). In case of the Cold Wind Cave, the permafrost ages are 79.7 ka, 104 ka and 180 ka BP subsequently. These old ages suggest that permafrost conditions deeply affected this cave, and possibly the whole surrounding terrain, during earlier glacial periods (Žák et al. 2009, 2012).

9.10 Greater Caucasus

The Greater Caucasus is a large region stretching between the Black Sea in the west and the Caspian Sea in the east (Fig. 9.48). The length of the mountain chain consisting of several subparallel ranges exceeds 1100 km, whereas the width is no more than 180 km. The highest point is Elbrus (5642 m). The Greater Caucasus is subdivided into four segments, namely, the Northwestern Caucasus (from the western edge to Fisht mountain), the Western Caucasus (from Fisht mountain to Elbrus mountain), the Central Caucasus (from Elbrus mountain to Kazbek mountain) and the Eastern Caucasus (from Kazbek mountain to the eastern edge); sometimes, the first two are labelled together as the Northwestern Caucasus or the Western Caucasus, i.e. these names can be used also as synonyms.

Glaciers are common in the Greater Caucasus, as well as epikarst and endokarst features. The exploration of both goes back to the nineteenth century, but the most serious advance was made in the twentieth century by Bush, Dinnik, Gvozdetskij, Panov, Podozerskij and many others. Among the most interesting areas for such studies is the Lagonaki Highland (height is generally over 1500 m) in the Western Caucasus, where glaciers and snowfields coexist with widespread karst. This area has been studied since the end of the nineteenth century with the main contribution made by Lozovoy. Another remarkable area is the Arabika Massif, also in the Western Caucasus, which is the place of the deepest known caves on Earth.

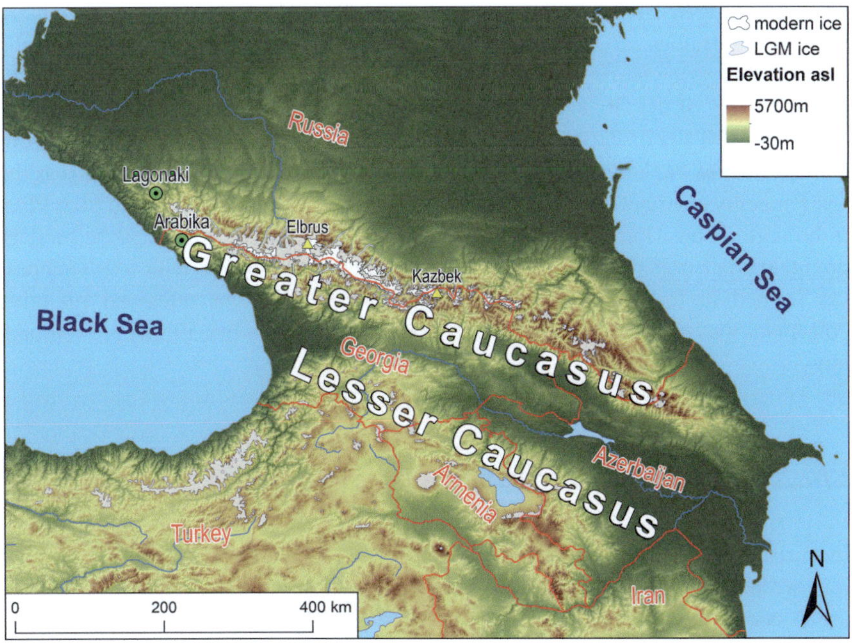

Fig. 9.48 Location map of the Greater Caucasus region. LGM data from Ehlers et al. (2011)

Unfortunately, the majority of the works devoted to the Greater Caucasus, including those of the above-mentioned scientists, were published in Russian, and this important region has remained poorly known to the broad international audience.

9.10.1 Geological Settings

The geology of the Greater Caucasus is similar to other Alpine orogenic belts. The crystalline core consists of Precambrian and Lower Paleozoic rocks. The Upper Paleozoic is represented by the Carboniferous coal-bearing sequences and the Permian red molassic complexes. The most widespread are the Mesozoic and Cenozoic deposits with the total thickness of several thousands of metres. Out of them, Jurassic and Cretaceous sedimentary rocks (sandstones, shales, carbonates and evaporates) are the most common. These are outcropped on the slopes and tops of many mountain ranges, including the lengthy Skalistyj Range and the Pastbitshnyj Range. The geological history of the Greater Caucasus was highly complex, and it is characterized by Tawadros et al. (2006), Ruban et al. (2007) and Ruban (2013). In the Precambrian and the Early Paleozoic, the Greater Caucasus was a part of the Afro-Arabian margin of Gondwana. In the mid-Paleozoic, it was separated (together with the other Galatian terranes) from this supercontinent and moved northwards where docked at the southern margin of Laurussia. This terrane was located near the Carnic Alps in the Late Paleozoic, where it was affected by the Hercynian orogeny. However, it was replaced to its present position at the southern periphery of Baltica already in the Triassic due to lateral displacements along the major shear zone. During the Jurassic, the Cretaceous and the Paleogene, the back-arc basin of the Greater Caucasus was occupied by the Caucasian Sea, which was a marginal sea of the Neo-Tethys Ocean. Finally, the uplift in the axial part of the region since the Paleogene led to the land growth. In the Neogene, there was the vast Paratethys Sea between the Caucasian island in the south and the Russian Plain in the north. The Alpine orogeny culminated in the Quaternary, when the modern mountain chain was formed. The fold-and-thrust belt (probably, with some nappes) was formed in the late Cenozoic, and tectonic activity has remained till now. Volcanism was active in the Quaternary, but the volcanoes like Elbrus are dormant now.

9.10.2 Climate and Vegetation

The Greater Caucasus is located in the temperate climate zone. The rainfall is higher in its western part, and the Lagonaki Highland, with its annual rainfall of 3600 mm/year is one of the wettest places in Russia. The mountains are covered with more or less dense forests, although steppe landscapes prevail at their toe and alpine meadows are typical for the most elevated places.

9.10.3 Glaciations

There are more than 2000 glaciers in the Greater Caucasus, which occupy an area of ~ 1500 km^2. The majority of glaciers are small in size, and up to 70% of them are located on the northern slopes of the Greater Caucasus. About a half of glaciers are located in the Central Caucasus, and a third are known in the Western Caucasus, and they are less frequent in the Eastern Caucasus (Panov 1993). The length of the largest glaciers exceeds 10 km and their area is more than 15 km^2. Morphologically, the glaciers of the Greater Caucasus are very diverse, and the simplest classification is to distinguish those growing in the valleys and others found near mountain summits (Panov 1993). The glaciation started at the beginning of the Quaternary, and its spatial extent peaked in the Late Pleistocene. In the modern time, glaciation has fluctuated rapidly together with the regional climate changes. There were glacial advances in the late sixteenth century, the second half of the seventeenth century and the first half of the nineteenth century (Solomina et al. 2016). Since the midst of the nineteenth century, a glacial retreat has started, and it has intensified in the last third of the twentieth century (Panov 1993; Stokes et al. 2006).

9.10.4 Glaciokarst Landforms

Karst of the Greater Caucasus is linked chiefly to the Devonian, Triassic, Jurassic, Cretaceous and some other carbonate rocks, from which the Upper Jurassic–Lower Cretaceous limestones and dolostones of the Skalistyj Range are the best karstified. A broad spectrum of small and big epikarst forms and caves are known (Gvozdetskij 1981). Many are Holocene in age, but some were formed in the Pleistocene and even before the glacial epoch (probably in the Pliocene). The existence of glaciokarst sensu *stricto* (after Veress 2016a, b) is not well documented, but it should be said that meltwater from the retreating glaciers (either in case of the modern retreat, or in case of the post-Pleistocene retreat) is a significant factor in the karstification of carbonate rocks. Moreover, this karstification is localized sometimes in the glacial landforms (Gvozdetskij 1981). Some karst depressions contain the material of glacier moraines (Kostin 1966), and, if so, melting of ice that filled these depressions could contribute significantly to both the development of these forms and water infiltration into carbonate rocks.

In the Greater Caucasus, the natural "laboratory" for karst studies (also in relation to glacial processes) is the above-mentioned Lagonaki Highland described comprehensively by Lozovoj (1984). The three highest mountains of this area are Fisht (2868 m), Oshten (2804 m) and Pshekha-Su (2743 m). Presently, two glaciers (Big Fisht glacier, 0.7 km^2, and Small Fisht glacier, 0.1 km^2) are established on Fisht and two glaciers (0.1 km^2 each) are established on Pshekha-Su. However, a much larger glacier was located there in the Late Pleistocene; its length reached

12 km, and the ice thickness was more than 150 m (Lozovoj 1984). And this is not the only paleoglacier of the Lagonaki Highland. Modern glaciers are the tiny little remnants of the Late Pleistocene glaciation. The rate of actual retreat of the Big Fisht glacier is more than 20 m/year. Upper Jurassic limestones and dolostones of the Lagonaki Highlands are karstified intensively everywhere, including the mentioned highest mountains. Lozovoj (1984) defined the landforms of the Fisht mountain as glaciokarstic (Fig. 9.49). The meltwater from the glaciers of Fisht and Pshekha-Su is not drained surficially, only by subsurface cavities. It cannot be excluded that the largest kamenitza on the slope of Fisht with 2 m diameter and 0.6 m depth (Lozovoj 1984) was formed due to glacial processes in the geological past. Of interest is the Ledjanaja Cave located on the southern flank of the Fisht massif and investigated in detail by Lozovoj (1984). Its length is only 50 m. There is a lot of ice and snow in this cave. Seasonal growth and melting of ice (some portion of the ice is apparently permanent) as well as frost weathering have contributed evidently to the cave development.

A great diversity of karst features (kamenitzas, grikes, rillenkarren, depressions, dolines, caves, etc.) are known from the Lagonaki Highland. Many are covered by the grass of alpine meadows. However, the most impressive are areas with non-covered grike fields (Fig. 9.50) on the Stonesea Range and near Oshten, which can be interpreted as glaciokarst sensu *lato* (after Veress 2016a, b). The degree of karstic denudation (up to 200 mm/ka according to Gvozdetskij 1981) and some observations on the state of kamenitzas imply that these features appeared in the Holocene. But the presence of large glaciers in the Late Pleistocene implies that meltwater from them could shape local landforms intensively. Three additional

Fig. 9.49 The Fisht—Oshten—Pshekha-Su group of peaks in the Western Caucasus, where small glaciers have contributed to karstification (photo by Ruban)

Fig. 9.50 Non-covered epikarst (grike "field") in the southern part of the Lagonaki Highland of the Western Caucasus (photo by Ruban)

mechanisms relevant to glaciokarst development in the Lagonaki Highland must be considered. First, this is the wide distribution of snowfields, which often fill karst depressions (Fig. 9.51). Meltwater from them enhance growth of surface and subsurface karst forms. At the times of climate cooling coincided with glacial advances, the spatial extent of snowfields increased, which facilitated karstification. Second, frost weathering is a significant local factor of karst growth (Kostin 1966; Lozovoj 1984), and, if so, this factor was even more significant at the times of glacial advances. Third, paleodolines are often the results of glaciokarst development (Veress 2016a, b). Such landforms are typical for the southern part of the Lagonaki Highlands, which experienced glaciation in the geological past. These landforms were created by glacial meltwater and snowmelt (Lozovoj 1984). The largest doline is named Tchashka, its long axis is 2 km in length, the width is 1 km, and the relative depth is up to 200 m (Lozovoj 1984).

The *Arabika Massif* is a high mountain karst plateau not far from the Black Sea, surrounded by deeply incised canyons, and towering up to 2705 m asl. It is a classical glaciokarst landscape with glacial valleys, cirques, arêtes, horn peaks and numerous dolines (Klimchouk et al. 2009). The main building rocks of the massif are Upper Jurassic and Lower Cretaceous limestones. The deepest known caves of the world (www.caverbob.com) are found just here, namely, Veryovkina (−2204 m), Krubera (−2197 m), Sarma (−1830 m) and Illuzia-Snezhnaja-Mezhonnogo (−1853 m). The caves are composed of vadose shafts and steep meandering passages, but at the deeper parts, phreatic morphology can be also observed. There are also old fossil passages, whose age is older than 500 ka based on speleothem dating (Klimchouk et al. 2009).

Fig. 9.51 A typical karst depression filled with snow in the Lagonaki Highland of the Western Caucasus (photo by Ruban)

In general, a part of the karst features of the Greater Caucasus and, particularly, of the Lagonaki Highland and Arabika Massif can be attributed to glaciokarst sensu *stricto*, and many of them can be attributed to glaciokarst sensu *lato*.

9.11 Patagonia (Chile)

The karst region in southern Chile is one of the least known and most unique glaciokarsts of the world. Research of the area in the modern sense began in the mid-1990s. A series of French expeditions revealed the special form of the archipelago (1997, 2000, 2006, 2008, etc.). In 2002, a smaller Hungarian expedition also studied the surface karstic forms of the island of Diego de Almagro (Veress et al. 2003, 2006). Its special features are the subpolar climatic environment, the extreme rainfall and the marble base rock. There are almost always strong winds on the Pacific islands in the southern foothills of the Andes. Due to extremely intense dissolution, giant forms were created both on the surface and in the endokarst. Two islands of the archipelago are built of calcareous rocks (Madre de Dios and Diego de Almagro). Their karstic formations are karren on the surface, and high mountain shaft caves under the surface. These are supplemented by caves related to eustatic and isostatic sea level changes (Jaillet et al. 2008). The geomorphology of the island has been shaped by karstification, glacial erosion and eustatic changes in sea levels. Because of the high intensity of the three processes, the morphology shows Quaternary or recent origin.

9.11.1 Geographical Location

The Region of Magallanes, Chile's southernmost region is of 132,000 km^2 area, with a significant part of the archipelago between 50° and 52°S (Fig. 9.52). The population density of the area is very low (1.1 p/km^2), and the two islands built up of calcareous rocks are completely uninhabited. Smaller geological explorations have been carried out on the islands (Biese 1956, 1957), mainly for mining purposes. Madre de Dios (50′20″) and Diego de Almagro (51′30″) are surrounded by a minor archipelago. The two karstic islands are 150 km away from each other.

9.11.2 Geological Features

The geological features of the archipelago look back to a long geological history. The tectonic activity of the area led to the complexity of geological formations. The rock formations, which build up the islands, along with the units of the Andes, are associated with the ancient Gondwana continent (Maire 1999). The Upper Paleozoic sedimentary and volcanic sediments carrying the karstic forms are arranged in a wide range. The sedimentary zone from the east is bordered by Mesozoic Patagonian batholite of granitic rocks. The age of the granite unit is 120–130 Ma (Forsythe and Mpodozis 1983).

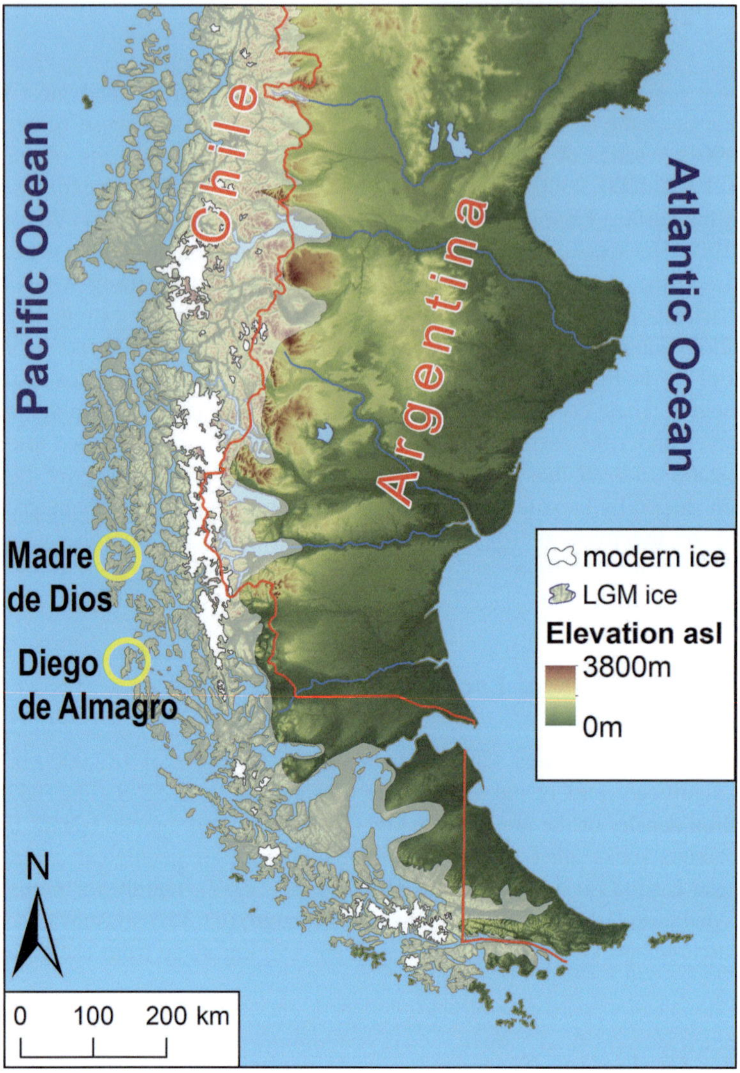

Fig. 9.52 Geographical location of Diego de Almagro and Madre de Dios islands with LGM and modern glacial extent

On the island of Madre de Dios, the Tarlton Formation (named after Tarlton Island in the vicinity of Madre de Dios) is the bearing rock of the karst landforms. It is a foraminiferous limestone accumulated in shallow, warm seawater environment and its thickness reaches 500 m (Forsythe and Mpodozis 1983). During the Carboniferous–Permian, the area was situated closer to the Equator, allowing coral reef development. The thickest unit is the Duke of York Formation, a volcanic rock mass. Denaro Formation is an ocean fragment, a silicate-cemented rock mass intercalated between the two dominant geological units.

The structure of Diego de Almagro is similar to the one described above, but, because of the different denominations of the subunits, we briefly present them. It is possible for karst to develop on the Pelantaro unit, exposed on the northeastern side of the island in an area of 25 km^2. Its thickness is hundreds of metres, but on the surface it only appears in a maximum of 2–3 km width. It also contains in marble, crystalline limestone and dolomite (Maire 1999). The other two formations of the island consist of non-karstic grauwacke and flysch (Huemul Formation) and greenschist (Escobar 1980).

In terms of karstification, Tarlton and Pelantaro Formations are of interest. These formations were generated as a classic atoll at the boundary of the Carboniferous and Permian. In the warm marine environment, the atolls formed on the surfaces of volcanic cones. The intensive tectonic movements during the formation period ordered the geological units into bands. The reduced distribution of calcareous rocks greatly facilitates the formation of vertical cave passages (Jaillet et al. 2008).

9.11.3 *Climate and Glaciation*

The current climate is described as temperate cold hyperhumid climate. According to Zamora and Santana (1979), the precipitation is between 7331 and 8495 mm/year. From the point of view of karstification, it is particularly important that 80% of precipitation comes as rain, while in the high mountains almost the same proportion is represented by snow (Maire 1990). The mean annual temperature is between 4 and 9 °C, and never exceeds 10 °C. The annual range of temperature is minimal. According to several studies, the most severe climatic factor in the area is the constant and strong wind. Average wind speed reaches 70 km/h. The decisive parameter, however, is that 150 km/h winds are not uncommon (Maire 1999). Jaillet et al. measured 140 km/h during the 2006 and 2008 expeditions (Jaillet et al. 2008). Under such circumstances, vegetation can only survive in wind shelters. Due to rainfall, however, it forms a completely impenetrable rainforest.

Patagonia's glaciation is approached from two directions in the literature. On one hand, it can be interpreted as the southern extension of the Andes' glaciation (Mercer 1976). On the other hand, it can be seen as part of the subpolar ice cap. According to (Sugden et al. 2005), the last glaciation peak was 27 ka ago BP, followed by glacier retreat. According to Ackert et al. (2008), 11 ka years ago there was a smaller glacial transgression, which is explained by a significant increase in precipitation.

Based on geomorphological observations, it is likely that the lowest level of glaciation reached the then sea level of −120 m and the glacier surface was at 600 m altitude. The peaks around 800 m certainly emerged from the ice as nunataks.

The southern Madre de Dios was covered with ice. Some nunataks indicate that the ice cover was incomplete. Diego de Almagro (150 km to the south) had a much larger ice cover. According to Hulton's model (1994), at the time of the LGM (Last Glacial Maximum), the average temperature was 3° lower than today. At the southern coast of the island, the former moraine chains can be discovered under the current water level. This proves that glaciers stretched far below the present sea level. Countless roches moutonnées prove the presence of glaciers. Because of the very significant postglacial dissolution, most of the glacial striae have been demolished by dissolution, which makes glacial reconstruction difficult. Erratic rocks (up to a few metres) can be found all over the island. They provide the ability to determine the degree of dissolution from the height of the karren pedestal (100 mm/ka, Maire 1999).

From the point of view of geomorphological processes, it is especially important to mention the lack of frost weathering. In case of subpolar karsts like around the Lena River (Veress et al. 2014), physical weathering virtually destroys all karstic form. In the Patagonian archipelago, due to the extreme marine climate, there is no frost weathering at all. Thanks to this, the karst of the archipelago is exempt from this destructive effect. The mean annual temperature fluctuation is around 4.5–5 °C, which is extremely low.

9.11.4 Glaciokarstic Forms

9.11.4.1 Glaciokarstic Macroforms

Well-known forms of European karstic mountain ranges are found in large numbers and huge in size on both islands. Glaciokarstic dolines are known mainly on Madre de Dios, with a diameter of several hundred metres. Glaciokarstic cirques are characteristic of both islands. In the zone below the peaks, they are located at 500–700 m elevations, often in rows. One of their types was entirely formed on calcareous rock. The other type is formed on the Duke of York Formation and at the bottom end, there are calcareous rocks. There is a sequence of shaft caves at the rock boundary. The entrance to the largest endokarstic systems is found at these sites. Glacial U-shaped valleys also formed in large numbers between 500 m altitude and the sea level. The main line of the U-shaped valleys is typically located in the non-karstic rock of the Duke of York Formation, while the upper margins of the valleys and ridges are karstic. In line with Quaternary climate changes, rock erosion has been highly variable. In cold periods, glaciers shaped their valleys on the less resilient sandstone, while marble and limestone were more resistant to glacial erosion. During warmer periods, karstification became dominant, and the highs of Tarlton and Pelantaro Formations suitable for dissolution have been destroyed. Some nunataks of marble show that the ice cover was not complete. Roches moutonnées also indicate the presence of ice. They may develop on any rock formations and they are abundant. Roches moutonnées of the Tarlton formation have undergone significant postglacial dissolutional transformation.

9.11.4.2 Karren Forms on the Islands of Diego de Almagro and Madre de Dios

The especially large and constant precipitation created a unique assemblage of karren. The basic system of karren forms was created by Bögli (1960), and then Tóth (2008) and Veress (2010) supplemented it with additional forms and subtypes. However, the karren forms found here differ not only in size from the types known in the Alps. The main reason for this is the permanent and strong wind (Maire 1999; Veress et al. 2006). The vegetation is concentrated in the protected depressions, while the rest of the island is completely bare. In the zones, where the Tarlton or the Pelantaro formations are exposed, several metres deep and wide karren forms are developed. Due to the fragmented surface, longitudinal forms are characteristic. Rillenkarren, rinnenkarren and wandkarren constitute the majority of the forms, but we can also encounter meandering karren with several metres in diameter. The central forms include heel print karren, which are often split into rows. Surface landforms created primarily by dissolution may reach extreme sizes. There are kamenitza-like forms, i.e. closed rock basins of some hundred m^2 in area, and solutionally enlarged pits of hundreds of metres deep, i.e. shafts.

Wind causes the solvent to flow on the surface and its solubility is almost constantly renewed (Veress et al. 2006). As a result, karren features unique on Earth are created. Such forms are microfolds, which are basically similar to blowouts. Plain karren are also known among high mountain karren types. However, the plain karren developed on the islands are perfectly straightened substrates and occasionally reach a size of 100 m^2. Further special forms are the relict ridges created by the wind. They are preserved behind insoluble (mostly sandstone) rocks in the leeward side. The French and Hungarian research expeditions described this phenomenon near Cerro Pelantaro peak (Maire 1999; Veress et al. 2003).

Also unique are those karren, which can be described as marinokarstic forms. The French expedition described them from the island of Guarello, belonging to Madre de Dios and the Hungarian research team from the shore of Abraham's fjord, Diego de Almagro. These solution benches show the postglacial uplift of the islands. After uplift, due to intense dissolution the wall karren crossed the abrasion platforms. The height of the abrasion platforms is less than 1 m on Diego de Almagro and 1–2 m on Guarello Island.

9.11.4.3 Subterranean Karst

Due to the extreme precipitation, the endokarstic forms are particularly well developed. As geological formations are arranged into bands, it is common to encounter karstic and non-karstic rocks side by side. At these junctions, inflow caves were created, which Maire (1999) classified into three types. Inflow caves fed by giant karren usually develop at the end of karren sculpted slopes in the Alps by the merging of pits. Here, in Patagonia, due to the almost constant precipitation, they receive continuous water supply with flow rates of several litres per second. Another inflow cave type is found at the border of the aquiclude, where water discharge is up to 2 m^3/s during intensive rainfalls. As a third type, shaft caves can be mentioned which are also fed by rinnenkarren from the surface and develop along fissures and faults. Unlike in the Alps, the role of snow in the formation of landforms is not significant as very little portion of the precipitation arrives as snowfall.

The development of endokarst did not stop during the cold periods of the Pleistocene. The highly intensive dissolution was only partly balanced by the erosional action of glaciers. The amount of glaciofluvial sediments that can be found in the underground karst shows that part of the subglacial meltwater percolated to the caves even in glacial periods. Near Cerro Pelantaro (Diego de Almagro), an inactive cave level has been cut by Pleistocene glaciers (Maire 1999).

9.11.4.4 Submarine Karst

French expeditions have described a number of submarine karren forms around the island of Madre de Dios. Their depth of occurrence is generally down to −20 m (Jaillet et al. 2008). As mentioned earlier, the sea level was 120 m lower during the LGM (27 ka). At that time, the glaciers separated the sea arms as ice shelves. Thus, in principle, karstic dissolution was hindered. The global sea level reached its current level by about 7 ka BP after the last deglaciation (Fleming et al. 1998). However, the local melting of ice was not coeval with the global sea level increase. There was a brief, a couple of thousand years long, interval when the local ice shelf had been melted, but the sea level was not yet risen to its actual level. During this period, karren forms that are currently at −20 m could develop. They are morphologically identical to the currently active surface karren. Heel print karren and large size rillenkarren positioned in steps can be found.

9.11.4.5 Glaciokarstic Model

As mentioned above, the karstic forms of the island can be considered unique, primarily due to extreme climate conditions. Perhaps due to the same reason, the area was discovered quite lately. The series of French expeditions revealed the geomorphological values of the island very effectively. The Hungarian expedition in 2002 was much smaller and focused on collecting data about karren (Veress et al. 2003, 2006). Since 1997, French expeditions conducted a systematic reconnaissance mapping and other instrumental geomorphological studies, in order to establish a glaciokarstic model of the islands (Maire 1999; Jaillet et al. 2008). This model is presented now. First, at the time of the LGM, glaciation from the South Pole reached the islands of Patagonia. At that time, the two studied islands were covered by glaciers with bottom at −120 m and top at 4–500 m above sea level. There was no surface karstification, but some of the shaft caves worked as melt-water drains under the glacier. In the second stage, glaciation was significantly reduced, only residual glaciers reached the sea. Most of the shafts had an active role in the drainage. In the third stage, significant surface and underground karstification took place. The glaciation completely disappeared. On the slopes of glacial valleys, karren development started. As the sea level rise had just begun, submarine karren could develop. The fourth stage is characterized by a powerful eustatic change in the sea level. Much of the previously formed littoral karsts (caves and karren) and glacial valleys were inundated by the emerging sea level. Lastly, in the final phase (Holocene), the isostatic uplift of the already ice-free islands began. Due to the uplift, some caves became vadose. Similarly, abrasion platforms created along the former sea level, emerged and became visible. Their altitudes of 3.5–5 m indicate the degree of sea level change (Jaillet et al. 2008).

9.12 Pyrenees and Cantabrian Mountains

The karstification in the Pyrenees is a particularly detailed repository of geological and paleoclimatic evolution. During the last major orogenic phase, in the Pliocene and the Quaternary, the Hercynian Basement and the Jurassic–Cretaceous sedimentary formations underwent a significant uplift. As a consequence, active erosion processes began to shape the mountains. The karst regions, which reached 1500 m altitude, suffered glacial transformation during the Pleistocene. Today, glaciokarsts are found in both the Eastern and Western Pyrenees. Predominant features are cirque dolines, poljes, karren fields and underground karst features. We can also classify the glacial karsts of the Pyrenees into two major types: active glaciokarsts and inherited glaciokarsts. Active glaciokarsts are very rare due to the limited present-day extent of glaciers. However, inherited glaciokarsts are more numerous. The *Arbas Massif* (Haute-Garonne) for instance, is known about its karstic glacial valleys and the 117-km-long Félix Trombe cave network. Another very special phenomenon is the world-famous underground system of *Pierre Saint Martin* (PSM) in the homonymous mountains with an overall length of 400 km.

9.12.1 Relief Settings

The Pyrenees (Fig. 9.53) are 450–490 km long in west to east direction along the northern boundary of the Iberian Peninsula, connecting the Atlantic Ocean and the Mediterranean Sea. The width of the ranges is 150–250 km with a total area of cca. 50,000 km^2. Together with the Cantabrian Mountains, they form a cca.

Fig. 9.53 Map of glaciokarst locations in the Pyrenees and the Cantabrian Mountains. LGM data from Ehlers et al. (2011)

1000-km-long mountain system. The highest point is Pic d'Aneto (3404 m) in the Maladeta group. The Aneto Glacier only extends over 0.6 km^2. The Maladeta Massif is the highest mountain block in the range. The Pyrenees have 20 peaks above 3000 m. According to the classical topographic division, three units are distinguished. The Atlantic Pyrenees (also known as the Low Pyrenees) at the western side of the mountain system, the Central Pyrenees with peaks above 3000 m. Finally, the Eastern Pyrénées (Catalonian Pyrenees) stretches to the Mediterranean Sea. In north to south cross section, the mountains are asymmetrical. The southern side is considerably more extensive than the north. Both are strongly dissected by north–south-aligned valleys.

9.12.2 Geologic Settings

The evolution of the Pyrenees took place parallel with the Alpine tectogenesis started about 40 million years ago. The deposits of the hemipelagic sedimentary basin between the Iberian Plate and the European Variscides were folded. Although the mountains are geologically young, their rocks are occasionally of much older geological age. They contain metamorphic, magmatic and volcanic formations. Mesozoic or younger sedimentary rock formations can be found in the northern and southern foreland of the mountain range. The Pyrenees show a symmetrical structure. The folding of the mountain runs parallel to the strike. The central highest unit is built up of different types of granite. In addition to granites, volcanic products are also present in small volumes, but we cannot find here rocks prone to karstification. The width of the northern zone ranges from 10 to 50 km, and its rocks are mainly of Jurassic and Cretaceous age. These were formed during tectonically quiet periods in neritic environment. The thicknesses of sedimentary formations locally reach 6000 m. We can find some metamorphic units as well in the northern zone. The southern zone of the Pyrenees consists of three major nappes, of which the Bosixols napple is the most significant with its thickness of 5000 m. Its Cretaceous limestone is locally covered by Eocene clays and sandstones.

9.12.3 Glaciations

The Pyrenees have attracted the interest of researchers already during early investigations into glaciation and glaciokarsts. The first important work concerning the division of the European Pleistocene was written by Albrecht Penck, who personally travelled to the Pyrenean glaciers in 1885 (Penck 1885). He returned there several times and published another work together with Brückner (Penck and Brückner 1901–1909). In the sense of modernity, Taillefer provided a conceptual work (1963, 1964, 1969, 1977, 1985). He was the first to reconstruct the extension of glaciers using morphometric methods. In his first works, he identified a single glacial period on the basis of moraine morphology in the valleys, but later on, he distinguished three periods.

It is widely accepted in the literature that in the Pyrenees the Last Glacial Maximum occurred several thousand years earlier than the global LGM (Gillespie and Molnar 1995; Calvet 2004; Pallás et al. 2006; Calvet et al. 2011). Most studies date the maximum glaciation at 70 ka and 45 ka BP (Vilaplana et al. 1989; Montserrat-Martí 1992; Bordonau 1992). On the northern slopes of the Pyrenees, we can find 900-m-thick and maximum 65-km-long ice covers at the peak of glaciation (Taillefer 1969). The glaciers in the Pau region reached 400 m above sea level in the Aquitaine basin. Due to higher insolation of the southern slopes, the thickness of glaciers in the southern side was only 600 m and the maximum length only 30 km (Martí Bono and García-Ruiz 1994). Due to the much shorter length, the glaciers on the Spanish side did not reach the Ebro Basin. For the period of deglaciation also a wider interval can be given, it occurred at cca. 26 ka BP in the low-lying valleys (Bordonau 1992), at 16 ka BP in the valleys of the high mountains, and finally between 16 and 13.5 ka, in the highest regions, depending on exposure and morphological situation (Delmas 2009). Similar results were obtained by Andricu-Poncl et al. (1988), who investigated the deglaciation of the northern side of the Pyrenees. According to him, the beginning of ice retreat after the most extended glaciation began before 38 ka BP, but with remarkable fluctuations. Due to the decreasing amount of precipitation, the disappearance of ice accelerated and the deglaciation was completed by 15 ka BP.

The Equilibrium Line Altitude (ELA) is currently located at 3000 m above sea level (Chueca et al. 1997). The areas of the current glaciers do not exceed 1 km^2 (Chueca et al. 2002).

9.12.4 Study Areas

9.12.4.1 Glaciokarst of Arudy

The *Arudy karst* on the northern side of the Western Pyrenees is a good example of the geomorphological complexity of the Pyrenees. The diverse geological evolution, the variations of faults and folds and the diverse climatic conditions are altogether responsible for richness in landforms. Throughout the history of karst formation, from tropical climates to Pleistocene glaciers, all kinds of climates affected the denudation of the surface (Auly 2008). The covered karst established on the Urgonian limestone extends from the peak region (Lazerque, Le Rey) to the glacier basin (Vallée d'Ossau).

9.12.4.2 Glaciokarst of Pierre Saint Martin Massif

The *Pierre Saint Martin Massif* is located in the Western Pyrenees. It is an outstanding site of glaciokarstic and speleological research, a symbolic place, where one of the deepest caves of the world is found. The mountain range of 140 km^2 area is a fine glaciokarst terrain at an altitude of 1800–2500 m asl. The surface is entirely covered by karren fields developed on tilted beds. Grikes, pits and shafts drain the surface water down to the karst interior. In the mountain range 450 km of underground network including cca. One hundred fifty caves have been explored (A.R.S. I.P. 1985). As a result of the early (Tertiary) erosion of the cover sediments, the Pierre Saint Martin has undergone continuous karstic development since then. Similarly to the neighbouring high mountains, Quaternary glaciokarst erosion was remarkable. The removal of 200–400 m of Cretaceous limestones opened up several old cave passages. Stalagmites formed in the Pliocene were also exposed to the surface (Bini et al. 1989). According to dated varve deposits in the Aranzadi Gallery, there was a significant fluvioglacial activity in the Middle Pleistocene (Maire 1990; Quinif and Maire 1998).

9.12.4.3 Glaciokarst of Ordesa and Monte Perdido

Monte Perdido (3355 m asl) is the third highest peak in the Pyrenees, whereas *Ordesa* is a classical U-shaped glacial valley (Fig. 9.54). The area is rich in spectacular glaciokarst landforms and it is also valuable for biological reasons, and thus it became protected as a national park already in 1918. Monte Perdido and its surroundings are mostly built up of upper Cretaceous and Paleocene–Eocene limestone and dolomites. Structurally, they are part of the southern Pyrenean fold-and-thrust belt (Fig. 9.55), formed by a set of imbricated thrust sheets (Lambán et al. 2015). Deep canyons, U-shaped valleys with vertical cliffs, cirques, dolines, shafts and karren are the characteristic glaciokarst features of the Perdido–Ordesa area (García-Ruiz et al. 2014; Fig. 9.56). Today, only a small glacier survives north of Monte Perdido, but likely, it will disappear in some decades due to global warming. Geochemical properties of this unique, high mountain karst aquifer with large vertical differences have been studied recently by Lambán et al. (2015).

Fig. 9.54 The glacially formed Ordesa valley with steep cliffs (photo by Telbisz)

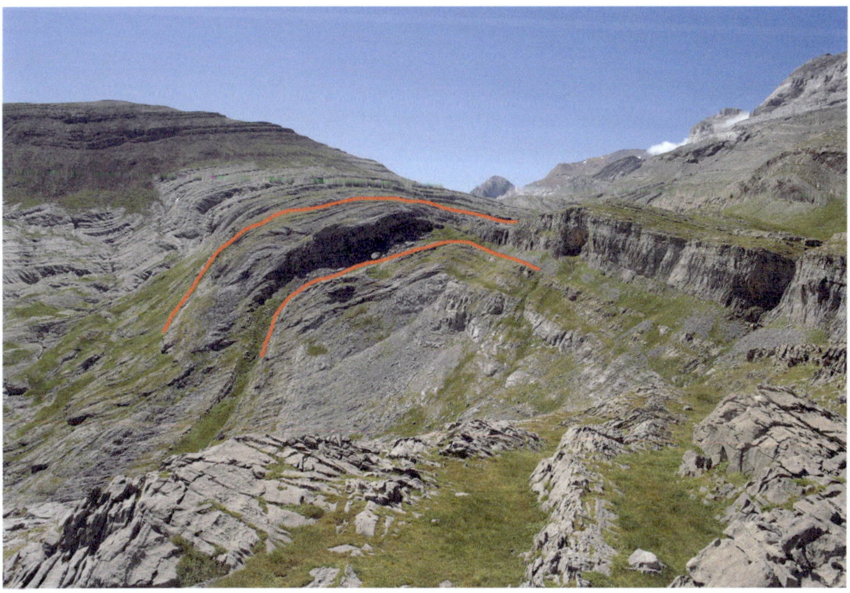

Fig. 9.55 Fold in the Monte Perdido thrust (photo by Telbisz)

Fig. 9.56 Connection of the glacially polished plateau and the U-shaped Ordesa valley (photo by Telbisz)

9.12.4.4 Glaciokarst of Picos de Europa

Picos de Europa is the best-known and most explored unit in the Cantabrian Mountains. The mountain range is built up of mostly Carboniferous limestones. Two of the 11 deepest caves in the world are found here (www.caverbob.com), but there are several other cave systems in the mountains. This spectacular karstic and glaciokarstic terrain can be explained by three reasons. First, the karstifiable sedimentary rocks have extremely large thicknesses. Second, during the Hercynian and Alpine tectogeneses, they experienced intensive tectonic effects, and large fold-and-fault systems as well as numerous small fissures have been created. Third, the variegated but usually wet climate favoured karstification. The tropical climate of the Pliocene was characterized by high temperatures and increased precipitation (Bertrand 1971). During Pleistocene cold periods, the mountains naturally received less precipitation than in the warm periods; however, snowfall was *relatively* high feeding the glaciers, which caused intensive erosion and glaciokarst development. Although the mountains became ice-free at around 15–16 ka BP, nival karstification still actively forms the glaciokarst of Picos de Europa (Smart 1986).

Picos de Europa are a limestone mountain range of coastal location, about 20 km from the Cantabrian Sea, separated from the rest of the Cantabrian Mountains by deep valleys. The area of the mountains is 650 km^2, and crest elevations generally exceed 2000 m. The highest peak is Torre Cerredo with 2648 m (Rodríguez-Rodríguez et al. 2014). There are 13 mountain peaks exceeding 2500 m elevation.

Three large rivers cut the mountain block into separate parts. Picos de Europa is a geologically complex mountain range, where most of the rock formations are of Paleozoic origin. The main mountain-building rock is a Carboniferous limestone dominant in the central and eastern parts of the mountain range (Cendrero and Saiz de Omenaca 1979). The rock, which is resistant to mechanical erosion, was deposited in several hundred metres of thickness. Later, tectonic processes caused intense fissuring of the rock mass. Due to its high permeability, it is characterized by well-developed cave systems. In addition, Paleozoic greywacke, quartzite and granites are also found in the area. In the Central and Western Massifs, a total area of cca. 200 km^2 was glaciated and the lengths of the glaciers reached 7–8 km (Serrano et al. 2012, 2013). Spectacular glacial features are characteristic in Urrieles, which is the central and highest unit of the mountain range. Everywhere above 1000 m asl, the dominant landforms are of glaciokarstic origin (Miotke 1968). On the surface, cuestas can be identified. The landform assemblage consists of shafts, dolines and karren fields. Several of the world's deepest caves are found here, including the Torca del Cerro del Cuevon and the Sima de la Cornisa, which are deeper than 1500 m. Typical surface morphological elements are the giant glaciokarst dolines.

9.13 Rocky Mountains

The Rocky Mountains form the longest continuous mountain chain in the North American Cordilleras. The mountain ranges stretch from Liard River at 59°N–35°N in New Mexico. The Canadian Rockies are found mostly in British Columbia, but smaller parts extend to Alberta as well, whereas the US Rockies lie in several states, namely, in Montana, Idaho, Wyoming, Utah and Colorado, and to a little extent in New Mexico and Washington. Although the mountain ranges are contiguous, the Canadian and US Rockies are remarkably different from several points of view, their topography, geologic settings, glaciations and spatial distributions of karst features have distinct characteristics. Interestingly, the natural boundary between the northern and southern parts roughly overlaps the international border between Canada and the USA.

Glaciokarst terrains exist in both parts, but they are more extended in the northern, Canadian Rockies. In this segment, there are glaciokarsts, which are actually covered by glaciers, or which are situated in the close vicinity of glacierized areas. Among these locations, *Castleguard Cave* and its environment has the highest scientific significance, because it was thoroughly described and studied by Derek Ford and his colleagues (Ford 1971b, c, 1979, 1983a, b, c; Ford et al. 1983; Atkinson 1983; Gascoyne et al. 1983; Schroeder and Ford 1983; Smart 1983), so it became a locus typicus of glaciokarsts.

Most parts of the Rocky Mountains are rarely populated, and certain places are still hard to access, and hence karst studies in the Rockies were scarcer and later in time than in the more easily approachable glaciokarsts (see Chap. 1). Nevertheless, difficult accessibility means an advantage if nature conservation is the goal. It is especially important today, when tourism is a quickly developing sector carrying some risks to the sensible glaciokarst terrains in terms of water resources, vegetation, fauna and soils (Werner 1979).

9.13.1 Geologic and Tectonic Settings

In the Paleozoic Era, the western part of what is North America today was largely covered by shallow seas, in which several kilometres thick limestone and dolostone formations were deposited. These rock formations of mostly Cambrian, Devonian and Mississippian ages are the main building rocks of the majority of the present-day karst terrains in the Rocky Mountains. The tectonic structure of the Rocky Mountains was formed during three orogenies, the last of them being the Laramide Orogeny 80–55 Ma ago, which resulted in the uplift of the mountain ranges (King 2015). A remarkable characteristic of this mountain-building period was that the subduction angles of the oceanic Kula and Farallon Plates, which constituted previously the bottom of the northern Pacific Ocean, were different. In the south, the subduction angle of the Farallon Plate was gentle, and therefore the US Rockies are relatively far from the plate boundary, and hence from the ocean, at 800–1500 km distance. Moreover, magmatic intrusions also played a significant role in the uplift of the southern parts. As a result, by the Early Tertiary, a Tibetan-type high and large plateau had been formed in the territory of the present-day Southern Rocky Mountains. Meanwhile, in the Canadian segment, the Kula Plate subducted at a relatively steeper angle; consequently, these ranges are closer to the Pacific Ocean, and the tectonic style is typical of accretionary wedges, i.e. there are folds, overthrusts, duplexes and imbricate thrust systems. Overthrusts and folds are present in the US Rockies as well, but there, in many cases, strata became dipped due to magmatic intrusions. Naturally, the above tectonic processes affected the carbonate layers as well, and thus dip angles are highly variegated in the karst terrains of the Rocky Mountains, and Paleozoic carbonate layers are usually rich in tectonic joints and fissures. Tectonism was active during the Quaternary as well; moreover, isostatic changes also contributed to vertical movements causing both uplift and subsidence according to interglacials and glacials. Logically, the isostatic reactions were more significant in the northern segment, where the magnitude of vertical isostatic motion could reach 200–300 m (Pierce 2003; Clague and Ward 2011).

In between the mountain-building periods, and afterwards, the amount of erosion was highly remarkable, and the related sediments were deposited in the intermountain basins, rifts and along the oceanic and continental margins of the mountains. The latest erosional phase also affected and dissected the previously deposited sediments (McMillan et al. 2006).

The spatial distribution of carbonate rocks is uneven (Fig. 9.57). While in the Canadian Rockies the limestones and dolostones are predominant, extending over

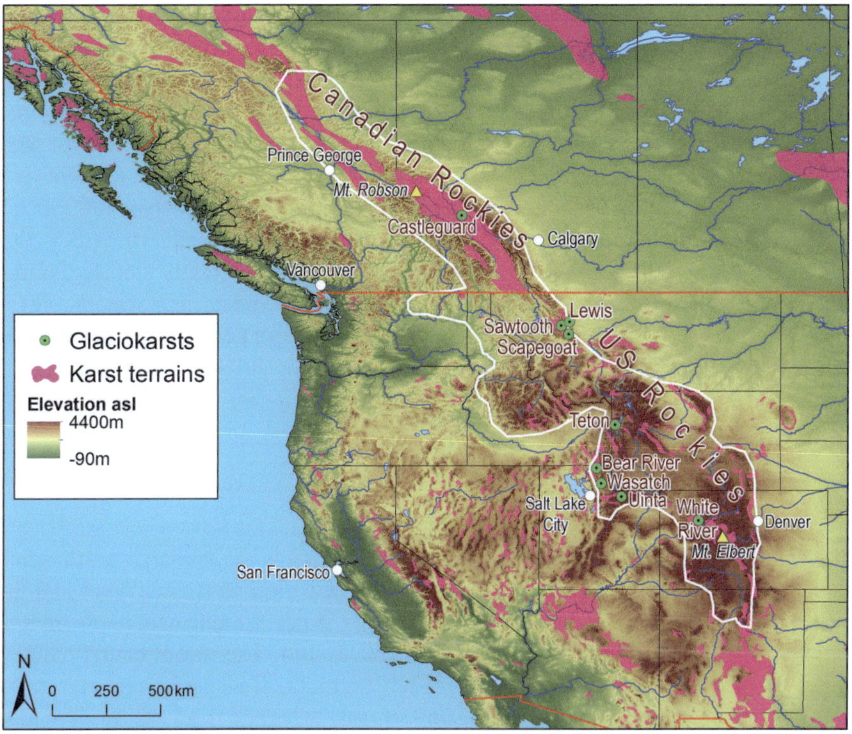

Fig. 9.57 Location of glaciokarst terrains in the Rocky Mountains. Data source of karst terrains: Ford and Williams (2007), and http://web.env.auckland.ac.nz/our_research/karst/

large areas contiguously, in the US Rockies, the carbonate rocks have a significantly lower proportion and they are more scattered and limited to smaller patches.

9.13.2 Relief Settings

The topography of the Rocky Mountains also mirrors the distinct geological characteristics of the northern and southern parts. The Canadian Rockies are narrower with 200–300 km width and they emerge from a lower base level. This base level is at cca. 500 m elevation asl in the intermountain basins and at the northwestern Prairie, which marks the eastern boundary of the Canadian Rockies. The orientation of the ranges is northwest–southeast, but it turns to north–south at 50.5°N. The mountains are contiguous here, and at the western side of the ranges, there is a remarkable topographic feature, the Rocky Mountain Trench, which was created by fault tectonics. The northern segment of the trench was formed by strike-slips, whereas the southern segment was created by normal faults (Gabrielse 1985).

To the contrary of the Canadian parts, the US Rockies consist of several tens of ranges, which are more separated from each other, and the strikes of the mountain ranges are more variegated. Moreover, the base level of the surroundings is significantly higher than in Canada, and it is at about 1000–1200 metres above sea level. Peak elevations are also somewhat higher in the south. While the highest peak of the Canadian Rockies, Mt. Robson is at 3954 m asl, the top of the US Rockies, Mt. Elbert is at 4401 m asl and some 50 other summits are higher than 4000 m asl.

Nevertheless, a common feature of both the northern and southern parts is that the jagged, dissected rocky peaks abruptly emerge from the gently sloping Prairie, when approaching the mountains from the east. This spectacular topography is mostly due to Quaternary glaciations. Cirques, arêtes, horn peaks, U-shaped troughs and hanging valleys are widespread in most part of the Rocky Mountains.

9.13.3 Climatic Settings

The climate-determining factors in the Rocky Mountains are the large north–south extension of the mountain chain, the variegated elevation above sea level and the distance from the Pacific Ocean. The climate is partly continental, partly mountainous, with precipitation decreasing from wet to arid. The mean annual temperature is 4.1 °C in Prince George (British Columbia, Canada), which is at the northern part of the mountains at 569 m asl, whereas 10.4 °C in Trinidad (Colorado, USA, at 1827 m asl) near the southern end of the US Rockies (https://en.climate-data.org). January mean temperature shows harsh winter in Prince George with −10.2 °C, while a mild winter in Trinidad with −0.1 °C. On the other hand, the summer is relatively warm even in the north (Prince George: 15.9 °C July mean temperature) and hot in the south (22 °C in Trinidad). Obviously, the higher mountains have respectively lower temperatures. Since the westerly winds are predominant at the latitudes of the whole Rocky Mountains, the precipitation is mostly a function of the distance from the Pacific Ocean. As the Canadian parts are closer to the ocean, the highest yearly precipitation values of 1500 mm are found just there, while the southern parts, farther from the ocean, are more arid, especially the valleys and basins, where even desert conditions prevail in some places with less than 200 mm yearly precipitation. Again, the mountains complicate this picture and get more precipitation from the eastward-blowing air masses. Most part of the precipitation falls as snow in the mountains. A further impact of the westerly winds is that the snow is swept away from the western slopes, and accumulated in the eastern, lee sides of the ridges, that influences the development of glacial forms like cirques, and karst processes as well (Medville et al. 1979; Wilson 1979; Bobrowsky and Rutter 1992).

9.13.4 Glaciation Characteristics and Phases

At present, extended areas covered by glaciers are found only in the Canadian Rockies. The largest of them is the Columbia Icefield with 325 km^2 area, and thickness between 100 and 365 m. The icefield has several outlet glaciers towards the neighbouring valleys. It is also the locality of *Castleguard Cave*. On the other hand, in the US Rockies, there are only alpine-type valley glaciers, half of them in Montana, more than a quarter of them in Wyoming and even less in Colorado. The present glaciers are all warm-based, and it was similar in most cases during the glacial periods as well (Ford 1971a, b, 1979). The formation of glaciers is restricted in the southern parts not only due to higher insolation, but also due to the lower amount of precipitation. Taken into account the recent global warming trend, it is predicted that most glaciers of the US Rockies may vanish in some decades. As a consequence of the above facts, larger karst areas covered actually by glaciers are present only in the Canadian Rockies.

The chronology of glacial periods in the Rocky Mountains has been summarized by Pierce (2003). Regional names are used for the different glacial periods based on thoroughly studied moraine locations. In the USA, *Bull Lake* glaciation is the counterpart of Illinoian, whereas *Pinedale* glaciation matches the Late Wisconsinan, which was the time of the global LGM. In Canada, the latter is called as *Fraser* glaciation. There is very limited information about Early or Middle Pleistocene glaciations, because later glaciations devastated almost all previous traces (Clague and Ward 2011). The first, relatively well-documented glacial period was the Illinoian (MIS 6) between 200 and 130 ka BP. The Early Wisconsinan (MIS 4) glaciation took place between 70 and 35 ka BP with a remarkable interstadial. Finally, the Late Wisconsinan (MIS 2), including the LGM, occurred here between 30 and 10 ka BP, with a maximum between 19 and 17 ka (Bobrowsky and Rutter 1992; Benson et al. 2005; Clague and Ward 2011). The above glacial periods had several alternating advancing and retreat phases (Ford 1983a). The last deglaciation took place relatively quickly between 11 and 10 ka BP (Clague and Ward 2011). The last advance, though extremely limited with respect to former advances, occurred during the Little Ice Age (1200–1900 AD). Glacial chronology of the Rocky Mountains has been partly supported by karst studies as well. U-series dating of travertines has been performed by Sturchio et al. (1994), while Harmon et al. (1977) and Harmon (1979) dated speleothems from three Canadian and one US study area, using also U-series methodology. They linked measured speleothem growth ages (320–280 ka, 220–180 ka, 155–99 ka and 10 ka to present) to interglacial periods.

During the glacials, the ice sheets grew from three significant centres. In the northeast, the Laurentide Ice Sheet extended from the Canadian plains, in the northwest, the Cordilleran Ice Sheet extended over the Pacific Ranges and the intermountain basins, and finally, glaciers were independently formed in the Rocky Mountains, gradually coalescing with the other huge ice masses (Bobrowsky and Rutter 1992; Clague and Ward 2011, Fig. 9.58). At maximum glaciations, the

Fig. 9.58 Extension of Late Wisconsinan (LGM) ice in the Rocky Mountains. Data source of glacial extension: Dyke (2004)

expanding ice masses from these three centres were joined, but at smaller glaciations and in the transitional periods, these ice masses were separated from each other by ice-free corridors (Bobrowsky and Rutter 1992). During the LGM, the southern boundary of the unified large ice sheet run roughly along the present-day USA–Canada international border (Clague and Ward 2011). The surface of the Cordilleran Ice Sheet reached 2000–3000 m asl, and thus only the highest peaks emerged from the ice as nunataks (Clague and Ward 2011).

During the LGM, the ELA was at about 3300 m asl in the south (Colorado), and at cca. 2100 m asl in the north (Montana) of the US Rockies. In Montana, the ice sheet extending from north and valley glaciers advancing from the mountains were joined. The ELA had a westward decreasing gradient due to higher precipitation closer to the ocean (Pierce 2003). Based on ELA values of LGM glaciers, the then climate could be either 8.5 °C cooler in case of precipitation similar to present values, or 10–13 °C cooler in case of precipitation decreased by 44% (Leonard 1989).

9.13.5 Study Areas

Glaciokarsts of the *southern part of the Canadian Rockies* have been studied in detail by Ford (1979). The main karstifying rocks are Cambrian, Devonian and Mississippian limestones. Among them, there is a particularly well-karstified formation, the extremely thick-bedded Cathedral Formation which has even 5-m-thick layers in some places.

Mount Castleguard was one of the first places, where geochemical analysis was applied to demonstrate that the dissolution capacity of subglacial waters is limited (Ford 1979). Like in many other alpine glaciokarsts, springs are often found in hanging positions, high in the glacial valley sides. To the contrary, low-level springs are buried by glacial drift in some other places (Ford 1979, 1983a). During full glacial periods, ice covered most part of the surface, hydraulic gradient was low and phreatic dissolution in caves was minimal; instead, some parts of the caves were filled by fine-grained sediments. However, during transitional early or late glacial periods, the glaciers increased the hydraulic gradient, and hence quick flows were generated and the boundary of the vadose and phreatic zones often moved up or down. Thus, in these transitional periods, some pre-existing passages were enlarged, while some new bypass passages were created where pre-existing passages were clogged by sediments. A remarkable example is the *Nakimu Cave*, which has 140 m vertical dimension, and 120 m out of it is filled with deposits. Boulders and gravels are typical at the bottom, whereas silt and clay are abundant above (Ford 1979). The longest cave of all Canada is also found here, the famous *Castleguard Cave* (21 km), which consists of long, sub-horizontal, phreatic passages, occasionally pierced by vertical shafts. The peculiarity of that cave is the iceplug at its upstream end, which was formed by the direct penetration of glacier ice into the cave. In addition, speleothems forming below an active glacier were also observed here (Atkinson 1983). Besides, Ford (1979, 1983a) demonstrated here, how the subsurface karst system had been disintegrated by glacial erosion, namely, by valley entrenchment and cirque formation.

Theoretically, Ford (1979) grouped sensu *lato* glaciokarst landforms of the Canadian Rockies into six types, though he noted that some of these theoretical types could not be easily discriminated from each other in practice. The first group contains the *postglacial* landforms. Different types of karren (pits, groovings, channels) belong to this group. Unlike in the Alps, rounded karren are sparse features in the Canadian Rockies. It is due to the fact that forest clearing and grazing was insignificant in the higher zones of the Canadian Rockies, and thus soil erosion was also limited. There are abundant solution dolines, and suffosion dolines formed on glacial drift, which are also principally postglacial forms. Collapse dolines, in turn, are uncommon here. The second group contains *subglacial* forms. Among them, the subglacial calcite precipitates can be mentioned, which have been described just from this area. They are ephemeral phenomena, formed on limestone surfaces below glacier ice, but in some decades after glacier withdrawal they are demolished by rain and meltwaters (Hallet 1976; Ford 1979). The third group, the *karstiglacial* forms, is originally created by ice and later transformed by karst processes. Mostly, depressions belong to this group, and they are very frequent in the Castleguard karst, often containing small lakes, but these lakes never (or only exceptionally) overspill because of their karstic drainage. The fourth group is called sensu *stricto glaciokarstic* forms, meaning that the primary form is of karstic origin transformed later by glacial processes. Only some larger depressions can be unambiguously mentioned in this group and even in those cases, it is not easy to demonstrate their primary karstic origin. The fifth type, the *mixed* landforms, was created during several alternating glacial and karstic periods. Likely, the relatively large, 6 km by 2 km area and 15-m-deep Medicine Lake belongs to this group. Finally, the truly *preglacial* forms, i.e. forms, which are older than the first glacial, are uncommon, probably some old cave passages can be mentioned, which avoided glacial modifications (Ford 1979, 1983a). As for the altitude zones, they are also different than in the Alps. Three zones can be observed here: first, the *boreal forest*, where soils are formed on glacial drift and the typical karst forms are suffosion and collapse dolines, while bare karren are rare phenomena. Second, the *glaciated tundra*, where karst forms are the most numerous, there are abundant depressions and karren, but the impact of frost shattering already affects this zone. Finally, the third zone is the *unglaciated tundra*, where frost shattering is the dominant process, and it hinders the evolution of large-scale karren forms. Nevertheless, there are no distinct elevation boundaries between these zones, and even they intermingle with each other in some places (Ford 1979).

The deepest cave of Canada, the *Arctomys Cave* (526 m depth) is found near Mount Robson. This cave is so deep due to the steep dipping of beds, which is in turn the result of the above-mentioned tectonic processes. In spite of the great depth of the cave, shafts are surprisingly few, there are only five of them and even they are not too deep.

In the northern part of the US Rockies, in Montana, glaciokarsts are found in the *Lewis Range*, *Sawtooth Range*, and in the *Scapegoat Mountains* (Campbell 1979). They are also developed on Paleozoic (Cambrian, Devonian and Mississippian) limestones. Tectonic control is very strong in case of these karsts, i.e. folds, faults and overthrusts complicate the structure (Bodenhamer 2007). Where dip angles are steep, surface landforms and even caves are few. Caves and surface landforms are usually constrained by layer dip and joints. Where the dip of strata is less than 10°, karst forms are more abundant both at the surface and underground. The effect of wind direction is clearly observable, at the eastern side of ridges, where snow can accumulate, sinkholes are more frequent. Most caves are found in syncline areas, and passages follow the axis of the syncline. Speleothems are rare phenomena, but ice is more often found in caves. Karren features are frequent at the surface, especially where soil is missing and the limestone is thick-bedded. In case of thin-bedded limestones, frost shattering more effectively obstructs the evolution of karren. Depressions are often filled with shattered rock debris. Late Pleistocene glaciation dissected the former, contiguous karst aquifer (Bodenhamer 2007).

In the *South Teton Range* (Wyoming), Paleozoic (Cambrian to Mississippian) carbonate formations are up 1000 m thickness, that provides a good basis for karstification (Medville et al. 1979). Structural control is evident here as well, and there are large bedrock slabs reaching even 10 km^2. Well-developed karren (meanderkarren, spitzkarren, trittkarren and kluftkarren) are typical everywhere on these surfaces, whereas sinking features, dolines and shafts are mostly distributed along the margins of these bedrock slabs. Depressions are often filled with shattered limestone debris. Due to the moderately dipping layers, typical schichttreppenkarst (stepped bedding planes) can be observed in many places. The largest shafts reaching down to 50 m depth were developed in these areas. Further on, some collapse dolines are also present. In addition, cirque-dolines can be also recognized, containing in some cases small tarns, and limestone pavements are also found near them. The so-called cirque-dolines are karstically drained, there are caves below them and they get significant amount of aggressive water from the surrounding non-karstic ridge slopes, and thus active karst evolution is possible here. The steep rock walls are dissected by dissolution features, natural bridges, niches and fossil cave entrances that are all present. Wind Cave is an actively developing cave, in which passages were formed partly by phreatic and partly by vadose invasion waters, which is reflected in the keyhole profile of passages.

In *Bear River Range* (Utah), the Paleozoic Formations are almost 1000-m-thick, but in this case, dolostones are present in a relatively large proportion beside limestones. As dolomite is more inclined to frost shattering, karstification is more confined here (Wilson 1979). A further limiting factor is that precipitation is very low; at present, it is only 155 mm a year. Here again, wind direction determines snow accumulation, and thus glacial cirques as well as karstic dissolution features prefer the eastern–northeastern slopes. This fact that has been statistically demonstrated by Wilson (1979). There are two larger cirques, which can be considered karstiglacial in terms of Ford (1979). Besides, there are smaller solution dolines, some collapse dolines, karst corridors formed along joints, stepped pavements and smaller dissolution features, i.e. linear karren, small pits, spongy karren, kamenitzas, spitzkarren and karr pedestals. Caves are few, small, mostly of vadose origin and located mainly along the margin of the formerly glacier-covered terrain (Wilson 1979). Surface drainage operates only during strong snowmelt periods, usually from late spring to early summer. Karst springs are well-developed and their discharges highly fluctuate according to snowmelt periods (Spangler 2001). Speleothem ages, fluvioglacial sediments and the deranged topography of Bear River Range all suggest that karstification and especially the caves are the remnants of a former (possibly preglacial) period, or at least caves started to evolve before the first glaciation (Spangler 2001).

The glaciokarst of *Wasatch Range* is also found in Utah (White 1979). The main bedrocks of the karst terrains are Mississippian carbonates, but they are tectonically distributed into smaller segments, and dolomites have a relatively high proportion. The highest part of the Wasatch Range is strongly dissected, with arêtes, cirques and deep valleys. Some of the valleys have typical U-shaped glacial form, but there are fluvial V-shaped canyons, too. The drainage is mostly superficial, while karstic drainage is subordinate. The most significant cave of the Wasatch Range is the *Timpanogos Cave*, located in a hanging position. The bare limestones are often scenes of pavements and other karren (e.g. rinnenkarren, rillenkarren, spitzkarren and kluftkarren), but larger dissolutional depressions are uncommon.

The *Uinta Mountains* are stretching in the close neighbourhood of the Wasatch Ranges. The west–east strike is a peculiar characteristic of Uinta within the Rocky Mountains. In fact, the Uinta Mountains were formed by an elliptical dome-like uplift. The central, higher parts are built up of crystalline rocks, which are surrounded by a relatively narrow, elongated limestone ring (Godfrey 1985). The higher parts have a remarkable, erosionally levelled surface, a broad rolling landscape, locally called as *summit flats* (Munroe 2006). The north–south extension of these summit flats increases from west to east. As they are preglacial surfaces, truncated gradually by glacial erosion, it is concluded that Pleistocene ice had gradually less extension from west to east, i.e. the ELA was higher in the east. It is likely the result of the eastward decreasing precipitation (Munroe 2006). Karren features are subordinate here, but closed depressions are more frequent and caves are larger than in the Wasatch Range (White 1979). Streams flowing from the higher, non-karstic terrains cross the narrow karst bands. There are some well-developed stream sink caves, the longest of them being the *Little Brush Creek Cave* (9.5 km). For half a year, during spring and summer, caves cannot be entered because of the high amount of water flowing into them due to snowmelt. There are phreatic passages in the caves, and remarkable 3D maze sections as well, which are characteristic of caves formed by flooding waters (Palmer 2003). The valleys, which avoided glaciation, are often dry valleys with karstic drainage. However, in the valleys, which were heavily glaciated, there is no measurable water loss. It means that during the 11 ka since glacier withdrawal, there was not enough time for the evolution of draining passages (Godfrey 1985).

The *White River Plateau* in Colorado is particularly interesting, because it is a 50 km by 60 km dome-like structure, which bears a relatively contiguous karst plateau, and thus it was a unique location in the US Rockies, where a large area cap ice could form during the Pleistocene (Maslyn and Davis 1979). The terrain can be divided into distinct karst zones, which are determined by their climatic and geologic settings. First, in the *lower* zone (up to about 2100 m asl), the structural control is dominant. Due to the dome-like structure, layer dips are steep, controlling cave evolution, but in general, either surface or underground karst forms are subordinate in this zone. Second, in the *middle* zone (from 2100 to 3000 m asl), there are alpine meadows, and the drainage is partly superficial, partly karstic. The largest caves are found in this zone, notably the *Groaning Cave*, which is the longest cave of the US Rockies with its 21.5 km length. Basically, it is a 2D maze cave as a relatively thin geological formation constrains its passages. The surface is partly covered by a shale cap, and during glacial periods, this shale cap hindered the water to flow into the caves, thus saving them from being clogged by debris. In the sides of canyons, there are several cave entrances marking a former higher water level which could be either due to glaciation or due to tectonic uplift. The *upper* or *alpine* zone (above 3000 m asl) is a well-developed alpine-type glaciokarst. The shale cap is missing here, and during most part of the year, snow covers the area that provides a permanent water recharge for karstification. There are variegated types of karren, and limestone pavements. Solution dolines are also abundant, and occasionally some really large features of 800 m length are also observable. Collapse dolines, shafts and dry valleys are present, too. However, the effect of frost shattering is significant in this zone, and the resulted debris often fill the depressions. The caves of this zone are mostly of vadose type, often with vertically elongated passages according to joints. Their evolution is active and they are related to present-day surface landforms, so they are possibly of postglacial origin (Maslyn and Davis 1979).

9.14 Scandinavia and Svalbard

In Scandinavia and Svalbard, surface karst terrains have generally small extension (Gunn 2004). The bedrock of karst terrains is almost exclusively marble (Fig. 9.59), which is usually found in narrow, elongated zones, and hence they are generally categorized as *stripe karsts* (Horn 1935; Skoglund et al. 2010; Lauritzen 1984, 2005). Since the whole area was glaciated during the cold periods of the Pleistocene, practically every karst terrain of Scandinavia and Svalbard falls in the *glaciokarst* category. Surface karst landforms are limited in numbers and dimensions, but caves are not negligible, both their number or their sizes are considerable. At present, some 2000 caves are known in Norway (Lauritzen and Skoglund 2013). Karst research in Scandinavia goes back to the beginning of the twentieth century (Oxaal 1914 in Lauritzen and Gascoyne 1980), and one of the main issues has always been the relationship of cave evolution and glacial phases, namely, whether the main cave forming phases occurred in the cold or in the warm periods, and whether speleogenesis was active during glaciations or not (Lauritzen and Gascoyne 1980; Skoglund and Lauritzen 2010, 2011). The karst of Svalbard merits special attention, because it is still a partly glaciated, partly periglacial area; hence, processes observed here can be thought of as models of Scandinavian karst evolution during glacial periods (Lauritzen 2006).

9.14.1 Geologic and Tectonic Settings

The majority of Scandinavian karsts are found in the Scandinavian Mountains, whose structure was formed during Caledonian Orogeny. The marble rocks, which bear most of the Scandinavian karsts, were also formed during this tectonic phase (Lauritzen 2005). As a result of metamorphosis, the former porosity (primary porosity, bedding planes, fractures) of the carbonate rocks were lost, and therefore later cave evolution was initiated along post-metamorphic fractures (Skoglund et al. 2010). Tectonic processes of the Caledonian Orogeny included intense folding and overthrusting that resulted variegated structural settings of the marble layers bearing today the karst features. Marble layers are present at the surface only in narrow, elongated stripes, and

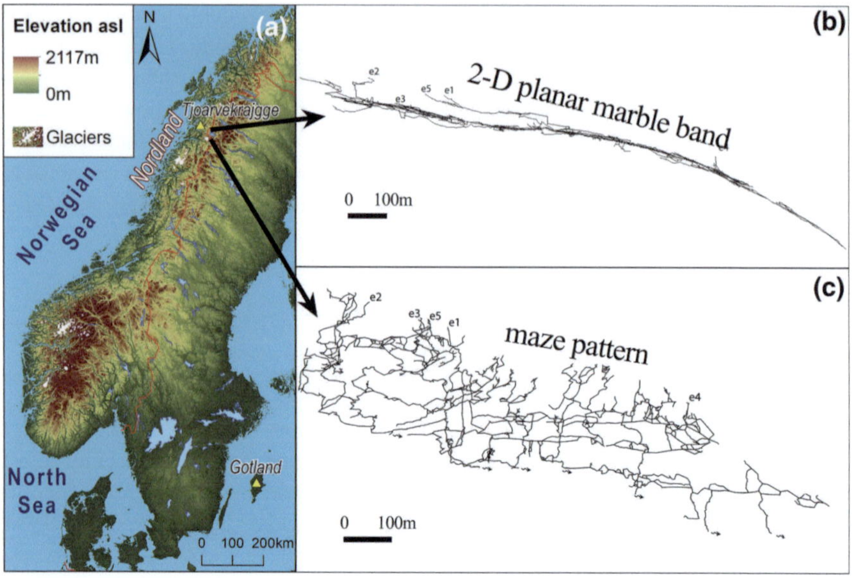

Fig. 9.59 Location of Tjoarvekrajgge Cave, Nordland and Gotland in Scandinavia (**a**), profile of Tjoarvekrajgge Cave (**b**) looking northeast (40°) and upslope (24°), and profile of the same cave looking north and horizontal (**c**). Observe that the cave is constrained in a 2d planar marble band, but the passages form a labyrinth within this band. Cave profiles from Finnesand and Curl (2009)

they have a contact with non-karstic rocks, for instance, with mica schists (Lauritzen 2005). The tilt of marble layers is also highly variable, from almost horizontal to near vertical. Cave passages formed in these marble layers are constrained by the tilt and extension of marble; thus, in many cases, they are approximately two-dimensional systems (Lauritzen 1986; Skoglund et al. 2010, 2011). The karst of Svalbard (which is not part of the Scandinavian Mountains) was also formed on marble rocks (Gunn 2004; Lauritzen 2006), but another karst terrain found in Götland Island (Baltic Sea) was created on Silurian reef limestones (Gunn 2004).

9.14.2 Relief Settings

The karst terrains of Scandinavia and Svalbard are generally found at low elevations, and therefore their evolution has been influenced in many cases by the changing sea level as well (Svendsen and Mangerud 1987). The preglacial relief of the Scandinavian Mountains could be characterized by a gently undulating surface, which is called locally as *paleic surface* (Lauritzen 2005). The remains of the paleic surface are recognizable at many places as plateau areas. However, in case of Scandinavia, these plateaus are not related to karst processes. These plateaus are dissected by deeply incised glacial valleys, in which lakes and fjords are found. The isostatic rebound following the deglaciation is remarkable throughout the area, and high uplift took place in Scandinavia since the terminus of the last glaciation. Nevertheless, the uplift rate was not uniform during this period. Today, rates of uplift range from 1 to 8 mm/a, but the total Holocene uplift reached some 100 m in some quickly uplifting areas (Fjeldskaar et al. 2000).

9.14.3 Climatic Settings

Nowadays, Norway is one of the wettest regions in Europe. In the middle coastal areas, notably in Nordland County, where most of the karst terrains are found, the annual precipitation is 1500–2000 mm (Skoglund and Lauritzen 2013). Temperatures are relatively high due to the North Atlantic Current; however, given the high latitudes, mean annual temperatures are around 2–3 °C (Skoglund and Lauritzen 2013). In Svalbard, which is even more extreme to the North, the precipitation is significantly lower, notably 400 mm a year in the western parts, and the mean annual temperature is only −4 °C, that implies permafrost conditions, i.e. groundwater is in a frozen state from 2 to 400 m depth, except taliks (Lauritzen 2006).

9.14.4 Glaciation Characteristics and Phases

During the Quaternary, several tens of glaciation phases affected Scandinavia and Svalbard. The first glaciations took place already at the end of the Pliocene, 2.75 Ma ago (Mangerud et al. 2011). The first glaciation, which extended over the shelf terrain, occurred 1.6 Ma ago in Svalbard, and 1.1 Ma ago off the Norwegian coasts (Sejrup et al. 2005). The MIS 12 glaciation, which occurred cca. 0.5 Ma ago, was really powerful, and some further retreat and advance phases took place afterwards. The last significant glaciation was the Weichselian, which began in MIS 5d 120 ka ago (Mangerud et al. 2011). The Weichselian largely reworked the sediments and landforms of earlier glaciations. The LGM occurred during MIS 2, started cca. 28–30 ka ago. The warming period following the LGM was shortly interrupted by the cold Younger Dryas at 13–12 ka BP. The melting of glaciers and the resulted sea level increase took place at a quicker pace than the uplift caused by the postglacial isostatic rebound; hence, Early Holocene marine levels and abrasion landforms are found in some places more than 50 m higher than the present-day sea level. As a result, certain caves, which are today at more than 50 m above sea level, could be at or below sea level in the Early Holocene (Skoglund and Lauritzen 2010, 2011, 2013).

9.14.5 Study Areas

Most of the Norwegian karst terrains are found in *Nordland County*, which is at the northern middle sections of the Atlantic Coast (Fig. 9.60). Surface karst landforms are rare features, and in general, only small size (less than 10 m diameter) sinkholes and stream sinks are present here, whereas karren are not mentioned at all in the literature (Skoglund and Lauritzen 2010, 2011).

On the other hand, there are a number of caves with a variegated morphology. Nevertheless, cave passages are usually narrow (Horn 1935). Relict phreatic caves are frequent phenomena, and vadose caves also occur, though they are not so common. In addition, compound caves with both phreatic and vadose passages also exist (Faulkner 2006; Øystese et al. 2005). Many caves have maze segments (Skoglund et al. 2010). The cave entrances are often found at hanging positions in the side of glacial valleys (Lauritzen 1986). Speleothems do exist in Nordland caves, but their number is limited, and they are found mostly in caves below the treeline (Lauritzen and Gascoyne 1980). Theories of local speleogenesis must explain the above observations. The role of static, geologic factors, i.e. the layer settings of marble rocks, tectonic fissures, younger isostatic fissures, is usually unambiguous, but the nature of dynamic, i.e. hydrologic factors in cave evolution can be modelled by several ways (Skoglund and Lauritzen 2011). An important fact

Fig. 9.60 Marble karst in Vadvetjåkka National Park (Kiruna, Sweden, photo by Egri)

is that the corrosion capacity of subglacial meltwaters is extremely small (Ford 1971c), and therefore if subglacial speleogenesis is considered sensu stricto, then passage enlargement rates are particularly small, even if it is taken into account that there was no permafrost below thick and more than 400-m-wide glaciers (Lauritzen and Gascoyne 1980).

Previously, it was argued that most Norwegian caves are of postglacial origin given the small size of passages (Horn 1935). Further on, it was supposed that cave passages could evolve quickly during deglaciation phases, when they got high amount of water from the retreating and melting glaciers. A novel version of these former theories emphasizes the role of ice-dammed lakes, which could recharge high amount of water into the caves (Faulkner 2006; Murphy et al. 2015). Since these lakes were short-lived with cca. 1000-year-long lifespan, the calculated total discharges and passage enlargement rates are not satisfactory enough to explain the formation of the present cave dimensions.

As for the age of caves, early ideas stated that there were no preglacial caves in Norway (Horn 1935). However, U-series speleothem data proved that some caves already existed at least before the last glacial period, the Weichselian. Lauritzen and Gascoyne (1980) were the first, who measured speleothem ages and their oldest speleothem sample turned out to be 145 ka old. Later on, speleothems older than 350 ka, and even older than 750 ka were also found in Norwegian caves (Lauritzen and Skoglund 2013). Consequently, the reason for the small size of cave passages is not the young age of caves, but the slow dissolution rates and the long inactive periods.

Characteristic features of phreatic caves are the elliptic cross sections, the ceiling half tubes, which were formed when the passages were partly clogged, and the scallops which were created by the flowing water. These latter forms can be used to estimate the velocity, direction and even the discharge of the former water flow in the given cave passage. Typical velocity values are surprisingly small, in the order of 10 cm/s (Skoglund and Lauritzen 2011; Lauritzen and Skoglund 2013). Slow flow velocities are also supported by the fact that fine-grained sediments are often deposited in the passages (Skoglund and Lauritzen 2010). In many cases, former flow directions are different from what could be expected based on the present relief settings. Taken into consideration all of the above observations, the formation of the phreatic cave passages, which are inactive at present, is generally explained by the *ice-contact speleogenesis* theory (Lauritzen 1986; Lauritzen and Skoglund 2013). The idea is that when glacier ice filled the valleys, at the contact of the ice and rock, surface meltwater flowed into the karst (Fig. 9.61). Dissolution capacity of surface meltwaters was much higher than that of the subglacial waters. Moreover, the glacier and the karstic rock formed a combined aquifer, and hence hydraulic gradients were small that explains the small flow velocities as well as the curious flow directions, which are not in agreement with the present-day topography. Certain cave passages are thought of as specific Nye-channels by this theory.

The formation of maze caves is generally explained by two reasons (Palmer 1991), by diffuse water recharge, and by floodwater recharge. In Scandinavia, maze caves are widespread, and the reasons for their formation have been intensively

Fig. 9.61 Contact of rock and ice in Vadvetjåkka National Park (Kiruna, Sweden, photo by Egri)

studied by field observations and theoretical models as well. It is concluded that glacier-related maze caves could be produced either by the water level increase due to the damming effect of the ice at the output side, or by the diffuse recharge from the ice at the input side (Skoglund et al. 2010). Beside ice-contact speleogenesis, in case of abundant water recharge, the widening of cave passages by dissolution and by vadose flow could be remarkable during interglacial periods, too.

Glacier exaration was also pronounced in Norway. On the average, 520 m surface lowering was calculated for Mid-Norway during the whole Quaternary period (Mangerud et al. 2011). Naturally, there are local deviations from this mean value, but the general magnitude of glacier erosion is well represented by that mean value. Many of the cave passages are found today in hanging positions. It is due to either the intense deepening or widening of glacial valleys. Some of the passages became open as a result of glacier erosion, whereas many others could be totally destroyed (Skoglund and Lauritzen 2011). The dominant surface landforms in Norway are related to glacier erosion or accumulation, and surface karst morphology is only subordinate.

The longest cave in Scandinavia, the *Tjoarvekrajgge* (Finnesand and Curl 2009), is a typical example for the above characteristics. Namely, it is a maze cave, formed in a two-dimensional marble stripe karst setting, in the side of a former glacier valley, near Bonå in Nordland County (Fig. 9.59). The actual explored length of the cave is 25.2 km (http://www.caverbob.com). Most of the larger cave systems in Norway were formed during several glacial–interglacial cycles, for instance, *Okshola Cave* (Skoglund and Lauritzen 2010), but there are also simple caves, whose formation is restricted to the last glacial cycle only. *Kvithola Cave* is a fine example for the second type (Lauritzen 1986).

The glaciokarst of *Götland Island* (Sweden) has different characteristics from the aforementioned terrains in almost all aspects, since topographic, lithologic and morphologic features are all different. This karstland has small relief differences, only in the order of some 50 m. The low-elevation "plateau" bears 33 dolines, and in the underground, the Lummelunda Cave has a branchwork structure with several fossil and active levels (Gunn 2004).

In *Svalbard*, karst terrains are found in the western parts, and *Blomstrandsoya Island* has been presented in detail by Lauritzen (2006). The island as a whole can be considered as a giant roche moutonnée of 32 km^2. Although permafrost is present in most of the area, there exist some small size, closed depressions, namely, collapse and suffosion dolines, as well as some *bogaz* features and caves. Bogaz forms were supposedly formed by subglacial water flows, i.e. they are Nye-channels. Caves are dominantly found along the shoreline, and they were formed mostly by abrasion; however, karstic dissolution may have contributed to their evolution. All caves situated at higher levels are clogged either by sediment or by ice, and thus it is impossible to penetrate them deeper than some 10 m. Nonetheless, some scallops are observable in these short passages demonstrating paleoflow conditions. Flowstones and cemented carbonates are also found in these caverns, and dating of these substances provided 32 ka and 23.9 ka BP, that suggest interstadial formation of these features. Moreover, some other karst terrains have also been described from Svalbard, namely, a "polje" with 200 m length and 6 m depth from *Sarsøyra* (Barbaroux and Besset 1968), ponors and outflows from *Sørkapp Land* (Baranowski 1975) and a 30-m-long cave from *Hornsund* (Pulina 1974). Salvigsen and Elgersma (1985) reported dolines with sinking water from *Vardeborgsletta*. They stated that these forms are basically of karstic origin, but their morphology has been modified by thermokarst processes as well. Cohen (2013) described the geomorphology, including dolines and karst lakes, of *Linnédalen*, a former glacial valley, which was deglaciated at 12.5 ka BP, but cooling periods at 4–5 ka BP and in the Little Ice Age led to the formation of small alpine and cirque glaciers.

9.15 Tien Shan and Pamir

The Tian Shan (also spelled Tien Shan) and the Pamir form together with some auxiliary ranges a big mountainous region ("node") in Central Asia that is linked closely to the Hindu Kush, the Karakoram and Tibet (Fig. 9.62). The highest point of the Tian Shan is Pobeda (Tomur) Peak (7439 m), and the highest points of the Pamir are Kongur Peak (7719 m) and Ismoil Somoni (Communism) Peak (7495 m). Further high summits are Abu Ali ibn Sina (Lenin) Peak (7134 m) and Korzhenevskaja Peak (7105 m) in the Pamir and famous Khan Tengri (6995 m) in the Tian Shan. The Tian Shan and the Pamir include numerous ranges, and the

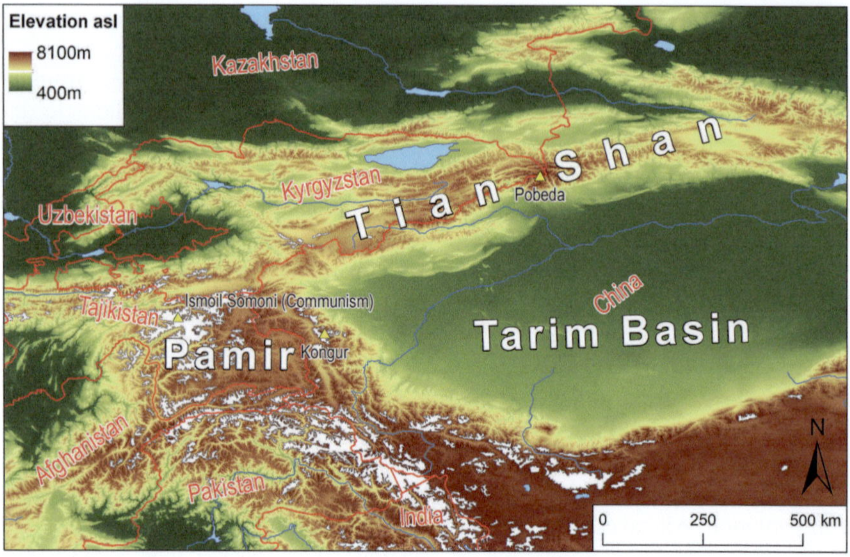

Fig. 9.62 Location map of the Pamir–Tian Shan region with modern glaciers

orographical architecture of the region is highly complex. For instance, the Tian Shan is subdivided into the Northern, Eastern, Western, Southwestern and Internal segments.

Glaciers are very common in the region. These have been studied actively since the nineteenth century by Russian and some other explorers, including Oshanin, Semenov-Tjan-Shanskij and many others. Serious advances were made by the specialists of the ex-USSR in the twentieth century, for instance, by Kalesnik, who hypothesized glaciokarst in the Tian Shan.

9.15.1 Geologic Settings

The geology of the Tian Shan and the Pamir is characterized by unprecedented degree of complexity, and it is yet to be fully understood. Crystalline and sedimentary rocks of all ages are represented there. Tectonically, the region comprises fold-and-thrust structures and nappes. It is formed of domains (former terranes) of different paleotectonic affinity. At least a part of this region was affected by the Hercynian orogeny, and the Alpine uplift started in the mid-Cenozoic has been continuing until nowadays.

9.15.2 Climate and Vegetation

The climate is temperate to subtropical with strong continental patterns with abnormally cold winter and warm (somewhere cool and short) summer. The annual rainfall varies significantly within the territory (from less than 100 mm/year to more than 1600 mm/year). Steppe, forests, alpine meadows and highland deserts are typical landscapes. Permafrost is developed locally.

9.15.3 Glaciations

Glaciers occupy significant area in the region: ~7800 glaciers (with more than 7300 km^2 area) are known in the Tian Shan and ~7100 glaciers (with ~7500 km^2 area) are known in the Pamir. In the latter, 10% of the entire territory is glaciated (Gvozdetskij and Golubtchikov 1987). The spatial extent of the glaciation has fluctuated; certain retreat has occurred in the past decades (Khromova et al. 2014; Jin et al. 2016); however, some advance of the Pamirian glaciers was also registered (Szulc-Rojan 1995). As shown by modern studies in the Chinese Pamir (Schoenbohm et al. 2014), development of glaciers interacts not only with climate, but also tectonics, river erosion and some other processes.

9.15.4 Glaciokarst Landforms

Karst is widespread in the Tian Shan and the Pamir: there are numerous paleokarst features developed prior to the Quaternary (in the Paleogene and the Miocene), whereas the present aridity of the region's climate precludes intense karstification (Gvozdetskij 1981). Presently, meltwater contributes to karst development in limestones of various ages. Additionally, frost weathering destroys original karstic landforms and produces peculiar features. Ice is found in the Syjkyrduu (Rangkul) Cave in the Eastern Pamir (Ridush 1993). This cave has been explored to a length of more than 2 km, and it has developed in Triassic limestones. The ice is supposed to be the relic of a paleoglacier. The cave is of hydrothermokarst origin, and it experienced glacial erosion.

Although the classical carbonate karst is known in both the Tian Shan and the Pamir, of larger interest is the widespread gypsum karst and salt karst that can be interpreted in the terms of glaciokarst sensu *stricto*. Some examples are described by Gvozdetskij (1981). First, glacial meltwater facilitates gypsum dissolution in many places. Second, intensively karstified gypsum layers are overlain by moraines of the past glaciers in several places in the Pamir. Third, on the right bank of the Kyzylsu River, karst is developed in large "exotic" blocks of gypsum detached and transported by the paleoglaciers.

Interesting features were described by Kalesnik (1935) and later Mavlyudov (2004) in the Tian Shan. Depressions and shafts at the edge of glaciers in the upper valley of the Naryn River were formed directly in the ice because of maximum "concentration" of meltwater in those places. Although this phenomenon was termed initially as *glacial karst*, it should be distinguished from typical glaciokarst (sensu Veress 2016a, b).

References

Ackert RP Jr, Becker RA, Singer BS, Kurz MD, Caffee MW, Mickelson DM (2008) Patagonian glacier response during the Late Glacial-Holocene transition. Science 321:392–395

Adamson KR, Woodward JC, Hughes PD (2014) Glaciers and rivers: pleistocene uncoupling in a Mediterranean mountain karst. Quatern Sci Rev 94:28–43

Adamson KR, Woodward JC, Hughes PD (2015) Middle Pleistocene glacial outwash in poljes of the Dinaric karst. Geol Soc Am Spec Pap 516:516–520

Andrieu-Ponel V, Hubschman J, Jalut G, Herail G (1988) Chronologie de la déglaciation des Pyrénées françaises. Bulletin de l'Association Française pour l'Etude du Quaternaire 34–35(2–3):55–67

A.R.S.I.P. (1985) Le Karst de la Pierre Saint-Martin en quelques chiffres. Karstologia 6(2):3–6

Atkinson TC (1983) Growth mechanisms of speleothems in castleguard cave, Columbia Icefields, Alberta, Canada. Arct Alp Res 15:523–536

Aucelli PP, Cesarano M, Di Paola G, Filocamo F, Rosskopf CM (2013) Geomorphological map of the central sector of the Matese Mountains (Southern Italy): an example of complex landscape evolution in a Mediterranean mountain environment. J Maps 9(4):604–616

Audra P (1994) Alpine karst speleogenesis: case studies from France (Vercors, Chartreuse, Ile de Crémieu) and Austria (Tennengebirge). Cave Karst Sci 21(3):75–80

Audra P (2000) Le karst haut alpin du Kanin (Alpes Juliennes, Slovénie-Italie). Karstologia 35(1):27–38

Auly T (2008) Quelques morphologies de rapport karst/glaciaire dans les Pyrénées (France). In: Tyc A, Stefaniak K (ed) Karst and cryokarst. University of Silesia Faculty of Earth Sciences, University of Wrocaw Zoological Institute, Sosnowiec-Wroclaw, Pologne, pp 129–154

Baker VR (1968) Limestone caves in glaciated areas. Nat Speleol Soc Bull 30(2):36–37

Baranowski S (1975) Report on the field work of the Polish scientific expedition to Spitsbergen in 1974. Uniwersytet Wroclawski, 25 p

Barbaroux L, Besset Y (1968) Le karst de Sarsøyra (RiveNord Avatsmarkbre) Vest Spitsbergen. Norois 57:97–103

Baroni C, Pieruccini P, Bini M, Coltorti M, Fantozzi PL, Guidobaldi GI, Nannini D, Ribolini A, Salvatore MC (2015) Geomorphological and neotectonic map of the Apuan Alps (Tuscany, Italy). Geografia Fisica e Dinamica Quaternaria 38(2):201–227

Bathrellos GD, Skilodimou HD, Maroukian H (2015) The significance of tectonism in the glaciations of Greece. Geological Society, London, Special Publications, p 433

Bauer F, Zötl J (1972) Karst of Austria. In: Herak M, Stringfi eld VT (eds) Karst, important Karst of the Northern Hemisphere. Elsevier, Amsterdam, pp 225–265

Bayari S, Özbek O (1995) An inventory of karstic caves in the Taurus Mountain Range (Southern Turkey): preliminary evaluation of geographic and hydrologic features. Cave Karst Sci 21 (3):81–92

Bayari S, Zreda M, Çiner A, Nazik L, Törk K, Özyurt N, Klimchouk A, Sarikaya AM (2003) The extent of Pleistocene ice cap, glacial deposits and glaciokarst in the Aladaglar massif: central Taurids range, southern Turkey. In: XVI INQUA Congress, Paper, 55360

Bayrakdar C, Çilğin Z, Döker MF, Canpolat E (2015) Evidence of an active glacier in the Munzur Mountains, eastern Turkey. Turk J Earth Sci 24(1):56–71
Bennett RA, Hreinsdóttir S, Buble G, Bašić T, Bačić Ž, Marjanović M, Casale B, Gendaszek MA, Cowan D (2008) Eocene to present subduction of southern Adria mantle lithosphere beneath the Dinarides. Geology 36(1):3–6
Benson L, Madole R, Landis G, Gosse J (2005) New data for late Pleistocene Pinedale alpine glaciation from southwestern Colorado. Quatern Sci Rev 24(1):49–65
Bertrand G (1971) Morphostructures cantabriques: Picos de Europa, "Montaña" de León et de Palencia. Revue Geographique des Pyrenees et du Sud-Ouest 42(1):49–70
Biese WB (1956) Uber Karstvorkommen in Chile. Die Höhle 7:91–96
Biese WB (1957) Auf der Marmor-Insel Diego de Almagro (Chile). Natur und Volk 87 (4):123–132
Bini A (1997) Problems and methodologies in the study of the Quaternary de- posits of the southern side of the Alps. Geologia Insubrica 2(2):11–20
Bini A (1998) Rapporti tra evoluzione dei versanti e endocarso: studio dei sedimenti della grotta La Nevera (2693 LO CO) sul M. Generoso (Svizzera, Italia), Il Quaternario
Bini A, Delannoy JJ, Maire R, Quinif Y (1989) Générations de cavités karstiques dans les chaînes alpines. C R Acad Sci Paris 309(2):1183–1190
Bini A, Tognini P, Zuccoli L (1998) Rapport entre karst et glaciers durant les glaciations dans les vallées préalpines du Sud des Alpes. Karstologia 32:7–26 (Fédération française de spéléologie, Paris & Association française de karstologie, Bordeaux)
Birkenmajer K (2008) Karst sink-holes in the Würm Glaciation deposits, subsurface drainage and extent of Triassic limestones in the Sucha Woda Valley, Polish Tatra Mts (West Carpathians). Stud Geol Pol 131:281–289
Blasi C, Di Pietro R, Pelino G (2005) The vegetation of alpine belt karst-tectonic basins in the central Apennines (Italy). Plant Biosyst Int J Dealing Aspects Plant Biol 139(3):357–385
Bobrowsky P, Rutter N (1992) The quaternary geologic history of the Canadian Rocky Mountains. Géog Phys Quatern 46(1):5–50
Bočić N, Faivre S, Kovačić M, Horvatinčić N (2012) Cave development under the influence of Pleistocene glaciation in the Dinarides–an example from Štirovača Ice Cave (Velebit Mt., Croatia). Z Geomorphol 56(4):409–433
Bočić N, Pahernik M, Mihevc A (2015) Geomorphological significance of the palaeodrainage network on a karst plateau: the Una-Korana plateau, Dinaric karst, Croatia. Geomorphology 247:55–65
Bodenhamer HG (2007) Preglacial development of caves at structural duplexes on the Lewis Thrust, Glacier National Park, Montana. J Cave Karst Stud 69(3):326–341
Bögli A (1960) Kalklösung und Karrenbildung. Zeitsch f Geomorph N E 2:4–21
Bognar A, Faivre S (2006) Geomorphological traces of the younger Pleistocene glaciation in the central part of the Velebit Mt. Hrvatski geografski glasnik 68(2):19–30
Bordonau J (1992) Els complexos gla'cio-lacustres relacionats amb el darrer cicle glacial als Pirineus. Geoformaediciones, Logrono
Bosák P, Ford DC, Glazek J, Horácek I (eds) (2015) Paleokarst: a systematic and regional review, vol 1. Elsevier
Bozkurt E, Mittwede SK (2001) Introduction to the geology of Turkey—a synthesis. Int Geol Rev 43(7):578–594
Braun DD (1989) Glacial and periglacial erosion of the Appalachians. Geomorphology 2(1–3):233–256
Butvilovskiy VV (1993) Paleogeografiya poslednego oledeneniya i golotsena Altaya: sobytiyno-katastroficheskaya model (Paleogeography of the last glaciation and the Holocene of Altai: catastrophic event model). Tomsk University, Tomsk, p 253
Calvet M (2004) The Quaternary glaciation of the Pyrenees. In: Quaternary glaciations, extent and chronology, pp 120–128

Calvet M, Delmas M, Gunnell Y, Braucher R, Bourlcs D (2011) Recent advances in research on Quaternary glaciations in the Pyrenees. Quaternary glaciations, extent and chronology, a closer look Part IV. Elsevier, Amsterdam, pp 127–139

Campbell N (1979) Alpine karst of the Scapegoat-Bob Marshall Wilderness and adjoining areas, North-Central Montana. NSS Bulletin 41:66–69

Carulli GB (2006) Carta Geologica del Friuli venezia Giulia, 1:150,000—Servizio Geologica, Firenze

Cendrero A, Saiz de Omenaca J (1979) Geology of the Picos de Europa; a brief outline. Mem Soc Bot Geneve 1:23–29

Chardon M (1984) Le role des héritages quaternaires dans les karsts alpins: le cas des Alpes du Nord. 1er semester Karst des Alpes occidentales. Karstologia 3:12–14

Chardon M (1989) Les karsts de l'avant-pays alpin au N des Alpes occidentales. Karstologia 13:21–30

Chueca J, Peña Monné JL, Lampre F, Julián A (1997) La Pequeña Edad del Hielo en el Pirineo Central y Meridional. Inferencias paleoambientales a partir de datos geomorfológicos. In: Gomez A, Ortiz et Perez A (eds) Las huellas glaciares de las montañas españolas. Universidade de Santiago de Compostela, Spain, pp 307–328

Chueca J, Julián A, Peña-Monné JL (2002) Comparación de la situación de los glaciares del Pirineo español entre el final de la Pequeña Edad del Hielo y la actualidad. Boletín Glaciológico Aragonés 3:13–41

Çılğın Z, Bayrakdar C, Oliphant JS (2014) An example of polygenetic geomorphologic development (Karst-Glacial-Tectonics) on Munzur Mountains: Kepir Cave-Elbaba spring karstic system. J Hum Sci 11(1):89–104

Çiner A, Sarıkaya MA, Yıldırım C (2015) Late Pleistocene piedmont glaciations in the Eastern Mediterranean; insights from cosmogenic ^{36}Cl dating of hummocky moraines in southern Turkey. Quatern Sci Rev 116:44–56

Clague JJ, Ward B (2011) Pleistocene Glaciation of British Columbia. In: Ehlers, Gibbard, Hughes (2011) Developments in quaternary sciences, vol 15. Elsevier, pp 563–573

Cohen SM (2013) Geomorphological studies of a karst system in a permafrost environment at Linnédalen, western Spitsbergen. Master's thesis

Cooper M (2014) Verification of post-glacial speleogenesis and the origins of epigene maze caves in New York. M.Sc., dissertation, Mississippi State University

Cooper MP, Mylroie JE (2015) Glaciation and speleogenesis. Springer

Costantini EA, Fantappié M, L'Abate G (2013) Climate and pedoclimate of Italy. In: The soils of Italy, Springer, pp 19–37

Coutterand S, Buoncristiani JF (2006) Paléogéographie du dernier maximum glaciaire du Pléistocene recent de la region du massif du Mont Blanc, France. Quaternaire 17(1):35–43

Cvijić J (1899) Glacijalne i morfološke studije o planinama Bosne, Hercegovine i Crne Gore. Glas Srpske Kraljevske Akademije Nauka, Beograd, vol 57, 196 p

Cvijić J (1900) L'époque glaciaire dans la péninsule des Balkans. Annales de Géographie 9:359–372

Delannoy JJ (1986) Contribution a l'étude des circulations aquifères dans le géosystème Coulmes-Choranche; Présentation du site experimental de la Grotte de Coufin. Revue de Géographie Alpine 74(1–2):83–92

Delmas M (2009) Chronologie et impact géomorphologique de glaciation quaternaires dans l'est des Pyrénées. Géomorphologie. Université Panthéon-Sorbonne, Paris

Dilek Y (2006) Collision tectonics of the Mediterranean region: causes and consequences. Geol Soc Am Spec Pap 409:1–13

Djurović P (2009) Reconstruction of the pleistocene glaciers of Mount Durmitor in Montenegro. Geografski Zbornik/Acta Geographica Slovenica 49(2):263–289

Djurović P, Petrović A (2007) Large Canyons in Dinaric and Prokletije Mountains Region of Montenegro. Geographica Pannonica 11(1):14–18

Drew DP (1983) Accelerated soil erosion in a karst area: the Burren, western Ireland. J Hydrol 61 (1–3):113–124

Ducić V, Luković J, Burić D, Stanojević G, Mustafić S (2012) Precipitation extremes in the wettest Mediterranean region (Krivošije) and associated atmospheric circulation types. Nat Hazards Earth Syst Sci 12(3):687–697

Dyke AS (2004) An outline of North American deglaciation with emphasis on central and northern Canada. Dev Quat Sci 2:373–424

Dyke AS, Andrews JT, Clark PU, England JH, Miller GH, Shaw J, Veillette JJ (2002) The Laurentide and Innuitian ice sheets during the last glacial maximum. Quatern Sci Rev 21 (1):9–31

Ehlers J, Gibbard PL, Hughes PD (eds) (2011) Quaternary glaciations—extent and chronology: a closer look. Elsevier, 1126 p

Ekmekci M (2003) Review of Turkish karst with emphasis on tectonic and paleogeographic controls. Acta Carsologica 32(2):205–218

Escobar F (1980) Mapa Geologico de Chile, escala 1/1,000,000 (feuille sud). Servicio Nacional de Geologia y Mineria, Departamento de Geologia General, Santiago

Faccenna C, Becker TW, Auer L, Billi A, Boschi L, Brun JP, Capitanio FA, Funiciello F, Horváth F, Jolivet L, Piromallo C, Royden L, Rossetti F, Serpelloni E (2014) Mantle dynamics in the Mediterranean. Rev Geophys 52(3):283–332

Farrant AR, Simms MJ (2011) Ogof Draenen: speleogenesis of a hydrological see-saw from the karst of South Wales. Cave Karst Sci 38(1):31–52

Farrant AR, Smith CJ, Noble SR, Simms MJ, Richards DA (2014) Speleogenetic evidence from Ogof Draenen for a pre-Devensian glaciation in the Brecon Beacons, South Wales, UK. J Quat Sci 29(8):815–826

Faulkner T (2006) The impact of the deglaciation of central Scandinavia on karst caves and the implications for Craven's limestone landscape. Re-thinking Craven's Limestone Landscape, p 4

Finnesand T, Curl RL (2009) Morphology of Tjoarvekrajgge, the longest cave of Scandinavia. In: 15th International congress of speleology proceedings, pp 878–883

Fjeldskaar W, Lindholm C, Dehls JF, Fjeldskaar I (2000) Postglacial uplift, neotectonics and seismicity in Fennoscandia. Quatern Sci Rev 19(14):1413–1422

Fleming K, Johnston P, Zwartz D, Yokoyama Y, Lambeck K, Chappell J (1998) Refining the eustatic sea-level curve since the Last Glacial Maximum using far- and intermediatefield sites. Earth Planet Sci Lett 163(1–4):327–342

Ford DC (1971a) Alpine Karst in the Mt. Castleguard-Columbia Icefield Area, Canadian Rocky Mountains. Arct Alp Res 15(4):239–252

Ford DC (1971b) Alpine Karst in the Mt. Castleguard-Columbia Icefield Area, Canadian Rocky Mountains. Arct Alp Res 239–252

Ford DC (1971c) Characteristics of limestone solution in the southern Rocky Mountains and the Selkirk Mountains, Alberta and British Columbia. Can J Earth Sci 8:585–609

Ford DC (1979) A review of alpine karst in the southern Rocky Mountains of Canada. NSS Bulletion 41:53–65

Ford DC (1983a) Alpine Karst Systems at Crowsnest Pass, Alberta-British Columbia, Canada. J Hydrol 61(1):187–192

Ford DC (1983b) Concluding discussion. Arct Alp Res 61(1):551–554

Ford DC (1983c) The physiography of the Castleguard karst and Columbia icefields area, Alberta, Canada. Arct Alp Res 61(1):427–436

Ford DC, Williams P (2007) Karst hydrogeology and geomorphology. Wiley, West Sussex, England

Ford DC, Smart PL, Ewers RO (1983) The physiography and speleogenesis of Castleguard Cave, Columbia Icefields, Alberta, Canada. Arct Alp Res 61(1):437–450

Forsythe R, Mpodozis C (1983) Geología del basamento pre-jurasico superior en el archipiélago Madre de Dios, Magallanes, Chile. Servicio Nacional de geologia y Mineria, Chile, Boletin, vol 39, 63 p

Gabrielse H (1985) Major dextral transcurrent displacements along the Northern Rocky Mountain Trench and related lineaments in north-central British Columbia. Geol Soc Am Bull 96(1):1–14

Gachev E, Stoyanov K, Gikov A (2016) Small glaciers on the Balkan Peninsula: State and changes in the last several years. Quatern Int 415:33–54
Gadek B, Litwin L (1999) Glaciokarst of subalpine and alpine zone of the Mala Laka Valley, Tatra Mts., Poland. Acta Carsologica 28(1):71–86
García-Ruiz JM, Valero-Garcés BL, Beguería S, López-Moreno JI, Martí-Bono C, Serrano-Muela P, Sanjuan Y (2014) The Ordesa and Monte Perdido National Park, Central Pyrenees. Landscapes and landforms of Spain. Springer, Netherlands, pp 165–172
Garrity CP, Soller DR (2009) Database of the Geologic Map of North America; adapted from the map by Reed JC Jr. and others (2005) U.S. Geological Survey Data Series 424
Gascoyne M, Latham AG, Harmon RS, Ford DC (1983) The antiquity of Castleguard Cave, Columbia Icefields, Alberta, Canada. Arct Alp Res 61(1):463–470
Gillespie A, Molnar P (1995) Asynchronous maximum advances of mountain and continental glaciers. Rev Geophys 33(3):311–364
Giraudi C (2005) Middle to Late Holocene glacial variations, periglacial processes and alluvial sedimentation on the higher Apennine massifs (Italy). Quatern Res 64(2):176–184
Giraudi C (2012) The Campo Felice Late Pleistocene Glaciation (Apennines, central Italy). J Quat Sci 27(4):432–440
Giraudi C (2015) The Upper Pleistocene deglaciation on the Apennines (Peninsular Italy). Cuadernos de investigación geográfica 41(41):337–358
Giraudi C, Giaccio B (2015) Middle Pleistocene glaciations in the Apennines, Italy: new chronological data and preservation of the glacial record. Geol Soc Lond Spec Publ 433 (1):161–178
Giraudi C, Bodrato G, Lucchi MR, Cipriani N, Villa IM, Giaccio B, Zuppi GM (2011) Middle and late Pleistocene glaciations in the Campo Felice Basin (central Apennines, Italy). Quatern Res 75(1):219–230
Godfrey AE (1985) Karst Hydrology of the South Slope of the Uinta Mountains, Utah. Geology and Energy Resources, Uinta Basin of Utah, pp 277–294
Goldie HS (2006) Mature intermediate-scale surface karst landforms in NW England and their relations to glacial erosion. In: Kiss A, Mezősi G, Sümeghy Z (2006) Landscape, environment and society, pp 225–238
Gremaud V, Goldscheider N (2010) Geometry and drainage of a retreating glacier overlying and recharging a karst aquifer, Tsanfleuron-Sanetsch, Swiss Alps. Acta carsologica 39(2):289–300
Gruber P, Gy Kovács, Sz Somlai (1998) Vertikális karsztformák vizsgálata az ausztriai Totes-Gebirgében. Karsztfejlődés 2:201–210
Grund A (1910) Beiträge zur Geomorphologie des Dinarischen Gebirges. Geographische Abhandlungen 7(H3):121–125
Grunewald K, Scheithauer J (2010) Europe's southernmost glaciers: response and adaptation to climate change. J Glaciol 56(195):129–142
Gunn J (2004) Encyclopedia of caves and karst science. Fitzroy Dearborn, New York
Gutak JM (2002) Evolyutsiya devonskikh basseynov sedimentatsii v Gornom Altae, Geologiya Devonskoy sistemy (Evolution of the Devonian sedimentation basins in the Altai Mountains Geology of the Devonian system). Materialy Mezhdunarodnogo simpoziuma, Syktyvkar, Respublika Komi, Rossiya, 9–12 iyulya 2002 g.: pp 18–20
Gutak JM (2015a) Evolyutsiya obstanovok sedimentogeneza v zapadnoy chasti Altae-Sayanskoy skladchatoy oblasti (ASSO) v pozdnem dokembrii-paleozoe, Evolyutsiya osadochnykh protsessov v istorii Zemli (Evolution of sediment genesis environments of in the western part of the Altai-Sayan folded area (ASFA) in the Late Precambrian-Paleozoic. Evolution of sedimentary processes in the Earth's history). Materialy 8-go Vserossiyskogo litologicheskogo soveshchaniya, Moskva 27–30 oktyabrya 2015 g., Moscow, Gubkin University) vol 1, pp 85–87
Gutak JM (2015b) Paleozoyskie rify zapadnoy chasti Altae-Sayanskoy skladchatoy oblasti (paleotektonicheskie obstanovki obrazovaniya), Geologiya rifov (Paleozoic reefs of the western part of the Altai-Sayan folded area (paleotectonic environment of formation) Geology

of reefs). Materialy Vserossiyskogo litologicheskogo soveshchaniya, Syktyvkar, Respublika Komi, Rossiya 15–17 iyunya 2015, pp 38–39
Gutak JM, Batyaeva SK, Lyakhnitskiy VN, Fedak SI (2001) Yurskie otlozheniya Gornogo Altaya, Aktual'nye voprosy geologii i mineragenii yuga Sibiri (Jurassic deposits of the Altai Mountains Actual problems of geology and mineralogy of southern Siberia, Novosibirsk, pp 49–57
Gutak JM, Bagmet GN, Valieva FL, Fedak SI (2004) Dokembriyskie otlozheniya basseyna r. Eskongo (Gornyy Altay), Priroda i ekonomika Kuzbassa (Precambrian sediments of the river Eskongo basin (the Altai Mountains) Nature and the economy of Kuzbass). Novokuznetsk, Geologiya i paleontologiya 9:8–13
Gutak JM, Antonova VA, Bagmet GN, Gabova MF, Savitskiy VR, Tolokonnikova ZA (2008) Ocherki po istoricheskoy geologii Kemerovskoy oblasti (Essays on historical geology of the Kemerovo region). Novokuznetsk, KuzSPA, 132 p
Gvozdetskij NA (1981) Karst (Karst). Mysl', Moscow (214 p)
Gvozdetskij NA, Golubtchikov YN (1987) Gory (Mountains). Mysl', Moscow (399 p)
Hallet B (1976) Deposits formed by subglacial precipitation of $CaCO_3$. Geol Soc Am Bull 87 (7):1003–1015
Harmon RS (1979) U-series dating of speleothems and a glacial chronology for western North America. NSS Bulletin 41:102–104
Harmon RS, Ford DC, Schwarcz HP (1977) Interglacial chronology of the Rocky and Mackenzie Mountains based upon ^{230}Th–^{234}U dating of calcite speleothems. Can J Earth Sci 14(11): 2543–2552
Hercman H, Bella P, Głazek J, Gradziński M, Lauritzen SE, Løvlie R (1997) Uranium-series dating of speleothems from Demanova Ice Cave: a step to age estimation of the Demanova Cave System (the Nizke Tatry Mts., Slovakia). In: Annales Societatis Geologorum Poloniae, vol 67, no 4, pp 439–450
Hinnov LA (2003) Lofer Cyclothems in the Dachstein Limestone of the Julian Alps. Geol Soc Am Abstr Programs 35(6):426
Horn G (1935) Űber die Bildung von Karsthöhlen unter einem Gletcher. Nor Geogr Tidsskr 5:494–498
Hughes P (2007) Recent behaviour of the Debeli Namet glacier, Durmitor, Montenegro. Earth Surf Process Land 32(10):1593–1602
Hughes P (2008) Response of a Montenegro glacier to extreme summer heatwaves in 2003 and 2007. Geogr Ann: Ser A, Phys Geogr 90(4):259–267
Hughes PD, Woodward JC (2017) Quaternary Glaciation in the Mediterranean Mountains: a new synthesis. Geol Soc Lond Spec Publ 433:1–23
Hughes PD, Gibbard PL, Woodward JC (2003) Relict rock glaciers as indicators of Mediterranean palaeoclimate during the Last Glacial Maximum(Late Wuermian) in northwest Greece. J Quat Sci 18(5):431–440
Hughes PD, Woodward JC, Gibbard PL, Macklin MG, Gilmour MA, Smith GR (2006) The glacial history of the Pindus Mountains, Greece. J Geol 114(4):413–434
Hughes PD, Gibbard PL, Woodward JC (2007) Geological controls on Pleistocene glaciation and Cirque form in Greece. Geomorphology 88(3):242–253
Hughes PD, Woodward JC, Van Calsteren PC, Thomas LE, Adamson KR (2010) Pleistocene ice caps on the coastal mountains of the Adriatic Sea. Quatern Sci Rev 29(27):3690–3708
Hughes PD, Woodward JC, Van Calsteren PC, Thomas LE (2011) The glacial history of the Dinaric Alps, Montenegro. Quatern Sci Rev 30(23):3393–3412
Hulton N, Sugden D, Payne A, Clapperton C (1994) Glacier modeling and the climate of Patagonia during the last glacial maximum. Quatern Res 42(1):1–19
Isotta FA, Frei C, Weilguni V, Perčec Tadić M, Lassègues P, Rudolf B, Pavan V, Cacciamani C, Antolini G, Ratto SM, Maraldo L, Micheletti S, Bonati V, Lussana C, Ronchi C, Panettieri E, Marigo G, Vertačnik G (2013) The climate of daily precipitation in the Alps: development and analysis of a high-resolution grid dataset from pan-Alpine rain-gauge data. Int J Climatol 34 (5):1657–1675

Ivy-Ochs S, Kerschner H, Reuther A, Preusser F, Heine K, Maisch M, Kubik PW, Schlüchter C (2008) Chronology of the last glacial cycle in the European Alps. J Quat Sci 23(6–7):559–573

Jaillet S, Lans B, Maire R, Tourte B, L'équipe Ultima Patagonia-2006 (2008) Héritage glaciaire et karstification de l'archipel calcaire de Madre de Dios - Patagonie, Chili. Actes colloque AFK, Sion, Edytem, Cahiers de géographie, vol 7, pp 39–50

Jin S, Tian X, Feng G (2016) Recent glacier changes in the Tien Shan observed by satellite gravity measurements. Glob Planet Change 143:81–87

Jurewicz E (2005) Geodynamic evolution of the Tatra Mts. and the Pieniny Klippen Belt (Western Carpathians): problems and comments. Acta Geol Pol 55(3):295–338

Kalesnik SV (1935) Ledniki verkhovev Bol'shogo Naryna, Tjan' Shan' (Glaciers of the upper Big Naryn, Tian Shan). Trudy lednokovykh epspeditsij 2:83–186

Karpunin AM, Mamonov SV, Mironenko OA, Sokolov AR (1998) Geologicheskie pamyatniki prirody Rossii: K 300-letiyu gorno-geologicheskoy. sluzhby Rossii, 1700–2000 (Russian geological sites. 300th anniversary of Russian Mining and Geological Service, 1700–2000). S-Petersburg, 200 p

Khromova T, Nosenko G, Kutuzov S, Muraviev A, Chernova L (2014) Glacier area changes in Northern Eurasia. Environ Res Lett 9:015003

King PB (2015) Evolution of North America. Princeton University Press

Klimchouk A, Nazik L, Bayari S, Tork K, Kasjan Y (2004) Kuzgun Cave and its Context: the first super-deep cave in the Aladaglar Massif, Turkey

Klimchouk A, Bayari S, Nazik L, Törk K (2006) Glacial destruction of cave systems in high mountains, with a special reference to the Aladaglar massif, Central Taurus, Turkey. Acta Carsologica 35(2):111–121

Klimchouk AB, Samokhin GV, Kasian Y (2009) The deepest cave in the world in the Arabika massif (Western Caucasus) and its hydrogeological and paleogeographic significance. In: Proceedings of 15th international congress of Kerrville, Texas, pp 898–905

Kostin PA (1966) Karst Peredovogo khrebta i polosy kuest Severo-Zapadnogo Kavkaza (Karst of the Peredovoj Range of the stripe of cuestas of the Northwestern Caucasus). Moscow, 25 p

Kotarba A, Hercman H, Dramis F (2001) On the age of Campo Imperatore glaciations, Gran Sasso Massif, Central Italy. Geogr Fis Dinam Quat 24:65–69

Kranjc A (2006) Some large dolines in the Dinaric karst. Speleogenesis Evol Karst Aquifers 4 (1):1–4

Krasnaya kniga Altayskogo kraya (2009) Osobo okhranyaemye prirodnye territorii (The Red Book of the Altai Territory. Protected areas), Barnaul (273 p)

Křížek M, Mida P (2013) The influence of aspect and altitude on the size, shape and spatial distribution of glacial cirques in the High Tatras (Slovakia, Poland). Geomorphology 198:57–68

Kuhlemann J, Milivojevic M, Krumrei I, Kubik PW (2009) Last glaciation of the Šara range (Balkan peninsula): increasing dryness from the LGM to the Holocene. Austrian J Earth Sci 102(1):146–158

Kuhlemann J, Gachev E, Gikov A, Nedkov S, Krumrei I, Kubik P (2013) Glaciation in the Rila Mountains (Bulgaria) during the last glacial maximum. Quatern Int 293:51–62

Kunaver J (1983) Geomorphology of the Kanin Mountains with special regard to the glaciokarst. Geografski zbornik 22:201–344

Kurter A (1991) Glaciers of Turkey. Satellite image atlas of glaciers of the world, US Geological Survey Washington, G1-G30

Lambán LJ, Jódar J, Custodio E, Soler A, Sapriza G, Soto R (2015) Isotopic and hydrogeochemical characterization of high-altitude karst aquifers in complex geological settings. The Ordesa and Monte Perdido National Park (Northern Spain) case study. Sci Total Environ 506:466–479

Lambeck K (1995) Late Pleistocene and Holocene sea-level change in Greece and south-western Turkey: a separation of eustatic, isostatic and tectonic contributions. Geophys J Int 122 (3):1022–1044

Lauritzen SE (1984) A symposium: arctic and alpine karst. Nor Geogr Tidsskr-Norw J Geogr 38 (3–4):139–143
Lauritzen SE (1986) Kvithola at Fauske; Northern Norway: an example of ice-contact speleogenesis. Nor Geol Tidsskr 66(2):153–161
Lauritzen SE (2005) Quaternary speleogenesis and landscape evolution in Scandinavia and Svalbard. In: 14th International congressof speleology, Athen, vol O-64, pp 1–4
Lauritzen SE (2006) Caves and speleogenesis at Blomstrandsøya, Kongsfjord, W. Spitsbergen. Int J Speleol 35(1):37–58
Lauritzen SE, Gascoyne M (1980) The first radiometric dating of Norwegian stalagmites–Evidence of pre-Weichselian karst caves. Nor Geogr Tidsskr 34:77–82
Lauritzen SE, Mylroie JE (2000) Results of a speleothem U/Th dating reconnaissance from the Helderberg Plateau, New York. J Cave Karst Stud 62(1):20–26
Lauritzen SE, Skoglund RØ (2013) Glacier ice-contact speleogenesis. Treatise Geomorphol 6:363–396
Leonard EM (1989) Climatic change in the Colorado Rocky Mountains: estimates based on modern climate at late Pleistocene equilibrium lines. Arct Alp Res 245–255
Lepirica A (2008) Geomorphological characteristics of the massif Prenj. Acta Carsologica 37(2–3):307–329
Lindner L, Dzierżek J, Marciniak B, Nitychoruk J (2003) Outline of Quaternary glaciations in the Tatra Mountains: their development, age and limits. Geol Q 47(3):269–280
Lindner L, Dzierżek J, Marciniak B, Nitychoruk J (2010) Outline of Quaternary glaciations in the Tatra Mts.: their development, age and limits. Geol Q 47(3):269–280
Litwin L, Andreychouk V (2008) Characteristics of high-mountain karst based on GIS and remote sensing. Environ Geol 54:979–994
Lóczy D, Stankoviansky M, Kotarba A (eds) (2012) Recent landform evolution: the Carpatho-Balkan-Dinaric region. Springer
Lozovoj SP (1984) Lagonakskoe nagor'e (Lagonaki Highland). Krasnodarskoe knizhnoe izdatel'stvo, Krasnodar, p 160
Maire R (1990) La haute montagne calcaire: karsts, cavités, remplissages, paléoclimats, Quaternaire. Thèse d'Etat, Univ. de Nice, Karstologia-mémoires 3, La Ravoire, 731 p
Maire R (1999) Les glaciers de marbre de Patagonie, Chili. Un karst subpolaire océanique de la zone australe. Karstologia 33:25–40
Makos M, Nitychoruk J, Zreda M (2013) Deglaciation chronology and paleoclimate of the Pięciu Stawów Polskich/Roztoki Valley, high Tatra Mountains, Western Carpathians, since the Last Glacial Maximum, inferred from ^{36}Cl exposure dating and glacier–climate modelling. Quatern Int 293:63–78
Makos M, Dzierżek J, Nitychoruk J, Zreda M (2014) Timing of glacier advances and climate in the High Tatra Mountains (Western Carpathians) during the Last Glacial Maximum. Quatern Res 82(1):1–13
Mangerud J, Gyllencreutz R, Lohne Ö, Svendsen JI (2011) Glacial history of Norway. In: Ehlers J, Gibbard PL, Hughes PD (2011) Quaternary glaciations-extent and chronology: a closer look. Elsevier
Marjanac L, Marjanac T (2004) Glacial history of the Croatian Adriatic and coastal Dinarides. In: Ehlers J, Gibbard PL (eds) Quaternary glaciations. Developments in quaternary science, vol 2, pp 19–26
Marjanac T, Marjanac L (2016) The extent of middle Pleistocene ice cap in the coastal Dinaric Mountains of Croatia. Quatern Res 85(3):445–455
Martí Bono C, García-Ruiz J (eds) (1994) El Glaciarismo surpirenaico: nuevas aportaciones. Geoforma Ediciones, Logroño, p 142
Martinetto E, Ravazzi C (1997) Plant biochronology of the Valle della Fornace succession (Varese) based on the Plio-Pleistocene record in northern Italy. Geologia Insubrica 2(2):81–98
Maslyn RM, Davis DG (1979) Karst development on the White River Plateau, Colorado. NSS Bulletin 41:95–101

Mavlyudov BR (2004) Lednikovyj karst (Glacial karst). In: Karstovedenie – XXI vek: teoretitcheskoe i praktitcheskope znatchenie, 25–30 May 2004. Perm', pp 69–74

McMillan ME, Heller PL, Wing SL (2006) History and causes of post-Laramide relief in the Rocky Mountain orogenic plateau. Geol Soc Am Bull 118(3–4):393–405

Medville DM, Hempel JC, Plantz C, Werner E (1979) Solutional landforms on carbonates of the Southern Teton Range Wyoming. NSS Bulletin 41:70–79

Menkovic L, Markovic M, Cupkovic T, Pavlovic R, Trivic B, Banjac N (2004) Glacial morphology of Serbia, with comments on the Pleistocene Glaciation of Monte Negro, Macedonia and Albania. Dev Quat Sci 2:379–384

Mercer JH (1976) Glacial history of southern most South America. Quatern Res 6:125–166

Merritt JW, Auton CA, Connell ER, Hall AM, Peacock JD (2003) Cainozoic geology and landscape evolution of north-east Scotland. British Geological Survey, Edinburgh

Mihevc A, Prelovšek M, Hajna NZ (eds.) (2010) Introduction to the Dinaric karst. Inštitut za raziskovanje krasa ZRC SAZU, 71 p

Milivojević M (2007) Glacial relief of Mts. Volujak, Bioć and Maglić. Geogr Inst Jovan Cvijić Spec Issue 68:1–132

Milivojević M, Menković L, Ćalić J (2008) Pleistocene glacial relief of the central part of Mt. Prokletije (Albanian Alps). Quatern Int 190(1):112–122

Miotke FD (1968) Karstmorphologische Studien in der glacial-überformten Höhenstufe der Picos de Europa, Nordspanien. Selbtverlag der Geografhischen Gessesllschaft Hannover, p 4

Moles NR, Moles RT (2002) Influence of geology, glacial processes and land use on soil composition and Quaternary landscape evolution in The Burren National Park, Ireland. Catena 47(4):291–321

Montserrat-Martí JM (1992) Evolución glaciar y postglaciar del clima y la vegetación en la vertiente sur del Pirineo: estudio palinológico. Monografías del Instituto Pirenaico de Ecología 6:1–147

Munroe JS (2006) Investigating the spatial distribution of summit flats in the Uinta Mountains of northeastern Utah, USA. Geomorphology 75(3):437–449

Murphy P, Westerman AR, Clark R, Booth A, Parr A (2008) Enhancing understanding of breakdown and collapse in the Yorkshire Dales using ground penetrating radar on cave sediments. Eng Geol 99(3):160–168

Murphy PJ, Faulkner TL, Lord TC, Thorp JA (2015) The caves of Giggleswick Scar-examples of deglacial speleogenesis? Cave and Karst Science 42(1):42–53

Orvošová M, Deininger M, Milovský R (2014) Permafrost occurrence during the Last Permafrost Maximum in the Western Carpathian Mountains of Slovakia as inferred from cryogenic cave carbonate. Boreas 43(3):750–758

Oxaal L (1914) Kalkstenshuler i Ranen. Norg Geol Unders 69:1–47

Øystese A, Johannesen F, Lauritzen SE (2005) Speleogenesis and landscape evolution in Tromsdalen karst, Verdal, Norway. In: 14th international congressof speleology, Athen, vol P-14, pp 1–2

Pallás R, Rodés A, Braucher R, Carcaillet J, Ortuno M, Bordonau J, Bourlcs D, Vilaplana JM, Masana E, Santanach P (2006) Late Pleistocene and Holocene glaciation in the Pyrenees: a critical review and new evidence from ^{10}Be exposure ages, south-central Pyrenees. Quatern Sci Rev 25(21–22):2937–2963

Palmer AN (1991) Origin and morphology of limestone caves. Geol Soc Am Bull 103(1):1–21

Palmer AN (2003) Speleogenesis in carbonate rocks. Speleogenesis Evol Karst Aquifers 1(1):2–11

Pamić J, Gušić I, Jelaska V (1998) Geodynamic evolution of the Central Dinarides. Tectonophysics 297(1):251–268

Panov DV (1993) Evoljutsija sovremennogo oledenenija Kavkaza (Evolution of the modern glaciation of the Caucasus). Sankt-Peterburg, Gidrometeoizdat, 432 pp. (in Russian)

Penck A (1885) La Période glaciaire dans les Pyrénées. Bulletin de la Societe d'histoire naturelle de Toulouse 19:105–200

Penck A, Brückner E (1901–1909) Die Alpen im Eiszeitalter, Christian-Herman Tauchnitz, Leipzig 1199 p

Petrović AS (2014) A Reconstruction of the Pleistocene Glacial Maximum in the Žijovo Range (Prokletije Mountains, Montenegro). Geografski Zbornik/Acta Geographica Slovenica 54(2)
Piccini L, Zanchetta G, Drysdale RN, Hellstrom J, Isola I, Fallick AE, Leone G, Doveri M, Mussi M, Mantelli F, Molli G, Lotti L, Roncioni A, Regattieri E, Meccheri M, Vaselli L (2008) The environmental features of the Monte Corchia cave system (Apuan Alps, central Italy) and their effects on speleothem growth. Int J Speleol 37(3):153–172
Pierce KL (2003) Pleistocene glaciations of the Rocky Mountains. Dev Quat Sci 1:63–76
Pope RJ, Hughes PD, Skourtsos E (2015) Glacial history of Mt Chelmos, Peloponnesus, Greece. Geol Soc Lond Spec Publ 433(1):211–236
Pulina M (1974) Preliminary studies on denudation in SW Spitsbergen. Bulletin de l'Academie Polonaise des Sciences-Serie des Sciences de la Terre 22(2):83–89
Quinif Y, Maire R (1998) Pleistocene deposits in Pierre Saint-Martin cave, French Pyrenees. Quatern Res 49:37–50
Ridge JC (2004) The Quaternary glaciation of western New England with correlations to surrounding areas. Dev Quat Sci 2:169–199
Ridush BT (1993) Petschera Syjkyrduu na Vostotchnom Pamire (The Syjkyrduu Cave in the Eastern Pamir). Svet 1–2:5–9
Rodríguez-Rodríguez L, Jiménez-Sánchez M, Domínguez-Cuesta MJ, Aranburu A (2014) Research history on glacial geomorphology and geochronology of the Cantabrian Mountains, North Iberia (43–42°N/7–2°W). Quatern Int 364:6–21
Ruban DA (2013) The Greater Caucasus—a Galatian or Hanseatic terrane? Comment on "The formation of Pangea" by G.M. Stampfli, C. Hochard, C. Verard, C. Wilhem and J. von Raumer (Tectonophysics 593 (2013) 1-19). Tectonophysics 608:1442–1444
Ruban DA, Al-Husseini MI, Iwasaki Y (2007) Review of Middle east Paleozoic plate tectonics. GeoArabia 12:35–56
Rudoy AN (2005) Gigantskaya ryab' techeniya, istoriya issledovaniy, diagnostika, paleo-geograficheskoe znachenie (Giant stream ripple, history of research, diagnostics, paleogeographic value). Tomsk, 224 p
Salvigsen O, Elgersma A (1985) Large-scale karst features and open taliks at Vardeborgsletta, outer Isfjorden, Svalbard. Polar Res 3(2):145–153
Sarıkaya MA, Zreda M, Çiner A, Zweck C (2008) Cold and wet Last Glacial Maximum on Mount Sandıras, SW Turkey, inferred from cosmogenic dating and glacier modeling. Quatern Sci Rev 27(7):769–780
Sarıkaya MA, Ciner A, Zreda M (2011) Quaternary glaciations of Turkey. Dev Quat Sci 15:393–403
Sarıkaya MA, Çiner A, Haybat H, Zreda M (2014) An early advance of glaciers on Mount Akdağ, SW Turkey, before the global Last Glacial Maximum; insights from cosmogenic nuclides and glacier modeling. Quatern Sci Rev 88:96–109
Schoenbohm LM, Chen J, Stutz J, Sobel ER, Thiede RC, Kirby B, Strecker MR (2014) Glacial morphology in the Chinese Pamir: Connections among climate, erosion, topography, lithology and exhumation. Geomorphology 221:1–17
Schroeder J, Ford DC (1983) Clastic sediments in Castleguard Cave, Columbia Icefields, Alberta, Canada. Arct Alp Res 61(1):451–461
Seefeldner E (1961) Salzburg und seine Landschaften. Eine geographische Landeskunde. Salzburg/Stuttgart, Das Bergland-Buch, 573 p
Sejrup HP, Hjelstuen BO, Dahlgren KT, Haflidason H, Kuijpers A, Nygård A, Praeg D, Stoker MS, Vorren TO (2005) Pleistocene glacial history of the NW European continental margin. Mar Pet Geol 22(9):1111–1129
Serrano E, González-Trueba JJ, González-García M (2012) Mountain glaciation and paleoclimate reconstruction in the Picos de Europa (Iberian Peninsula, SW Europe). Quatern Res 78:303–314
Serrano E, Gomez-Lende M, Gonzalez-Trueba JJ, Turu V, Ros S (2013) Fluctuaciones glaciares pleistocenas en las Monta ~ nas Pasiegas (Cordillera Cantabrica). Cuaternario y Geomorfología 27:91–110

Skoglund RØ, Lauritzen SE (2010) Morphology and speleogenesis of Okshola (Fauske, northern Norway): example of a multi-stage network cave in a glacial landscape. Norw J Geol 90:123–137

Skoglund RØ, Lauritzen SE (2011) Subglacial maze origin in low-dip marble stripe karst: examples from Norway. J Cave Karst Stud 73(1):31–43

Skoglund RØ, Lauritzen SE (2013) Characterisation of a post-glacial invasion aquifer: the Grønli-Seter karst system, northern Norway. Norw J Geol 93(1):61–73

Skoglund RØ, Lauritzen SE, Gabrovšek F (2010) The impact of glacier ice-contact and subglacial hydrochemistry on evolution of maze caves: a modelling approach. J Hydrol 388(1):157–172

Smart CC (1983) The hydrology of the castleguard karst, columbia icefields, Alberta, Canada. Arct Alp Res 61(1):471–486

Smart PL (1986) Origin and development of glaciokarst closed depressions in the Picos de Europa, Spain. Z Geomorphol 30:423–443

Smith GW, Nance RD, Genes AN (2006) Pleistocene glacial history of Mount Olympus, Greece: Neotectonic uplift, equilibrium line elevations, and implications for climatic change. Geol Soc Am Spec Pap 409:157–174

Solomina O, Bushueva I, Dolgova E, Jomelli V, Alexandrin M, Mikhalenko V, Matskovsky V (2016) Glacier variations in the Northern Caucasus compared to climatic reconstructions over the past millennium. Glob Planet Change 140:28–58

Spangler LE (2001) Delineation of recharge areas for karst springs in Logan Canyon, Bear River Range, northern Utah. In: US Geological Survey Karst Interest Group Proceedings. Water-Resources Investigations Report, 01-4011

Stea RR (2004) The Appalachian glacier complex in maritime Canada. Dev Quat Sci 2:213–232

Stepišnik U, Žebre M (2011) Glaciokras Lovčena, E-Geograff 2. Univerza v Ljubljani, Filozofska faklulteta, Ljubljana

Stepišnik U, Ferk M, Kodelja B, Medenjak G, Mihevc A, Natek K, Žebre M (2009) Glaciokarst of western Orjen, Montenegro. Cave Karst Sci 36(1):21–28

Stepišnik U, Grlj A, Radoš D, Žebre M. (2016) Geomorphology of Blidinje, Dinaric Alps (Bosnia and Herzegovina). J Maps 12(sup1):163–171

Stokes CR, Gurney SD, Shahgedanova M, Popovnin V (2006) Late-20th-century changes in glacier extent in the Caucasus Mountains, Russia/Georgia. J Glaciol 52:99–109

Sturchio NC, Pierce KL, Murrell MT, Sorey ML (1994) Uranium-series ages of travertines and timing of the last glaciation in the northern Yellowstone area, Wyoming-Montana. Quatern Res 41(3):265–277

Styllas MN, Schimmelpfennig I, Ghilardi M, Benedetti L (2015) Geomorphologic and paleoclimatic evidence of Holocene glaciation on Mount Olympus, Greece. The Holocene 26(5):709–721

Sugden DE, Bentley MJ, Fogwill CJ, Hulton NRJ, McCulloch RD, Purves RS (2005) Late-glacial events in Southernmost South America: a blend of "Northern" and "Southern" hemispheric climatic signals? Geogr Ann Ser A Phys Geogr 87(2):273–288

Svendsen JI, Mangerud J (1987) Late Weichselian and Holocene sea-level history for a cross-section of western Norway. J Quat Sci 2(2):113–132

Szczygieł J (2015) Cave development in an uplifting fold-and-thrust belt: case study of the Tatra Mountains, Poland. Int J Speleol 44(3):341–359

Szczygieł J, Gaidzik K, Kicińska D (2015) Tectonic control of cave development: a case study of the Bystra Valley in the Tatra Mts., Poland. Ann Soc Geol Pol 85(2):387–404

Szulc-Rojan E (1995) The contemporary glaciers of Pamirs. Czasopismo Geograficzne 66:303–315

Taillefer F (1963) La carte de Morphologie glaciaire des Pyrénées au 1/50,000, Feuilles de Foix et de Vicdessos. Revue Géographique des Pyrénées et du Sud-Ouest 34:5–10

Taillefer F (1964) Glaciaire pyrénéen: versant nord et versant sud. R.G.P.S.O. 28(3):221–243

Taillefer F (1969) Les Glaciations des Pyrénnées. In: Actes 8ème congrès international INQUA, Supplément du Bulletin de l'Association Française pour l'Etude du Quaternaire, pp 19–32

Taillefer F (1977) Le glacier de l'Ariège dans le bassin de Tarascon. Revue Géographique des Pyrénées et du SudOuest 48:269–286
Taillefer F (1985) Idées actuelles sur les glaciations dans les Pyrénées de l'Ariège. Revue Géographique des Pyrénées et du Sud-Ouest 56:323–338
Talour B (1976) Hydrogéologie karstique du Massif du Grand Som (Chartreuse, Isere) – Grenoble. Institut de Géologie, These de 3eme cycle, 166 p
Tawadros E, Ruban D, Efendiyeva M (2006) Evolution of NE Africa and the Greater Caucasus: common patterns and petroleum potential. In: The Canadian Society of Petroleum Geologists, the Canadian Society of Exploration Geophysicists, the Canadian Well Logging Society Joint Convention, 15–18 May 2006. Calgary, pp 531–538
Telbisz T (2010a) Morphology and GIS-analysis of closed depressions in Sinjajevina Mts (Montenegro). Karst Dev 1(1):41–47
Telbisz T (2010b) Glacio-karst features of the Sinjajevina Mts (Montenegro): an overview and DEM-analysis. Karst Dev 1(1):17–22
Telbisz T, Dragušica H, Nagy B (2009) Doline Morphometric Analysis and Karst Morphology of Biokovo Mt (Croatia) Based on Field Observations and Digital Terrain Analysis. Croat Geogr Bull 71(2):5–22
Telbisz T, Mari L, Szabó L (2011) Geomorphological characteristics of the Italian side of Canin massif (Julian Alps) using digital terrain analysis and field observations. Acta Carsologica 40 (2):255–266
Tóth G (2003) Karrenmorphologische Forschungen im Dachstein und im Toten-Gebirge. Gmundner Geo-Studien 2, Beiträge zur Geologie des Salzkammerguts, pp 191–198
Tóth G (2004) Karrenmorphologische Forschungen im Dachstein und im Toten-Gebirge. In: Weidinger JT, Lobitzer H, Spitzbart I. Beiträge zur Geologie des Salzkammerguts, Gmundner Geo-Studien 2. Erkudok-Institut; Museum Gmunden, Gmunden, pp 191–198
Tóth G (2008) Une nouvelle approche du systeme des lapiés alpins nus. Karsts de montagne. Géomorphologie, patrimoine et ressource. Chambéry, Université de Savoie, pp 147–156
Tóth G (2009) Karren features in the Dachstein mountain. In: Ginés A, Knez M, Slabe T, Dreybrodt W (eds) Karst rock features: Karren sculpturing. Postojna, Ljubljana, Zalozba ZRC, pp 313–322
Tóth G, Reynard E (2011) Developpement et spécificités des cellules lapiazées sur Lapiaz de Tsanfleuron (Alpes Bernoises, Suisse). Z Geomorphol 55(2):231–245
Tronov MV (1948) Ocherki oledeneniya Altaya (Glaciers and climate). Gidrometeoizdat, Leningrad (407 p)
Tronov MV (1966) Ledniki i klimat (Essays on Altai glaciation). Geografiz, Moscow (376 p)
Turkin YA, Fedak SI (2008) Geologiya i veshchestvennye kompleksy Gornogo Altaya (Geology and compositional complexes of the Altai Mountains). Tomsk, 460 p
Uggeri A, Felber M, Bini A, Bignasca C, Ravazzi C (1994) La successione della Val Fornace, I depositi Plio-Quaternari e l'evoluzione del territorio varesino, Milano, pp 63–92
Uggeri A, Felber M, Bini A, Ravazzi C, Bignasca C, Heller F (1995) Pliocene-Pleistocene environmental evolution in the Varese region (NW Lombardia, Northern Italy): evidence of a Pliocene glaciation. In: Abstracts XIV international congress INQUA, Berlin, Terra Nostra, 2/95, 280 p
Urdea P (2004) The Pleistocene glaciation of the Romanian Carpathians. Dev Quat Sci 2:301–308
van der Meer JJ, van Tatenhove FG (1992) Drumlins in a full alpine setting: some examples from Switzerland. Geomorphology 6(1):59–67
Van Husen D (2000) Geological processes during the Quaternary. Mitt Österr Geol Ges 92:135–156
Velić J, Velić I, Kljajo D (2011) Sedimentary bodies, forms and occurrences in the Tudorevo and Mirovo glacial deposits of northern Velebit (Croatia). Geologia Croatica 64(1):1–16
Veress M (2010) Karst environments: Karren formation in high mountains. Springer, Netherlands, p 230
Veress M (2016a) Covered karsts. Springer, Dordrecht, p 536

Veress M (2016b) Postglacial evolution of paleodepressions in glaciokarst areas of the Alps and Dinarides. Z Geomorphol 60(4):343–358

Veress M (2017) Solution doline development on glaciokarst in alpine and dinaric areas. Earth-Sci Rev 173:31–48

Veress M, Tóth G, Zentai Z, Czöpek I (2003) Vitesse de recul d'un escarpement lapiazé (Ile Diego de Almagra, Patagonia, Chili). Karstologia 41(1):23–26

Veress M, Szunyogh G, Zentai Z, Tóth G, Czöpek I (2006) The effect of wind on karren formation on the Island of Diego de Almagra (Chile). Z Geomorphol 50(4):425–445

Veress M, Zentai Z, Péntek K, Döbröntei L (2014) A Léna pillérei. Földrajzi Közlemények 138 (1):17–36

Vilaplana JM, Montserrat J, Schlüchter C (1989) Recent progress in Quaternary stratigraphy: the Lake Llauset séquence in the Spanish Pyrenees. In: Rose J, Schlüchter C (eds) Quaternary type sections: imagination or reality?. Balkema, Rotterdam, pp 113–124

Viles HA (2003) Conceptual modelling of the impacts of climate change on karst geomorphology in the UK and Ireland. J Nat Conserv 11(1):59–66

Vlahović I, Tišljar J, Velić I, Matičec D (2005) Evolution of the Adriatic carbonate platform: palaeogeography, main events and depositional dynamics. Palaeogeogr Palaeoclimatol Palaeoecol 220(3):333–360

Waltham AC (1978) The caves and karst of Astraka, Greece. Trans Br Cave Res Assoc 5:1–12

Waltham T, Lowe D (eds) (2013) Caves and Karst of the Yorkshire Dales, vol 1. British Cave Research Association, Buxton, 264 p

Waltham AC, Simms MJ, Farrant AR, Goldie HS (1997) Karst and caves of Great Britain. In: Geological conservation review, vol 12, 358 p

Washburn AL (1979) A survey of periglacial processes and environments. Geocryology, 406 p

Weary DJ (2008) Preliminary Map of potentially karstic carbonate rocks in the central and southern Appalachian states, No. 2008-1154, Geological Survey (US)

Weingartner H (1983) Geomorphologische Studien im Tennengebirge. Im Selbstverlag des Institutes für Geographie der Universität Salzburg, 196 p

Weremeichik JM, Mylroie JE (2014) Glacial Lake Schoharie: an investigative study of glaciolacustrine lithofacies in caves, Helderberg Plateau, Central New York. J Cave Karst Stud 76(2):127–138

Werner E (1979) Alpine karst in the Rocky Mountains—introduction to the symposium. NSS Bulletin 41:51–52

White WB (1979) Karst landforms in the Wasatch and Uinta Mountains, Utah. NSS Bulletin 41:80–88

Williams PW (1966) Limestone pavements with special reference to western Ireland. Trans Inst Br Geogr 155–172

Wilson JR (1979) Glaciokarst in the Bear River Range, Utah. NSS Bulletin 41:89–94

Wilson P, Lord T, Rodés Á (2013) Deglaciation of the eastern Cumbria glaciokarst, northwest England, as determined by cosmogenic nuclide (^{10}Be) surface exposure dating, and the pattern and significance of subsequent environmental changes. Cave Karst Sci 40(1):22–27

Woodward JC, Hamlin RHB, Macklin MG, Hughes PD, Lewin J (2008) Glacial activity and catchment dynamics in northwest Greece: long-term river behaviour and the slackwater sediment record for the last glacial to interglacial transition. Geomorphology 101(1):44–67

Yilmaz Y, Tüysüz O, Yigitbas E, Can Genç S, Sengör AMC (1998) Geology and tectonic evolution of the Pontides. Memoirs-American Association of Petroleum Geologists, pp 183–226

Žák K, Hercman H, Orvošová M, Jačková I (2009) Cryogenic cave carbonates from the Cold Wind Cave, Nízke Tatry Mountains, Slovakia: Extending the age range of cryogenic cave carbonate formation to the Saalian. Int J Speleol 38(2):139–152

Žák K, Richter DK, Filippi M, Živor R, Deininger M, Mangini A, Scholz D (2012) Cryogenic cave carbonate—a new tool for estimation of the Last Glacial permafrost depth of the Central Europe. Clim Past Discuss 8(3):2145–2185

Zamora E, Santana A (1979) Características climáticas de la costa occidental de la Patagonia entre las latitudes 46° 40′ y 56° 30′S. Anales del Instituto Patagonia. Punta Arenas (Chile) 10:109–154

Zasadni J, Kłapyta P (2014) The Tatra Mountains during the last glacial maximum. J Maps 10 (3):440–456

Žebre M, Stepišnik U (2014) Reconstruction of Late Pleistocene glaciers on Mount Lovćen, Montenegro. Quatern Int 353:225–235

Žebre M, Stepišnik U (2015a) Glaciokarst geomorphology of the Northern Dinaric Alps: Snežnik (Slovenia) and Gorski Kotar (Croatia). J Maps

Žebre M, Stepišnik U (2015b) Glaciokarst landforms and processes of the southern Dinaric Alps. Earth Surf Proc Land 40(11):1493–1505

Žebre M, Stepišnik U, Fabekovič G, Grlj A, Koblar S, Kodelja B, Pajk V, Štefanić K (2013) Pleistocenska poledenitev Biokova. Dela 39:141–155

Žebre M, Stepišnik U, Colucci RR, Forte E, Monegato G (2016) Evolution of a karst polje influenced by glaciation: the Gomance piedmont polje (northern Dinaric Alps). Geomorphology 257:143–154

Zreda M, Çiner A, Sarıkaya MA, Zweck C, Bayarı S (2011) Remarkably extensive glaciation and fast deglaciation and climate change in Turkey near the Pleistocene-Holocene boundary. Geology 39(11):1051–1054

Index

M. Veress et al., *Glaciokarsts*, Springer Geography,
https://doi.org/10.1007/978-3-319-97292-3

MIX
Papier aus verantwortungsvollen Quellen
Paper from responsible sources
FSC® C105338
FSC
www.fsc.org

If you have any concerns about our products,
you can contact us on
ProductSafety@springernature.com

In case Publisher is established outside the EU,
the EU authorized representative is:
Springer Nature Customer Service Center GmbH
Europaplatz 3, 69115 Heidelberg, Germany

Printed by Libri Plureos GmbH
in Hamburg, Germany